The Role of VLBI in Astrophysics, Astrometry and Geodesy

NATO Science Series

A Series presenting the results of scientific meetings supported under the NATO Science Programme.

The Series is published by IOS Press, Amsterdam, and Kluwer Academic Publishers in conjunction with the NATO Scientific Affairs Division

Sub-Series

I. Life and Behavioural Sciences	IOS Press
II. Mathematics, Physics and Chemistry	Kluwer Academic Publishers
III. Computer and Systems Science	IOS Press
IV. Earth and Environmental Sciences	Kluwer Academic Publishers
V. Science and Technology Policy	IOS Press

The NATO Science Series continues the series of books published formerly as the NATO ASI Series.

The NATO Science Programme offers support for collaboration in civil science between scientists of countries of the Euro-Atlantic Partnership Council. The types of scientific meeting generally supported are "Advanced Study Institutes" and "Advanced Research Workshops", although other types of meeting are supported from time to time. The NATO Science Series collects together the results of these meetings. The meetings are co-organized bij scientists from NATO countries and scientists from NATO's Partner countries – countries of the CIS and Central and Eastern Europe.

Advanced Study Institutes are high-level tutorial courses offering in-depth study of latest advances in a field.
Advanced Research Workshops are expert meetings aimed at critical assessment of a field, and identification of directions for future action.

As a consequence of the restructuring of the NATO Science Programme in 1999, the NATO Science Series has been re-organised and there are currently Five Sub-series as noted above. Please consult the following web sites for information on previous volumes published in the Series, as well as details of earlier Sub-series.

http://www.nato.int/science
http://www.wkap.nl
http://www.iospress.nl
http://www.wtv-books.de/nato-pco.htm

Series II: Mathematics, Physics and Chemistry – Vol. 135

The Role of VLBI in Astrophysics, Astrometry and Geodesy

edited by

Franco Mantovani
Istituto di Radioastronomia,
Consiglio Nazionale delle Ricerche, Bologna, Italy

and

Andrzej Kus
Radio Astronomy Department,
Nicolaus Copernicus University, Torun, Poland

Kluwer Academic Publishers

Dordrecht / Boston / London

Published in cooperation with NATO Scientific Affairs Division

Proceedings of the NATO Advanced Study Institute on
The Role of VLBI in Astrophysics, Astrometry and Geodesy
Bologna, Italy
September 17–29, 2001

A C.I.P. Catalogue record for this book is available from the Library of Congress.

ISBN 1-4020-1876-2 (PB)
ISBN 1-4020-1875-4 (HB)
ISBN 1-4020-2406-1 (e-book)

Published by Kluwer Academic Publishers,
P.O. Box 17, 3300 AA Dordrecht, The Netherlands.

Sold and distributed in North, Central and South America
by Kluwer Academic Publishers,
101 Philip Drive, Norwell, MA 02061, U.S.A.

In all other countries, sold and distributed
by Kluwer Academic Publishers,
P.O. Box 322, 3300 AH Dordrecht, The Netherlands.

Printed on acid-free paper

Printed in the Netherlands.

TABLE OF CONTENTS

Preface

The NATO Advanced Study Institute Series 'The Role of VLBI in Astrophysics, Astrometry and Geodesy' appears 13 years after the pioneering summer school on 'VLBI Techniques and Applications' edited by M. Felli and R.E. Spencer. Its purpose is to update and expand our knowledge on the scientific potential of Very Long Baseline Interferometry (VLBI) and its development. In particular, the VLBI technique is reviewed in the light of the most advanced astronomical observations. The scientific issues in the field of galactic and extragalactic astronomy, astrometry and geodesy, where this technique is recognised to play a crucial role, are emphasized. The book includes such important aspects as a high-resolution approach to non-thermal emission from extragalactic radio sources and the principles of synchrotron emission in an astrophysical context. The basic concepts of the theory of relativistic jets are presented. The evolution of young, powerful radio sources is explained, while the structure of radio sources is treated in the context of the unified scheme models. Important topics are discussed, such as scintillation of extragalactic radio sources, radio polarimetry, investigation of the faint sub-mJy and micro-Jy radio sky, evolution of radio emission from Supernovae, and astrophysical masers. In addition the readers will find more technical issues such as radio and optical interferometry and tropospheric and ionospheric phase calibration. Applications and results of VLBI for Geodesy and Geodynamics are also extensively discussed. This book is intended to be a support for teachers and students in Astrophysics and Geodesy at the undergraduate and graduate levels, and an aid for scientists planning their own investigations.

Acknowlegments

The Editors wish to thank the NATO Science Committee of the Advanced Study Institute programme, the European VLBI Consortium Directors for their continued encouragements, the Consiglio Nazionale delle Ricerche (CNR) and the EC FP5 Project Infrastructure Cooperation Network in Radio Astronomy - RadioNET (Contract No. HPRI-CT-1999-40003) for the funds made available to the students.

The Editors also wish to thank the members of the Local Organizing Committee for making the Institute a success, and the staff of the Computer Centre of the Istituto di Radioastronomia for providing the computer facilities for tutorials and internet connections.

The Mayor of Castel San Pietro Terme, the Mayor of Medicina, the Chairman of the 'Pro Loco' Castel San Pietro Terme, are thanked for their kind collaboration, and the Manager and the staff of the Albergo delle Terme, where the Institute was held, for their contribution to a pleasent stay for all the participants.

Finally, The Editors wish to thank the lecturers for their stimulating contributions and the students for their active participation, in particular those who have given oral presentations at the Institute.

Participants

Directors of Institute

Franco Mantovani
Istituto di Radioastronomia
Consiglio Nazionale delle Ricerche
Bologna, Italy

and
Andrzej Kus
Radio Astronomy Department
Nicolaus Copernicus University
Torun, Poland

Local Organizing Committee

Marco Bondi
Miguel Angel Perez-Torres
Franco Mantovani
Tiziana Venturi

Istituto di Radioastronomia
Bologna, Italy

Speakers

Prof. B. Anderson – University of Manchester, Jodrell Bank Observatory
Dr. M. Bondi – Istituto di Radioastronomia, Bologna
Dr. G. Brunetti – Department of Astronomy, University of Bologna
Prof. R. Booth – Chalmers University, Onsala Space Observatory
Prof. I. Browne – University of Manchester, Jodrell Bank Observatory
Prof. J. Campbell – Geodaetisches Institut, University of Bonn
Dr. W. Cotton – National Radio Astronomy Observatory, Charlottesville
Dr. P. Diamond – University of Manchester, Jodrell Bank Observatory
Prof. R. Fanti – Department of Physics, University of Bologna
Dr. H. Falcke – Max-Planck-Institut fuer Radioastronomie, Bonn
Dr. S. Frey – Institute of Geodesy, Penc
Dr. M. Garrett – Joint Institute for VLBI in Europe, Dwingeloo

Prof. A. Kus – Department of Astronomy, University of Torun
Dr. J-F. Lestrade – Observatoire de Paris, Paris
Dr. F. Mantovani – Istituto di Radioastronomia, Bologna
Prof. J. Marcaide – Department of Astronomy, University of Valencia
Dr. C. O'Dea – Space Telescope Institute, Baltimore
Dr. R. Porcas – Max-Planck-Institut fuer Radioastronomie, Bonn
Dr. A. Zensus – Max-Planck-Institut fuer Radioastronomie, Bonn

Tutorials

W. Cotton – Introduction to AIPS I

W. Cotton – Introduction to AIPS II

M. Garrett – Difmap

W. Cotton, M. Garrett, A. Richards – Preparing for a VLBI experiment

W. Cotton, M. Garrett – Phase-referencing

W. Cotton – Polarization Calibration

W. Cotton – FITS Header

A. Richards – Spectral lines

Participants

ASI STUDENTS FROM NATO COUNTRIES

CANADA
Scott William – Physics and Astronomy Department, University of Calgary, Calgary,
GERMANY
Bach Uwe – Max Planck Institut fuer Radioastronomy, Bonn
Cimo' Giuseppe – Max Planck Institut fuer Radioastronomy, Bonn

Fuhrmann Lars – Max Planck Institut fuer Radioastronomy, Bonn
Kadler Matthias – Max Planck Institut fuer Radioastronomy, Bonn
Koerding Elmar – Max Planck Institut fuer Radioastronomy, Bonn
Medici Alessio – Max Planck Institut fuer Radioastronomy, Bonn
Middelberg Enno – Max Planck Institut fuer Radioastronomy, Bonn
Hagiwara Yoshiaki – Max Planck Institut fuer Radioastronomy, Bonn
GREECE
Gonidakis Ioannis – Jodrell Bank Observatory, University of Manchester, U.K.
Papageorgiou Andreas – University of Central Lancashire, Preston , U.K.
Rovilos Emmanouel – Jodrell Bank Observatory, University of Manchester, U.K.
HUNGARY
Frey Sandor – FOMI Satellite Geodetic Observatory, Budapest
Mosoni Laszlo – FOMI Satellite Geodetic Observatory, Budapest
Paragi Zsolt – FOMI Satellite Geodetic Observatory, Budapest
ITALY
Baldacci Lara – Istituto di Radioastronomia, Bologna
Buemi Carla – Istituto di Radioastronomia, Noto, Siracusa
Crapsi Antonio – Istituto di Radioastronomia, Bologna
Giroletti Marcello – Istituto di Radioastronomia, Bologna
Goddi Ciriaco – Osservatorio Astronomico, Cagliari
Leto Paolo – Istituto di Radioastronomia, Noto, Siracusa
Nagar Neil – Osservatorio Astrofisico di Arcetri, Firenze
Schwegmann Wolfgang – Istituto di Radioastronomia, Bologna
Tinti Sara – Istituto di Radioastronomia, Bologna
NETHERLANDS
Mack Karl-Heinz – Astron-Netherlands Foundation for Research in Astronomy, Dwingeloo
Reynolds Cormac – Joint Institute for VLBI in Europe, Dwingeloo
Sjouwerman Lorant – Joint Institute for VLBI in Europe, Dwingeloo
POLAND
Gawronski Marcin P. – Torun Centre for Astronomy, Torun
Hrynek Grzegorz – Torun Centre for Astronomy, Torun
Jamrozy Marek – Astronomical Observatory, Krakow
Katarzynski Krzysztof – Torun Centre for Astronomy, Torun
Kunert Magdalena – Torun Centre for Astronomy, Torun
Niezurawska Anna – Torun Centre for Astronomy, Torun
Zajaczkowski Radomil – Torun Centre for Astronomy, Torun
PORTUGAL

Goncalves Angelino Sabino Lira da Silva Universidade de Madeira, Funchal

SPAIN

Agudo-Rodriguez Ivan – Instituto de Astrofisica de Andalucia, Granada

de Gregorio Monsalvo Itziar – Autonoma University of Madrid, Madrid

de los Reyes Lopez Raquel – Complutense University of Madrid, Madrid

Perucho Pla Manuel – Valencia University, Valencia

Rebeca Soria – Autonoma University of Madrid, Madrid

UNITED KINGDOM

Bains Indra – University of Hertfordshire, Hatfield

Richards Anita – Jodrell Bank Observatory, University of Manchester, Manchester

USA

Avruch Ian – Joint Institute for VLBI in Europe, Dwingeloo, Netherlands

ASI STUDENTS ELIGIBLE PARTNER COUNTRIES

LATVIA

Berzins Karlis – Ventspils International Radio Astronomy Center, Riga

RUSSIAN FEDERATION

Ivanov Dimitri – Institute of Applied Astronomy, St. Petersburg

Skurikhina Elena – Institute of Applied Astronomy, St. Petersburg

Syrovoi Serguei – Institute of Applied Astronomy, St. Petersburg

Zheltykov Oleg – Institute of Applied Astronomy, St. Petersburg

Kanevsky Boris – Astro Space Centre, Moscow

Smirnov Alexander – Astro Space Center, Moscow

Chuprikov Andrey – Astro Space Center, Moscow

Kozlova Iraida Alexandrovna – Institute of Applied Astronomy, St. Petersburg

SLOVENIA

Slosar Anze – Cavendish Laboratory, Cambridge, U.K.

UKRAINE

Vorobjov Daniel – National Taras Shevchenko University, Kiev

ASI STUDENTS MEDITERREAN DIALOGUE COUNTRIES

ALGERIA

Ghezali Boualem – National Centre of Spatial Techniques, Arzew

Taibi Habib – National Centre of Spatial Techniques, Arzew

JORDAN

Khassawneh Awni – Royal Jordanian Geographic Center, Amman

ASI STUDENT NON ELIGIBLE PARTNER COUNTRIES

AUSTRIA
Bohem Johannes – Institute of Geodesy and Geophysics, Vienna
FINLAND
Kononen Prisse – Metsahovi Radio Observatory, Metsahovi
SWEDEN
Pestalozzi Michele R. – Onsala Space Observatory, Onsala
Yi Jiyune – Onsala Space Observatory, Onsala

Contributed Papers

M. Pelucho – Phisical parameters and evolution models for relativistic jets in CSOs
G. Cimo' – A very rapid estreme scattering event in IDV source 0954+65
I. Agudo – A mechanism for generation of moving components in relativistic jets
M. Kadler – A multi-frequency VLBA study of the pc-scale twin jets in NGC 1052
L. Fuhrmann – Annual modulation in the IDV properties of 0917+624
A. Chuprikov – A new software for VLBI data processing
M. Pestalozzi – Onsala Methanol blind Survey of the Galactic Plane
A Slozar – VSA: Very Short Array

RADIO ASTRONOMY:

A Short Historical Introduction

FRANCO MANTOVANI
Istituto di Radioastronomia, Consiglio Nazionale delle Ricerche
Via P. Gobetti, 101 - 40129 Bologna, Italy

1. Introduction

This lecture is the first of a short series given as an introduction at the Advanced Study Institute 'The Role of VLBI in Astrophysics, Astrometry and Geodesy'. The aim of these lectures is mainly to provide the students with an historical background for the various branches of radio astronomy which will be developed in the following lessons of the course. This lecture deals with the birth of radio astronomy following the work of Karl Guthe Jansky on electrical disturbances in telephone radio-linked communications, and with the enormous progress made by radio astronomers in the years after the discovery.

For this contribution, I'm deeply indebted to the work of Woodruff Turner Sullivan III on the history of radio astronomy. His two interesting books (see References) have been a strong stimulus and a great help for the preparation of this lecture. The papers quoted in the present contribution are available in the books listed in the references.

2. The discovery of Karl Guthe Jansky

On 27 April 1933 Karl Guthe Jansky gave a contribution before the U.R.S.I. (International Scientific Radio Union) meeting in Washington entitled: 'Electrical Disturbances of Extra-terrestrial Origin'. That date is officially considered the beginning of radio astronomy. The public was made aware of the discovery by an article in the 'The New York Times' (Friday, May 5th, 1933) following a press release issued by the Bell Telephone Laboratories. Karl Jansky was, in fact, an employee of Bell Telephone Laboratories, a branch of the Atlantic Telephone & Telegraph company (AT&T). Bell Labs had a strong international reputation in scientific research. Their research staff included C.J. Davisson and L.H. Germer who investigated electron diffrac-

F. Mantovani and A. Kus (eds.), The Role of VLBI in Astrophysics, Astrometry and Geodesy, 1–12.

tion by crystals, and H. Nyquist and J.B. Johnston, who worked on the problems of noise generated by electronic components. In the years 1925-28, Bell Labs ranked amongst the top ten institutions in terms of number of articles published in Physical Review.

The first AT&T research connected with radio telephony was begun by C.R. Englund in 1914 on long wave signal and static levels. In 1919, R. Bown and Friis (later, the supervisor of Jansky) joined him in investigating problems of radio propagation, measurement methodology, and receiver and antenna design. Radio communication research had began more than a decade before thanks to the innovative research of an Italian scientist, Guglielmo Marconi. It was December 12th, 1901 when Marconi first sent radio signals across the Atlantic [1]. The existence in the upper atmosphere of an electrified layer (the Kennelly-Heaviside layer) was later proposed to explain the propagation of radio signals around the curve of the Earth. However, strong evidence for the layer was found only in 1925 by E.V. Appleton and M.A. Barnett.

Prior to 1920, intercontinental radio circuits worked at frequencies lower than 100 kHz. Later, it was found that 'shortwave' (wavelengths of 200 m and shorter; frequencies > 1.5 MHz) worked well for long-distance contacts. The introduction of the high-vacuum, oxide-coated triode allowed a more intense development and use of shortwave. Radio communications rapidly evolved from radiotelegraphy (Morse code) to radiotelephony, the transmission of voice. The latter required increased bandwidth, a simultaneous two-way service, and greater reliability and fidelity 24 hours-a-day. The first radiotelephone service between New York and London, transmitted on 60 kHz, was introduced by AT&T in 1927 and cost 75 US Dollars for a 3 minute call. In 1929, shortwave services operating at frequencies between 9 MHz and 21 MHz, using smaller antennas and transmitters to achieve the same signal levels and permitting many voice channels, were introduced. Those services were less affected by interference from atmospheric noise. However, new sources of interference, automobile ignitions, intrinsic noise in the electronics and magnetic storms, now hindered radio communications.

In the Spring of 1930 Jansky and the Bell Labs group moved to Holmdel, New Jersey, where they built an array of quarter-wavelength sections 29 m in length, which automatically rotated every 20 minutes. This array observed the 'quiet' band centered on the wavelength of 14.5 m. Only signals coming from a single direction, perpendicular to the array, were detected by the antenna. The equipment and techniques were described by Jansky

[1] To celebrate the centennial of that event, the Istituto di Radioastronomia and the Fondazione Guglielmo Marconi have 'broadcast' radio signals from the Medicina 32-m radio telescope at 1300 MHz using the reflection from the Moon on December 12th., 2001.

(1933) in the report 'Directional Studies of Atmospherics at High Frequencies'. The data analysis suggested that three sources of static were present: (a) static received from local thunderstorms, (b) very steady static coming from thunderstorms some distance away, most likely by way of ionosphere refractions and (c) a very steady hiss of unknown origin.

3. Connection to the stars

The idea of the static arriving from a 'fixed region of the sky' has an uncertain genesis. The Jansky's paper in the *Proceedings of the Institute of Radio Engineers* has the relatively cautious title 'Electrical Disturbances of Extra-terrestrial Origin'. In it, Jansky derives the direction of arrival of the radio waves (R.A. = $18^h\,00^m \pm 30^m$; Dec. = $-10° \pm 30°$) and points out that this direction is not far from both the direction in Hercules towards which the Sun moves with respect to nearby stars and the direction in Sagitarius towards the centre of our Milky Way. Reading Jansky's papers, we see clearly that he had to learn classical astronomy to conclude that the 'interference' originated from a fixed point of the sky. Moreover, he also had to address astrophysics to suggest a possible emission mechanism responsible for the detected radio signals.

The communications by Jansky had little reaction from his colleagues. Bell Labs issued a press release which made Jansky an instant public celebrity. However, Jansky effectively ceased working on the 'star static' soon after he wrote an article for *Popular Astronomy* (September 1933) in an effort to communicate with the astronomical community.

The reaction of the scientific community to Jansky's work was rather unenthusiastic. The only observations of any depth were carried out by G.W. Potapenko and D.F. Folland from the Caltech Institute of Technology. Their antenna, observing at 14.6 m, consisted of a two one-meter diameter loops at a spacing of 2.5 m on an equatorial mount. In 1936, Folland presented a seminar in which he summarized Jansky's results and then worked out a rough estimate of the flux densities implied by Jansky's published data. Comparison with that expected for blackbody emission at a temperature of 10,000 K then revealed that the star noise was of much higher intensity and therefore undoubtedly not arising from thermal processes.

More observations were needed (as usual!), and Potapenko and Porter designed a 180 ft by 90 ft rhombic antenna mounted on a rotating assembly. However, R. Millikan, head of the Physics Department and President of Caltech, was not convinced it was worth to spend the 1000 US Dollars needed to build it. CIT became interested again in radio astronomy only much later.

4. Grote Reber: First map of the Galaxy

One of the few people to recognise the importance of Jansky's work was Grote Reber, a radio engineer during the day in Chicago, but at night a radio astronomer in his backyard. He realised that further progress would only be possible with better equipment. Reber designed and built a steerable 31 ft (9 metres) paraboloid reflector which was able to observe at different frequencies.

Reber's first observations were made in the Spring of 1938 at 3300 MHz, a frequency chosen so as to achieve a higher angular resolution than before. He also expected a much stronger signal if the radiation was similar to that of a black body. He did not detect any signal at 3300 MHz, nor at 910 MHz in a second attempt. On one hand, this finding was quite interesting since it established that the extraterrestrial radiation was not thermal in origin. However, the non-detection of radio emission was quite discouraging to Reber. Later, he used the lower frequency of 160 MHz and, in 1939, he was finally able to confirm Jansky's discovery.

Reber submitted his results to *Astrophysical Journal.* The editor, Otto Struve, decided to publish the experimental part of the paper, leaving out the theoretical interpretation since he was not convinced it was correct. Reber was the first to suggest that the radiation mechanism is due to free-free transitions by electrons in a hot plasma, but he over-corrected for the effects of stimulated emission and ignored the effects of opacity.

Reber described his equipment in a paper in the *Proceedings of I.R.E.* In the summary, he wrote:

> It is suggested that cosmic static is the equivalent of thermal agitation in which all space is the conductor and the input terminals of the detecting equipment are projected by means of an antenna system to some far-distant part of space.

Thus, Reber introduced for the first time the concept of far-field pattern.

Reber realised the need for an all-sky survey. To this end, he automated his data recording system and improved the receiver stability and sensitivity ($T_{sys} \approx 11{,}000\,K$). He was finally able to present the first radio contour maps of the sky, in which the central disk of the Milky Way is seen. In other papers, Reber also pointed out that regions of more intense emission can be recognised, superimposed on the diffuse galactic emission, in the direction of several constellations, for instance, Sagitarius, Cygnus and Cassiopea.

5. Astrophysical interpretation

Using Reber's data, L.G. Henyey and P.C. Keenan (1940) checked the calculations of the expected free-free emission from interstellar gas at radio wavelegths. They extrapolated the theory of radiative transfer and atomic recombination to wavelengths several orders of magnitude longer than previous. The agreement of computed spectrum with Reber's 160 MHz measurements was a mere coincidence since only a portion of the radiation at low frequencies is thermal by nature. Henyey and Keenan reached the important and correct conclusion that Jansky's observed brightness temperatures of about 10^5 K at 20.6 MHz cannot be reconciled with the thermal emission from a gas at a temperature of 10^4 K. A different interpretation of the observations was required.

A possible interpretation involving the synchrotron mechanism was suggested in 1950 by Kiepenheuer. He was the first to recognise the link between cosmic rays and the general galactic emission. It took several years before his proposal, an alternative to the interpretation of cosmic radio noise as the integrated emission from unresolved radio stars, became generally accepted.

The synchrotron theory was developed almost exclusively in the Soviet Union by Ginzburg, Shklovsky and their collaborators, who realised the importance of the link between cosmic ray physics and radio astronomy. Ginzburg developed the theory for the Galactic background while Getmansev (1952) derived the relation between the radio spectral index α and the electron energy spectral index γ, i.e. $\alpha = (1 - \gamma)/2$. Shklovsky suggested the existence of a spherical galactic corona of cosmic rays, accounting for the radio brightness found at high galactic latitudes. The primary electrons, which were necessary in the theory, were definitely established to be a component of cosmic rays only in 1960 (Earl 1961, Meyer and Vigt 1961).

6. Prediction of the 21 cm H–line and of molecular line emission

During the Second World War, Oort in Holland, went through the Reber's paper in *Astrophysical Journal.* He soon understood the importance of these dust-penetrating radio waves for the investigation of galactic structure. The second speaker in a meeting he organised was H.C. Van de Hulst. In the second part of Hulst's talk on the 'Origin of radio waves from space', he discussed of the possibility of discrete lines in the spectrum of interstellar hydrogen and considered the detectability of the 21 cm transition arising from the hyperfine splitting of the hydrogen atom, despite the fact that the line had yet to be detected in laboratory. He derived a frequency of 1411 MHz (the actual value is 1420.4 MHz) and suggested that, using the expected physical conditions for the interstellar medium, the line should be easily observable if the probability of spontaneous emission is greater than 8×10^{-17} sec^{-1}. This represents the beginning of radio spectroscopy.

A few years later, independent calculations of the 21 cm line were made by Shklovsky, and he was able for the first time to calculate the probability of the hyperfine transition. In a series of papers, he also discussed the possible emission of radio lines from interstellar molecules. Shklovsky's prediction for the wavelength of the OH line was not correct due to some approximations that he used. However, he anticipated the fundamental field of galactic microwave spectroscopy.

Radio spectroscopy has some advantages over continuum work. In particular, the line profile can reveal the temperature and density of the region from which the line is detected and radial velocities can be estimated. The observed line frequency indicates which atomic or molecular species is present.

In 1951, *Nature* simultaneously published three papers by Ewen and Purcell (Harvard University), Muller and Oort (Holland), and Christiansen and Hindman (Australia). These authors were able to measure the frequency shift of the peak neutral Hydrogen emission at several longitudes on the Galactic plane. Using the Oort dynamic model of the Galaxy, they suggested that the majority of the Neutral Hydrogen lies in a thin layer 400 pc thick. Some time later, van de Hulst, Muller and Oort published a map of the physical conditions and density of Neutral Hydrogen along the spiral structure of the Galaxy.

Dutch and Australian started surveys of opposite sides of the Milky Way, which allowed to assemble the 'Leiden-Sydney' map presented by Oort, Kerr and Westerhout in 1958. At the beginning of 1950s there was also the basic finding of the 21-cm line absorption against strong background source of radio continuum (Williams and Davis, 1954; Hagen, Lilley and McClain, 1955).

7. The development of radio astronomy

The overall development of radio astronomy can be view as a continuous quest for greater sensitivity and better angular resolution. The improved sensitivity came from a combination of lower-noise receivers and larger collecting areas. Greater resolution at a given wavelength can only be achieved building larger antennas or longer baseline interferometers. In the '50s, it was the problem of resolution which was of greater concern for radio astronomers who developed several new techniques. One of the most important among them was the Mills Cross, which employed a type of phase-switching with the phase connection between two intersecting line arrays. A second technique, the concept of aperture synthesis mapping, became practical as a result of technological progress and the advent of large digital computers. Improvements in receiver phase stability allowed the relative phase to be followed over periods of several hours. Large parabolic reflectors were also designed and constructed. Hanbury Brown and collaborators developed the intensity interferometer; this, in turn, promoted new theoretical and experimental work in correlation interferometry.

In the paper 'A new radio interferometer and its application to the observation of weak radio stars' (1951, Proc. of the Royal Society), Martin Ryle wrote:

> A new type of radio interferometer has been developed which has a number of important advantages over earlier systems. Its use enables the radiation from weak 'point' source such as radio star to be recorded independently of the radiation of much greater intensity from an extended source. ... Besides improved sensitivity, the new system has a number of other advantages, particularly for the accurate determination of the position of the radio source. Unlike earlier systems the accuracy of position finding is not seriously affected by rapid variations in the intensity of radiation. It also has important applications to the measurement of the angular diameter and polarization of weak source of radiation.

8. Discrete sources

Since the initial discovery of the phenomenon in 1946, discrete radio sources have been central to the development of radio astronomy. The first recognition of a discrete source came from the chance discovery by Hey, Parsons and Phillips in 1946 of rapid intensity variations in Cyg A. They accurately measured the position of the source and established that the 'source of disturbance' was $< 2°$. Their paper is notable for the deduction that the fast time variations imply that the origin lies in 'a small number of discrete sources'. Bolton and Stanley (1948) used a sea-cliff interferometer (a form of Lloyd's mirror interferometer) to determine the size of Cyg A to be less than 8 arcmin in extent. In a follow-up paper, they published the first overlay of a radio source position superimposed on a photograph of the region in the optical.

9. Optical identification

The first catalogues of radio sources had an almost complete lack of reliable optical identifications. Typical errors in the radio source position were of the order of 1 to 3 degrees. Ryle *et al.* inferred radio emission from the Andromeda Nebula; this was later confirmed by Hanbury Brown and Hazard (1950, 1951) who were able to reduce the positional errors to 0.5° using a 218 ft fixed paraboloid at Jodrell Bank Observatory. Using an interferometer aerial system on a wavelength of 3.7 m, F.G. Smith (1951) was able to determine the positions of Cyg A and Cas A with an accuracy of 1 second in R.A. and 1 arcmin in Declinations. In 1954, Baade and Minkowsky identified the radio source Cyg A with an extragalactic object.

As new sources were discovered and the accuracy of their position improved, several identifications with optical objects become possible. The four strongest sources of small angular size were each found to correspond to different classes of optical objects, i.e. a supernova remnant (Tau A), a bright elliptical galaxy with a peculiar jet (Vir A), an apparent pair of colliding galaxies at a distance of 300 million light years (Cyg A) and a system of rapidly moving filaments in our own galaxy (Cas A). This made the correlation between radio sources and their optical counterparts quite perplexing.

In 1960, the small diameter radio source 3C295 (number 295 in the Third Cambridge Catalogue) was identified with a 20th-magnitude galaxy at redshift 0.46 (Minkowski 1960). Then radio source 3C48 was identified with what appeared to be a 16th-magnitude 'stellar object'. The subsequent discovery of night-to-night variations in the light intensity led to the reasonable conclusion that 3C48 was indeed a true 'radio star' in our galaxy. Soon 3C196 and 3C286 were also found to appear stellar and it appeared that as many as twenty percent of all high-latitude radio sources were of this type.

Lunar occultation measurements (Hazard *et al.* 1963) gave an accurate position of the strong, compact radio source 3C273. Maarten Schmidt (1963) identified 3C273 with a 13th-magnitude stellar object, and he noted that the relatively simple optical line spectrum could be interpreted as a redshifted (z=0.16) Balmer series with MgII lines. A reinspection of the 3C48 spectrum indicated that its red-shift would be 0.37. Spectra of other similar objects led to the identification of CIII, CIV and Lyα lines, permitting much larger redshifts to be measured. The discovery of radio sources at cosmological distances was astounding since this demonstrated the existence of highly-luminous, galactic objects. Almost immediately, it was understood by Zel'dovich, Lynden-Bell and Salpeter that the mechanism powering these objects was gravitational in nature.

The word 'quasar' was coined and it is now used to describe the entire class of highly-redshifted 'quasi-stellar objects' or QSOs, some of which are not radio-bright. Bright radio quasars now allowed cosmologists to probe the Universe to deeper redshifts than had been possible before.

10. New windows in cosmology

Several surveys of the entire sky were undertaken in the mid 1950s: 2C (Shakeshaft *et al.* 1955; Ryle and Scheuer 1955), MSH (Mills and Slee 1957).

The first catalogues of radio sources established the isotropic distribution of 'radio stars'. The two surveys were found to show poor agreement, thus giving rise to the so called 'log N - log S' controversy. Later, the discrepancy was resolved when it was understood that these surveys were "confusion-limited"; below a certain flux density threshold, the number of sources in the radio sky becomes so large that low-resolution radio telescopes blend them together. However, the important contribution of these surveys, in addition to systematically cataloguing radio sources, is mainly in the introduction of a new window on cosmology.

11. Source structure

The observed finite angular extent of the most intense sources made the nomenclature of 'radio stars' inappropriate for the sources of radio emission. Mills (1952) was the first to make angular size measurements of the strong sources; these measurements were also useful for optical identification. He found angular widths at 100 MHz of about 20 arcmin for both Cen A and For A and of about 33 arcmin for Pup A. With a radio-linked interferometer with a 10 km long baseline, Mills also observed Cyg A, Tau A, Vir A and Cen A and accurately measured their east-west extent. Hanbury Brown, Jennison and Das Gupta (1952) presented the results of the first observations done at Jodrell Bank with the intensity interferometer. They pointed out that the angular size of Cyg A is highly non-circularly-symmetric. Using data taken at close intervals over a wide range of spacings, Jennison and Das Gupta later found that Cyg A can be modelled with two well separated components of approximately equal intensity. The model used to fit the visibility amplitudes consisted of two, almost equal components $51'' \times 30''$ in size, separated by about $88''$.

The discovery of the first double radio source immediately raised the question of why the optical image and the radio brightness distribution should be so different — a question astronomers are still asking themselves today.

12. Parallel developments

Radio waves from pulsars (pulsating radio sources) were discovered in 1967. The event can be taken as a prime example of 'serendipitous' discovery, similar to Jansky's discovery more than thirty years before. These discoveries are not due merely to fortune but also to sagacity of the researcher.

In a survey conducted at Cambridge University, some of the newly-discovered sources showed rapid changes of brightness if observed close to the direction of the Sun, a phenomenon called 'scintillation'. The smallest-diamenter radio sources scintillate while larger-diameter radio sources shine steadily. From the temporal and spatial evolution of the scintillation, source diameters can be estimated. A cheap radio telescope using kilometres of wire supported on wooden posts was built for scintillation studies. The detected signals were recorded using automatic pen recorders. Jocelyn Bell, a graduate student at the time, studied miles of paper chart records. In August 1967, she and Antony Hewish detected pulses of radio emission from the costellation of Vulpecula. The pulses repeated every 1.3373 seconds. The announce of the pulsating radio source opened a new window for the study of the last phases of stellar evolution.

In 1968, Thomas Gold concluded that pulses must come from highly-magnetized, rapidly-rotating, neutron stars. He based this conclusion on the millisecond intensity fluctuation of the signals, which implied that the emitting regions were smaller than 300 km. The stability of their rotation periods required the inertia of a stellar mass. The discovery of pulsars arrived after a thirty-year-long period of speculation on the possibility of neutron stars.

The detailed story of the discovery and later results from pulsar investigations can be found in Don Backer's contribution to the comprehensive book edited by Verschuur and Kellermann (1988).

13. Very Long Baseline Interferometry

The rotation of the earth changes the projected interferometer baseline and this allows measurements of a range of spatial frequencies. The Cambridge One-Mile radio telescope was designed to take advantage of this technique. Other telescope arrays were built following the success of the Cambridge One Mile. In addition to the capability to image structural details of the radio sources not seen with more primitive instruments, these interferometers also found quasars and other objects that appear nearly point-like in structure.

Following the observations done on 'big sources' the interest of the Jodrell Bank group moved to 'small sources'. To increase the resolution of the interferometer longer baselines were required. Directly-connected telescopes were not practicable so the completely new technique of radio linking was introduced. This approach imposed many technical demands like slowing down the visibility fringe rate and delay compensation and new techniques had to be developed. The longest baseline was increased to 115 km or 60,000 wavelengths; even at this separation, seven sources were found to be unresolved with angular sizes less than 1 or 2 arcsec. The full story can be found in the contribution by Hanbury Brown 'The development of Michelson and Intensity Long Baseline Interferometry' in the proceedings of the workshop 'Serendipitous Discoveries in Radio Astronomy (Kellermann and Sheets

1983). During the discussion that followed Marshall Cohen mentioned that

> It was those seven sources and the 120 km, of whatever it was, that was one of the main factors in going to Very Long Baseline Interferometry, which began very shortly after Henry Palmer stopped extending the baseline... .

By the mid-1960s it was well known that scintillation and time variability of the radiation from quasars implied angular sizes of < 0.01 arcsec. Moreover, maser emission from OH molecules at 18 cm was unresolved at 0.1 arcsec resolution. The motivation to develop VLBI came from the realisation that many radio sources have structures that cannot be resolved by interferometers with baselines of a few hundred kilometres. The technical requirements for VLBI were widely discussed in the early 1960s (*e.g.* Matveenko, Kardashev and Solomitskii 1965). Many groups contributed to the development of the technique. The first successful experiment was done in January 1967 by a group from the University of Florida. They detected fringes from the burst radiation of Jupiter at 18 MHz (Brown, Carr and Block 1968). A Canadian group built an analog recording system with a bandwidth of 1 MHz based on television tape recorders (Broten *et al.* 1967). They obtained fringes at a frequency of 448 MHz on baselines of 183 km and 3074 km from several quasars in April 1967. An American group (NRAO and Cornell Univ.) developed a computer-compatible digital system with a bandwidth of 360 kHz (Bare *et al.* 1967). They obtained fringes from several quasars at 610 MHz on a baseline of 220 km. A third group from MIT joined in the development of the NRAO-Cornell system and obtained fringes at a frequency of 1665 MHz on a baseline of 845 km from several OH masers in June 1967 (Moran *et al.* 1967). A notable early achievement was the measurement made by Burke *et al.* (1972); they obtained a resolution of 0.2 milli arcsecond using antennas in Westford (Massachussetts) and Simeiz (Crimea) operating at 1.3 cm.

During the mid-1970s, several groups of astronomers agreed to combine their facilities in order to increase the number of simultaneous baselines available for VLBI observations. The use of non-synchronised local oscillators precluded the calibration of the phase of the fringe patterns. The earlier idea of "closure phase" first introduced by Jennison at Jodrell Bank in the late 50's for their long baseline observations was revisited by Rogers *et al.* (1974). This allowed the development of so-called "hybrid mapping" and this became the adopted method to recover the source brightness distribution.

The technical developments of VLBI were paralleled by the introduction of relatively cheap, fast digital computers. These made the computationally-intensive technique of VLBI possible. Similarly, the improved receiver technology for commercial and military communications was readily adopted by the radio astronomy community.

The great potential of VLBI in Astrometry and Geodesy was soon recognized. NASA and other agencies in the USA started cooperative programmes of geodetic measurements. The results obtained in years of observations of compact radio sources and the impact of VLBI in Astrophysics will be discussed in the following lessons.

References

Classics in Radio Astronomy, W.T. Sullivan, III, Reidel Publishing Company, Dordrecht, 1982

Serendipitous Discoveries in Radio Astronomy, Proceedings of a workshop held at the National Radio Astronomy Observatory, Green Bank, WV on May 4,5,6, 1983, edited by K.I. Kellermann and B. Sheets

The Early Years of Radio Astronomy, edited by W.T. Sullivan, III, Cambridge University Press, Cambridge, 1984

Galactic and Extragalactic Radio Astronomy, edited by G.L. Verschuur & K.I. Kellermann, Springer-Verlag, New York, 1988

PRINCIPLES OF SYNCHROTRON EMISSION IN AN ASTROPHYSICAL CONTEXT

T. HUEGE AND H. FALCKE
Max-Planck-Institut für Radioastronomie
Auf dem Hügel 69, 53121 Bonn, Germany

Abstract. In these lecture notes, we give a short introduction to the very general topic of synchrotron radiation and its relevance in an astrophysical context. We deduce the power radiated by a single highly relativistic charged particle moving in a homogeneous B-field and the characteristics of the synchrotron spectrum it emits. On this basis, aspects of synchrotron radiation from multi-particle systems such as the spectrum generated by a power-law particle distribution, synchrotron absorption effects and synchrotron self-Compton effects, are discussed.

1. Introduction

"Synchrotron radiation" occurs whenever the presence of highly relativistic charged particles and the occurence of magnetic fields coincide. It was first proposed as a concept by Schwinger (1949) and soon thereafter suggested as the source of radio emission from the then newly discovered cosmic radio sources by Alfvén and Herlofson (1950) (for an account of the "history" of synchrotron radiation see Shklovsky 1960). Synchrotron radiation is non-thermal radiation, meaning that the emitting particles are not in thermal equilibrium. Its most outstanding property is that it is inherently strongly polarised, a trait that makes it distinguishable from thermal and other nonthermal radiation mechanisms and allows the observer to deduce the configuration of the magnetic fields at the source. It can occur over large frequency regimes from radio frequencies up to X-rays, and since the only necessary prerequisites are the presence of high energy particles and magnetic fields, is a very universal phenomenon that can be observed in many classes of astrophysical objects and contexts such as:

– jets of radio galaxies

F. Mantovani and A. Kus (eds.), The Role of VLBI in Astrophysics, Astrometry and Geodesy, 13–27.

– radio through X-ray emission from BL Lacs
– supernovae and supernova remnants
– non-thermal radiation from stars
– magnetic filaments (Galactic center)
– galaxy and cluster halos.

Synchrotron radiation from charged particles gyrating in the earth's magnetic field ("geo-synchrotron radiation") may also be able to explain the radio pulses that have been observed from cosmic ray air showers in the earth's atmosphere (Falcke and Gorham 2002, Huege and Falcke 2002).

Of the many books that cover the physics of synchrotron radiation, we can especially recommend Jackson (1975) and Shu (1991) for a detailed and thorough discussion. For this short introductory article, however, we follow closely the less mathematical argumentation of Rybicki and Lightman (1979) to which the reader is referred for more information.

In the following section we derive the power radiated by a single charge moving highly-relativistically in a homogeneous B-field. The spectrum it emits is discussed in section 3. Multi-particle distributions and the effects associated with them are reviewed in sections 4 (power-law particle distributions), 5 (absorption effects) and 6 (synchrotron self-Compton).

2. Radiated power

The total power P radiated by an accelerated charge q with velocity $\vec{v}$ is given by the "Larmor Formula", which is derived from the Poynting vector through integration over solid angle:

$$P = \frac{2q^2}{3c^3}\,|\dot{\vec{v}}|^2. \tag{1}$$

The acceleration of the moving particles is governed by the Lorentz force

$$\vec{F} = q\left(\vec{E} + \frac{\vec{v}}{c} \times \vec{B}\right). \tag{2}$$

In case of synchrotron radiation, we usually deal with purley magnetic fields, i.e. $\vec{E} = 0$, so that, using the Lorentz factor γ and charge mass m, we can write

$$\vec{F} = \frac{\mathrm{d}}{\mathrm{d}t}(\gamma m\vec{v}) = \frac{q}{c}\,\vec{v} \times \vec{B}. \tag{3}$$

Since $\vec{v} \cdot (\vec{v} \times \vec{B}) = 0$, the Lorentz force is always perpendicular to the instantaneous velocity vector. This implies $|\vec{v}| = \mathrm{const}$ (and therefore $\gamma = \mathrm{const}$) while the direction of $\vec{v}$ changes. In other words, we have conservation of particle energy:

$$\frac{\mathrm{d}W}{\mathrm{d}t} = \frac{\mathrm{d}}{\mathrm{d}t}(\gamma mc^2) = \vec{v} \cdot \vec{F} = 0 \tag{4}$$

and the equation of motion becomes

$$\gamma m \frac{\mathrm{d}}{\mathrm{d}t}\vec{v} = \frac{q}{c}\,\vec{v} \times \vec{B}. \tag{5}$$

Keep in mind, however, that we neglect the energy losses through radiation which we are actually about to calculate. (They would enter the equations through the E-field generated by the radiation itself.) As the acceleration is also perpendicular to $\vec{B}$, its components parallel and perpendicular to the B-field lines are

$$\frac{\mathrm{d}}{\mathrm{d}t}\vec{v}_{\|} = 0, \quad \frac{\mathrm{d}}{\mathrm{d}t}\vec{v}_{\perp} = \frac{\mathrm{d}}{\mathrm{d}t}\vec{v} = \frac{q}{\gamma mc}\,\vec{v} \times \vec{B}, \tag{6}$$

and therefore, given a *homogeneous B-field* with $B = |\vec{B}|$, and

$$v_{\perp} = |\vec{v}_{\perp}| = \sqrt{|\vec{v}|^2 - |\vec{v}_{\|}|^2} = |\vec{v}| \sin\alpha = \mathrm{const}, \tag{7}$$

where α is the (constant) pitch-angle between the direction of $\vec{B}$ and $\vec{v}$, we have

$$|\dot{\vec{v}}| = \left|\frac{\mathrm{d}}{\mathrm{d}t}\vec{v}_{\perp}\right| = \frac{q}{\gamma mc}\,v_{\perp}B = \omega_B\,v_{\perp} = \mathrm{const}. \tag{8}$$

This corresponds to uniform linear motion along the B-field lines superposed by circular ($\omega_B = \mathrm{const}$) motion in the perpendicular plane with frequency

$$\omega_B = \frac{qB}{\gamma mc}, \tag{9}$$

which toghether leads to helical particle motion. Transformation from the particle's to the observer's frame of reference introduces an additional factor γ^2 in $\dot{v}_{\perp}$ (cf. Rybicki and Lightman 1979 eq. 4.92) so that

$$|\dot{\vec{v}}| = \gamma^2 \omega_B v_{\perp} \tag{10}$$

and using the Larmor Formula (1) the total radiated power becomes

$$P = \frac{2q^2}{3c^3}\gamma^4 \frac{q^2B^2}{\gamma^2 m^2 c^2} v_{\perp}^2 = \mathrm{const}. \tag{11}$$

Note that synchrotron radiation from protons is negligible compared to radiation by electrons/positrons due to the much lower q/m ratio. Using the classical electron radius

$$r_{\mathrm{ce}} = \frac{e^2}{mc^2} \tag{12}$$

and $\beta = v/c$, for electrons we can also write

$$P = \frac{2}{3} r_{\rm ce}^2 c \beta_\perp^2 \gamma^2 B^2. \tag{13}$$

For particles with an isotropic distribution of constant velocity β, we can average over pitch-angle to derive the average power radiated per particle:

$$\left\langle \beta_\perp^2 \right\rangle = \frac{1}{4\pi} \int \beta^2 \sin^2 \alpha \, \mathrm{d}\Omega = \frac{\beta^2}{4\pi} \int_0^{2\pi} \int_0^{\pi} \sin^3 \alpha \, \mathrm{d}\alpha \, \mathrm{d}\phi = \frac{2}{3} \beta^2 \tag{14}$$

so that

$$P = c \left(\frac{2}{3} r_{\rm ce} \gamma \beta B \right)^2 . \tag{15}$$

Note that this is equivalent to adopting a pitch-angle α of

$$\alpha_{\rm iso} = \arcsin \sqrt{\frac{2}{3}} \approx 54.7^\circ. \tag{16}$$

The radiation process can also be interpreted as the scattering of a virtual photon originating from the B-field off the electron. With the Thomson cross section

$$\sigma_{\rm T} = \frac{8\pi}{3} r_{\rm ce}^2 \tag{17}$$

and the magnetic field energy density

$$U_B = \frac{B^2}{8\pi} \tag{18}$$

the average total power radiated per particle can then be written as

$$P = \frac{4}{3} \sigma_{\rm T} c \left(\gamma\beta\right)^2 U_B. \tag{19}$$

3. Spectrum

A detailed derivation of the synchrotron spectrum is beyond the scope of this article and can be found in Jackson (1975) or Shu (1991). We can, however, already deduce a lot of information about the general form of the spectrum from a few basic considerations.

For non-relativistic particles, the radiation purely consists of line emission at the fundamental frequency $\omega_B/2\pi$ with a dipole radiation pattern and is named "cyclotron radiation". For relativistic particles, however, it is heavily beamed in the forward direction, into a cone of semi-angle $\sim 1/\gamma$, which causes the spectrum to spread to and be dominated by frequencies

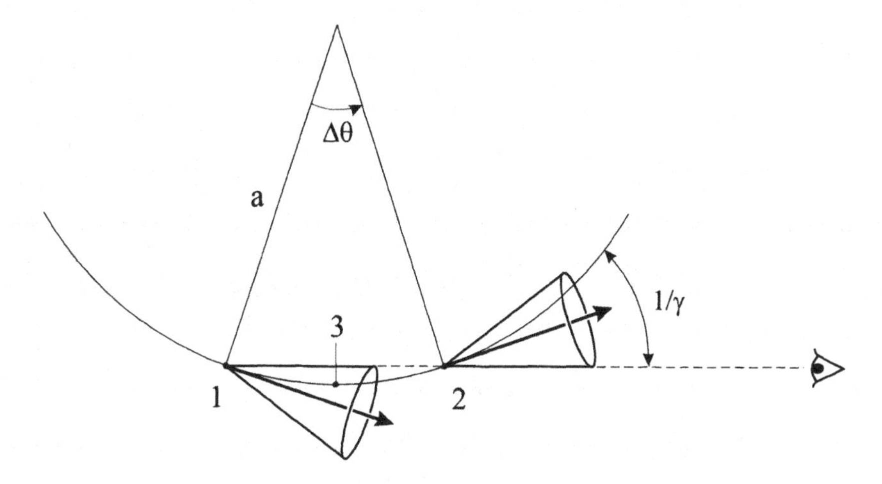

Figure 1. The beaming cone sweeping accross the observer's line of sight.

$\gg \omega_B/2\pi$: The helical motion corresponds to a trajectory with curvature radius

$$a = \frac{v}{\omega_B \sin\alpha}, \tag{20}$$

so that for linear uniform motion, i.e. $\alpha = 0$, we have no curvature at all as $a = \infty$. (Caution: a is not to be confused with the radius of the circular projection of the helix onto the plane normal to the magnetic field lines, $v \sin\alpha/\omega_B$.) As illustrated in Fig. 1, an observer will see a short flash of radiation each time the beaming cone sweeps across him or her, similar to a lighthouse beam. The flash starts when the beaming cone reaches the observer's line of sight at point 1 and ends when it leaves his or her line of sight at point 2. This corresponds to an angular distance of $\Delta\theta \approx 2/\gamma$, and using the radius of curvature a corresponds to a travelled distance

$$\Delta s = a\,\Delta\theta \approx \frac{2a}{\gamma} = \frac{2v}{\gamma\,\omega_B\,\sin\alpha} \quad \Rightarrow \quad \Delta t = \frac{\Delta s}{v} \approx \frac{2}{\gamma\,\omega_B\,\sin\alpha}. \tag{21}$$

The time difference $\Delta t_{\rm obs}$ that the observer perceives, however, is shorter by an amount $\approx \Delta s/c$, the time it takes the radiation emitted at point 1 to propagate to point 2. We therefore have

$$\Delta t_{\rm obs} \approx \frac{2}{\gamma\,\omega_B\,\sin\alpha}\left(1 - \frac{v}{c}\right). \tag{22}$$

For $v \to c \;\Rightarrow\; \beta \to 1$ and $\gamma \to \infty$, we can Taylor expand

$$\left(1 - \frac{v}{c}\right) = \left(1 - \sqrt{1 - \frac{1}{\gamma^2}}\right) \approx \frac{1}{2\gamma^2} \tag{23}$$

and get

$$\Delta t_{\mathrm{obs}} \approx \frac{1}{\gamma^3 \omega_B \sin\alpha}. \tag{24}$$

The measured pulse frequency is therefore larger by approximately a factor γ^3 than the gyration frequency. From Fourier theory we know that there is an inverse relation between the time length of a signal and the width of the associated spectrum: The spectrum of a δ-peak signal spreads to infinite frequencies, whereas the spectrum of an infinite sine wave signal is a single δ-peak at the corresponding frequency. In other words,

$$\Delta\omega\Delta t > 1, \tag{25}$$

so that a short pulse of length Δt must have a very broad-band spectrum with the highest frequencies of order $1/\Delta t$. Let us conveniently define a characteristic frequency

$$\omega_{\mathrm{c}} = \frac{3}{2}\gamma^3\omega_B \sin\alpha = \frac{3}{2}\gamma^2\frac{eB}{mc}\sin\alpha, \tag{26}$$

at which, as it turns out, the spectrum approximately peaks (more precisely, it peaks at $\approx 0.3\,\omega_{\mathrm{c}}$), and which can be converted from angular to "normal" frequency via

$$\nu_{\mathrm{c}} = \frac{\omega_{\mathrm{c}}}{2\pi}. \tag{27}$$

One can get a feeling for the frequencies involved by the parametrisation

$$\nu_{\mathrm{c}} \approx 3.4\ \mathrm{MHz}\cdot\gamma^2\left(\frac{B}{\mathrm{G}}\right) \quad \text{for} \quad \alpha = \alpha_{\mathrm{iso}}. \tag{28}$$

But what about the form of the spectrum? We can derive some of its properties by an analysis of the the properties of $E(t)$ and its Fourier transform $\hat{E}(\omega)$. The beaming cone, as mentioned earlier, has a semi-opening angle of order $1/\gamma$. This sets a critical angle θ_{c}, because an observer off axis by an angle $\theta > \theta_{\mathrm{c}}$ from the center of the cone does not receive significant radiation. The time depencence of the E-field an observer measures is then governed only by the time evolution of the ratio $\theta(t)/\theta_{\mathrm{c}} = \theta(t)\gamma$ (the other quantities are constant as seen in section 2). We can thus write

$$E(t) \propto f(\theta(t)\,\gamma), \tag{29}$$

t being the observer's time. Setting $s = 0$ and $t = 0$ for the moment where $\theta = 0$, i.e. for position 3 in Fig. 1, we see that

$$\theta(s) \approx \frac{s}{a} = \frac{s}{v}\,\omega_B \sin\alpha \tag{30}$$

and, using (23),

$$t(s) \approx \frac{s}{v}\left(1 - \frac{v}{c}\right) \approx \frac{s}{v}\frac{1}{2\gamma^2} \tag{31}$$

and therefore, using (26),

$$\theta(t)\gamma \approx (2\gamma^3\omega_B \sin\alpha)\, t \propto \omega_c t. \tag{32}$$

Thus, we can write

$$E(t) \propto g(\omega_c t), \tag{33}$$

and via Fourier transform and with $\xi = \omega_c t$ we get for the spectrum of the E-field

$$\hat{E}(\omega) \propto \int_{-\infty}^{\infty} E(t) \mathrm{e}^{i\omega t}\, \mathrm{d}t \propto \int_{-\infty}^{\infty} g(\xi) \mathrm{e}^{i\omega\xi/\omega_c}\, \mathrm{d}\xi \propto H\left(\frac{\omega}{\omega_c}\right). \tag{34}$$

The energy radiated per unit solid angle per unit frequency interval is proportional to the square of $\hat{E}_\omega$,

$$\frac{\mathrm{d}^2W}{\mathrm{d}\omega \mathrm{d}\Omega} \propto |\hat{E}(\omega)|^2, \tag{35}$$

and to get the time-averaged radiated power P, we have to average over an orbital period T and integrate over solid angle Ω, both of which are independet of observing frequency ω. The general frequency dependence is therefore unchanged by the integration and we get

$$P(\omega) = T^{-1} \int_\Omega \frac{\mathrm{d}^2W}{\mathrm{d}\omega \mathrm{d}\Omega}\, \mathrm{d}\Omega = T^{-1}\, \frac{\mathrm{d}W}{\mathrm{d}\omega} = C_\mathrm{x} \cdot F\left(\frac{\omega}{\omega_c}\right). \tag{36}$$

An integration over frequency yields the total radiated power

$$P = \int_0^\infty P(\omega) \mathrm{d}\omega = C_\mathrm{x} \int_0^\infty F\left(\frac{\omega}{\omega_c}\right) \mathrm{d}\omega = \omega_c\, C_\mathrm{x} \int_0^\infty F(x)\, \mathrm{d}x \tag{37}$$

where we have made a change of integration variable to $x = \omega/\omega_c$. Direct comparison with (26) and P as derived in Sec. 2,

$$P = \frac{2q^4B^2\gamma^2\beta^2 \sin^2\alpha}{3m^2c^3} \tag{38}$$

yields

$$C_\mathrm{x} \int_0^\infty F(x)\, dx = \left(\frac{2}{3}\right)^2 \frac{e^3B\beta^2 \sin\alpha}{mc^2}. \tag{39}$$

Adopting $F(x)$ (and thus $\int F(x)\mathrm{d}x$) as dimensionless quantity whose normalisation factor we can arbitrarily set, we can derive the dependence of C_{x} on all the physical parameters,

$$C_{\mathrm{x}} = \frac{\sqrt{3}}{2\pi}\frac{e^3 B \sin\alpha}{mc^2}, \tag{40}$$

where we have used our freedom of arbitrary normalisation to accomodate the notation usually applied in the literature and also set $\beta \approx 1$, since we are only interested in highly relativistic electrons. With (36) we then have the result

$$P(\omega) = \left(\frac{\sqrt{3}}{2\pi}\right)\frac{e^3 B \sin\alpha}{mc^2} F\left(\frac{\omega}{\omega_{\mathrm{c}}}\right) \tag{41}$$

and

$$P(\nu) = 2\pi P(\omega). \tag{42}$$

We have therefore established that $P(\omega)$ is basically a function of $\omega/\omega_{\mathrm{c}}$. A more quantitative analysis reveals the exact solution for $F\left(\frac{\omega}{\omega_{\mathrm{c}}}\right)$ (plotted in Fig. 2) whose asymptotical values are (cf. Rybicki and Lightman 1979 eq. 6.34):

$$F(x) \sim \frac{4\pi}{\sqrt{3}\Gamma(1/3)}\left(\frac{x}{2}\right)^{1/3}, \qquad x \ll 1 \tag{43}$$

$$F(x) \sim \frac{\pi}{2}^{1/2} e^{-x} x^{1/2}, \qquad x \gg 1. \tag{44}$$

4. Power-law particle distribution

In case of synchrotron radiation, the emitting particles are usually *not* in thermal equilibrium, but can rather be described with a power-law distribution for a certain range of $\gamma_1 < \gamma < \gamma_2$:

$$N(\gamma)\,\mathrm{d}\gamma = C\gamma^{-p}\,\mathrm{d}\gamma \tag{45}$$

The normalisation constant C is related to the number (and therefore energy) density of the particles. An assumption often employed is that of "minimum energy", i.e. the configuration that produces the measured flux density with a minimum in the sum of particle kinetic energy and B-field energy as given by (18). This energy minimum turns out to be close to equipartition of the energy between particles and magnetic field, and allowing for a slight deviation through the equipartition factor k, we get

$$\int_{\gamma_1}^{\gamma_2} N(\gamma)(\gamma m_e c^2)\,\mathrm{d}\gamma = \int_{\gamma_1}^{\gamma_2} C\gamma^{-p}(\gamma m_e c^2)\,\mathrm{d}\gamma \equiv k\frac{B^2}{8\pi}, \tag{46}$$

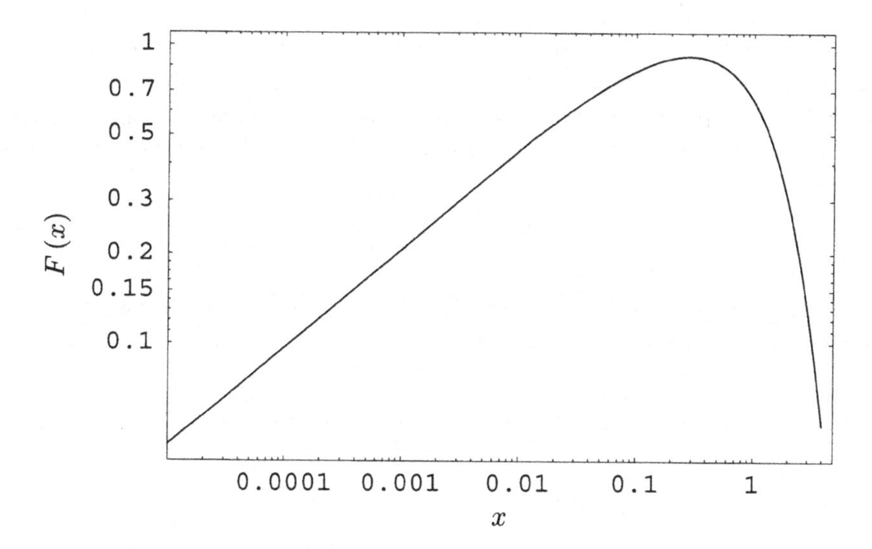

Figure 2. The form of the function $F(x)$. The spectrum peaks at $\approx 0.3\ \omega_c$.

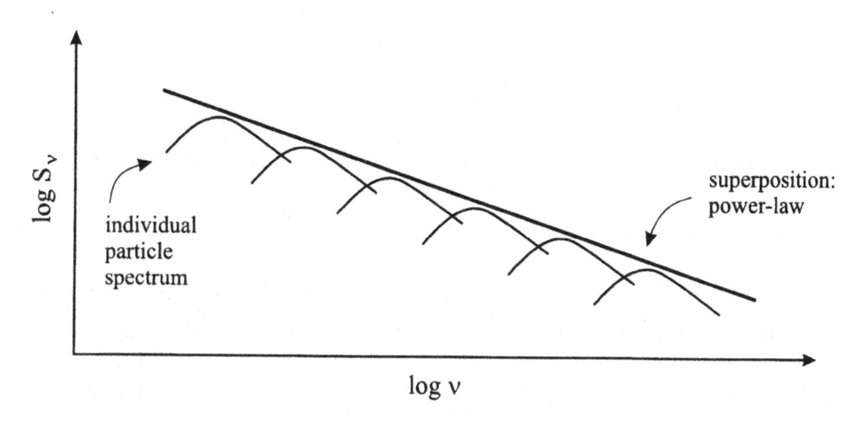

Figure 3. The superposition of the single-particle synchrotron spectra following a power-law distribution is again a power-law.

so that

$$C = \frac{kB^2}{8\pi m_e c^2 \int_{\gamma_1}^{\gamma_2} \gamma^{-p+1}\,\mathrm{d}\gamma}, \tag{47}$$

or in the special case $p = 2$ that arises from Fermi shock-acceleration of a non-relativistic mono-atomic gas with adiabatic index $5/3$,

$$C = \frac{kB^2}{8\pi m_e c^2 \ln\left(\frac{\gamma_2}{\gamma_1}\right)}. \tag{48}$$

The synchrotron spectrum of an individual particle is dominated by contributions close to the characteristic frequency ω_c. If the energy range $[\gamma_1, \gamma_2]$ over which the power-law holds is broad enough, which we can approximate by adopting $\gamma_1 \approx 0$ and $\gamma_2 \approx \infty$, the superposition of the individual particle spectra

$$P(\omega) \propto \int_0^\infty F\left(\frac{\omega}{\omega_c}\right) \gamma^{-p} \, d\gamma \tag{49}$$

will again have the form of a power-law. From (26) we see that $\omega_c \propto \gamma^2$, so that with $x = \omega/\omega_c \propto \omega/\gamma^2$ we get

$$\gamma \propto \sqrt{\frac{\omega}{x}} \quad \Rightarrow \quad \frac{d\gamma}{dx} \propto \omega^{1/2} x^{-3/2}. \tag{50}$$

Application of these substitutions yields

$$\begin{aligned} P(\omega) &\propto \int_0^\infty F(x)\, \omega^{-p/2} x^{p/2} \omega^{1/2} x^{-3/2} \, dx \\ &\propto \omega^{\frac{(1-p)}{2}} \int_0^\infty F(x)\, x^{\frac{p-3}{2}} \, dx. \end{aligned} \tag{51}$$

The last integral is a constant and therefore the synchrotron spectrum is again a power-law with spectral index $\kappa = \frac{1-p}{2}$. Typical values are

$$\kappa = -0.5 \text{ to } -1.0 \quad \Rightarrow \quad p = 2 \text{ to } 3. \tag{52}$$

The spectrum then changes as the electron population "ages". Since the power radiated by a particle is $\propto \gamma^2 B^2$ as shown in (15), the highest energy electrons drop out of the power-law first. This leads to a "knee" in the spectrum whose location can be used to estimate the age of the electrons since their acceleration (see Fig. 4).

A detailed calculation (cf. Rybicki and Lightman (1979) eq. (6.36)) gives for the radiated power per unit volume per unit frequency:

$$\begin{aligned} \frac{dW}{dt d\omega dV} &= \frac{\sqrt{3} e^3 C B \sin\alpha}{2\pi m c^2 (p+1)} \Gamma\left(\frac{p}{4} + \frac{19}{12}\right) \\ &\times \Gamma\left(\frac{p}{4} - \frac{1}{12}\right) \left(\frac{m c \omega}{3 e B \sin\alpha}\right)^{-(p-1)/2}, \end{aligned} \tag{53}$$

which for $p = 2$, $\alpha = \alpha_{\rm iso}$ and conversion to $d\nu$ can be parametrised as

$$\frac{dW}{dt d\nu dV} \approx 5.7 \cdot 10^{-19} \frac{\rm erg}{\rm s\ cm^3\ Hz} \frac{k}{\ln(\gamma_2/\gamma_1)} \left(\frac{B}{\rm G}\right)^{3.5} \left(\frac{\nu}{\rm GHz}\right)^{-0.5}. \tag{54}$$

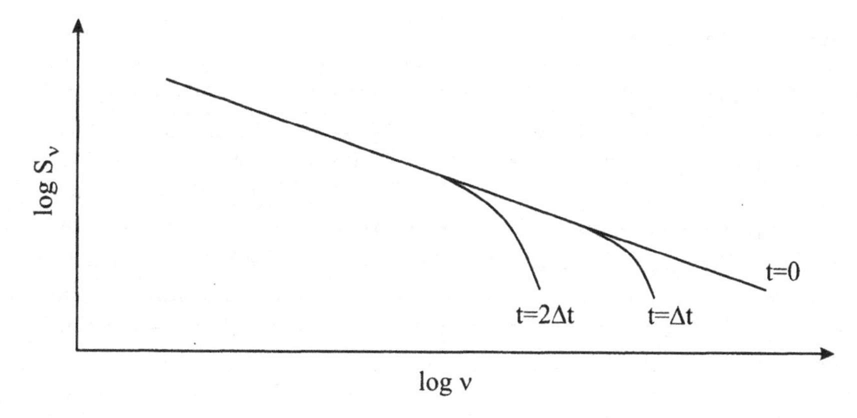

Figure 4. As the power-law synchrotron spectrum ages, the "knee" moves to lower frequencies.

Adopting a value of $\gamma_2/\gamma_1 \sim 10^4$, we can then calculate the expected flux density F_ν measured from a spherical source with radius R at a distance D:

$$\begin{aligned} F_\nu &= \frac{\mathrm{d}W}{\mathrm{d}t\mathrm{d}\nu\mathrm{d}V} \cdot \frac{\frac{4}{3}\pi R^3}{4\pi D^2} \\ &\approx 200\,\mathrm{Jy}\left(\frac{B}{\mathrm{mGauss}}\right)^{3.5}\left(\frac{R}{\mathrm{kpc}}\right)^{3}\left(\frac{\nu}{\mathrm{GHz}}\right)^{-0.5}\left(\frac{D}{\mathrm{Gpc}}\right)^{-2}. \end{aligned} \qquad (55)$$

This is obviously very strongly dependent on the magnetic field strength and the size of the emitting region. It is worth noticing that the flux density is not proportional to the energy density of the magnetic field ($U_B \propto B^2$), but $\propto U_B^{1.75}$.

Synchrotron radiation is strongly linearly polarised (the elliptical components cancel given a smooth distribution of pitch angles) up to a maximum of 75% for monoenergetic particles. For a power-law distribution of particles the degree of polarisation corresponds to $\Pi = \frac{p+1}{p+\frac{7}{3}}$ (69 % for $p = 2$). Tangling of B-fields, however, reduces this to much lower levels of typically a few percent. In radio galaxies only the hotspots are sources of highly polarised synchrotron emission. From the polarisation vectors one can deduce valuable information about the structure of the magnetic fields at the source — if one knows how much the polarisation vector has been rotated through Faraday rotation on the way from source to observer. This so-called "rotation measure" can be determined through multi-frequency observations because of its characteristic λ^2-dependence.

5. Absorption effects

The emitting particles not only produce synchrotron radiation but can also absorb it. If, for a given frequency, the particle density is high enough, i.e. in the optically thick case, these effects significantly alter the spectrum an observer receives. A rigid derivation is possible using the Einstein emission/absorption coefficients. We here, however, take a more phenomenological approach comparing the case of synchrotron radiation with that of thermal radiation. For thermal radiation, the source function S_ν (the specific intensity in W m^{-2} Hz^{-1} sr^{-1} that one receives from a source in case of high optical depth) is simply the Planck function

$$S_\nu = \left(\frac{2\nu^2}{c^2}\right)\left(\frac{h\nu}{e^{h\nu/kT}-1}\right). \tag{56}$$

The first factor represents a universal phase-space factor, while the second factor corresponds to the mean energy of an oscillator emitting radiation of frequency ν. In the low-frequency Rayleigh-Jeans limit $h\nu \ll kT$, it approximates kT. What would the second factor become in case of nonthermal synchrotron radiation? By analogy to the aforementioned, this should be the mean energy of a synchrotron electron emitting radiation of frequency ν. From our earlier discussion we know that particles with energy γmc^2 radiate dominantly at $\nu \approx \nu_c \propto \gamma^2$ as defined in (27). Combined this yields

$$\gamma \propto \nu^{1/2} \tag{57}$$

and substituting the second factor in (56) with (γmc^2) leads to

$$S_\nu \propto \left(\frac{2\nu^2}{c^2}\right)\left(\nu^{1/2}mc^2\right) \quad \Rightarrow \quad S_\nu \propto \nu^{5/2}. \tag{58}$$

Note that the spectral index $\kappa = 5/2$ does not depend on the power law index p at all and differs from the index of blackbody radiation in the Rayleigh-Jeans limit $\kappa = 2$.

In the optically thin case, on the other hand, e.g. at high frequencies, the spectrum follows the unaltered power-law synchrotron spectrum with index $\kappa = (1-p)/2$, i.e. $\kappa = -0.5$ for $p = 2$. Fig. 5 shows a "textbook example" of an astrophysical source with a self-absorbed synchrotron spectrum.

For frequencies lower than the frequency $\nu_c(\gamma_1)$ corresponding to the lower cutoff γ_1 of the particle power-law distribution, the radiation is dominated by particles with $\gamma = \gamma_1$ and its spectrum therefore approximates a monoenergetic synchrotron spectrum with $\gamma = \gamma_1$ and spectral index $\kappa = 1/3$ as given in (43). For optically thick emission, the spectrum then approaches that of a blackbody with $\kappa = 2$.

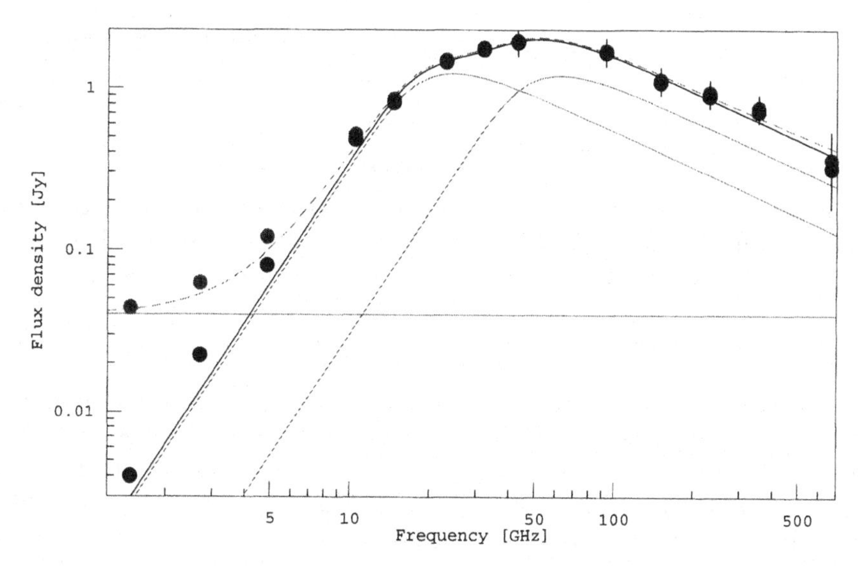

Figure 5. Synchrotron self-absorbed spectrum of III Zw 2, taken from Falcke *et al.* (1999). After subtraction of a constant background from the measured flux densities (grey points), the resulting data (black points) can be reproduced very well with a superposition of two self-absorbed synchrotron spectra with spectral indices $\kappa = 2.5$ in the low-frequency optically thick and $\kappa = -0.75$ in the high-frequency optically thin regime.

The absorption coefficient α_ν (the fractional loss of intensity per unit length) of a synchrotron source is given by eq. (6.53) of Rybicki and Lightman (1979), which, however misses a factor $(mc^2)^{p-1}$:

$$\begin{aligned} \alpha_\nu &= \left(mc^2\right)^{p-1} \frac{\sqrt{3}q^3}{8\pi m} \left(\frac{3q}{2\pi m^3 c^5}\right)^{p/2} C(B\sin\alpha)^{(p+2)/2} \\ &\times \Gamma\left(\frac{3p+2}{12}\right) \Gamma\left(\frac{3p+22}{12}\right) \nu^{-(p+4)/2}. \end{aligned} \quad (59)$$

For a typical synchrotron source with $p = 2$ and a pitch-angle corresponding to that of an isotropic distribution of particle velocities, $\alpha = \alpha_{\rm iso}$, we can parametrise this as

$$\alpha_\nu \approx 4.5 \cdot 10^{-12}\ \mathrm{cm}^{-1} \frac{\mathrm{k}}{\ln(\gamma_2/\gamma_1)} \left(\frac{\mathrm{B}}{\mathrm{G}}\right)^4 \left(\frac{\nu}{\mathrm{GHz}}\right)^{-3} \quad (60)$$

and estimate the optical thickness τ as a function of the length R of the line of sight through the source region to

$$\tau = \alpha_\nu R \approx 1.5 \cdot 10^{-3}\, k \left(\frac{B}{\mathrm{mGauss}}\right)^4 \left(\frac{R}{\mathrm{kpc}}\right) \left(\frac{\nu}{\mathrm{GHz}}\right)^{-3}. \quad (61)$$

The synchrotron self-absorption frequency ν_{ssa} is defined as the frequency for which $\tau = 1$,

$$\nu_{\mathrm{ssa}} \approx 115\,\mathrm{MHz}\; k^{1/3} \left(\frac{B}{\mathrm{mGauss}}\right)^{4/3} \left(\frac{R}{\mathrm{kpc}}\right)^{1/3}. \tag{62}$$

Calculating B from (55) and applying this to (62), we can relate the synchrotron self-absorption frequency and the corresponding flux-density to the size and distance of the source:

$$\nu_{\mathrm{ssa}} \approx 5.7\,\mathrm{MHz}\; k^{-1/17} \left(\frac{F_{\nu_{\mathrm{ssa}}}}{\mathrm{Jy}}\right)^{8/17} \left(\frac{D}{\mathrm{Gpc}}\right)^{16/17} \left(\frac{R}{\mathrm{kpc}}\right)^{-1}. \tag{63}$$

For observed flux densities around 1 Jy for a typical strong source and observation frequencies in the GHz regime it is then obvious that synchrotron self-absorption can only play a role for very compact sources where $R/D \sim 1$ milli-arcsecond. Lobes of radio galaxies are therefore optically thin, while their VLBI cores are optically thick.

6. Synchrotron self-Compton

Another important effect is the up-scattering of synchrotron photons by the high-energy synchrotron electrons through inverse Compton processes, which can lead to photons with energies as high as TeV. The derivation of the energy transfer from electron to photon is done most easily in the rest frame of the electron, in which a photon of (observer frame) energy $h\nu$ will have an energy $\sim \gamma h\nu$. As long as $\gamma h\nu \ll m_e c^2$, the cross section of the scattering process is then simply given by the Thomson cross section (17). The up-scattered photon will have an (observer frame) energy $\sim \gamma^2 h\nu$. The ratios of the photon energy in the observer frame, in the electron rest frame, and in the observer frame after the scattering event therefore correspond approximately to

$$1 : \gamma : \gamma^2. \tag{64}$$

For isotropic photon and electron distributions, the spectral form is basically unchanged from the original synchrotron power-law spectrum apart from an obvious stretching by a factor $\sim 4\gamma^2$. There is, however, a maximum attainable energy for the emitted photons that arises from the Klein-Nishima effect which reduces the cross section from its classical value for high photon energies. Photon energies higher than $\sim \gamma mc^2$ can therefore not be attained, i.e. the maximum is governed directly by the maximum electron energy.

The total power radiated by inverse Compton scattering can be shown to approximate (cf. Rybicki and Lightman 1979 eq. 7.16)

$$P_{\mathrm{compt}} = \frac{4}{3}\sigma_T c (\gamma\beta)^2 U_{\mathrm{ph}}, \tag{65}$$

where $U_{\rm ph}$ is the source photon energy density. Direct comparison with (19) shows that

$$\frac{P_{\rm synch}}{P_{\rm compt}} = \frac{U_B}{U_{\rm ph}}. \tag{66}$$

An example for a spectrum including synchrotron self-Compton emission is shown in Koerding and Falcke (2003, this volume) for the TeV-blazar Mrk 421.

References

Alfvén, H. and Herlofson, N. (1950), *Phys. Rev.* **78**, p. 616

Falcke, H., Bower, G.C., Lobanow, A.R. et al. (1999), *ApJ* **514**, p. L17

Falcke, H. and Gorham, P. (2002), *Astropart. Phys.* in press, astro-ph/0207226

Huege, T. and Falcke, H. (2002), *Proceedings of the 6th European VLBI Network Symposium, Ros, E., Porcas, R.W., Zensus, J.A. (eds.), MPIfR Bonn, Germany*, p. 25, astro-ph/0207647

Jackson, J.D. (1975) Classical Electrodynamics, *John Wiley & Sons, New York*, pp. 655–679

Koerding, E. and Falcke, H. (2003), *this volume*

Rybicki, G.B. and Lightman, A.P. (1979) Radiative Processes in Astrophysics, *John Wiley & Sons, New York*, pp. 167–194

Schwinger, J. (1949), *Phys. Rev.* **75**, p. 1912

Shklovsky, I.S. (1960) Cosmic Radio Waves, *Harvard Univ. Press, Cambridge, Ma.*, p. 191ff

Shu, F.H. (1991) The Physics of Astrophysics **Vol. I**, *University Science Books, Mill Valley, Ca.*, pp. 173–200

NON–THERMAL EMISSION FROM EXTRAGALACTIC RADIO SOURCES: A HIGH RESOLUTION – BROAD BAND APPROACH

GIANFRANCO BRUNETTI
Istituto di Radioastronomia del CNR
via P. Gobetti 101, I-40129 Bologna, Italy

1. ABSTRACT

In the framework of the study of extragalactic radio sources, we will focus on the importance of the spatial resolution at different wavelengths, and of the combination of observations at different frequency bands. In particular, a substantial step forward in this field is now provided by the new generation X-ray telescopes which are able to image radio sources in between 0.1–10 keV with a spatial resolution comparable with that of the radio telescopes (VLA) and of the optical telescopes. After a brief description of some basic aspects of acceleration mechanisms and of the radiative processes at work in the extragalactic radio sources, we will focus on a number of recent radio, optical and X–ray observations with arcsec resolution, and discuss the deriving constraints on the physics of these sources.

2. INTRODUCTION

This contribution is focussed on some basic concepts regarding the study of the non–thermal emission from extragalactic radio sources based on a broad band, multifrequency approach. The principal difficulty in this study arises by the different sensitivity and spatial resolution of the instruments in different bands. Radio telescopes easily reach sub–arcsec spatial resolutions and can image very faint sources by relatively short observations. On the other hand, optical telescopes generally have only arcsec spatial resolution so that combined radio – optical studies are limited by the lower spatial resolution of the optical telescopes. Although, the problem is alleviated by making use of optical HST observations, it still remains to a certain degree. In order to extend this approach to higher energies, X–ray telescopes with arcsec resolution (at least) and good sensitivity are required. Neverth-

F. Mantovani and A. Kus (eds.), The Role of VLBI in Astrophysics, Astrometry and Geodesy, 29–82.

less the poor spatial resolution of the past X–ray observatories, pioneeristic studies in this direction have been attempted in the last 10-20 years with combined radio (and optical) and *Einstein* or ROSAT X–ray observations. More recently, a substantial step forward has been achieved thanks to the

TABLE 1. X-ray Observatories

Satellite	Instrument	Flux→1 cts/s (erg/s/cm^2)	Energy Band (keV)	Resolution (arcsec)
Einstein	HRI	1.6E-10	0.5-4.0	4
	IPC	2.8E-11	0.5-4.0	
ROSAT	HRI	3.7E-11	0.5-2.4	3*
	PSPC	1.4E-11	0.5-2.4	20
ASCA	SIS	3.3E-11	0.4-12	180
	GIS	3.8E-11	0.4-12	180
BeppoSAX	MECS	8.1E-11	1.3-10	60
	PDS	9.5E-11	13-80	
Chandra	HRC-I	2.8E-11	0.4-10	0.5
	ACIS-I	1.1E-11	0.4-10	0.5

Notes: **Column 3**, Flux→ is the flux (erg/s/cm^2) necessary to have 1 cts/s in the detector. **Column 5**: (*) means that the spatial resolution results affected by errors in the aspect solution associated with the wobble of the space craft.

advent of the new generation X–ray satellites: *Chandra* and XMM-*Newton*. In Tab.1 we report the main capabilities of a selection of past X–ray observatories compared with those of *Chandra*: the abrupt increase of the spatial resolution combined with the high effective area of *Chandra* represent a 'new revolution' in astrophysics (data taken from Cox, 1999). For the first time it is now possible to investigate the X–ray properties of extragalactic radio sources performing spatially resolved spectroscopy over a relatively large energy band (about 0.4–10 keV). *Chandra* allows us to obtain images with $\leq$arcsec spatial resolution, comparable with the typical resolution of the VLA radio images and of the optical telescopes and thus it is a tool to perform, for the first time, multiwavelength studies from the radio band to the X–rays.

As we will show in this contribution, the high resolution broad band study of non–thermal emission from the extragalactic radio sources allows us to sample the emission due to very different portions of the spectrum of the relativistic electrons or to study the emission due to different emitting mechanisms at work in the observed regions. After the first 3 years of new *Chandra* observations, it is now clear that the scenarios describing these objects should now be partially revisited.

In Sect.3 we discuss some theoretical aspects related to the relativistic plasma in extragalactic radio sources. More specifically, in Sects. 3.1–3.3 we report the basic concepts of particle acceleration and discuss the resulting spectrum of the relativistic electrons in the framework of the standard shock acceleration scenario. In Sect. 3.4 we introduce the non–thermal emitting mechanisms. In particular, we focus on the sampling of the spectrum of the emitting electrons provided by the observations at different frequencies. In Sect. 3.5 we review the basic methods to derive the energetics of the relativistic plasma in the extragalactic radio sources. Finally, in Sect.4 we concentrate on the most recent high resolution multiwavelength studies of extragalactic radio sources. Our 'biased' review is mainly focussed on the new *Chandra* observations of radio lobes, radio jets and radio hot spots. Combining the data with the methods showed in Sect. 3, we discuss on the new constraints on the shape of the low and high energy end of the electron spectrum, on the kinematics of radio jets, and on the energetics of extragalactic radio sources.

In this lecture I will focus on standard-leptonic models. For secondary and hadronic models the reader may refer to Mannheim et al. (1991) and Aharonian (2002).

3. PARTICLE ACCELERATION AND RADIATIVE MECHANISMS

In this Section we give the basic concepts of the particle acceleration and of the time evolution of the energy of the particles. We focus our attention on the case of electrons which are responsible for the observed non–thermal emission. The acceleration of electrons to high energies in various environments is a problem of widespread interest in astrophysics. Indeed, the synchrotron emission of jets in a large number of extragalactic radio sources (e.g., Miley, 1980; Fomalont, 1983; Venturi, this proceedings) requires in situ acceleration of electrons because of the short synchrotron life–times (e.g., Rees, 1978; Bicknell & Melrose, 1982). Also the radio luminosity of supernova remnants and their broad band spectrum (e.g., Clark & Caswell, 1976; Dickel, 1983) suggest that electrons are accelerated to high energy in these cases. Finally, the diffuse Mpc–size radio emission discovered in an increasing number of clusters of galaxies (e.g., Feretti & Giovannini 1996) requires large scale in situ acceleration of protons and electrons (e.g., Jaffe 1977; Tribble 1993; Sarazin 1999; Blasi 2001; Brunetti et al. 2001a; Petrosian 2001).

The basic goal of this Section is simply to show, based on general theoretical arguments, that the energy distribution of the relativistic electrons in extragalactic radio sources cannot be a simple power law. Although a power law energy distribution is commonly assumed in the literature, here we will point out that this approximation cannot be adopted in a broad band, multiwavelength (radio to X–ray observations) approach, and further that the observations can be used to constrain the energy distribution of the emitting electrons. In this way we will test the acceleration theories.

3.1. COMPETING MECHANISMS AT WORK

The relativistic electrons in extragalactic radio sources are subject to several inescapable processes that change their energy with time. For a general description of the leading mechanisms at work we refer the reader to classical books (e.g., Ginzburg 1969; Pacholczyk 1970; Rybicki & Lightman 1979; Melrose 1980).

Here, it is important to underline that the efficiency of these processes is related to the energy of the electrons in a way that depends on the particular process. As a consequence, the time evolution of relativistic electrons at different energies is expected to be dominated by different processes. In the following, we give some basic relationships:

• Relativistic electrons with Lorentz factor γ mixed with a thermal plasma lose energy via ionization losses and Coulomb collisions; for relativistic electrons one has :

$$\left(\frac{d\gamma}{dt}\right)^{-}_{\rm ion} = -1.2 \times 10^{-12} n \left[1 + \frac{\ln(\gamma/n)}{75}\right] \quad (1)$$

where n is the number density of the thermal plasma.

• If the relativistic electrons are confined into an expanding region, they lose energy via adiabatic expansion:

$$\left(\frac{d\gamma}{dt}\right)^{-}_{\rm ex} = -\frac{1}{R}\frac{dR}{dt}\gamma \sim 10^{-11}\gamma \frac{v_{\rm ex}(t)}{c}\frac{1}{R(t)_{\rm kpc}} \quad (2)$$

where $v_{\rm ex}$ and R are the expansion velocity and dimension of the region, respectively.

• Relativistic electrons in a magnetic field B lose energy via synchrotron radiation :

$$\left(\frac{d\gamma}{dt}\right)^{-}_{\rm syn} = -1.9 \times 10^{-9}\gamma^2 B^2 \sin^2\theta \quad (3)$$

and via inverse Compton scattering with photons (of energy density $\omega_{\rm ph}$); in the Thompson regime, one has:

$$\left(\frac{d\gamma}{dt}\right)^{-}_{\rm ic} = -3.2 \times 10^{-8}\gamma^2\omega_{\rm ph} \tag{4}$$

• Finally, the electrons can be accelerated/re–accelerated by several mechanisms. The most commonly considered in the literature are the Fermi and Fermi–like processes. It should be stressed that, in general, such mechanisms are stochastic processes, i.e., given an initial monoenergetic electron distribution, they not only produce a net increase of the energy of the electrons but also a broadening of the energy distribution. Thus the effect of a reacceleration mechanism on an initial energy distribution can be described by the combination of a systematic acceleration of the electrons and of a 'statistical' broadening of the resulting energy distribution.
Two relevant cases are:

a) acceleration due to MHD turbulence; this is discussed in detail in a number of papers and classical books (e.g., Melrose 1980). If the resonance scattering condition (e.g., Eilek & Hughes 1991) is satisfied (this in general requires already relativistic electrons, e.g. Hamilton & Petrosian 1992), turbulent Alfven waves can accelerate electrons via resonant pitch angle scattering. For a power law energy spectrum of the Alfven waves:

$$P(k) = b\frac{B^2}{8\pi}\frac{s-1}{k_o}\left(\frac{k}{k_o}\right)^{-s} \tag{5}$$

in the range $k_o < k < k_{\rm max}$, where k is the wave number ($k_o << k_{\rm max}$) and b is a normalization factor indicating the fractional energy density in waves, the systematic energy gain is (e.g., Isenberg 1987; Blasi 2000; Ohno et al. 2002):

$$\left(\frac{d\gamma}{dt}\right)^{+}_{\rm A-tur} \sim (s+2)\xi_{\rm A}\gamma^{s-1} \tag{6}$$

where

$$\xi_{\rm A} = \frac{s-1}{s(s+2)}\frac{\pi b k_o {v_{\rm A}}^2}{c}\left(\frac{eB}{m_e c^2 k_o}\right)^{2-s} \tag{7}$$

and where $v_{\rm A}$ is the Alfven velocity.
MHD turbulence can also accelerate relativistic particles in radio sources via Fermi–like processes (e.g., Lacombe 1977; Ferrari et al. 1979). Under the simple assumption of a quasi–monochromatic turbulent scale (e.g., Gitti et al. 2002) the systematic energy gain is :

$$\left(\frac{d\gamma}{dt}\right)^{+}_{\rm F-tur} \simeq 4 \times 10^{-11}\gamma\frac{{v_{\rm A}}^2}{l}\left(\frac{\delta B}{B}\right)^2 \tag{8}$$

where l is the distance between two peaks of turbulence and $\delta B/B$ is the fluctuation in a peak of the field intensity with respect to the average field strength.

b) acceleration in a shock; (e.g., Meisenheimer et al. 1989); one has :

$$\left(\frac{d\gamma}{dt}\right)^{+}_{\rm sh} \simeq \gamma \frac{U_-{}^2 c}{r} \left(\frac{r-1}{r\lambda_+ + \lambda_-}\right) \tag{9}$$

where U_- is the velocity of the plasma in the region before the shock discontinuity (measured in the shock frame and in unit of c), r is the shock compression ratio, and $\lambda_\pm$ is the mean diffusion length of the electrons in the region before ($-$) and after ($+$) the shock. Assuming a similar B–field configuration and strength before and after the shock (i.e. at some distance from the shock in the pre– and post–shock region) Eq.(9) can be generalized as :

$$\left(\frac{d\gamma}{dt}\right)^{+}_{\rm sh} \simeq \gamma \frac{(cU_-)^2}{r} \left(\frac{r-1}{r+1}\right) \frac{1}{3\mathcal{K}(\gamma)} \tag{10}$$

where $\mathcal{K}(\gamma)$ is the spatial diffusion coefficient which depends on the physical conditions.

In the relevant astrophysical cases the time evolution of the energy of the electrons is given by the combination of several mechanisms, i.e. :

$$\left(\frac{d\gamma}{dt}\right) = \sum_i^{..} \left(\frac{d\gamma}{dt}\right)^{-}_i + \sum_i^{..} \left(\frac{d\gamma}{dt}\right)^{+}_i \tag{11}$$

In Fig.1 we report $d\gamma/dt$ due to the above processes as a function of γ for representative physical conditions. It is clear that the different processes dominate in different energy regions with the Coulomb losses being especially important at low energies. In addition, due to its dependence on the energy ($\propto \gamma$), the Fermi–like acceleration can be an efficient process in a restricted energy band, whereas Coulomb and radiative losses in general, prevent electron acceleration at lower and higher energies, respectively.

3.2. THE BASIC EQUATIONS OF PARTICLE EVOLUTION

In order to consider the general evolution of the particle spectrum, a kinetic theory approach is suitable (e.g., Melrose 1980; Livshitz & Pitaevskii 1981; Blandford 1986; Blandford & Eichler 1987; Eilek & Hughes 1991, and reference therein). Given $f(\mathbf{p},..)$, the distribution function of the electrons (i.e. $f(\mathbf{p},..)d\mathbf{p}$ is the number of electrons in the element $d\mathbf{p}$ of the momentum space), the description of the behaviour of this distribution function is given by a Boltzman equation:

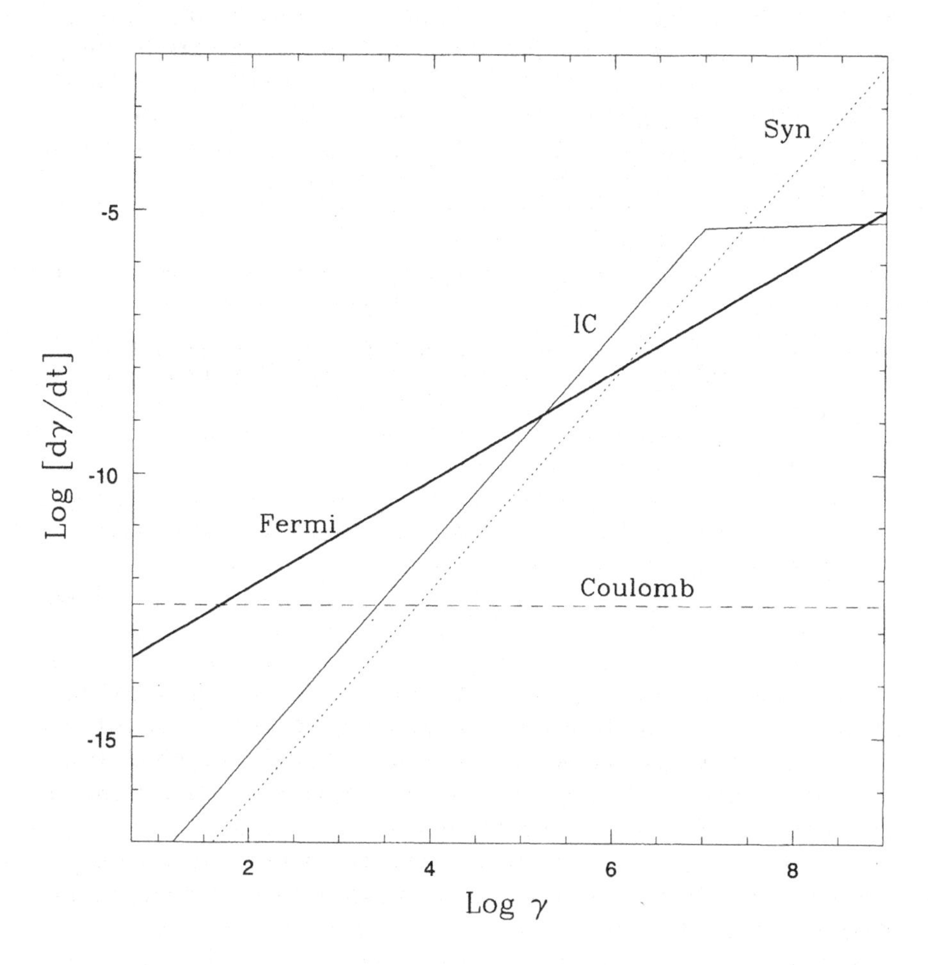

Figure 1. A sketch of the efficiency of acceleration (solid line) and loss (dashed lines) processes is shown, as a function of the Lorentz factor of the electrons. The artificial knee introduced in the IC losses at $\gamma > 10^7$ simulates the effect of the KN cross section.

$$\frac{df(\mathbf{p})}{dt} = \left(\frac{\partial f}{\partial t}\right)_{\rm coll} + \left(\frac{\partial f}{\partial t}\right)_{\rm diff} \tag{12}$$

where, the total time derivative is to be interpreted according to:

$$\frac{d}{dt} \rightarrow \frac{\partial}{\partial t} + \mathbf{v} \cdot \frac{\partial}{\partial \mathbf{r}} + \mathbf{F} \cdot \frac{\partial}{\partial \mathbf{p}} \tag{13}$$

where $\mathbf{v}$ is the particle velocity, and $\mathbf{F} = d\mathbf{p}/dt$ the force acting on the particles.

The *diffusion* term in Eq.(12) describes spatial diffusion, while the *collision* term accounts for all the physics of collisions and scattering (e.g., radiative losses, interaction with waves and shocks, Coulomb collisions). A general, stochastic acceleration may be thought of as a diffusion in the momentum space given by a diffusion coefficient, D_{pp}, so that in case of no spatial diffusion and no losses, Eq.(12) can be written as:

$$\frac{df(\mathbf{p})}{dt} = \frac{\partial}{\partial p_i}[(D_{pp})_{ij}(\mathbf{p})\frac{\partial}{\partial p_i}f(\mathbf{p})] \tag{14}$$

If isotropy of the scattering waves and electrons is assumed, $f(\mathbf{p})d\mathbf{p} = 4\pi p^2 f(p)dp$ and Eq.(12) can be written as (e.g., Melrose 1980; Eilek & Hughes 1991):

$$\frac{\partial}{\partial t}f(\mathbf{r},p,t) + \frac{\partial}{\partial x_i}[\mathcal{K}_{ij}\frac{\partial}{\partial x_j}f(\mathbf{r},p,t)] = \frac{1}{p^2}\frac{\partial}{\partial p}[\mathcal{S}p^4 f(\mathbf{r},p,t)$$
$$+p^2 D_{pp}\frac{\partial}{\partial p}f(\mathbf{r},p,t) + p^2 I(p)f(\mathbf{r},p,t)] + Q(\mathbf{r},p,t) \tag{15}$$

where we have considered the effect of the synchrotron and inverse Compton losses ($dp/dt = -\mathcal{S}p^2$), of the Coulomb losses ($dp/dt = -I(p)$) and considered a general form for the spatial diffusion of the electrons ($\mathcal{K}$ being the spatial diffusion coefficient). A term accounting for electron injection (injection rate $= Q$) has also been introduced.
Eq.(15) can be also written in the energy space. For an isotropic distribution of the momenta of the relativistic electrons, the electron spectrum is $N(\epsilon) = 4\pi p^2 f(p)dp/d\epsilon$ and one has :

$$\frac{\partial}{\partial t}f(\mathbf{r},p,t) = \frac{c}{4\pi p^2}\frac{\partial}{\partial t}N(\mathbf{r},\epsilon,t) \tag{16}$$

$$\frac{\partial}{\partial p}[\mathcal{S}p^4 f(\mathbf{r},p,t)] = \frac{c}{4\pi}\frac{\partial}{\partial p}[\mathcal{S}p^2 N(\mathbf{r},\epsilon,t)] \tag{17}$$

$$\frac{\partial}{\partial p}[p^2 I(p)f(\mathbf{r},p,t)] = \frac{c}{4\pi}\frac{\partial}{\partial p}[I(p)N(\mathbf{r},\epsilon,t)] \tag{18}$$

$$\frac{\partial}{\partial p}[p^2 D_{pp}\frac{\partial}{\partial p}f(\mathbf{r},p,t)] = \frac{c}{4\pi}\frac{\partial}{\partial p}[D_{pp}\frac{\partial}{\partial p}N(\mathbf{r},\epsilon,t) - \frac{2}{p}D_{pp}N(\mathbf{r},\epsilon,t)] \tag{19}$$

thus Eq.(15) can be written as:

$$\frac{\partial}{\partial t}N(\mathbf{r},\epsilon,t)+\frac{\partial}{\partial x_i}[\mathcal{K}_{ij}\frac{\partial}{\partial x_j}N(\mathbf{r},\epsilon,t)]=\frac{\partial}{\partial p}[\mathcal{S}p^2N(\mathbf{r},\epsilon,t)$$
$$I(p)N(\mathbf{r},\epsilon,t)++D_{pp}\frac{\partial}{\partial p}N(\mathbf{r},\epsilon,t)-\frac{2}{p}D_{pp}N(\mathbf{r},p,t)]+Q(x,\epsilon,t) \quad (20)$$

Let us assume that the reacceleration processes are Fermi–like, i.e. $D_{pp}=\frac{\chi}{2}p^2$ and that the Coulomb losses are negligible, i.e. $I(p)\sim 0$. In terms of Lorentz factors and neglecting the dependence on the spatial coordinates, it is possible to rewrite Eq.(20) as :

$$\frac{\partial}{\partial t}N(\gamma,t)=-\frac{\partial}{\partial\gamma}\left[\frac{d\gamma}{dt}N(\gamma,t)\right]+\frac{\chi}{2}\frac{\partial}{\partial\gamma}\left[\gamma^2\frac{\partial N(\gamma,t)}{\partial\gamma}\right]-\frac{N(\gamma,t)}{T_{\rm es}}+Q(\gamma,t) \quad (21)$$

where

$$\frac{d\gamma}{dt}=\chi\gamma-\mathcal{S}\gamma^2\mathrm{mc} \quad (22)$$

for simplicity we have replaced the general spatial diffusion term with a simple term $N/T_{\rm es}$, accounting for an energy independent diffusion of the electrons from the system. For a general treatment of the solution of Eq.(21) we refer the reader to Kardashev (1962). Some additional more recent applications can be found in a number of papers (e.g., Borovsky & Eilek 1986; Sarazin 1999; Brunetti et al. 2001a; Petrosian 2001 and references therein).

3.3. SHOCK ACCELERATION AND ELECTRON SPECTRUM

Shock acceleration is a well known astrophysical process (see Drury 1983 and Blandford & Eichler 1987 for a review). Energetic charged particles can be efficiently accelerated in shock waves either by drifts in the electric fields at the shock (e.g., Webb et al., 1983 and references therein), or by scattering back and forth across the shock in the magnetic turbulence present in the background plasma (e.g., Bell 1978a,b; Blandford 1979; Drury 1983). In the simplest case, i.e. that of a steady state shock without losses and with a seed monoenergetic particle distribution, the accelerated spectrum is a simple power law (e.g., Bell 1978a,b; Blandford & Ostriker, 1978). On the other hand, the effect of synchrotron and inverse Compton losses in diffusive shock acceleration causes the development of a cut–off in the accelerated spectrum and a break in the electron spectrum integrated over the post shock region (e.g., Meisenheimer et al. 1989). Detailed studies of the diffusive shock acceleration including energy losses under steady state conditions, find the development of both a high energy cut–off, and humps in the accelerated spectrum which depend on the physical conditions (Webb et al. 1984). Finally, more recent studies have investigated non steady state diffusive shock acceleration (Webb & Fritz 1990), multiple shocks acceleration (e.g., Micono et al. 1999; Marcowith & Kirk 1999), and particle acceleration via relativistic shocks (e.g., Kirk et al. 2000).

3.3.1. *A simplified semi-analytic approach*

In this Section we discuss a relatively simple analytic method to derive the energy distribution of the relativistic electrons accelerated in a shock region. In what follows we assumed that :

a) relativistic electrons, continuously injected in the shock region with a power law energy distribution are (re-)accelerated in this region subject to radiative losses. The final spectrum in the shock region is thus given by the competition between acceleration and loss terms;

b) the accelerated electrons diffuse from the shock region and continuously fill a post–shock region in which most of the observed emission is produced. As an additional simplification, we assume that the electrons are transported throughout the post–shock region with a constant velocity so that older electrons are located at larger distances from the shock. In the post–shock region the electrons are not (re-)accelerated but they are only subjects to radiative losses so that the energy of the electrons decreases with time.

As in Kirk et al.(1998), the electrons are assumed to be reaccelerated by Fermi–I like processes in the shock region from which they typically escape in a time $T_{\rm es}$. To calculate the electron spectrum in the shock region we can solve Eq.(21) with a time independent approach (i.e. $\partial N/\partial t = 0$):

$$\left(\frac{d}{d\gamma}N(\gamma)\right)\frac{d\gamma}{dt} + N(\gamma)\left[\frac{d}{d\gamma}\frac{d\gamma}{dt} + \frac{1}{T_{\rm es}}\right] - Q(\gamma) = 0 \tag{23}$$

where

$$\frac{d\gamma}{dt} = \chi\gamma - \beta\gamma^2 \tag{24}$$

$\beta = \mathrm{mc}\mathcal{S}$ in Eq.(22), and $Q(\gamma) = Q\gamma^{-s}$ (for $\gamma < \gamma_*$). The solution is :

$$N(\gamma) = \frac{Q}{\chi}\left(1 - \frac{\gamma}{\gamma_{\rm c}}\right)^{-\alpha_-} \gamma^{-\alpha_+} \int_{\gamma_{\rm low}}^{\tilde{\gamma}} \frac{y^{\alpha_+ - (1+s)}dy}{[1 - y/\gamma_{\rm c}]^{1-\alpha_-}} \tag{25}$$

where $\alpha_\pm = 1 \pm [\chi T_{\rm es}]^{-1}$, $\tilde{\gamma} = \min(\gamma_*, \gamma)$, and $\gamma_{\rm low}$ is an *artificial* cut–off introduced in the integration of the equations corresponding to the minimum energy of the electrons which can be accelerated by the shock. Indeed, it is well known that only particles with a Larmor radius larger than the thickness of the shock are actually able to 'feel' the discontinuity at the

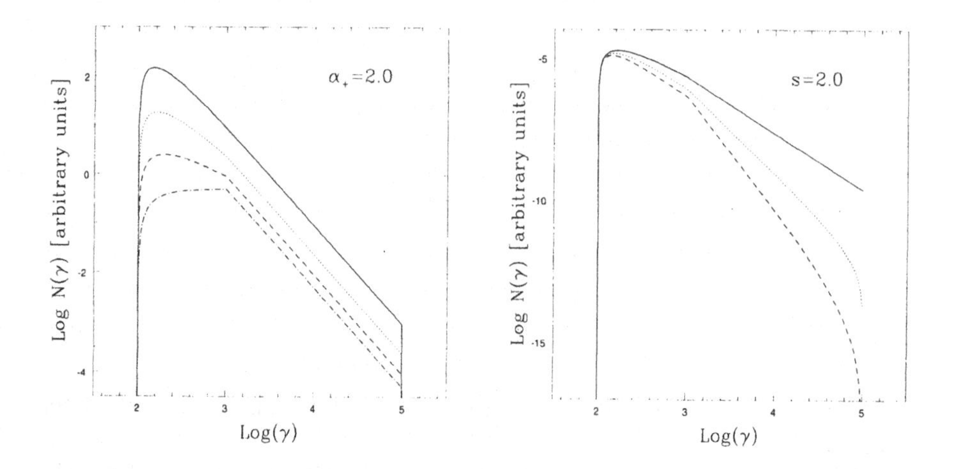

Figure 2. The time independent electron spectra in the shock region are reported as a function of the energy of the electrons for different values of the relevant parameters. Left panel: from the bottom we have: s=0, 1, 2, 3; $\alpha_+ = 2.0$. Right panel: from the bottom we have: α_+=4, 3, 2; $s = 2.0$.

shock (e.g., Eilek & Hughes 1991 and ref. therein). The shock thickness is of the order of the Larmor radius of thermal protons so that, as a first approximation, we might use $\gamma_{\rm low} \geq 10$ in the case of electrons.

The electron spectrum in the shock region is reported in Fig.2: a low energy cut–off is formed around $\gamma_{\rm low}$, whereas a high energy cut–off is formed at $\gamma_{\rm c} = \chi/\beta$, i.e. the energy at which the radiative losses outweight the acceleration efficiency. An additional break in the spectrum is obtained around γ_*; this break is due to the fact that, for $\gamma > \gamma_*$, all the electrons injected in the shock region can be accelerated at higher energies, whereas for $\gamma < \gamma_*$, only the electrons injected with energy $< \gamma$ can contribute to the accelerated spectrum.

As already anticipated, during the step b), once the electrons are re–accelerated in the shock region, they travel towards the post shock region subject to the radiative losses only. The evolution of the spectrum (25) is obtained solving the *continuity* equation (Eq.21) taking into account only the effect of the radiative losses (i.e., $d\gamma/dt = -\beta\gamma^2$) :

$$\frac{\partial N(\gamma,t)}{\partial t} = -\frac{\partial}{\partial \gamma}\left[\frac{d\gamma}{dt}N(\gamma,t)\right] \tag{26}$$

assuming an approximately constant magnetic field strength (i.e., β =const.), and a constant diffusion velocity $v_{\rm D}$, the resulting spectrum of the electrons at a distance $r = v_{\rm D}t$ is :

$$N(\gamma, t) = \frac{Q}{\chi}(1-\beta\gamma t)^{\alpha_+-2}\left[1-\frac{\gamma/\gamma_c}{1-\beta\gamma t}\right]^{-\alpha_-}\gamma^{-\alpha_+}\times$$
$$\int_{\gamma_{low}}^{\tilde{\gamma}(t)} y^{\alpha_+-(s+1)}\left[1-\frac{y}{\gamma_c}\right]^{\alpha_--1}\left[1-\frac{y}{\gamma_*}\right]^{s-2} dy \qquad (27)$$

where

$$\tilde{\gamma}(t) = \min\left\{\gamma_*, \frac{\gamma}{1-\gamma\beta t}\right\} \qquad (28)$$

Fig.3 shows the time/spatial evolution of the spectrum in the post–shock region.

In order to further simplify the scenario, we might assume that the spectrum of the injected electrons and that of the accelerated electrons have the same slope, i.e. $s = \alpha_+$ (as roughly expected if the electrons are continuously reaccelerated and released by a series of similar shocks in the jets) and that the acceleration time in the shock roughly equals the escape time, i.e. $\chi \sim \frac{1}{T_{es}}$. In this case Eq.(27) becomes :

$$N(\gamma, t) \simeq Q\gamma^{-2}\int_{\gamma_{low}}^{\tilde{\gamma}(t)} y^{-1}\left[1-\frac{y}{\gamma_c}\right]^{-1} dy \qquad (29)$$

In general, the lack of spatial resolution better than 1 arcsec over a wide energy band does not allow the resolution of the post–shock region in extragalactic radio sources at different frequencies. In fact, only information on the integrated emission from this region can be obtained. As a consequence, in order to obtain theoretical results to be compared with observations, one should deal with the electron spectrum integrated over the post–shock region. The size of this region, L, is determined by the diffusion length of the electrons in the largest considered time, T_L, so that the volume integrated spectrum of the electron population, $N_i(\gamma)$, is given by the sum of all the electron spectra in this region, i.e. it is obtained by integrating Eq(27, or 29) over the time interval $0 - T_L$ (or in an equivalent way over the distance interval from the shock $0 - L$ taking into account the relationship linking time and space). Assuming Eq.(29), the resulting integrated spectrum is reported in Fig.4 with the breaks and cut–offs indicated in the panel.

3.4. RADIATIVE PROCESSES AND ELECTRON SPECTRUM

One of the basic motivations for this contribution is the possibility to study extragalactic radio sources at different wavelength with comparable spatial resolution which has come about in the last 2–3 years. In this Section we

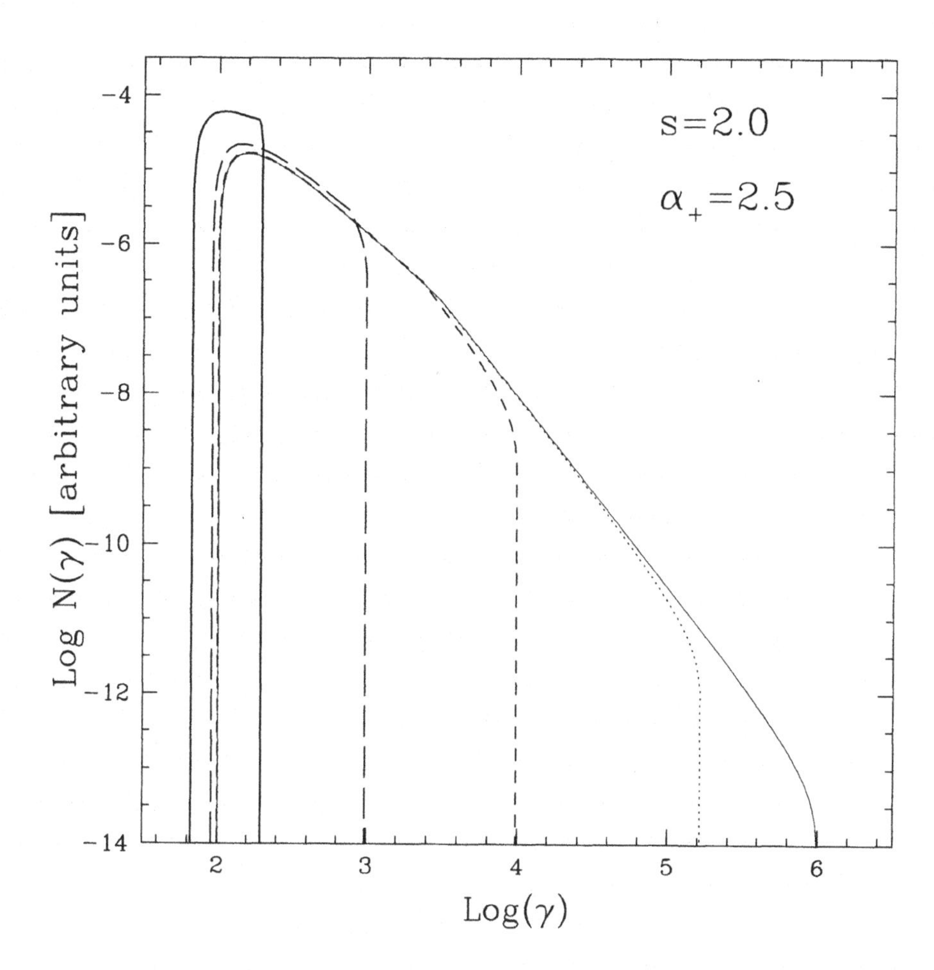

Figure 3. The time (or spatial) evolution of the electron spectrum in the post–shock region is reported for $s = 2.0$, $\alpha_+ = 2.5$, $\chi = 10^{-12}\mathrm{s}^{-1}$ and $\beta = 10^{-18}\mathrm{s}^{-1}$. The different spectra correspond to a time after injection in the post-shock region $\tau = 0$ (solid line), $5/\chi$ (dotted line), $10^2/\chi$ (dashed line), $10^3/\chi$ (long dashed line), $5 \times 10^3/\chi$ (thick solid line).

will point out that broad band (possibly from radio to X–rays) studies of the non–thermal emission from radio sources can allow us to derive unique information on the electron energy distribution.

Based on theoretical argumentations, in the preceding Section we have shown that the power law approximation for the electron spectrum, usually adopted to calculate the synchrotron and inverse Compton spectrum from extragalactic radio sources, is now inadequate. In particular we have shown that cut–offs at low and high energies, $\gamma_{\rm low}$ and $\gamma_{\rm c}$ respectively, might be

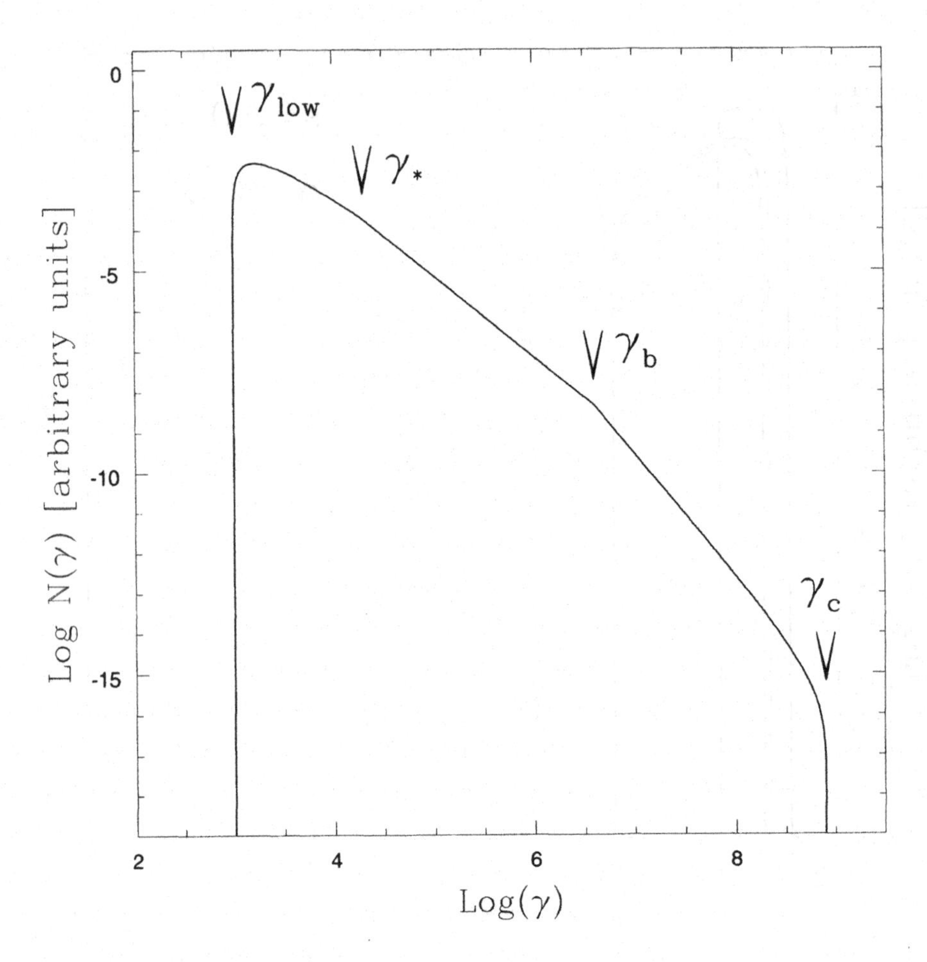

Figure 4. The electron spectrum integrated over the post–shock region is reported as a function of the energy for a given set of parameters. The energies corresponding to the relevant breaks and cut–offs are indicated.

expected if the relativistic electrons are accelerated in a shock region. In addition, a break at intermediate energies, γ_b, is expected if the accelerated electrons age in a post–shock region where acceleration processes are not efficient. Finally, a low energy break, γ_*, or a flattening of the spectrum, might be expected depending on the energy distribution of the electrons (already relativistic) when injected by the jet flow into the shock region. The presence and the location of all the above breaks and cut–offs depend on the relative importance of the mechanisms at work. As a consequence, broad band studies of non–thermal emission from the extragalactic radio sources, allow us not only to constrain the shape of the spectrum of the emitting electrons, but they can also constrain the efficiency of the different mechanisms.

In what follows we concentrate on deriving the energy of the electrons emitting in different frequency bands synchrotron and inverse Compton photons from compact regions (jets and hot spots) and from extended regions (radio lobes). For a general treatment of these emitting mechanisms, again, we refer the reader to classical books (e.g., Ribicky & Lightman 1979; Melrose 1980) or to seminal papers (e.g., Blumenthal & Gould 1970).

3.4.1. *Radio Lobes*

The most important radiative processes active in the radio lobes are the synchrotron emission and the IC scattering of external photons. As the radio lobes are generally regions extended tens or hundreds of kpc, the IC scattering of the synchrotron radiation produced by the same electron population (SSC) is not in general an efficient process.

SYNCHROTRON EMISSION: the typical energy of the electrons responsible for synchrotron emission observed at radio, optical and X–ray frequencies is given by:

$$\gamma_{\rm syn} \sim \frac{1}{2}\frac{10^3}{{B_{\mu G}}^{1/2}}{\nu_{\rm MHz}}^{1/2} \sim 5000 \\ \sim 1\times10^6 \\ \sim 4\times10^7 \tag{30}$$

where we have considered a magnetic field strength of the order of $\sim 30\mu$G, typical of the lobes of relatively powerful FRII radio galaxies.

IC EMISSION: the relativistic electrons in the radio lobes can IC scatter external photons to higher frequencies. The nature of the external photons depends on the astrophysical situation we are considering. Here we focus on the IC scattering of the CMB photons and on the IC scattering of the nuclear (quasar, QSO) photons.

- The IC scattering of CMB photons into the X–ray band is a well known process (e.g., Harris & Grindlay 1979) and it has been successfully revealed in a few radio galaxies (Section 4.1).

Optical emission from IC scattering of CMB photons by lower energetics electrons ($\gamma \sim 20$) was also tentatively proposed in order to account for the *optical-UV alignment* discovered in high–z radio galaxies (e.g., Daly 1992). So far, to our knowledge, there are no cases of detection of this effect. This may possibly be related to the expected flattening of the electron spectrum at $\gamma < 50$ due to the Coulomb losses (Sect.5).

- More recently, Brunetti et al. (1997) have proposed that the IC scattering of nuclear photons into the X–ray band might be an efficient process in powerful and relatively compact FRII radio galaxies and quasars. Powerful nuclei can isotropically emit up to $\sim 10^{47}$erg/s in the far-IR to optical band so that their energy density typically outweight that due to the CMB for ≤ 100 kpc distance from the nucleus.

The energy of the electrons emitting in the optical and X–ray band due to IC scattering is:

$$\begin{aligned}\gamma_{\rm ic} \sim \left(\frac{\nu}{\nu_{\rm E}}\right)^{1/2} &\sim 1000 \qquad ({\rm CMB} \rightarrow {\rm X-ray}) \\ &\sim 20 \qquad ({\rm CMB} \rightarrow {\rm optical}) \\ &\sim 100 \qquad ({\rm QSO} \rightarrow {\rm X-ray}) \qquad (31)\end{aligned}$$

$\nu_{\rm E}$ and ν being the frequency of the external (seed) and of the scattered photons, respectively.

Both IC/CMB and IC/QSO can produce detectable X–rays from the radio lobes which might introduce some degeneracy in the interpretation of the observed fluxes. However, it should be noticed that the contribution from the above IC processes can be easily disentangled based on the morphology of the observed X–ray emission. Indeed, in the IC/QSO model the photons propagate from the nucleus and their momenta are not isotropically distributed when they scatter with the electrons in the lobes: i.e., the IC scattering process is anisotropic. It follows that at any given energy of the scattered photons there will be many more scattering events when the velocities of the relativistic electrons point to the nucleus, (i.e. the direction of the incoming photons) than when they point in the opposite direction. The resulting IC emission will be enhanced towards the nucleus and will be essentially absent in the opposite direction. As a consequence, if the radio axis of the radio source does not lie on the plane of the sky, the X–rays from IC/QSO will be asymmetric: the smaller the angle between the axis and the line of sight, the greater the difference in IC emission from two identical lobes (Brunetti et al. 1997; Brunetti 2000).

In Fig.5 we report a compilation of the energy ranges selected by the different emitting processes in the different bands superimposed on representative electron spectra produced in different acceleration scenarios. As a net result, the combination of studies of the spectra emitted by different processes at different frequencies, will allow us to derive information/constraints on the electron spectrum which cover all the relevant energies.

In particular, it should be stressed that :

• In general, the detection of synchrotron radiation from the radio lobes at high frequencies (optical or X–rays) requires $\gamma > 10^5$ emitting electrons. Due to the radiative losses, these electrons have life times which are at

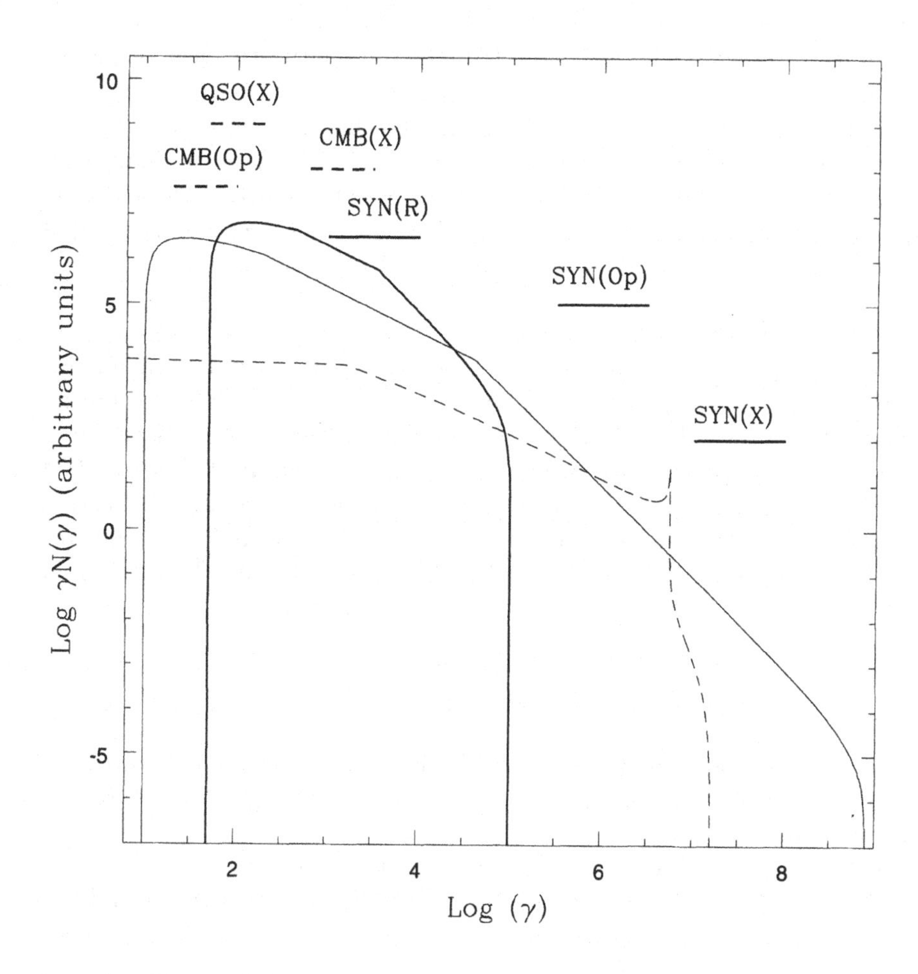

Figure 5. Scheme of the sampling of the spectrum of the emitting electrons with the different processes (SYN= synchrotron, CMB= IC/CMB, QSO= IC/QSO) in different bands (R= Radio, Op= Optical, X= X-ray) in the case of extended regions. The different electron spectra are shown for a set of different acceleration models.

least one order of magnitude shorther than the dynamical time/age of the radio lobes. As a consequence, the detection of such emission would prove the presence of effective in situ reacceleration (or injection) mechanisms distributed over whole the volume of the radio lobes. To our knowledge, so far, there are no clear detection of synchrotron optical (or even X-ray) emission from the lobes of radio galaxies and quasars.

• the X-ray emission from the IC/CMB is contributed by electrons in an energy range not far from that of the electrons emitting synchrotron radiation in the radio band ($\gamma \sim 10^3 - 10^4$). As it will be pointed out in the next Section, this helps constraining the number density of the relativistic electrons in the case where the spectrum of the relativistic electrons is not well known *a priori.*

• the X–ray emission produced by the IC/QSO is produced by low energy electrons ($\gamma \sim 100$). This is very important as the detection and study of the spectrum of this emission allows us to constrain the energy distribution of the emitting electrons at very low, and still unexplored energies, which might contain most of the energetics of the radio lobes.

3.4.2. *Jets and hot spots*

Although radio hot spots are believed to advance in the surrounding medium at non–relativistic velocities (e.g., Arshakian & Longair 2000), there are several indications that radio jets are moving at relativistic speeds out to several tens of kpc distance from the nucleus (e.g., Bridle & Perley, 1984; see also T. Venturi, these proceedings). As a consequence, in this Section, we assume that the photons emitted by the jet are observed at a frequency $\mathcal{D}\nu_o$, where ν_o is the emitted (i.e., jet frame), frequency and $\mathcal{D}$ the Doppler factor, i.e. :

$$\mathcal{D} = \frac{1}{\Gamma(1 - \frac{v_{\rm jet}}{c}\cos\theta_{\rm jet})} \tag{32}$$

where Γ is the Lorentz factor of the jet, $v_{\rm jet}$ is the velocity of the jet and $\theta_{\rm jet}$ is the angle between the direction of the velocity of the jet and the line of sight.

• In the case of a jet moving close to the direction of the observer (we assume $\mathcal{D} \sim \Gamma$), the typical energy of the electrons responsible for synchrotron emission observed at X–ray, optical and radio frequencies is given by:

$$\begin{aligned}\gamma_{\rm syn} \sim \frac{2\cdot 10^8}{\Gamma^{1/2}}\left(\frac{\epsilon({\rm keV})}{B_{\mu G}}\right)^{1/2} &\sim 3\times 10^6 \left(\frac{10}{\Gamma}\frac{500}{B_{\mu G}}\right)^{1/2}\\ &\sim 10^5 \left(\frac{10}{\Gamma}\frac{500}{B_{\mu G}}\right)^{1/2}\\ &\sim 400 \left(\frac{10}{\Gamma}\frac{500}{B_{\mu G}}\right)^{1/2}\end{aligned} \tag{33}$$

In Fig.6 we report the detailed calculation of the energies of the electrons responsible for the synchrotron emission observed in different frequency bands as a function of the angle $\theta_{\rm jet}$.

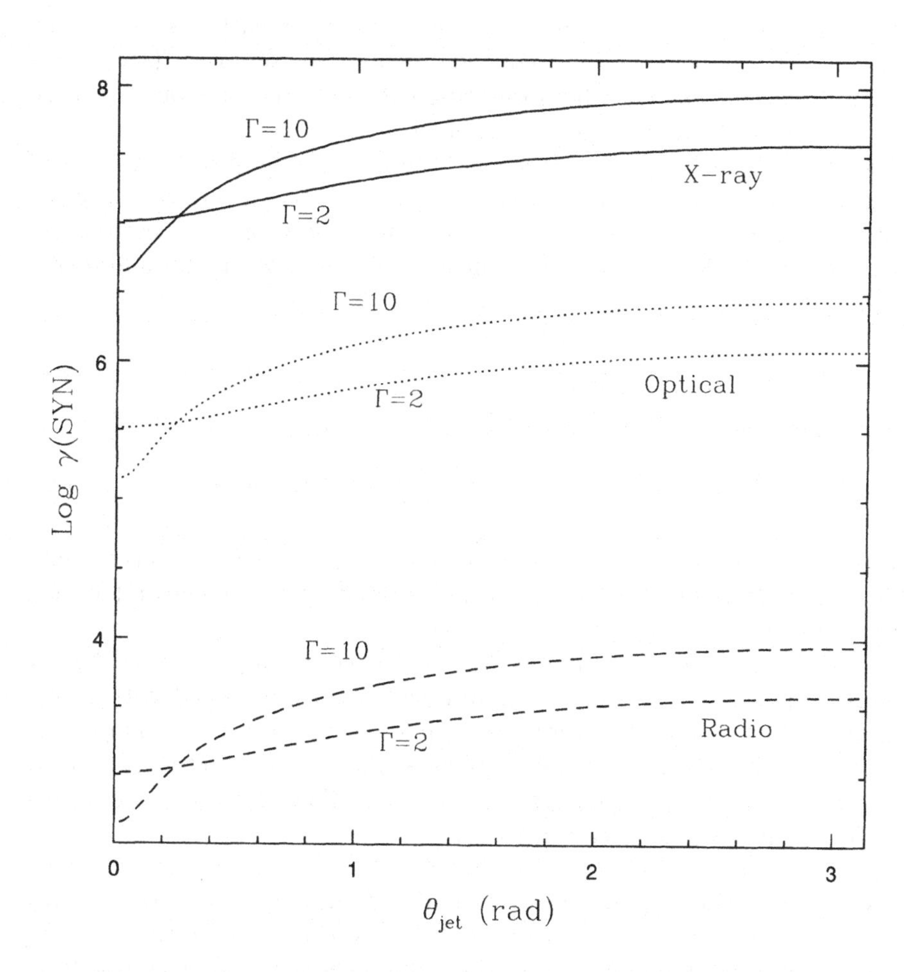

Figure 6. The energy of the electrons responsible for the synchrotron emission observed at radio (solid lines), optical (dotted lines) and X–ray (dashed lines) frequencies is reported as a function of the angle $\theta_{\rm jet}$ and for relevant values of the bulk Lorentz factor Γ. The plots are calculated assuming $B = 100\mu$G in the jet.

• A relevant radiative process in the case of compact regions is the inverse Compton scattering of the synchrotron photons emitted by the same electron population (SSC, e.g., Jones et al. 1974a,b; Gould 1979). In this case, the energy of the electrons giving the SSC emission observed at X–ray and optical frequencies is:

$$\gamma_{\rm ssc} \sim \left(\frac{\nu}{\nu_{\rm R}}\right)^{1/2} \sim 1.5 \times 10^4 \qquad\qquad \sim 500 \tag{34}$$

which does not depend on the motion of the jet with respect to the observer. This is because the frequency of both the observed (seed) synchrotron photons $\nu_{\rm R}$, and the observed (scattered) inverse Compton photons ν, scale with the Doppler factor of the jet.

• An additional radiative process is the inverse Compton scattering of external photons (e.g., Dermer 1995). In the case of a jet moving approximately in the direction of the observer, the energy of the electrons responsible for the inverse Compton emission observed at optical and X–ray frequencies is:

$$\gamma_{\rm eic} \sim \frac{1}{\Gamma}\left(\frac{\nu}{\nu_{\rm E}}\right)^{1/2} \sim 100 \times \left(\frac{10}{\Gamma}\right)\left(\frac{\nu_{\rm cmb}}{\nu_{\rm e}}\right)^{1/2}$$
$$\sim 3 \times \left(\frac{10}{\Gamma}\right)\left(\frac{\nu_{\rm cmb}}{\nu_{\rm e}}\right)^{1/2} \quad (35)$$

where we have parameterized the frequency of the external photons $\nu_{\rm e}$, with that of the CMB photons $\nu_{\rm cmb}$, whose energy density becomes relevant if the jet is moving at high relativistic speed.

In Fig.7 we report a compilation of the energy ranges selected by the different emitting processes in the different bands compared with representative electron spectra produced in different acceleration scenarios. As found in the case of the extended regions, studies of the spectra emitted by the different processes at different frequencies allow us to derive information/constraints on the electron spectrum.
In particular, it should be noticed that :

• The detection of synchrotron radiation from large scale jets at X–ray frequencies requires $\gamma > 10^6$ emitting electrons which, due to the radiative losses, have a relatively short life time and thus their presence points to the need for reacceleration (or injection) mechanisms in these jets. This requirement is even stronger in the case of high frequency synchrotron emission from jets moving at small angles with respect to the plane of the sky (e.g., in case of narrow line radio galaxies or FR I with quasi–symmetric jets). In the case of the hot spots, which are believed to advance at non relativistic speeds, the detection of synchrotron radiation in the optical band is very important as it can be considered as a prove for the presence of efficient in situ reacceleration mechanisms in these regions.

• If the emitting region is roughly homogeneous (in B field), the X–ray emission due to SSC process is produced by roughly the same electrons ($\gamma \sim 10^3 - 10^4$) emitting synchrotron radiation in the radio band. As it will be pointed out in the next Section, this would help in constraining the number density of the relativistic electrons in the case in which the energy distribution of the relativistic electrons is unknown *a priori*.

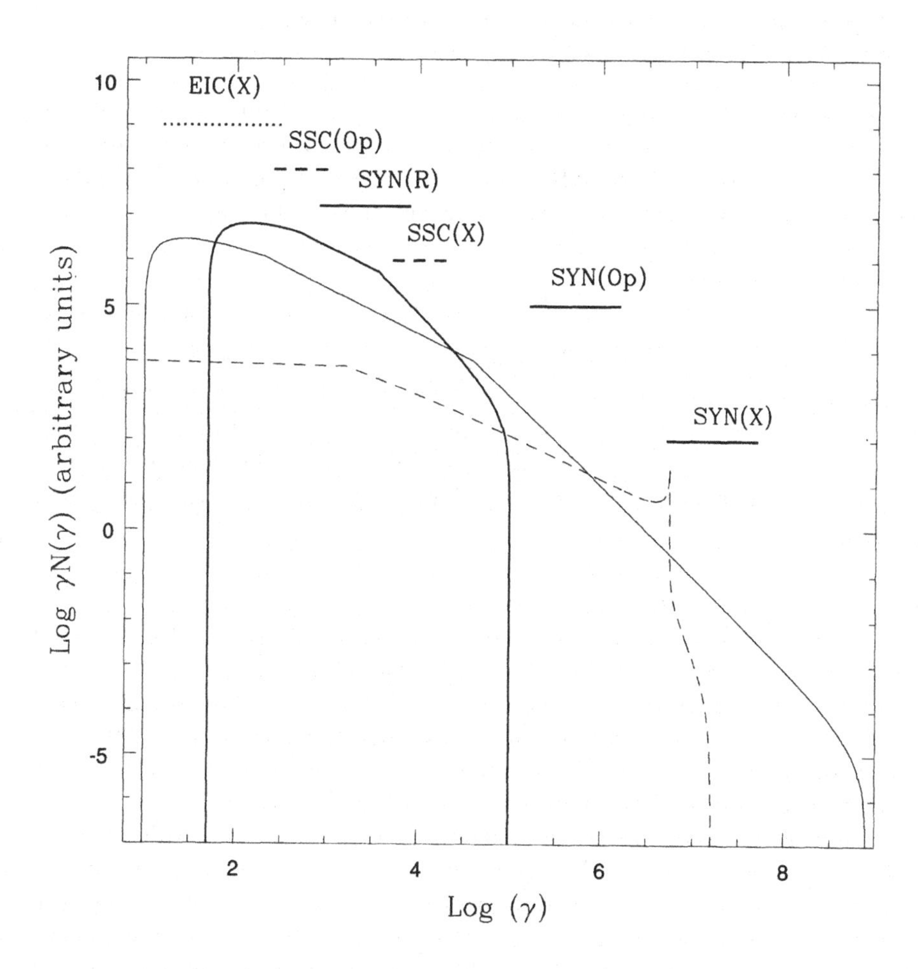

Figure 7. Scheme of the sampling of the spectrum of the emitting electrons with the different processes (SYN= synchrotron, EIC= IC/CMB assuming $\Gamma = 10$ and a jet moving in the direction of the observer, SSC= Syncro–self–Compton) in different bands (R= Radio, Op= Optical, X= X–ray) in the case of compact regions. The different electron spectra are shown for a set of different possible acceleration models.

• As the IC scattering of the external (e.g., CMB) photons is expected to be efficient only if the jets is moving in the direction of the observer at relativistic speed (e.g., Harris & Krawczynski 2002), the relative X–ray emission should be produced by low energy electrons ($\gamma \sim 100$). This is very important as the spectrum of the observed X–rays provides constraints to the energy distribution of the emitting electrons at very low and unexplored energies. This is similar to the case of the IC/QSO from the radio lobes.

3.5. METHODS TO CONSTRAIN THE ENERGETICS OF EXTRAGALACTIC RADIO SOURCES

One of the basic goals of radio astrophysics is to constrain the energetics of the extragalactic radio sources and the ratio between the energy density of particles and fields. Assuming for simplicity an electron energy distribution $N(\gamma) = K_e \gamma^{-\delta}$, the energy density of the relativistic plasma is :

$$\omega(e+p+B) = \mathrm{mc}^2 K_e (1+k) \int_{\gamma_{\rm low}}^{\gamma_{\rm max}} d\gamma \gamma^{-\delta+1} + \frac{B^2}{8\pi} \tag{36}$$

where k is the ratio between the energy of protons and electrons.

In what follows we describe the two most usually adopted arguments to constrain the energy density of the emitting plasma: the minimum energy assumption and the IC method. We refer the reader to classical books (e.g., Verschur & Kellermann, 1988) for additional methods (e.g., synchrotron self absorption).

3.5.1. *Minimum energy assumption*

As it is well known, the synchrotron properties result from a complicated convolution of magnetic field intensity and geometry with the electron spectrum. As a consequence, radio synchrotron observations alone are generally insufficient to derive the relevant physical parameters (e.g., Eilek 1996; Eilek & Arendt 1996; Katz–Stone & Rudnick 1997, Katz–Stone et al. 1999); this problem is known as the *synchrotron degeneracy*.

Hence, to have an idea of the energetics of extragalactic radio sources, radio astronomers are forced to calculate the minimum energy conditions: i.e., the minimum energy of the relativistic plasma required to match the observed synchrotron properties. For the details of this classical argument we refer the reader to Packolzyick (1970) and Miley (1980). Here, we will briefly focus on a critical review of the minimum energy formulae as usually adopted. The magnetic field strength yielding the minimum energy (equipartition field) is:

$$B_{\rm eq} = C_{\rm Pa}(\alpha)(1+k)^{2/7} \left(\frac{\int_{\nu_1}^{\nu_2} L_{\rm syn}(\nu) d\nu}{V} \right)^{2/7} \tag{37}$$

where the constant $C_{\rm Pa}$ can be derived from Packolzyik (1970). Eq.(37) is widely used by radio astronomers to calculate the minimum energy conditions in a radio source. For historical reasons the frequency band usually adopted to calculate the equipartition field is ν_1 =10 MHz – ν_2 =100 GHz, i.e. roughly the frequency range observable with the radio telescopes. From a physical point of view, the adoption of this frequency band means that, in the calculation of the minimum energy, it is assumed that only

electrons emitting between 10 MHz – 100 GHz, i.e. with energy between $\gamma_{\rm low} = a(\nu_1/B_{\rm eq})^{1/2}$ and $\gamma_{\rm max} = a(\nu_2/B_{\rm eq})^{1/2}$, are present in the radio source.
This generates a physical bias because :

i) There are no physical reasons which point to the absence of electrons at lower energies and a substantial fraction of the energetics might be associated to $\gamma < 1000$ electrons which usually emit synchrotron radiation below 10 MHz.

ii) As the energy of the electrons which emit synchrotron radiation at a given frequency depends on the magnetic field intensity, the low frequency cut–off, ν_1, in Eq.(37) corresponds to a low energy cut–off, $\gamma_{\rm low}$, in Eq.(36) which depends on the magnetic field strength $B_{\rm cq}$. As a net result, in radio sources with different $B_{\rm eq}$, the classical minimum energy formulae select different energy bands of the electron population.

Although point i) can be recovered introducing a very low frequency cut–off ($\sim 1 - 100$ kHz) in Eq.(37), it is not possible to recover the ii) with the classical equations (e.g., Myers & Spangler, 1985; Leahy 1991). Based on these considerations, in this contribution, we follow a different approach to calculate the minimum energy conditions which does not use a fixed emitted frequency band.

Expressing K_e in Eq.(36) in terms of the monochromatic synchrotron luminosity, $L_{\rm syn}(\nu)$, and of the magnetic field intensity, Eq.(36) can be immediately minimized as a function of the magnetic field. For $\alpha > 0.5$, the value of the magnetic field yielding the minimum energetics is (Brunetti et al., 1997):

$$B_{\rm eq} = \left[C(\alpha)(1+k)\frac{L_{\rm syn}(\nu)\nu^{\alpha}}{V} \right]^{\frac{1}{\alpha+3}} \gamma_{\rm min}{}^{\frac{1-2\alpha}{\alpha+3}} \tag{38}$$

which does not depend on the emitted frequency band but directly on the low energy cut–off of the electron spectrum. The ratio between the equipartition magnetic field, $B_{\rm eq}$ (Eq.38), and the classical equipartition field, $B'_{\rm eq}$ (Eq.37), is given by:

$$\frac{B_{\rm eq}}{B'_{\rm eq}} \sim 1.2 \left(B'_{\rm eq} \right)^{\frac{1-2\alpha}{2(\alpha+3)}} \gamma_{\rm min}{}^{\frac{1-2\alpha}{\alpha+3}} \tag{39}$$

As Eq.(38) selects also the contribution to the energetics due to the low energy electrons, the deriving intensity of the equipartition field is greater than that of the classical field. In addition, such a difference increases with decreasing magnetic field intensity of the radio sources and in the case of steep electron energy distributions (Fig.8).
Finally, it can be shown that, the ratio between particle and field energy densities is:

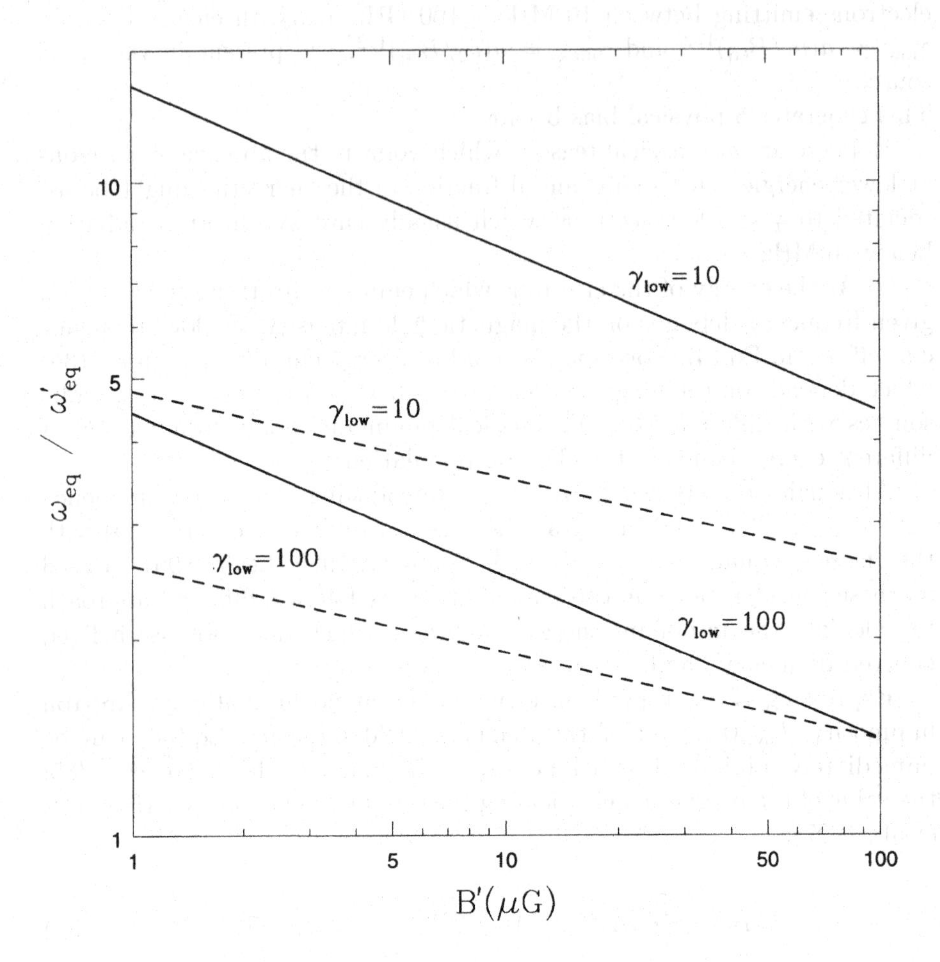

Figure 8. The ratio between the energy density of the equipartition field (Eq.38) and of the classical equipartition field (Eq.37) is shown. The ratio is reported as a function of the strength of the classical field $B'_{\rm eq}$, for different values of $\gamma_{\rm low}$ and of δ=2.5 (dashed lines) and 3.0 (solid lines).

$$\frac{\omega_{\rm eq}(e+p)}{\omega_{\rm eq}(B)} = \frac{2}{\alpha+1} \tag{40}$$

which is 4/3, i.e. the constant ratio obtained with the classical formulae, in the case $\alpha = 0.5$.

It should be reminded that in Section 4.2 and 4.3 we have shown that the electron energy distribution is not expected to be a simple power law and thus Eq.(38) is in general only an approximate solution. In order to calculate the equipartition field for a general electron energy distribution a numerical approach is required.

3.5.2. *Inverse Compton method*

Already many years ago it was pointed out (e.g., Felten & Morrison 1966) that the *synchrotron degeneracy* in determining the physical properties of radio sources could be broken by measurements of the X–rays produced by inverse Compton scattering.

It is well known (e.g., Blumenthal & Gould 1970) that the IC emissivity depends on the number and spectrum of the scattering electrons and of the incident photons. For a power law energy distribution of the electrons, one has:

$$L_{\mathrm{ic}}(\nu_{\mathrm{ic}}) = K_e V C_{\mathrm{ic}}(\alpha, n(\nu_{\mathrm{ph}}), ..)\nu_{\mathrm{ic}}^{-\alpha} \tag{41}$$

C_{ic} can be derived in a number of general cases (isotropica case: e.g., Blumenthal & Gould 1970, anisotropic case: e.g., Aharonian & Atoyan 1981; Brunetti 2000). Given the energy (and angular) distribution of the incident photons $n(\nu_{\mathrm{ph}})$, the measure of the IC flux and spectrum allows to constrain the electron number density and energy distribution.

For a power law energy distribution the synchrotron luminosity is given by (e.g., Ribickyi & Lightman 1979):

$$L_{\mathrm{syn}}(\nu_{\mathrm{syn}}) = K_e V C_{\mathrm{syn}}(\alpha) B^{\alpha+1} \nu_{\mathrm{syn}}^{-\alpha} \tag{42}$$

so that combining Eqs.(41) and (42) one can derive the intensity of the magnetic field :

$$B = \left[\frac{L_{\mathrm{syn}}}{L_{\mathrm{syn}}} \frac{C_{\mathrm{ic}}}{C_{\mathrm{syn}}}\right]^{1/(\alpha+1)} \left(\frac{\nu_{\mathrm{syn}}}{\nu_{\mathrm{ic}}}\right)^{\alpha/(\alpha+1)} \tag{43}$$

Once B is estimated, the ratio between particle and field energy density in the emitting region is given by :

$$\frac{\omega(e+p)}{\omega(B)} = \frac{2}{\alpha+1}\Delta^{\alpha+3} \tag{44}$$

where $\Delta = B_{\mathrm{eq}}/B$. Under equipartition conditions (i.e. $\Delta = 1$), Eq.(44) is equivalent to Eq.(40).

If the assumption of a power law energy distribution is relaxed, Eq.(43) should be replaced with a more complicated formula. Nevertheless, an approach to the determination of the magnetic field intensity is always possible. The most famous applications of this method are those of the IC scattering of CMB photons (e.g., Harris & Grindlay 1979) whose energy density and spectrum $n(\nu_{\rm ph})$, are well known, and of the SSC emission from compact regions (e.g., Jones et al. 1974a,b; Gould 1979). In these cases, as noticed in the previous Section, the radio synchrotron and the X–ray IC photons are emitted by about the same electrons and the value of the B field poorly depends on the electron spectrum (i.e., Eq.(43) is always applicable). A new application of the IC method has been proposed in the case of IC scattering of nuclear photons from the radio lobes (Brunetti et al., 1997).

4. THE NEW OBSERVATIONS

In this Section we will try to give a possibly 'unbiased' review of some of the most recent observations which are helping us to better understand the physics of extragalactic radio sources. Especially in the first part of this review, where we focus on recent detections of IC scattering from the lobes of radio galaxies and quasars, the number of observations is still very small. Consequently, unbiased considerations on the physics of radio lobes will be only obtained in the future thanks to the expected improvement of the statistics.

4.1. X–RAY OBSERVATIONS OF IC/CMB EMISSION FROM RADIO LOBES

Despite the poor spatial resolution and sensitivity, non–thermal IC/CMB X–ray emission from the radio lobes has been discovered by ROSAT and ASCA in a few nearby radio galaxies, namely Fornax A (Feigelson et al. 1995; Kaneda et al. 1995; Tashiro et al. 2001), Cen B (Tashiro et al. 1998), 3C 219 (Brunetti et al. 1999) and NGC 612 (Tashiro et al., 2000). By combining X–ray, as IC scattering of CMB photons, and synchrotron radio flux densities it was possible to derive magnetic field strengths (averaged over the total radio volume) 0.3–1 times the equipartition fields. In general, these observations have been complicated due to the weak X–ray brightness, relatively low count statistics and insufficient angular resolution of the instruments.

After approximately three years of observations with *Chandra* and XMM-*Newton* no clear evidence for diffuse emission from IC scattering of CMB photons from the lobes of extragalactic radio sources has been published on a referred journal. However, future to our knowledge there are preliminary results with *Chandra* and with XMM–*Newton* which appear very promising and to which the reader is referred at the epoch of the publication of this contribution.

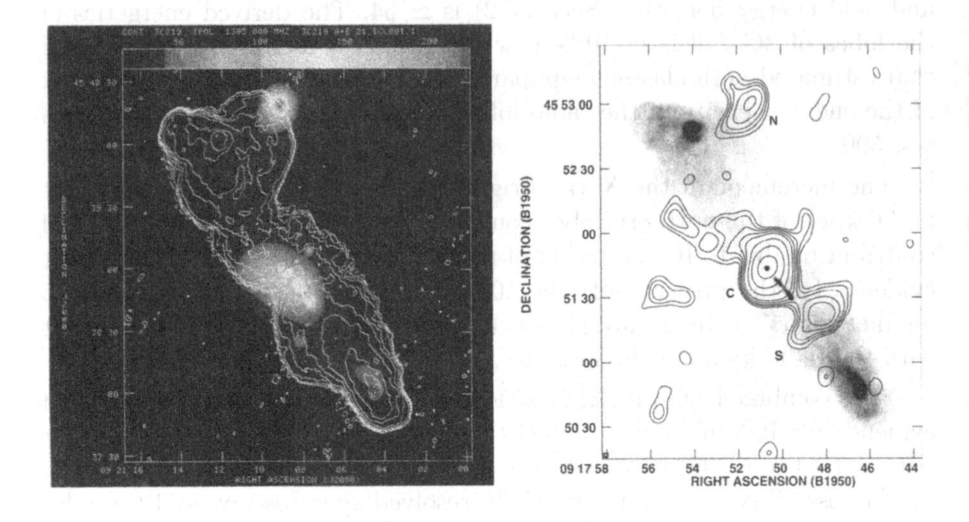

Figure 9. **Panel a**: Radio VLA contours at 1.4 GHz of 3C 219 superimposed on the *Chandra* 0.3–8 keV image (**grays**). The X–ray image is smoothed after removal of the nuclear source which affects only the innermost 3-5 arcsec. **Panel b**: X–ray ROSAT HRI image (**contours**) obtained after the subtraction of the nuclear source is overlaid with a VLA radio image (**grays**). The subtraction of the X–ray nucleus strongly affects the X–ray contours in the innermost 20 arcsec.

4.1.1. *3C 219*

So far, the only public (on electronic preprint archive) case of IC/CMB detection is 3C 219 (Brunetti et al.2002a, astro-ph/0202373). 3C 219 is a nearby (z=0.1744) powerful FRII radio galaxy extending for $\sim$ 180 arcsec corresponding to a projected total size of $\sim$ 690 kpc. The *Chandra* (17 ksec) 0.3–8 keV image is shown in Fig.9a superimposed on the radio contours from a deep 1.4 GHz VLA observation. Thanks to the arcsec resolution, it was possible to disentangle the nuclear emission (which affects only the innermost 3–5 arcsec) from the other components, and to identify the bright clump visible on the north–west with a background cluster at z=0.39. Diffuse emission coincident with the radio lobes also showing a brightness increment in the innermost part of the northern lobe is clearly detected.

The combined imaging and spectral analysis of this emission ($\sim$ 400 net counts) point to a non–thermal, IC/CMB origin of the large scale diffuse emission. Following the procedure described in Section 3.5.2, the comparison between radio and X–ray IC emission yields a precise measurement of the magnetic field intensity (averaged over the total radio volume) which results a factor of $\simeq$ 2.9 times lower than the equipartition value (assuming $\gamma_{\rm low}$ = 50 in Eq.(38). Under these conditions, the ratio between particle

and field energy densities (Sect.3.5.2) is $\simeq 64$. The derived energetics of the lobes of 3C 219 is $\simeq 10^{60}$erg which results a factor $\simeq 7$ larger than that estimated with classical equipartition formulae (Sect.3.5.1). The bulk of the energy density of the radio lobes is associated to the electrons with $\gamma < 500$.

The increment in the X–ray brightness present in the innermost part (~ 70 kpc) of the northern lobe (counter lobe) may indicate an additional contribution due to IC scattering of nuclear photons, thus providing direct evidence for the presence of $\gamma \sim 10^2$ electrons in the lobes. Finally, two distinct knots at 10–25 arcsec south of the nucleus, spatially coincident with the radio knots of the main jet, are visible in the *Chandra* image.

Past combined `ROSAT PSPC`, `HRI` and `ASCA` observations did also find evidence for IC emission in 3C 219 lobes out of equipartition conditions (Brunetti et al. 1999). However, the presence of the strong nuclear source, the impossibility to perform spatially resolved spectroscopy and the relatively poor sensitivity of `ROSAT HRI` required a follow up observation with *Chandra*. The 0.1–2 keV image from the 30 ksec `ROSAT HRI` observation is shown in Fig.9b: the emission within ~ 20 arcsec from the nucleus is strongly affected by the subtraction of the nuclear source. The comparison of the two images in Fig.9 is very instructive and shows well the real breakthrough in X–ray imaging provided by *Chandra*.

4.2. X–RAY OBSERVATION OF IC SCATTERING OF NUCLEAR PHOTONS FROM RADIO LOBES

The detection of X–ray emission from IC/QSO has been recently achieved in at least three objects (3C 179, 3C 207, 3C 295) whereas possible evidences have been suggested in other few cases (e.g., 3C 294: Fabian et al. 2001; 3C 219, Fig.9). A positive detection of this effect with *Chandra* was a specific prediction of this model in the case that a substantial fraction of the energetics of the radio lobes is associated to the low energy end of the electron spectrum. X–ray emission from IC scattering of nuclear photons with the relativistic electrons in the radio lobes is expected to be particularly efficient in the case of relatively compact (i.e., ≤ 100 kpc) and strong FRII radio galaxies and steep spectrum radio quasars. This is due to the dilution of the nuclear flux with distance from the nucleus.

4.2.1. *Radio galaxies: 3C 295*

This is a classical FRII at the center of a rich cluster (z=0.461). The X-ray data obtained with previous instruments (*Einstein* Observatory: Henry & Henriksen 1986; ASCA: Mushotzky & Scharf 1997; ROSAT: Neumann 1999) only allowed the study of the cluster emission.

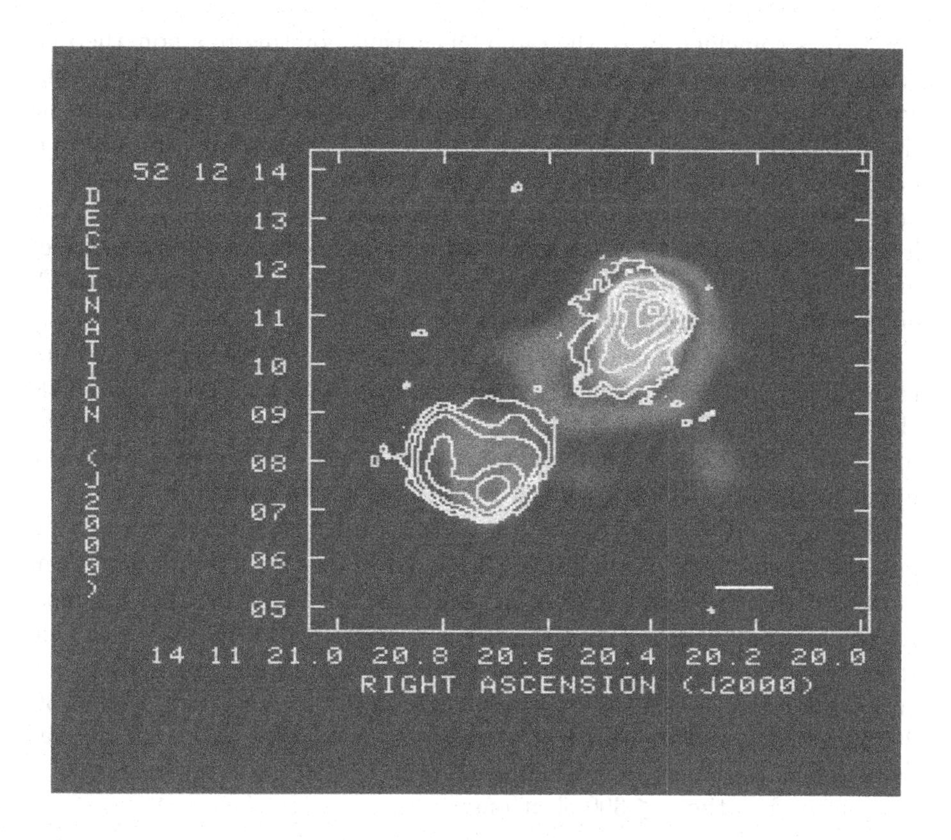

Figure 10. The GHz VLA radio map (contours) of 3C 295 is shown superimposed on the 0.1–2 keV *Chandra* image (grays). The X–ray image is obtained after the subtraction of the cluster emission.

3C 295 was the first FRII source observed by *Chandra.* Harris et al. (2000) obtained the 0.3–7 keV image of this radio galaxy in which the hot spots and nuclear emission were well separated from the surrounding cluster contribution. The presence of possible diffuse X–ray emission related to the radio lobes was first addressed by these authors. However, the morphology and intensity of this emission resulted particularly uncertain due to the presence of the bright nuclear source and of the northern hot spot at ~ 2 arcsec distance. Stimulated by the results of Harris et al. (2000), Brunetti et al. (2001b) performed a more detailed analysis in order to maximize the information on the X–rays from the radio lobes.

In particular, these authors performed the spectrum of the nuclear source which cames out to be highly absorbed by a column density of $\sim 10^{23}$cm^{-2} and thus almost absent in the 0.2–2 keV image. This image is shown in Fig.10: the morphology of the diffuse X–ray emission is double lobed with the X–rays coincident with the radio lobes, thus pointing to a non–thermal origin. In addition the asymmetry in the X–ray brightness (with the northern lobe a factor $\sim 2-4$ brighter than the southern one) appears to be the signature of the IC/QSO model. In order to reproduce the observed brightness ratio with this model an angle between radio axis and the plane of the sky of 6–13° is required with the northern lobe being further away from us. This geometry was confirmed by the discovery of a faint radio jet in the southern lobe (i.e., the near one) by P.Leahy with a deep MERLIN observation (private communication).

As stated in Sect.3, the spectrum from IC scattering of nuclear photons is a unique tool to constrain the energy distribution of $\gamma \sim 100-300$ electrons. Fig.11 shows the 0.5–2 keV spectral index predicted by the model as a function of a low energy cut–off in the electron spectrum : an upper limit $\gamma_{low} < 100$ is obtained from the *Chandra* data.

The IC scattering of the nuclear photons has been also used to calculate the magnetic field strength in the lobes of 3C 295. ISO measurement of 3C 295 flux (Meisenheimer et al. 2001) fix the far–IR nuclear luminosity to $\sim 3 \times 10^{46}$erg s^{-1} and the deriving value of the IC magnetic field strength results consistent with the value calculated under minimum energy assumption (Fig.12). The derived energetics of the radio lobes of 3C 295 is $\simeq 4.4 \times 10^{59}$erg and results a factor ~ 2 larger than that derived with classical equipartition formulae (Sect.3.5.1); a significant fraction of it is associated to the $\gamma \leq 300$ electrons.

4.2.2. *Lobe dominated quasars: 3C 179 and 3C 207*

The effect of the asymmetry in the X–ray distribution from the anisotropic IC scattering of the nuclear photons is maximized in the case of the steep spectrum quasars, which typically make an angle of 10–30 degrees between the radio axis and the line of sight. This provides an unambiguous identification of the process responsible for the X–ray emission as only the X–rays from the counter lobe are expected to be efficiently amplified and thus detected. In addition, in this case the far–IR to optical flux from the nuclear photons can be directly measured thus allowing a prompt estimate of the energy density of the scattering electrons (and magnetic field) in the radio lobes as in the case of the IC scattering of CMB photons (Sect.3.5.2).

So far there are two radio loud quasars observed with *Chandra* in which extended X–ray emission from the counter–lobe has been successfully detected, and for which no diffuse emission from the near lobe was detected: 3C 179 (Sambruna et al. 2002) and 3C 207 (Brunetti et al. 2002b). Both these sources are relatively compact and luminous, with prominent radio lobes making them ideal candidates to detect IC scattering of the nuclear photons in the radio lobes.

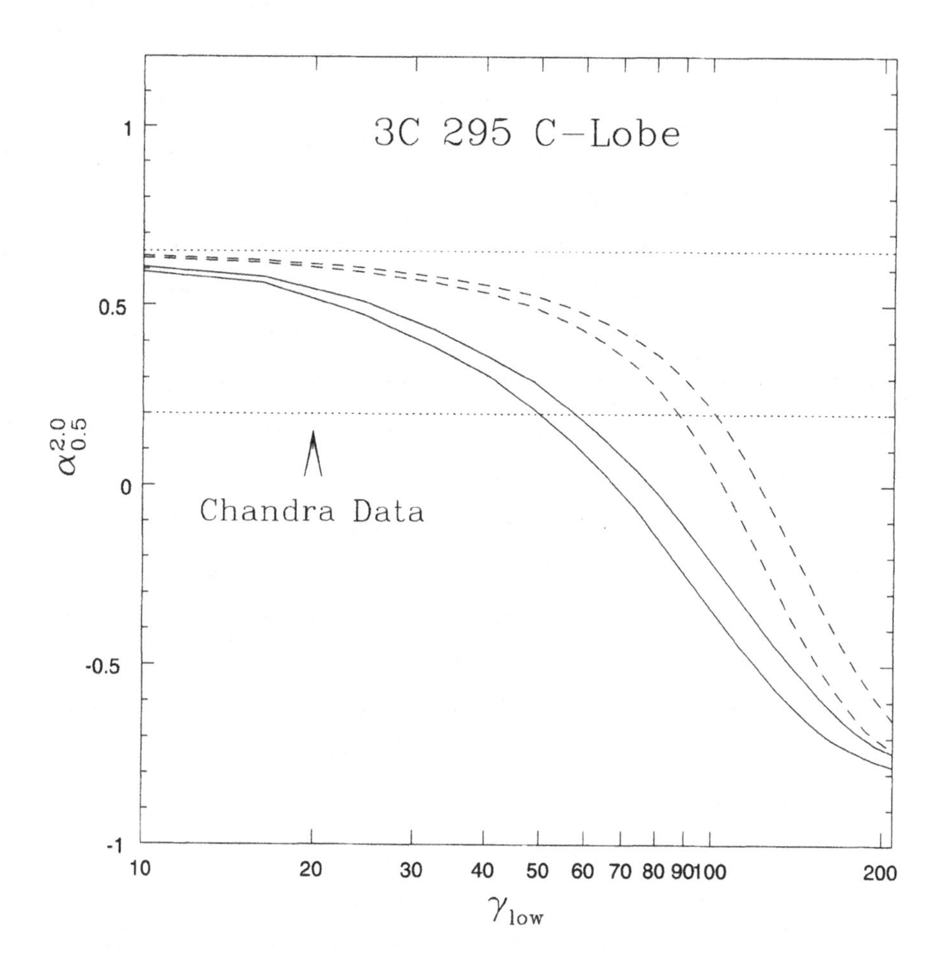

Figure 11. The 0.5–2 keV spectral index expected in the case of the IC scattering of nuclear photons is reported in the case of 3C 295. In the calculation we have assumed the Sanders et al. (1989) (solid lines) average SED of quasars and the SED of 3C 48 (dashed lines). For each SED we have reported two different curves giving uncertainties on the spectral index due to different set of parameters assumed in the model calculation. The limits on the observed 0.5–2 keV spectral index from our *Chandra* data analysis are also reported (dotted lines).

The 0.2–8 keV images of 3C 207 ($\sim$ 36 ksec exposure) is reported in Fig.13 superimposed on the VLA radio contours. The allowed regions of the values of the magnetic field strengths and of $\gamma_{\rm low}$ (Sect. 3.3) as inferred by the combined radio and X–ray fluxes and spectrum of 3C 207 are reported in Fig.14: the magnetic field strengths are lower, but within a factor of ~ 2, from the equipartition values. The resulting energetics of the radio lobes of 3C 207 is $\simeq 3 \times 10^{60}$erg; a large fraction of it is associated

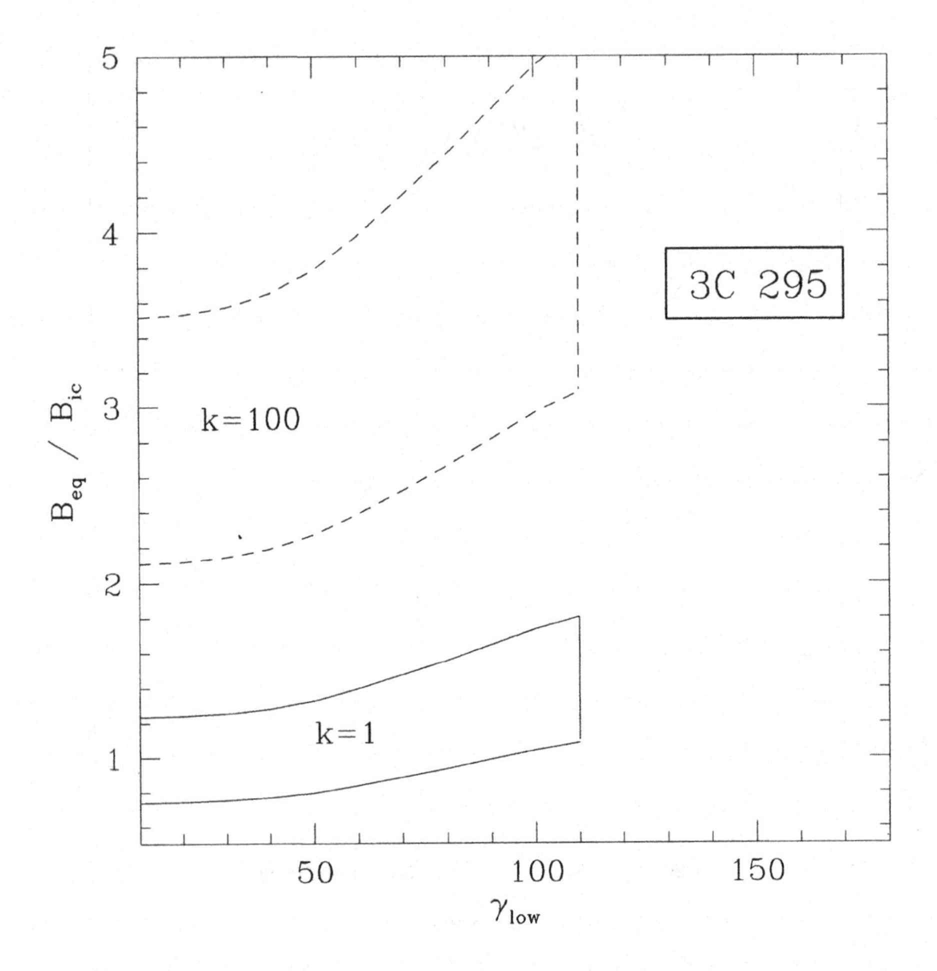

Figure 12. The ratio between the equipartition magnetic field intensity ($B_{\rm eq}$) and that estimated from the IC scattering is reported as a function of the low energy cut–off $\gamma_{\rm low}$. The calculation are performed for a bolometric far–IR/optical luminosity of the hidden quasar = 2–4$\times10^{46}$erg s^{-1}; k is the ratio between proton and electron energy density in the radio lobes.

to the $\gamma \leq 300$ electrons. The above value of the energetics is a factor ~ 4.5 larger than that obtained for 3C 207 with classical minimum energy formulae (Sect. 3.5.1). The shorter *Chandra* exposure ($\sim$ 9 ksec) in the case of 3C 179 makes difficult to constrain the energetics of the radio lobes. However, also in this case, the detection of IC/QSO proves that a consistent fraction of the energetics of the electrons component in the radio lobes is associated to low γ electrons.

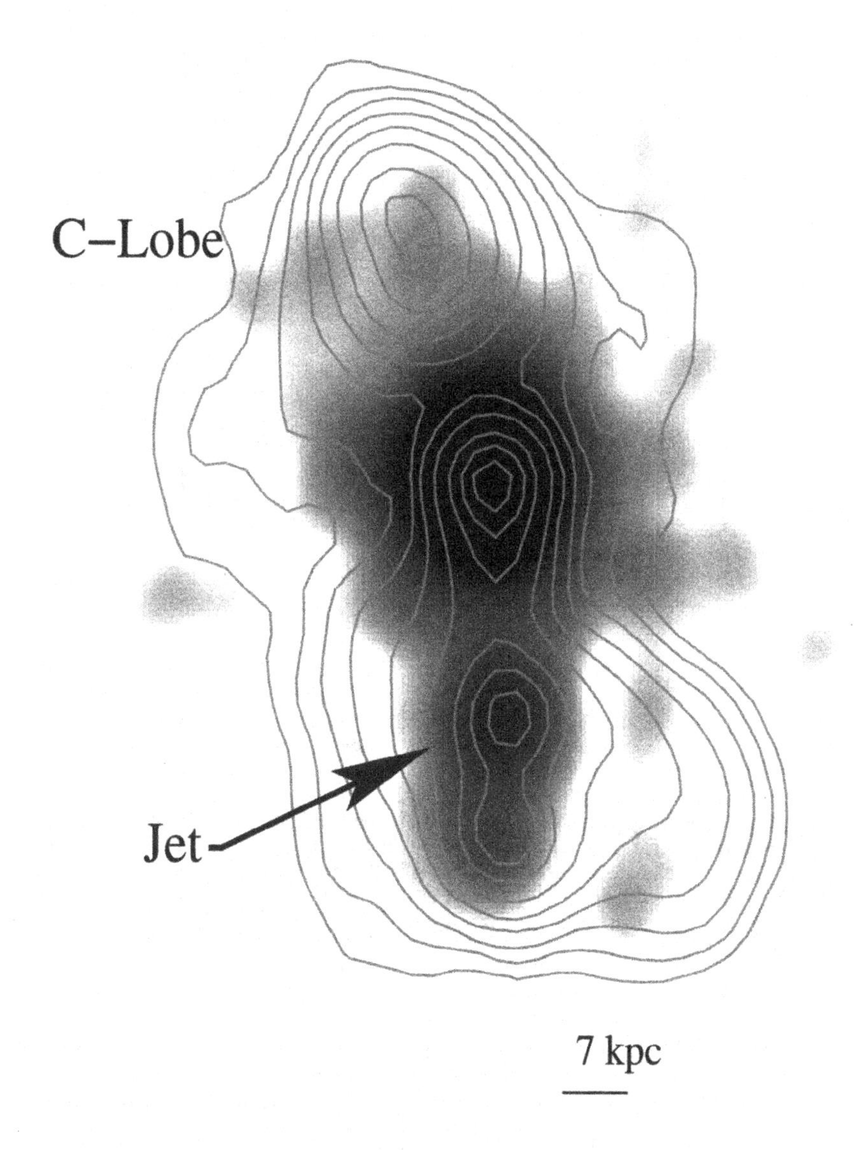

Figure 13. *Chandra* 0.2–8 keV image of 3C 207 (grays) superimposed on the 1.4 GHz VLA contours. The X–ray jet and counter lobe are indicated in the figure. The scale bar gives the resolution of the X–ray image (0.9 arcsec = 7 kpc); the resolution of the radio image is 1.4 arcsec.

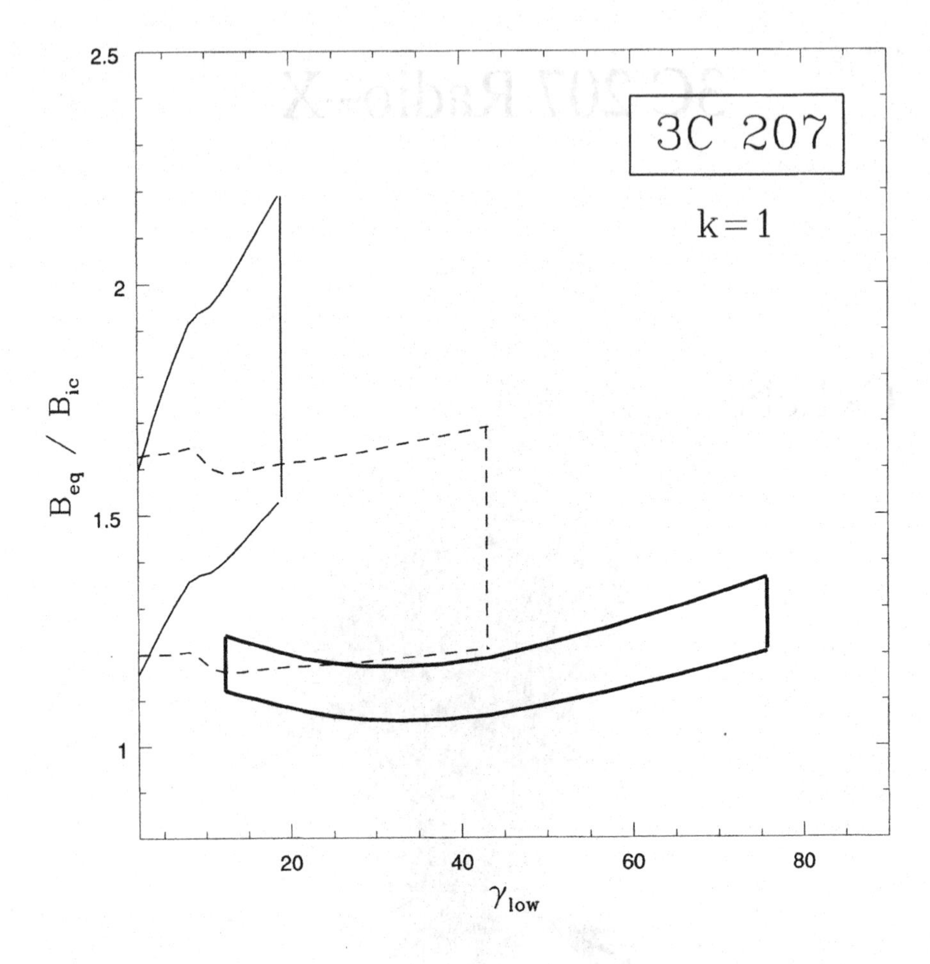

Figure 14. The allowance regions for the ratio between IC–magnetic field and equipartition field strength and for γ_{low} in the case of 3C 207 are reported for different assumed energy distributions of the electrons: power law $\delta = 2.8$ down to γ_{low} (thick solid region), accelerated spectrum with γ_*/γ_{low}=2 (dashed region) and =50 (thin solid region). In the case of reaccelerated electron spectra, the regions are calculated by assuming γ_b in the range 500–1000 and $\gamma_c >> 1000$.

4.3. JETS AND HOT SPOTS

RADIO OBSERVATIONS: Radio telescopes have imaged a large number of jets and hot spots of radio sources with arcsec or subarcsec spatial resolution (e.g., T.Venturi, this proceeding). The radio studies suggested the basic modelling of radio jets and hot spots. They provided evidence for relativistic motions of the radio jets from pc to kpc distances from the nucleus (e.g., Garrington et al. 1988; Bridle et al. 1994). The study of the polarization from jets and hot spots have suggested the presence of shocks and/or strong interactions with the surrounding IGM/ICM in which magnetic field amplification and particle reacceleration can take place. Finally, the study of the spectral synchrotron ages, combined with the direct (or statistical) measurement of the advancing motion of the radio lobes/hot spots have allowed a first order estimate of the dynamical age of extragalactic radio sources. This in turn allowed the measurement of the jet kinetic power under the assumption of minimum energy conditions (Rawlings & Saunders 1991).

Although the improvement of the radio telescopes and interferometers, and the advent of the future radio instruments (e.g., SKA) will allow to address a number of additional/substantial improvements in our understanding of the physics of radio sources, a multiwavelength approach is by far the most efficient tool to provide the next step in this topic. This is due to the *synchrotron degeneracy* (Sect.3.5.1).

OPTICAL OBSERVATIONS: although the search for optical emission from radio jets and hot spots has a long history (e.g., Saslaw et al. 1978; Simkin 1978; Crane et al. 1983), relatively few jets and hot spots have been detected as sources of optical emission so far (Tab.2, Tab.3, Meisenheimer et al. 1997). This is not only due to the power law decay with frequency of the synchrotron spectrum emitted from these regions, but also due to the presence of breaks and/or exponential cut-offs in the synchrotron spectrum below the optical band (Sect.3.3). The advent of the *Hubble Space Telescope* (HST) and more recently, of the 10 mt. generation of ground based telescopes (e.g., VLT, *Gemini*), has considerably improved the possibility to detect and study the optical emission from these regions.

X-RAY OBSERVATIONS: the study of X-ray emission from jets and hot spots has been considerably improved by the recent advent of the *Chandra* observatory. Before *Chandra* only a few cases of X-ray counterparts of radio jets and hot spots was discovered (see Tab. 2 and 3). The most spectacular result being the famous ROSAT HRI detection of both hot spots of the powerful radio galaxy Cygnus A (Harris et al. 1994). It was immediately clear from these past observations that the detected emission was of non-thermal origin with the best interpretation provided by synchrotron and SSC mechanisms under approximate minimum energy conditions. Likewise *Chandra* is really providing a significant progress on the study of the X-ray emission from jets and hot spots (see Tab. 2 and 3). Although the analysis of the increasing number of successful detections of X-ray emission from compact hot spots and jets has unambiguously confirmed the non-thermal nature of the X-rays from these sources, it is not clear whether the SSC and synchrotron model can provide or not a general interpretation of the data (e.g. Harris 2001). In addition the possibility to derive spectral analysis of the X–ray counterparts allows, for the first time, to constrain the spectrum of the emitting electrons in these regions.

In this Section we especially focus on the information on the low energy end and high energy end of the electron spectrum which are becoming available with the most recent multiwavelength studies.

4.3.1. *Constraining the LOW energy end of the electron spectrum*
As discussed in Sect.3.4, information on the low energy end of the electron spectrum can be provided by the detection of optical SSC fluxes from the hot spots and by the detection of X–ray emission via IC/CMB from the jets.

a) Optical SSC emission from hot spots – 3C 295–N and 196–N – : The northern hot spot of 3C 295 has been recently detected in the B–band with the HST telescope (Harris et al. 2000). Taking into account the radio, optical and X–ray data, Brunetti (2000) has shown that the radio to optical emission is not easily accounted for by a simple synchrotron model, whereas a synchrotron plus SSC model can account very well for the broad band spectrum with the radio matched by the synchrotron radiation, and the optical and X–rays matched by the SSC (Fig.15). The synchrotron radiation at $\geq$ GHz frequency is emitted by $\gamma > 1000$ electrons (Sect.3.4), whereas the SSC optical radiation is emitted by $\gamma \sim 500$ electrons. In particular, the model in Fig.15 is calculated assuming that the spectrum of the synchrotron emitting electrons can be extrapolated at lower energies ($\gamma_{\rm low} < 500$) and the data constrained the low energy break in the electron spectrum (if any) at $\gamma_* < 800$ (Sect. 3.3).

The optical counterpart of the northern hot spot of 3C 196 has been recently discovered by Hardcastle (2001) with the HST telescope. The high frequency radio fluxes show a prominent steepening so that a synchrotron model accounting for the radio spectrum is too steep and falls well below the optical flux. As in the case of 3C 295-N, a viable explanation for the optical emission is provided by the SSC mechanism under the assumption of approximate equipartition conditions in the hot spots (Hardcastle 2001). As in the case of 3C 295, the detection of optical SSC emission from the hot spot of 3C 196 points to the presence of $\gamma \sim 500$ electrons in the hot spot volume without a significant flattening of the electron spectrum at these energies.

b) X–ray emission from radio jets – external IC scattering –: one of the most impressive results from *Chandra* is the unexpected high detection rate of the radio jets in the powerful radio sources (Tab.2). The first object with a prominent X–ray jet discovered by *Chandra* was the flat spectrum quasar PKS 0637-752 at a redshift of z=0.653 (Chartas et al. 2000; Schwartz et al. 2000). The combined radio, optical and X–ray data exclude the possibil-

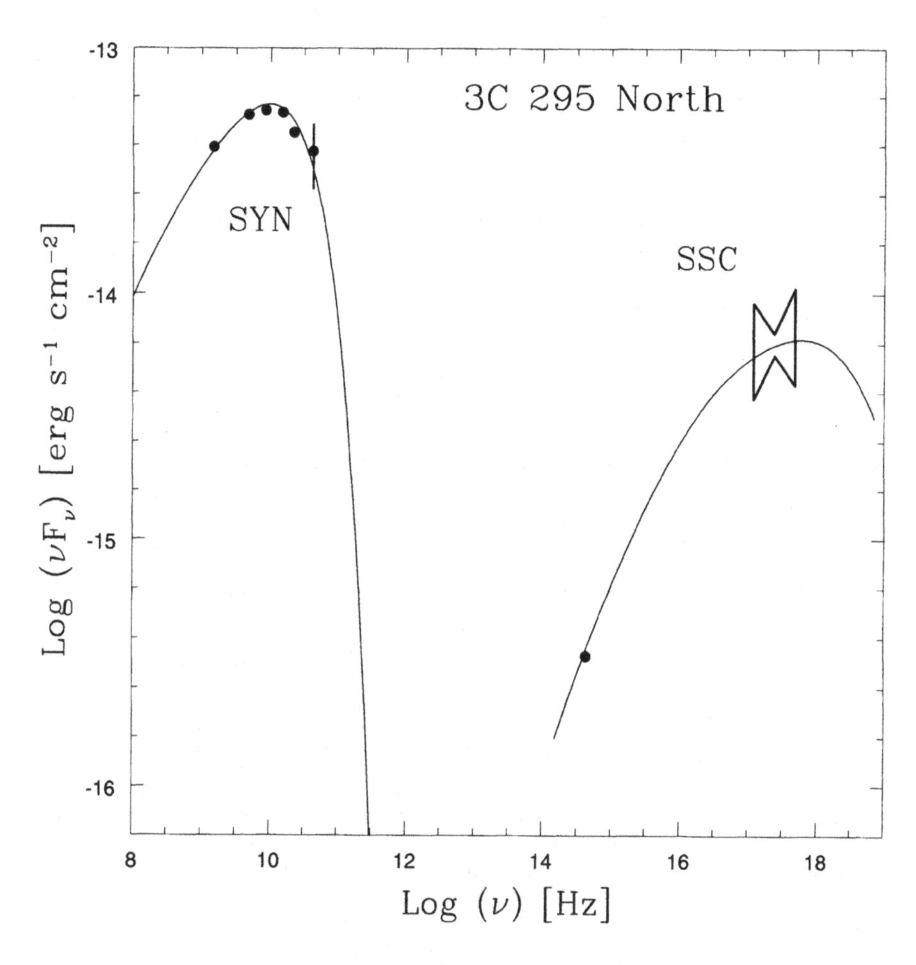

Figure 15. The radio to X–ray data points of the northern hot spots of 3C 295 are fitted with a synchrotron and SSC model: the SSC accounts for both optical and X–ray data.

ity of synchrotron X–ray emission. In addition Tavecchio et al.(2000) and Celotti et al.(2001) have shown that an SSC origin of the X–rays should require huge departures from equipartition, and an extremely high kinetic power of the jet. These authors were the first to successfully make use of the IC scattering of CMB photons by the relativistic electrons of the jet (a mechanism previously considered only for the jets on pc scales, e.g., Schlickeiser 1996) to reproduce the X–ray data of this object. In order to successfully fit the X–ray spectrum with this model they derive a high relativistic velocity ($\Gamma_{\rm bulk} \sim 10$) of the jet up to hundreds of kpc distance from the nucleus. In order to maintain such velocities, small radiative efficiencies in the jets are required with most of the energy extracted from the central

TABLE 2. High power objects

Name	z	Type	Assoc. Radio	Assoc. Optical	Chandra	Reference
3C 123	0.2177	RG	HS	N	Y	Ha01
Pictor A	0.0350	RG	HS	Y	N	W01
			Knots	N	Y	
PKS 0637-752	0.653	FSQ	Knots	Y	Y	C00,T00,Sc00,Ce01
3C 179	0.846	FSQ	Knot	N	Y	S02
			HS	N	Y	
3C 207	0.684	FSQ	Knot	N	Y	B02b
			HS	N	Y	
Q 0957+561	1.41	FSQ	Knots	N	Y	C02
PKS 1127-145	1.187	FSQ	Knots	N	Y	Si02
PKS 1136-135	0.554	FSQ	Knots	Y	Y	S02
4C 49.22	0.334	FSQ	Knots	Y	Y	S02
3C 273	0.1583	FSQ	Knots	Y	N	M01,S01
4C 19.44	0.720	FSQ	Knots	Y	Y	S02
3C 295	0.45	RG	HS	Y	Y	H00,B01b
3C 351	0.3721	SSQ	HS	Y	Y	B01c
3C 390.3	0.0561	RG	HS	Y	N	P97,H98
Cyg A	0.0560	RG	HS	N	N	H94,W00

References: Ha01=Hardcastle et al. 2001a, W01=Wilson et al. 2001, C00=Chartas et al. 2000, T00=Tavecchio et al. 2000, Sc00=Schwartz et al. 2000, Ce01=Celotti et al. 2001, S01=Sambruna et al. 2002, B02b=Brunetti et al. 2002b, C02=Chartas et al. 2002, Si02=Siemiginowska et al. 2002, M01=Marshall et al. 2001, S01=Sambruna et al. 2001, H00=Harris et al. 2000, B01b=Brunetti et al. 2001b, B01c=Brunetti et al. 2001c, P97=Prieto 1997, H98=Harris et al. 1998, H94=Harris et al. 1994, W00=Wilson et al. 2000. See http://hea-www.harvard.edu/XJET/index.html for an updated list.

black hole stored in the bulk motion of the plasma. In addition, as pointed out by Ghisellini & Celotti (2001), it can be considered that the radiative efficiency of the jet decreases with increasing the jet luminosity.

Additional evidences in favour of the IC/CMB and thus that high relativistic velocities are maintained by the jets up to tens or hundreds of kpc from the nucleus, is coming from other objects (e.g., Sambruna et al. 2002; Tab.2). In Fig.16a-c we report a compilation of radio to X-ray spectral energy distributions of some of these objects.

One of the most striking cases is the X-ray knot of the quasar 3C 207 at a redshift z=0.684 (Brunetti et al. 2002b). The resulting X-ray spectrum is considerably harder than the radio spectrum ($\Delta\alpha \sim 0.6$) so that

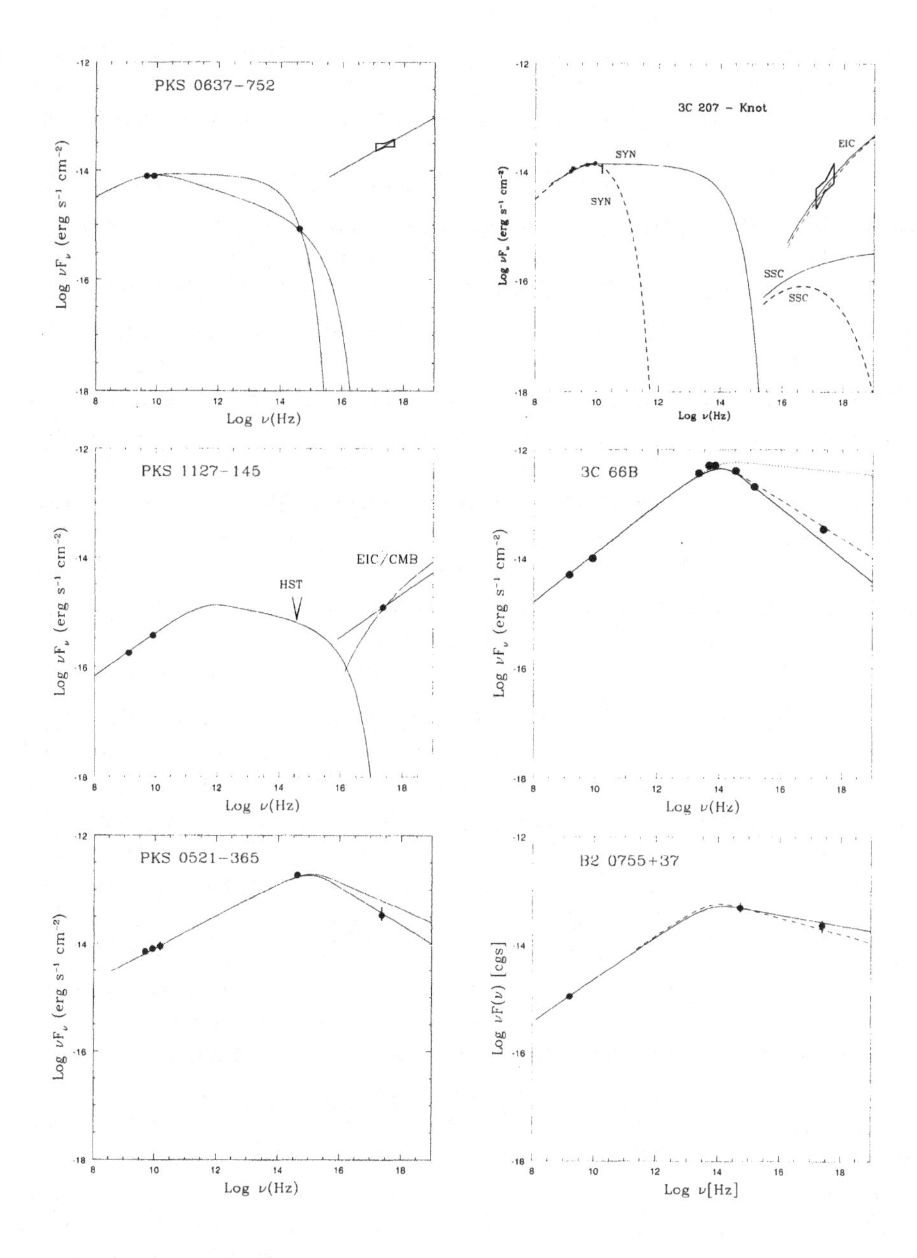

Figure 16. A compilation of SEDs of X–ray jets fitted with synchrotron and IC models. **Panel a)–c)**: SEDs in the case of jets of high power radio objects. Radio and optical data are fitted with synchrotron models, while X–ray data are accounted for via IC/CMB models. The SSC predictions are also reported in the case of 3C 207 (Panel b). **Panel d)–f)**: SEDs in the case of jets of low power radio objects. Radio, optical and X–ray data can be fitted with a synchrotron model which includes a non standard electron transport or adiabatic losses in the post–shock region. Standard synchrotron models (e.g., dotted line in 3C 66B, upper curves in PKS 0521-365 and B2 0755+37) cannot well account for the data.

it cannot be reproduced by the SSC spectrum even releasing the assumption of minimum energy conditions in the jet. On the other hand, as discussed in Sect.3.3, the electron spectrum of the low energy electrons emitting X–rays via IC/CMB ($\gamma \sim 100$) might be harder than that of the higher energy radio synchrotron electrons ($\gamma > 1000$), thus providing a natural explanation for the difference between X–ray and radio spectrum of this knot. It should be noted that, despite the poor statistics, a similar difference between radio and X–ray spectrum is also found in the jet of PKS 1127-145, which is the most luminous IC/CMB jet discovered so far (Siemiginowska et al. 2002).

If the IC/CMB interpretation is correct for these X–ray jets, then, for the first time, the modelling of the radio to X–ray spectrum allows the low energy end of the electron spectrum to be constrained in the regions where these electrons are (re)accelerated. These studies, however, are relatively complicated as the energy of the electrons giving the observed X–rays (Sect.3.4) depends on both the Lorentz factor of the bulk motion, and on the angle between the jet velocity and the line of sight. In the case of 3C 207 it can be shown that for substantial boosting (i.e., $\Gamma_{\rm bulk} > 4$ and $\theta < 10^o$) $\gamma_{\rm low} \lesssim 50 \leq \gamma_*$, whereas in the case of PKS 0637-752 $\gamma_{\rm low} \lesssim 30$.

4.3.2. *Constraining the HIGH energy end of the electron spectrum*

a) synchrotron optical emission from radio hot spots: As already discussed in Sect.3.4, synchrotron optical emission from radio hot spots is mainly due to $\gamma \geq 10^5$ electrons, which are probably close to the high energy end of the spectrum of the electrons accelerated in these regions. These electrons have a radiative life time about 300 times shorther than that of the electrons emitting the synchrotron radio spectrum of the same hot spots. Hence the optical detection of hot spots generally implies the in situ production of such energetic electrons (e.g., Meisenheimer et al., 1989). An important confirmation that the optical emission from the hot spots is of synchrotron nature is provided by the detection of optical linear polarization in a number of cases (3C 33: Meisenheimer & Röser, 1986; 3C 111, 303, 351, 390.3: Lahteenmaki & Valtaoja 1999; Pictor A West: Thomson et al. 1995). So far there are only about 15 hot spots detected in the optical band (see Gopal-Krishna et al. 2001 and ref. therein) and the radio to optical spectrum of these hot spots is well fitted by synchrotron radiation emitted by electrons accelerated in a shock region (e.g., Meisenheimer et al. 1997). In Fig.17 we report the radio to optical data of a few representative cases fitted with synchrotron models.

In principle, if this scenario is correct, the theory of shock acceleration allows us to get an independent estimate of the field strength at the hot spot by measuring the break and cut–off frequencies of the synchrotron spectrum and the hot spot length (e.g., Meisenheimer et al., 1989). This

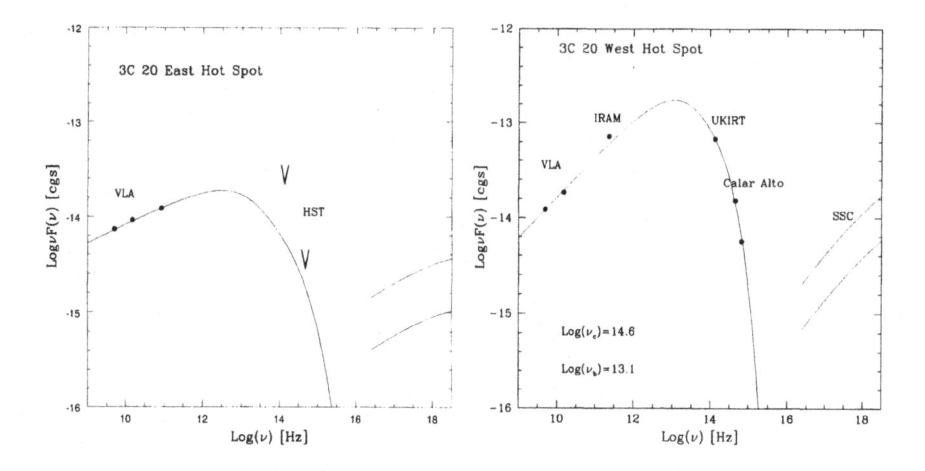

Figure 17. Synchrotron fits to the radio – optical SEDs of the radio hot spots 3C 20 East and West are reported. The spectrum of the emitting electrons is calculated under shock acceleration assumptions (Sect.3.3, Figs.3-4). In the case of 3C 20 West the values of the break and cut–off frequency are also given. The expected X–ray emission from both the hot spots due to SSC process are reported under equipartition conditions (lower curves) and assuming a magnetic field 2 times smaller than the equipartition (higher curves).

allows the estimate of both the maximum energy of the emitting electrons, and of the acceleration efficiency of the shock. In general, the estimated magnetic field strength is consistent with that estimated under minimum energy conditions within a factor of 2–3 (e.g., Meisenheimer et al., 1997). With these values, we have that the Lorentz factors of the electrons at the cut–off is in the range $\gamma_c = 10^5 - 10^6$, and that the acceleration time in the shock region is in the range $\sim 10^2 - 10^3$yrs.

An alternative scenario to that of the shock acceleration might be an extremely efficient transport – minimum energy losses – of the ultra relativistic electrons all the way from the core to the hot spots (e.g., Kundt & Gopal-Krishna 1980). Based on the evidences for relativistic jet bulk motion out to 100–kpc scales, Gopal-Krishna et al. (2001) have recently reconsidered a minimum loss scenario in which the relativistic electrons, accelerated in the central active nucleus, flow along the jets losing energy only due to the inescapable IC scattering of CMB photons. Under these assumptions, comparing the electron radiative life time with the travel time to the hot spots, these authors find that in situ electron re–acceleration is in general not absolutely necessary to explain the optical synchrotron radiation from the hot spots. In the framework of this minimum loss scenario, in Fig.18 we report the maximum distance that synchrotron optical electrons can cover as a function of the velocity of the jet flow, and for two different magnetic

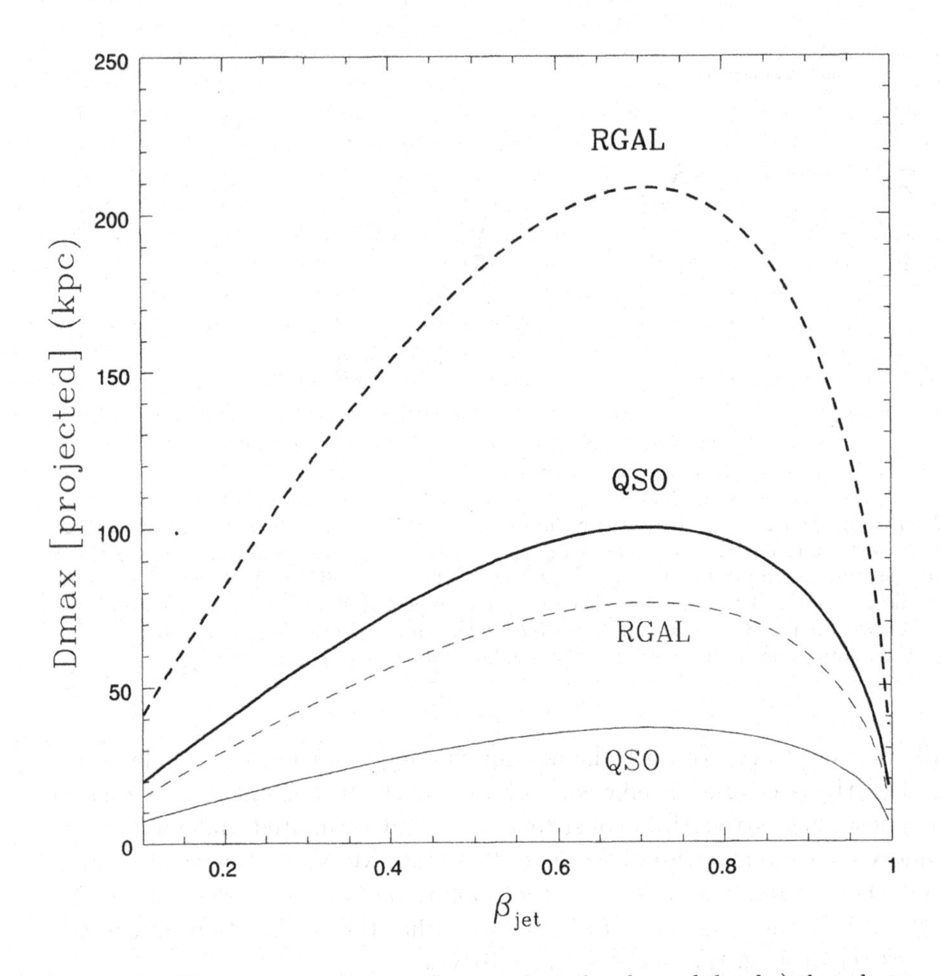

Figure 18. The maximum distance (projected on the plane of the sky) that electrons emitting synchrotron radiation at $\nu = 5 \cdot 10^{14}$Hz can cover is reported as a function of the velocity (v/c) of the jet. The calculation are performed for different inclination angles of the jet: quasar–like (solid lines) and radio galaxy–like (dashed lines). The results are shown in the case of z=0.2 (thin lines) and z=0.5 (thick lines).

field strengths in the hot spot region. Such distance is in general less than 100 kpc except in the case in which the magnetic field strength in the hot spot is very large. On the other hand, a number of optically detected radio hot spots are found at larger distances from the nucleus. In addition, the double hot spot in 3C 351 represents a clear counter example to the minimum energy loss scenario. Indeed, in this case the hot spot magnetic field ($B \leq 100\mu$G) is constrained matching the X–ray flux by the SSC process, and the distance of the hot spots from the nucleus is > 180 kpc. As both the hot spots emit synchrotron radiation at optical wavelengths, in situ reacceleration appears to be inescapable (Brunetti et al. 2001c).

b) synchrotron X–ray emission from jets: One of the most interesting findings of *Chandra* is that X–ray jets are relatively common also in the case of low power radio sources (Worral et al., 2001). These X–ray jets are usually interpreted as synchrotron emission from the very high energy end ($\gamma \sim 10^7$) of the electron population. Such interpretation is mainly supported by the observed radio to X–ray spectral distributions. In addition, it is supported by the fact that, contrary to the case of high power radio sources, the jets of low power objects are believed to move at sub/trans–relativistic speeds at kpc distances from the nucleus (e.g., T.Venturi, this proceedings and ref. therein) and thus the X–ray emission from IC/CMB is expected to be negligible. If so, the jets of low power radio sources can be considered laboratories to study the electron acceleration. Indeed, combining radio, optical and X–ray data it is possible to study the synchrotron spectrum from radio to X–ray frequencies and thus to sample the spectrum of the emitting electrons over more then 4 decades in energy. In particular, relatively deep multiwavelength observations with adequate frequency coverage of a few objects (M 87: Boehringer et al. 2001; 3C 66B: Hardcastle et al. 2001; PKS 0521-365: Birkinshaw et al. 2002) have shown that the radio to X–ray spectrum can be well fitted by a double power law model of slope α and $\alpha + (0.7 - 0.9)$ in the radio and in the optical to X–ray band, respectively. If further confirmed, this point is crucial as it generates problems in the interpretation of the data with acceleration models including standard electron diffusion in the post shock region (e.g. Bell, 1978a,b; Heavens & Meisenheimer, 1987) which, indeed, would predict a steepening of the synchrotron spectrum of only 0.5 in the optical to X–ray band.

More recently Dermer & Atoyan (2002) have proposed that synchrotron emission can successfully fits the X–ray data also in the case of some of the detected X–ray jets of high power radio sources usually interpreted via IC/CMB scattering. These authors have investigated the evolution of the spectrum of electrons accelerated up to very high energies ($\gamma > 10^8$) under the hypothesis that the radiative losses of the electrons are largely dominated by IC scattering rather than synchrotron. Under these conditions, if the photon energy density in the jet frame is dominated by boosted CMB photons (as in the case of the IC/CMB process), the energy dependence of the IC losses of the electrons with $\gamma > 10^7$ changes due to the effect of the Klein–Nishina cross section and the radiative losses for these electrons result alleviated (Fig.1). It can be shown that, under these conditions, the spectrum of the electrons may become harder for $\gamma > 10^7$ and that the resulting synchrotron emission may present a bump in the X–ray band similarly to that observed by *Chandra*. Assuming a transverse velocity structure in the jets, with a fast central spine surrounded with a boundary layer with a velocity shear (Sect. 4.3.5), it has been proposed that turbulence may also accelerate high energy electrons at such boundary layers (Owstroski, 2000; Stawarz & Ostrowski, 2002a). If this happens, X–ray synchrotron radiation from large scale jets is expected and it may account for some of the observed *Chandra* jets (Stawarz & Ostrowski, 2002b).

c) synchrotron X–ray emission from hot spots ?:

The effect of electron radiative cooling and the presence of a high energy cut–off in the electron spectrum produce an abrupt steepening of the spectrum of the hot spots below that extrapolated from the lower frequency power–low. This makes X–ray detection of synchrotron emission very difficult. In addition the photons emitted by competing processes particularly efficient at high frequencies (e.g., SSC) might completely hide those contributed by the synchrotron emission. So far, there are only two relatively secure cases of hot spots in which the synchrotron spectrum is given by a power law from the radio to the UV or even X–rays (3C 303 : Keel 1988, Meisenheimer et al. 1997 ; 3C 390.3 : Prieto 1997, Harris et al. 1998), indicating the continuation of the synchrotron spectrum at higher frequencies. An immediate implication is the presence of relativistic electrons with $\gamma \geq 10^7$ which, due to their short life time (considering typical hot spots' magnetic field strength $\geq 100\mu$G), require very efficient acceleration processes (acceleration time ≤ 100 yrs) and/or magnetic field strengths in the acceleration regions well below that calculated under equipartition conditions.

4.3.3. *Energetics: X–ray SSC emission from radio hot spots in FR II*

Until the advent of *Chandra* clear evidence for SSC emission had only been detected in the case of the hot spots of Cygnus A (Harris et al., 1994) in which case the magnetic field results close to the equipartition value.

Chandra has enabled significant progress in this field, with a number of successful detections in the first three years of observations (3C 295: Harris et al. 2000; Cyg A: Wilson et al. 2000; Pictor A: Wilson et al. 2001; 3C 123: Hardcastle et al. 2001; 3C 351: Brunetti et al. 2001c).

In the majority of the detected hot spots (Cygnus A–W and E, 3C 295–N, 3C 123) the magnetic fields derived comparing the radio and X–ray fluxes (Sect.3.5.2) result within a factor of 2 from the equipartition value. This has further motivated the usually adopted assumption of approximate equipartition between magnetic field and electron energy densities.

On the other hand, in the case of the double northern hot spots of 3C 351 (J and L components), if the SSC interpretation is correct, the magnetic field would result in both cases from a factor 3.5 to 5 smaller than the equipartition value (in case of ordered or isotropic field configuration,

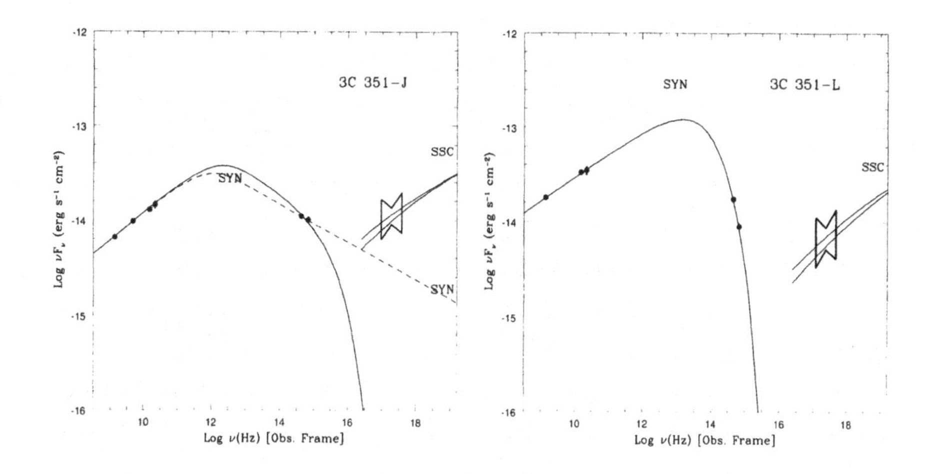

Figure 19. The radio, optical and X–ray data of the hot spots 3C 351–J (left panel) and 3C 351–L (right panel) are reported together with the synchrotron and SSC spectra. The spectrum of the emitting electrons is calculated assuming shock acceleration (Sect.3.3, Figs.3-4). Radio and optical fluxes are accounted for by synchrotron emission, whereas the X–rays are matched via SSC. In both cases, the reported SSC models are calculated assuming a substantial departure from equipartition conditions (see text for details).

respectively). Here we stress that this departure from equipartition implies an energy density of the electrons in the hot spots a factor > 60 larger than that of the magnetic field. Such a relatively strong departure from equipartition in these hot spots is further suggested by the modelling of the broad band synchrotron spectrum (Fig.19). Indeed, the magnetic field intensity, derived combining the optical synchrotron cut–off and break frequencies with the hot spots' lengths along the jet direction, lead to an independent estimate of the magnetic field strengths which is in good agreement with those obtained with the SSC argument (Brunetti et al. 2001c).

A particular intriguing case is the west hot spot of Pictor A (Wilson et al. 2001). The synchrotron spectrum shows an abrupt cut–off clearly indicated by a large number of optical data points and it falls orders of magnitude below the X–ray flux. On the other hand, the SSC interpretation would require a magnetic field strength about a factor 14 below the equipartition value to match the observed X–ray flux. In addition, the *Chandra* X–ray spectrum of the hot spot is relatively steep ($\alpha \simeq 1.1$) and it is poorly fitted by the spectrum expected in case of SSC emission ($\alpha \simeq 0.74$).

4.3.4. *Multiple electron populations in jets and hot spots ? : Pic A west*

TABLE 3. Low power objects

Name	z	Type	Assoc. Radio	Assoc. Optical	Chandra	Reference
3C 31	0.0167	RG	Knots	N	Y	Ha02
B2 0206+35	0.0368	RG	Knots	N	Y	Wo01
3C 66B	0.0215	RG	Knots	Y	Y	Ha01
3C 120	0.0330	RG	Knot	N	N	H99
B2 0755+37	0.0428	RG	Knots	Y	N	Wo01
3C 270	0.00737	RG	Knots	Y?	Y	Ch02
M 87	0.00427	RG	Knots	Y	N	Bi91,M02,W02
Cen A	0.001825	RG	Knots	?	Y	F91,K02
3C 371	0.051	BL	Knots	Y	Y	P01

References: Ha02=Hardcastle et al. 2002, Wo01=Worrall et al. 2001, Ha01=Hardcastle et al. 2001b, H99=Harris et al. 1999, Ch02=Chiaberge et al. 2002, Bi91=Biretta et al. 1991, M02=Marshall et al. 2002, W02=Wilson & Yang 2002, F91=Feigelson et al. 1991, K02=Kraft et al. 2002, P01=Pesce et al. 2001. See also http://hea-www.harvard.edu/XJET/index.html for an updated list.

and 3C 273 jet
One possibility to match the radio, optical and X–ray data of Pictor A west is to have two electron populations (Wilson et al. 2001). The first population of electrons (with injection spectral index $\delta \simeq 2.5$) is assumed to be in a region of relatively strong magnetic field (e.g. of the order of the equipartition field) and would be the responsible for the observed radio to optical spectrum via synchrotron emission. As already noticed, such a population would produce a SSC emission about two orders of magnitude below the observed X–ray flux. Thus it might be assumed a *second population* of relativistic electrons in the hot spot (with injection spectral index $\delta \simeq 3.3$), spatially separated from the first population and in low field regions. This *second population* would emit negligible synchrotron radiation and it may produce efficient IC scattering of the synchrotron radio–optical photons (from the first population) matching the observed X–ray spectrum. Alternatively, the electron spectrum of the second population might extend up to very high energies (with $\nu_c > 10^{17}$Hz) matching the X–rays via synchrotron radiation. In this last case a value of the injection spectral index $\delta \simeq 2.15$ and a break frequency in between 10^{11} and 10^{17}Hz is required in order to not overproduce the synchrotron spectrum emitted by the first population (Wilson et al. 2001). Both these possibilities are *ad hoc* and, so far, they are not well physically motivated so that the case of Pictor A west remains poorly understood and future radio and *Chandra* observations are

still required.

A second interesting case of possible multiple electron populations is the well studied jet of 3C 273. Recently, Marshall et al.(2001) and Sambruna et al.(2001) performed a detailed radio, optical (HST) and X-ray (*Chandra*) study of this jet. These authors provide different interpretations of the stronger knots in this jet. In particular, Sambruna et al.(2001) have suggested an IC/CMB interpretation for the X-ray spectrum of the knots since a synchrotron model from a single electron population cannot fit the radio to X-ray spectra. On the other hand, making use of a different data set, Marshall et al.(2001) were able to fit the observed spectra with a single synchrotron model. The most important difference in the two data sets is given by the different slopes of the optical spectra; additional observations and more detailed data analysis are probably requested to understand this discrepancy. A detailed optical-UV study of the jet of 3C 273 has been performed by Jester et al.(2001). These authors have measured the optical-UV spectral index along the jet discovering that it is only slowly changing without showing a clear trend along the jet. This has been interpreted as the signature of continuous reacceleration processes (Fermi I and II - like) of the relativistic electrons active along the jet which, in principle, might yield multiple electron populations or a single electron population with a complex spectrum. It should be noticed that, indeed, Sambruna et al.(2001) do not exclude the possibility to fit the radio to X-ray spectrum of the knots with synchrotron emission by *ad hoc* multiple (or complex) electron populations. A further step forward in the study of the jet of 3C 273 has been recently obtained by Jetser et al.(2002). These authors discovered the presence of an UV excess in the bright knots with respect to that expected from standard synchrotron models. This strengthens the possibility of a complex spectrum of the electrons or the presence of coexisting multiple electron populations. Jester et al.(2002) have also pointed out that the UV excess in the knots A and D2+H3 is consistent with the contribution due to the extrapolation at UV frequencies of the X-ray spectrum thus suggesting a common origin for the UV excess and for the X-rays. At the light of these results, we might conclude that the UV to X-ray spectrum can be produced by IC/CMB radiation emitted by the electron population giving the radio to optical emission via synchrotron process (as in Sambruna et al.2001). On the other hand, a synchrotron origin for the UV to X-ray spectrum of the bright knots cannot be excluded if a second - high energy - electron population is assumed (Jester et al. 2002). Additional multifrequency observations are required to test the different scenarios.

4.3.5. *Velocity structure in radio jets*

Komissarov (1990) suggested that the emission minima observed near the starting point of some FR I jets could result from Doppler dimming and that the appearance of the jet might be due to the presence of a slow moving boundary *layer* which would be less dimmed than the faster internal *spine*. Laing (1993) made the connection to polarization structure of FR I jets proposing a two component model consisting of an internal high velocity *spine* containing a magnetic field which has no longitudinal component but is otherwise random and a lower velocity external *layer* with an entirely longitudinal field structure. If FR I jets have field and velocity structure of this type and if they are launched at relativistic speeds decelerating away from the nucleus, the emission of the *spine* and *layer* components will suffer different effects due to beaming. Consequently, which component would dominate the emission properties of the jets depends on the angle with the line of sight and on the distance from the nucleus. This model provides a natural explanation for the tendency of the apparent magnetic field direction to be longitudinal close to the core and transverse further out. Detailed applications of models with velocity structure have been performed in the case of individual radio galaxies (e.g., 3C 31: Laing 1986; 3C 296: Hardcastle et al. 1997) and also for a small sub–sample selected from the B2 sample (Laing et al. 1999). More recently, Chiaberge et al.(2000) have explored the viability of the unification of BL Lacs and FR I radio galaxies by comparing the core emission of radio galaxies with those of BL Lacs of similar extended radio power in the radio-optical luminosity plane. In agreement with the Komissarov & Laing findings, these authors conclude that velocity structures in the jet are necessary to reconcile the observations with the unification scheme.

A possibility to study the velocity structure of radio jets (of both high and low power objects) on kpc scales is to compare their radio and X–ray emission properties. Assuming a transverse velocity structure, the emission of the jets pointing in the direction of the observer should be dominated by the contribution of the fast moving *spine*, whereas that from a misaligned jet should be dominated by the emission of the slow moving *layer* which is less dimmed by transverse Doppler boosting. In the case of high power radio objects, this scenario might be easily tested by *Chandra*. Indeed, these jets are highly collimated up to tens or hundreds kpc distance from the nucleus and thus indicating that their velocity structure (if any) should be preserved on large scales. As stated in Sect.4.3.1, the X–ray emission from the jets of core dominated quasars indicate a highly relativistic motion up to tens of kpc distance from the core. These velocities might be associated with a fast *spine* with a low radiative efficiency. On the other hand, future *Chandra* observations of the jets of the misaligned parent population (FR II radio galaxies) might constrain the contribution from any slow *layer*.

5. SUMMARY

In this contribution we have tried to summarize new insights on the physics of extragalactic radio sources from recent studies based on combined observations in different bands. We concentrated ourselves on the case of the non–thermal emission produced in kpc–scale regions (radio lobes, kpc–scale jets, hot spots) and on the spectrum and energetics of the emitting relativistic electrons.

We have shown that the spectrum of the relativistic electrons in radio sources can be approximated with a power law only in a relatively narrow energy range. In general, the shape of the spectrum depends on the acceleration mechanisms active in the emitting regions and on the competition between such mechanisms and both the processes responsible for the energy losses and the spatial diffusion of the relativistic electrons. Measurements of the flux and spectrum produced by different non–thermal emitting processes in different frequency bands can allow to trace the spectrum of the emitting electrons and thus to constrain the physics of the acceleration in these remote regions. Such measurements are possible only now by combining radio, optical and X–ray observations with arcsec resolution.

Here, a 'biased' summary of some of the most promising recent findings:

• *Low energy end of the electron spectrum*: An advance in constraining the energetics associated to the low energy electrons has coming from the new detections of extended X–ray emission from IC scattering of nuclear photons in powerful FR II radio galaxies and quasars. This effect opens a new window on the study of electrons with $\gamma \sim 100$ in radio lobes which are invisible in the radio band as, in general, they would emit synchrotron radiation in the 0.01 to 1 Mhz frequency range. We have stressed that, making use of the classical minimum energy formulae, these electrons are not taken into account in the calculation of the energetics of the radio lobes. On the other hand, the first X–ray detections of IC scattering of nuclear photons indicate that the bulk of the energetics of the radio lobes is contained by these low energy electrons. Additional evidences that the spectrum of the relativistic electrons extends down to low energies ($\gamma <$ 500) is provided by the possible detection of SSC optical emission from radio hot spots and by the X–ray emission from powerful radio jets due to boosted IC scattering of CMB photons.

• *High energy end of the electron spectrum*: The maximum energy of the electrons in extragalactic radio sources depends on the balance between acceleration and losses mechanisms. The discovery of optical synchrotron emission from hot spots in the last decades has pointed out the presence of efficient accelerators active in these regions and able to accelerate electrons up to very high energies ($\gamma \geq 10^5$). More recently, the discovery of synchrotron X–ray emission from an increasing number of jets of low power radio galaxies suggest the presence of even more energetic electrons ($\gamma \geq 10^{6-7}$) possibly accelerated in low B–field regions. These findings and the recent suggestion about synchrotron X–ray emission also from jets of powerful sources, would indicate the presence of extremely efficient accelerators in the emitting regions. The extremely short acceleration time–scales (down to $\sim 10^2$ yrs) requested by these findings are actually starting to put interesting limits on the spatial diffusion coefficient of the electrons and thus on the microphysics prevailing in these regions. The recent additional evidence for continuous reacceleration of relativistic electrons in radio jets might finally suggest the presence of energetic turbulence and thus that part of the kinetic power of the jets is dissipated into the developing of plasma instabilities and in the re–acceleration of relativistic particles. This might match with the proposed scenario in which a slow *layer*, where a fraction of the kinetic power is dissipated, surrounds a fast moving *spine*.

• *Energetics of radio lobes and hot spots*: A crucial point in the study of the extragalactic radio sources is the calculation of the energetics associated to these objects. As we have shown, the advent of *Chandra* makes it possible to extensively apply the IC method and to derive the energy density of the relativistic electrons and of the magnetic field in the emitting regions. This has been done in the case of a number of hot spots and radio lobes : the derived energetics are usually larger (1 to 30 times) than that calculated with the classical minimum energy formulae. This is a direct consequence of the presence of low energy electrons which contain most of the energetics, but it is also due to moderate departures from the minimum energy conditions found in a number of cases. Additional statistics is requested to better address this point. We further claim that deep X–ray follow up of the most unambiguous detections of diffuse IC emission from radio lobes will probably allow to derive the first maps of the magnetic field intensity in the extragalactic radio sources providing unvaluable information on the prevailing physics.

• *Kinematics of the radio jets*: One of the most interesting findings obtained combining the radio, optical and *Chandra* X–ray data of powerful radio jets is the recent discovery of highly relativistic speeds of the jets up to several tens of kpc of distance from the nucleus. Bulk Lorentz factors $\Gamma_{\mathrm{bulk}} \sim 3$–10 up to these distances limit the radiative efficiency of the jets that should be low in order to preserve the kinetic power of the jet itself. Again, a scenario with a velocity structured jet with a fast *spine* surrounded by a slow *layer* might help to better understand this findings.

Acknowledgements I am grateful to all my collaborators, in particular to M.

Bondi, A. Comastri and G. Setti for help and discussions. I am indebted to F. Mantovani who invited me to give these lectures and to M. Marcha for a careful reading of the manuscript and for useful comments.

References

Aharonian, F.A., 2002, *MNRAS*, **332**, 215
Aharonian, F.A., Atoyan, A.M., 1981, *Ap&SS*, **79**, 321
Arshakian, T.G., Longair, M.S., 2000, *MNRAS*, **311**, 846
Bell, A.R., 1978a, *MNRAS*, **182**, 147
Bell, A. R., 1978b, *MNRAS*, **182**, 443
Bicknell, G. V., Melrose, D. B., 1982, *ApJ*, **262**, 511
Birkinshaw, M., Worrall D. M., Hardcastle, M. J., 2002, *MNRAS*, in press; astro-ph/0204509
Biretta, J. A., Stern C. P., Harris, D. E., 1991, *AJ*, **101**, 1632
Blandford, R. D., 1986, in *Magnetospheric Phenomena in Astrophysics*, eds. R.I. Epstein & W.C. Feldman (New York: AIP), p.1
Blandford, R. D., Ostriker, J. P., 1978, *ApJ*, **221**, L29
Blandford, R. D., Eichler, D., 1987, *Physics Rep.*, **154**, 1
Blasi, P., 2000, *ApJ*, **532**, L9
Blasi, P., 2001, *APh*, **15**, 275
Blumenthal, G. R., Gould, R. J., 1970, *RvMP*, **42**, 237
Böhringer, H., Belsole, E., Kennea, J., et al., 2001, *A&A*, **365**, L181
Borovsky, J. E., Eilek, J. A., 1986, *ApJ*, **308**, 929
Bridle, A. H., Perley, R. A., 1984, *Ann. Rev. Astr. Ap.*, **22**, 319
Bridle, A. H., Hough, D. H., Lonsdale, C. J., Burns, J. O., Laing, R. A., 1994, *AJ*, **108**, 766
Brunetti, G., 2000, *Astroparticle Physics*, **13**, 107
Brunetti, G., Setti, G., Comastri, A., 1997, *A & A*, **325**, 898
Brunetti, G., Comastri, A., Setti, G., Feretti, L., 1999, *A & A*, **342**, 57
Brunetti, G., Setti, G., Feretti, L., Giovannini, G., 2001a, *MNRAS*, **320**, 365
Brunetti, G., Cappi, M., Setti, G., Feretti, L., Harris, D. E., 2001b, *A & A*, **372**, 755
Brunetti, G., Bondi, M., Comastri, A., et al., 2001c, *ApJ*, **561**, L 157
Brunetti, G., Comastri, A., Dallacasa, D., Bondi, M., Pedani, M, Setti, G. 2002a, *Proc. Symposium New Visions of the X-ray Universe in the XMM-Newton and Chandra Era'*; astro-ph/0202373
Brunetti, G., Bondi, M., Comastri, A., Setti, G., 2002b, *A & A*, **381**, 795
Celotti, A., Ghisellini, G., Chiaberge, M., 2001, *MNRAS*, **321**, L1
Chartas, G., Worrall, D. M., Birkinshaw, M., et al., 2000, *ApJ*, **542**, 655
Chartas, G., Gupte V., Garmire G., et al., 2002, *Apj*, **565**, 96
Chiaberge M., Celotti, A., Capetti, A., Ghisellini, G., 2000, *A&A*, **358**, 104
Chiaberge M., Gilli, R., Macchetto, F. D., Sparks, W. B., Capetti, A., 2002, *ApJ*, submitted; astro-ph/0205156
Clark, D. H., Caswell, J. L., 1976, *MNRAS*, **174**, 267
Cox, A. C., 1999, *Allen's Astrophysical Quantities* (Springer-Verlag)
Crane, P., Tyson, J. A., Saslaw, W. C., 1983, *ApJ*, **265**, 681
Daly, R. A., 1992, *ApJ*, **386**, L9
Dermer, C. D., 1995, *ApJ*, **446**, L63
Dermer, C. D., Atoyan, A. M., 2002, *ApJ*, **568**, L81
Dickel, J.R., 1983, in *Supernova remnants and their X-ray emission, IAUS*, **101**, 213
Eilek, J. A., Hughes, P. A., 1991, in *Beams and Jets in Astrophysics*, eds. P.A.Hughes (Cambridge astrophysics series), p.428
Eilek, J. A., 1996, in *Extragalactic radio sources, IAUS*, **175**, 483

Eilek, J. A., Arendt, P. N., 1996, *ApJ*, **457**, 150
Fabian A. C., Crawford, C. S., Ettori, S., Sanders, J. S., 2001, *MNRAS*, **322**, L11
Feigelson, E. D., Schreier E. J., Delvaille J. P., et al., 1981, *ApJ*, **251**, 31
Feigelson, E. D., Laurent-Muehleisen, S. A., Kollgaard, R. I., Fomalont, E. B., 1995, *ApJ*, **449**, L149
Felten, J. E., Morrison, P., 1966, *ApJ*, **146**, 686
Feretti, L., Giovannini G., 1996, in *Extragalactic radio sources, IAUS*, **175**, 333
Ferrari, A., Trussoni, E., Zaninetti, L., 1979, *A&A*, **79**, 190
Garrington, S.T., Leahy, J. P., Conway, R. G., Laing, R. A., 1988, *Nature*, **331**, 147
Ghisellini, G., Celotti, A., 2001, *MNRAS*, **327**, 739
Ginzburg, V. L., 1969, *Elementary Processes for Cosmic Ray Astrophysics* (Gordon & Breach Science Publishers: New York)
Gitti, M., Brunetti, G., Setti, G., 2001, *A&A*, **386**, 456
Gopal-Krishna, Subramanian, P., Wiita, P. J., Becker, P. A., 2001, *A&A*, **377**, 827
Gould, R. J., 1979, *A&A*, **76**, 306
Hamilton, R. J., Petrosian, V., 1992, *ApJ*, **398**, 350
Hardcastle, M. J., 2001, *A & A*, **373**, 881
Hardcastle, M. J., Alexander, P., Pooley, G. G., Riley, J. M., 1997, *MNRAS*, **288**, L1
Hardcastle, M. J., Worrall, D. M., 2000, *MNRAS*, **319**, 562
Hardcastle, M. J., Birkinshaw, M., Worrall, D. M., 2001a, *MNRAS*, **323**, L 17
Hardcastle, M. J., Birkinshaw, M., Worrall, D. M., 2001b, *MNRAS*, **326**, 1499
Hardcastle, M. J., Worrall, D. M., Birkinshaw, M., Laing R. A., Bridle A. H., 2002, *MNRAS*, in press; astro-ph/0203374
Harris, D. E., 2001, in "Particles and Fields in Radio Galaxies", R. A. Laing and K. M. Blundell (Eds.), ASP Conf. Series, in press; astro–ph/0012374
Harris, D. E., Grindlay, J. E., 1979, *MNRAS*, **188**, 25
Harris, D. E., Stern, C. P., 1987, *ApJ*, **313**, 136
Harris, D. E., Carilli, C. L., Perley, R. A., 1994, *Nature*, **367**, 713
Harris, D. E., Leighly, K. M., Leahy, J. P., 1998, *ApJ*, **499**, L 149
Harris, D. E., Hjorth, J., Sadun, A. C., Silverman, J. D., Vestergaard, M., 1999, *ApJ*, **518**, 213
Harris, D. E., Nulsen, P. E. J., Ponman, T. J., et al., 2000, *ApJ*, **530**, L 81
Harris, D. E., Krawczynski, H., 2002, *ApJ*, **565**, 244
Heavens, A.F., Meisenheimer, K., 1987, *MNRAS*, **225**, 335
Henry, J. P., Henriksen, M. J., 1986, *ApJ*, **301**, 689
Jaffe, W. J., 1977, *ApJ*, **216**, 212
Jester, S., Röser, H.-J., Meisenheimer, K., Perley, R., Conway, R., 2001, *A&A*, **373**, 447
Jetser, S., Röser, H.-J., Meisenheimer, K., Perley, R., 2002, *A&A*, **385**, L27
Jones, T. W., O'Dell, S. L., Stein, W. A., 1974a, *ApJ*, **188**, 253
Jones, T. W., O'Dell, S. L., Stein, W. A., 1974b, *ApJ*, **192**, 259
Isenberg, P. A., 1987, *JGR*, **92**, 1067
Kaneda, H., Tashiro, M., Ikebe, Y., et al., 1995, *ApJ*, **453**, L13
Kardashev, N. S., 1962, *Soviet Astronomy - AJ*, **6**, 3
Katz–Stone, D. M., Rudnick L., 1997, *ApJ*, **488**, 146
Katz–Stone, D. M., Rudnick L., Butenhoff, C., O'Donoghue, A. A., 1999, *ApJ*, **516**, 716
Keel, W. C., 1988, *ApJ*, **329**, 532
Kirk, J. G., Rieger, F. M., Mastichiadis, A., 1998, *A&A*, **333**, 452
Kirk, J. G., Guthmann, A. W., Gallant, Y. A., Achterberg, A., 2000, *ApJ*, **542**, 235
Komissarov, S. S., 1990, *Sov. Astronomy Lett.*, **16**, 284
Kraft R. P., Forman W. R., Jones C., et al., 2002, *ApJ*, **569**, 54
Kundt, W., Gopal-Krishna, 1980, *Nature*, **288**, 149
Lacombe, C., 1977, *A&A*, **54**, 1
Lähteenmäki, A., Valtaoja, E., 1999, *ApJ*, **521**, L493
Laing, R. A., 1993, in *Astrophysical Jets*, D.Burgarella, M.Livio,C.P.O'Dea eds. (Cambridge University Press, Cambridge), p.95

Laing, R. A., 1986, in *Energy transport in radio galaxies and quasars, ASP Series*, **100**, 241
Laing, R. A., Parma, P., de Ruiter, H. R., Fanti, R., 1999, *MNRAS*, **306**, 513
Leahy, J. P., 1991, in *Beams and Jets in Astrophysics*, eds. P.A.Hughes (Cambridge astrophysics series), p.100
Livshitz, E. M., Pitaevskii, L. P., 1981, *Physical Kinetics* (Pergamon: Oxford)
Mannheim, K., Krülls, W. M., Biermann, P. L., 1991, *A&A*, **251**, 723
Marcowith, A., Kirk, J. G., 1999, *A&A*, **347**, 391
Marshall, H. L., Harris, D. E., Grimes, J. P., et al., 2001, *ApJ*, **549** L167
Marshall, H. L., Miller, B. P., Davis, D. S., et al., 2002, *ApJ*, **564**, 683
Meisenheimer, K., Roser, H.-J., 1986, *Nature*, **319**, 459
Meisenheimer, K., Roser, H.-J., Hiltner, P. R., et al., 1989, *A&A*, **219**, 63
Meisenheimer, K., Yates, M. G., Roeser, H.-J., 1997, *A&A*, **325**, 57
Meisenheimer, K., Haas, M., Müller, S. A. H., Chini, R., Klaas, U., Lemke, D., 2001, *A&A*, **372**, 719
Melrose, D. B., 1980, *Plasma Astrophysics, Nonthermal Processes in Diffuse Magnetized Plasmas* (Gordon & Breach Science Publishers)
Micono, M., Zurlo, N., Massaglia, S., Ferrari, A., Melrose, D. B., 1999, *A&A*, **349**, 323
Miley, G., 1980, *ARA&A*, **18**, 165
Myers, S. T., Spangler, S. R., 1985, *ApJ*, **291**, 52
Mushotzky, R. F., Scharf, C. A., 1997, *ApJ*, **482**, L13
Neumann, D. M., 1999, *ApJ*, **520**, 87
Ostrowski, M., 2000, *MNRAS*, **312**, 579
Pacholczyk, A. G., 1970, *Radio Astrophysics* (W.H. Fereeman and Company: San Francisco)
Pesce, J. E., Sambruna R. M., Tavecchio, F., et al., 2001, *ApJ*, **556**, L79
Petrosian, V., 2001, *ApJ*, **557**, 560
Prieto, M. A., 1997, *MNRAS*, **284**, 627
Rawlings, S., Saunders, R., 1991, *Nature*, **349**, 138
Rees, M. J., 1978, *Nature*, **275**, 35
Rybicki, G. B., Lightman A. P., 1979, *Radiative Processes in Astrophysics* (Wiley, New York)
Sambruna, R. M., Urry, C. M., Tavecchio, F., 2001, *ApJ*, **549**, L161
Sambruna, R. M., Maraschi, L., Tavecchio, F., et al., 2002, *ApJ*, **571**, 206
Sanders, D. B., Phinney, E. S., Neugebauer, G., Soifer, B. T., Matthews, K., 1989, *ApJ*, **347**, 29
Sarazin, C. L., 1999, *ApJ*, **520**, 529
Saslaw, W. C., Tyson, J. A., Crane, P., 1978, *ApJ*, **222**, 435
Schlickeiser, R., 1996, *A&AS*, **120**, 481
Schwartz, D. A., Marshall, H. L., Lovell, J. E. J., et al., 2000, *ApJ*, **540**, L69
Siemiginowska, A., Bechtold, J., Aldcroft, T. L., et al., 2002, *ApJ*, **570**, 543
Simkin, S. M., 1978, *ApJ*, **222**, L55
Stawarz, L., Ostrowski, M., 2002a, *PASA*, **19**, 22
Stawarz, L., Ostrowski, M., 2002b, *ApJ* in press; astro-ph/0203040
Tashiro, M., Kaneda, H., Makishima, K., et al., 1998, *ApJ*, **499**, 713
Tashiro, M., Makishima, K., Kaneda, H., 2000, *AdSpR*, **25**, 751
Tashiro, M., Makishima, K., Iyomoto, N., Isobe, N., Kaneda, H., 2001, *ApJ*, **546**, L19
Tavecchio, F., Maraschi, L., Sambruna, R. M., Urry, C. M., 2000, *ApJ*, **544**, L23
Thomson, R. C., Crane, P., Mackay, C. D., 1995, *ApJ*, **446**, L93
Tribble, P. C., 1993, *MNRAS*, **263**, 31
Verschur, G. L., Kellermann, K. I., 1988, *Galactic and Extragalactic Radio Astronomy* (Springer–Verlag)
Webb, G. M., Drury, L. O'C., Biermann, P., 1984, *A&A*, **137**, 185
Webb, G. M., Fritz, K. D., 1990, *ApJ*, **362**, 419
Wilson, A. S., Young, A. J., Shopbell, P. L., 2000, *ApJ*, **544**, L27

Wilson, A. S., Young, A. J., Shopbell, P. L., 2001, *ApJ*, **547**, 740
Wilson, A. S., Yang Y., 2002, *ApJ*, **568**, 133
Worrall, D. M., Birkinshaw, M., Hardcastle, M. J., 2001, *MNRAS*, **326**, L 7

THE OBSERVATIONAL PROPERTIES OF BLAZARS; SELECTION, RELATIONSHIPS AND UNIFICATION

I.W.A. BROWNE[1]
University of Manchester
Jodrell Bank Observatory,
Macclesfield,
Cheshire, SK11 9DL,
UK

1. Introduction

The term Blazar is in some ways too specific giving the (possibly mistaken) impression that there exists a single well-defined class of "blazars" all having the same overall properties. In other ways it is too restrictive, leading to the exclusion from discussion of objects which have the same physical properties but which do not happen to meet some strict (but arbitrary) observational selection criteria. The problem is that those who observe like to adopt some convenient selection criteria to create neat and tidy samples while those who like to interpret the population statistics are tempted to ignore the selection criteria and may sometimes reach conclusions that have more to do with selection than with astrophysics. Samples may be well-selected in that they satisfy well-defined criteria but, in terms of containing a single physical class of object, they are rarely complete or 100% reliable.

What I will aim to do in this chapter is to try and strip away the observational fog and discuss what we really know about the broad-band non-thermal emission from jets in AGN. These are the jets which, when observed with the right sensitivity and from the right direction, can lead to an object being recognized and classified as a blazar. I will review the history of the subject, describe how blazar samples are selected, list the main samples and discuss broad-band spectral energy distributions (SEDs) with particular emphasis on the position of the peak in the synchrotron emission and correlations with total luminosity. Finally, I will take a brief look at unification, emphasizing that it makes little astrophysical sense to distinguish blazars from other radio-loud AGN.

F. Mantovani and A. Kus (eds.), The Role of VLBI in Astrophysics, Astrometry and Geodesy, 83–91.

2. History

In the late 1960s and early 1970s it became clear that a few of the optical objects associated with strong flat-spectrum radio sources exhibited very unusual behaviour; they were optically highly variable and often had high and variable optical linear polarizations. The variability and high polarizations were naturally explained in terms of optical synchrotron emission arising in some very compact region. Some of the objects were quasars with strong and broad emission lines but others were more enigmatic, having no emission lines. The prototype of this latter group was BL Lac[1] which displayed the most extreme behaviour. For a time there was even a dispute about whether BL Lac was a galactic or an extragalactic object and this was only finally resolved when the spectrum of the faint fuzz associated with BL Lac was shown to have the features of an elliptical galaxy at a redshift of 0.07 (Oke and Gunn, 1974). The term BL Lac object was first used by Stritmatter et al (1972)[2]. The term Blazar was coined by Ed Speigel in 1978 during the Pittsburgh Conference on BL Lac Objects to cover both the BL Lacs and those quasars that showed similar optical behaviour.

Some key points are worth emphasizing:

1. All objects displaying blazar behaviour have strong flat-spectrum radio cores and are also strong X-ray sources.
2. Not all objects with strong flat spectrum radio cores show blazar behaviour (only 50% of flat-spectrum radio-quasars are classified as blazars), neither are all strong extragalactic X-ray sources.
3. Blazars are believed to be the Doppler beamed cores of common types of radio sources, FR1 and FR2 radio galaxies.
4. The spectral energy distributions (SEDs) of Blazars are very broad with peaks (in $\nu f(\nu)$ space) in the IR/optical/UV/X-ray and in the X-ray/Gamma-ray part of the spectrum.

3. How Blazars are found

The ways in which blazar samples are put together are particularly important because they define the selection effects and, like it or not, these play a major role in the subject. Because the SEDs are very broad (see Figure 1) it turns out that there are currently two reasonably efficient ways of

[1]BL Lacertae had been classified as an irregular variable star by Hoffmeister (1929), hence its unusual name for an extragalactic object. There are two other well-known BL Lac objects that also appear in early lists of irregular variable stars, AP Lib and W Com. A comprehensive list of extragalactic objects which were first discovered as variable stars can be found at http://www.klima-luft.de/steinicke/AGN/vargal/vargal2000.htm

[2]The first listing of six objects, four of which are now recognized as BL Lacs, is contained in Browne (1971).

constructing Blazar samples: one either starts with a radio survey or an X-ray survey. In addition, some combination of optical monitoring, optical polarimetry and/or spectroscopy is required; though it is common to refer to radio-selected or X-ray-selected samples, this is a misnomer since optical confirmation is always required. It is important to emphasize that, only when the optical non-thermal emission is bright enough compared to other optical emissions, will the object be classified as a blazar. If this is not the case the object will end up in a sample of other objects, e.g. quasars, elliptical galaxies, etc., depending on the dominant source of optical emission.

The standard optical characteristics that an object must display in order to be classified as a blazar are one or more of the following:

1. Be an OVV; i.e. display >0.5 magnitude changes of optical flux density on time-scales of days.
2. Be highly polarized. Plane polarization >3% is the adopted criterion mainly because transmission through the Galactic interstellar medium can produce polarizations of up to ~2% so only objects with polarization of >3% can be recognized reliably as intrinsically polarized.
3. Have a featureless (emission line equivalent width <5 Å) optical spectrum.

N.B. These traditional criteria are entirely empirical, set mainly for optical observing convenience rather than any natural break in the distributions of intrinsic properties between blazars and non-blazars.

4. Blazar samples

Creating samples of blazars is hard work because of the optical observations involved. At present there are less than 1000 objects catalogued as blazars. Much of the systematic work on compiling blazar samples has been done from X-ray surveys. Some of the main blazar samples are are:

1. EMSS sample obtained with the Einstein X-ray satellite (Stocke et al., 1991)
2. Einstein Slew survey – not as sensitive as EMSS but covers a larger area (Perlman et al., 1996).
3. ROSAT-based surveys mostly made by cross correlation with radio surveys. These include REX (Caccianiga et al., 1999), DXRBS (Landt et al., 2001) and RASS (Bade et al., 1998).
4. The 1Jy BL lac sample selected at 5 GHz and covering the whole sky (Stickel et al., 1993).
5. The 200mJy sample (Marchã et al., 1996), selected at 5 GHz.
6. CLASS samples down to 30 mJy (Marchã et al., 2001: Caccianiga et al., 2002), again selected at 5 GHz.

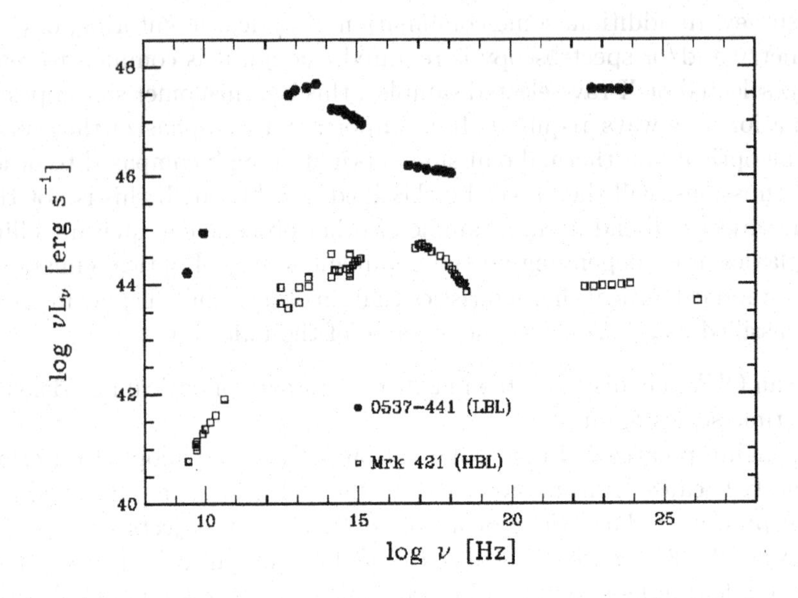

Figure 1. The spectral energy distributions (SEDs) of two BL Lac objects, 0537-441 and Mrk 421. The former is an LBL and the latter an HBL.

5. Blazar SEDs

Blazars are the broadest band emitters of electromagnetic radiation in the Universe (see Figure 1). All the SEDs, however, appear to follow a similar template with two broad peaks, the lower frequency one believed to be synchrotron emission and the higher frequency one believed to be inverse Compton emission (IC). This IC peak may arise either from synchrotron-self-Compton, where the electrons are scattered off synchrotron photons generated in the jet itself, or from scattering of external photons, possibly provided by the AGN. One of the most intriguing aspects of the whole blazar phenomenon is that, while the basic template remains the same from object to object, the positions of the synchrotron and IC peaks can differ by many orders of magnitude. Looking for a physical explanation for this diversity seems to me to be a key problem for blazar researchers to focus upon.

Over the years various sub-classifications based on the SEDs have been in use. How useful they are is an open question but, since they are widely used in the literature, it is appropriate to give a brief summary. Objects with their synchrotron peaks in the near infrared are generally classified as LBLs (Low frequency BL Lacs) while those with peaks in the UV/X-ray part of the spectrum are designated HBLs (High frequency BL Lacs). Most LBLs have been identified from radio surveys and most HBLs from X-ray surveys. (An older classification scheme was in terms of selection method; RBLs for radio-selected and XBLs for X-ray selected.) The different mixture of LBLs and HBLs depending on selection method should be no surprise since radio selection favours strong radio emission and X-ray selection favours strong X-ray emission[3]. The fact that joint selection by radio plus X-ray emission produces many objects with intermediate type SEDs (e.g. REX; Caccianiga et al., 1999) perhaps reinforces the view that LBL/HBL classification is telling more about selection methods than about astrophysics. The astrophysics lies in explaining why there is a continuum of SED properties rather than in explaining the existence of dubious sub-classes.

5.1. CORRELATIONS OF SEDS WITH LUMINOSITY

Donato et al (2001) is the most recent of a series of papers based on compilations of blazar SEDs. They have now compiled SEDs for a total 268 blazars, those that have sufficiently good observations to enable the synchrotron and IC peaks to be identified. Fossati et al. (1998) and Donato et al. (2001) claim that it is possible to describe the overall SED of a blazar by a model with one parameter, the bolometric luminosity. In Figure 2 we show one of their plots. The five different curves are obtained by averaging the SEDs of blazars in different radio luminosity ranges. Within this sample there is a unambiguous trend for the higher luminosity objects to have lower frequency synchrotron (and IC) peak frequencies; LBLs tend to be high luminosity objects while HBLs are low luminosity ones. This is potentially a very exciting result, particularly since there is an elegant physical explanation has been offered for the effect by Ghisellini (1999). He suggests that higher luminosity objects have more luminous AGN which produce sufficient ambient photons for the IC scattering of these to take over from synchrotron-self-Compton as the dominant loss mechanism for the high energy synchrotron electrons. This will have the effect of moving the synchrotron cut-off to lower frequencies in objects with strong AGN.

Have we really got away from selection effects? I worry that we have not since in the heterogenious sample available to Donato et al., the highest luminosity objects tend to be radio-selected and the lowest luminosity ob-

[3]This is all very familiar to radio astronomers who, for example, are not surprised by the 178 MHz 3CR survey containing mostly steep spectrum sources and the 5 GHz GB6 survey containing a high proportion of flat spectrum sources. See Kellermann, et al. (1968) for a quantitative analysis of how the spectral mix varies with the frequency of sample selection.

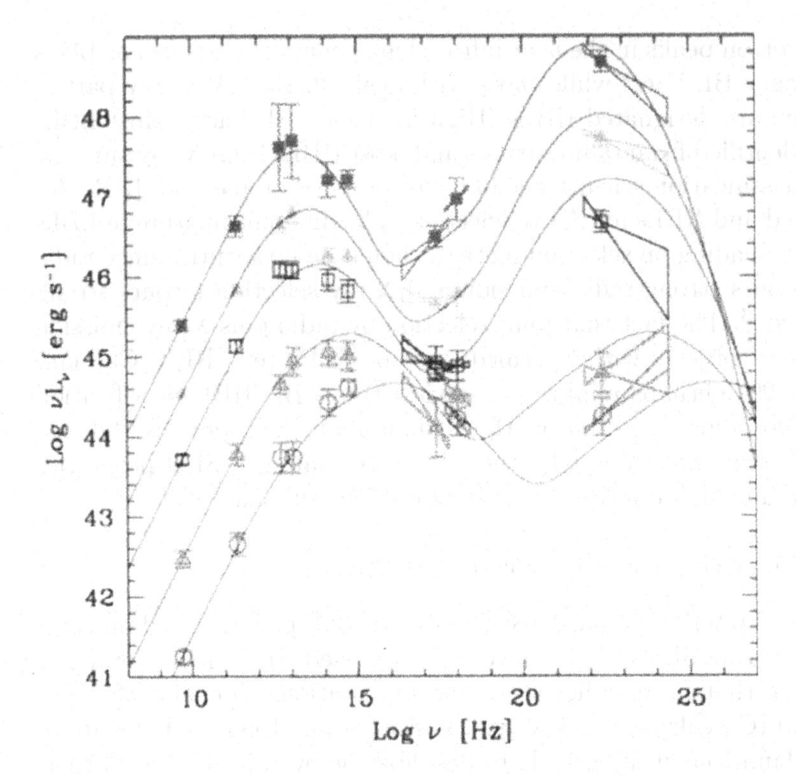

Figure 2. The averaged SEDs of blazars divided by radio luminosity, after Donato et al., 2001.

jects, X-ray selected. A critical investigation is to find out what the SEDs of low luminosity radio-selected objects look like? We do not have as much information as we would like but Anton (2002) has been looking at the radio-to-optical SEDs of low-redshift flat-spectrum radio sources from the Marchã et al. 200 mJy sample. Sub-mm observations with SCUBA and far-infrared observations with ISO have been included in her analysis. The preliminary conclusion from this work is that a significant number of the low radio luminosity objects may have an even lower break frequency (one might classify these as VLBLs!) than some of the high luminosity LBLs. The SEDs for two such examples are illustrated in Figure 3 and Figure 4. The Anton (2002) results need to be confirmed and extended, but appear to indicate that the neat picture of a monotonic relation between the synchrotron peak frequency and luminosity represented by Figure 2 is not the whole story.

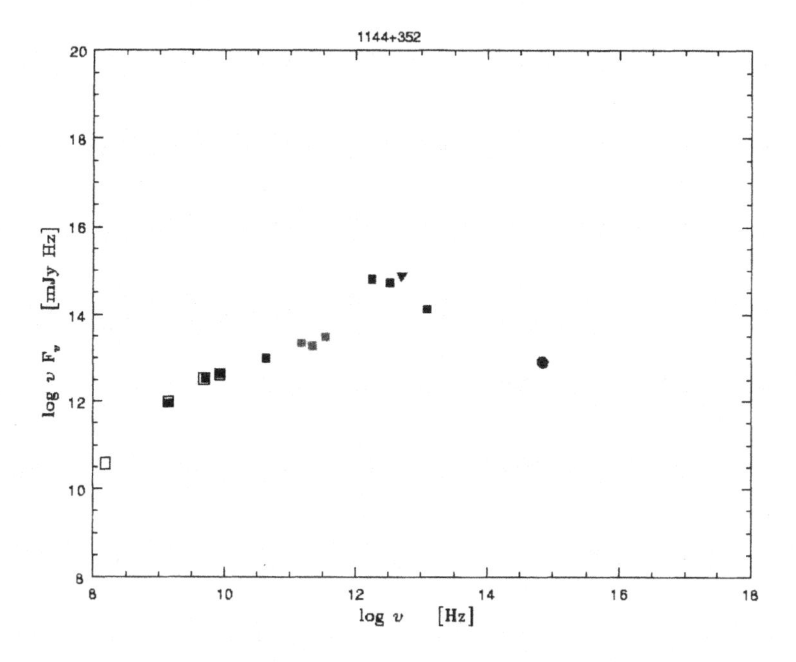

Figure 3. SED of the low radio-luminosity flat radio-spectrum source B1144+352 from Anton (2002). Not that the peak in the SED is at infrared wavelengths

6. Blazars and unification

Blazars are believed to the highly Doppler beamed cores of common types of radio sources. BL Lac objects are thought to be beamed Fanaroff and Riley type 1 sources (FR1s) and the broad emission-line objects beamed FR2s. The former sources have relatively low intrinsic luminosity (discounting beamed emission), relaxed extended structures, no prominent hotspots weak emission lines and show little cosmological evolution. FR2s in contrast have high radio luminosities, highly collimated radio structures, strong emission lines and undergo strong cosmological evolution. The OVV quasars and HPQs show strong evolution while BL Lac objects show little cosmological evolution and have (almost by definition) weak emission lines. Thus a reasonably consistent picture emerges. More details are given in Chapter?? which contains a review of unified models.

6.1. COSMOLOGICAL EVOLUTION OF BL LAC OBJECTS

One possible flaw in the unification picture is the claimed difference in cosmological evolution for BL Lac objects selected from radio surveys and

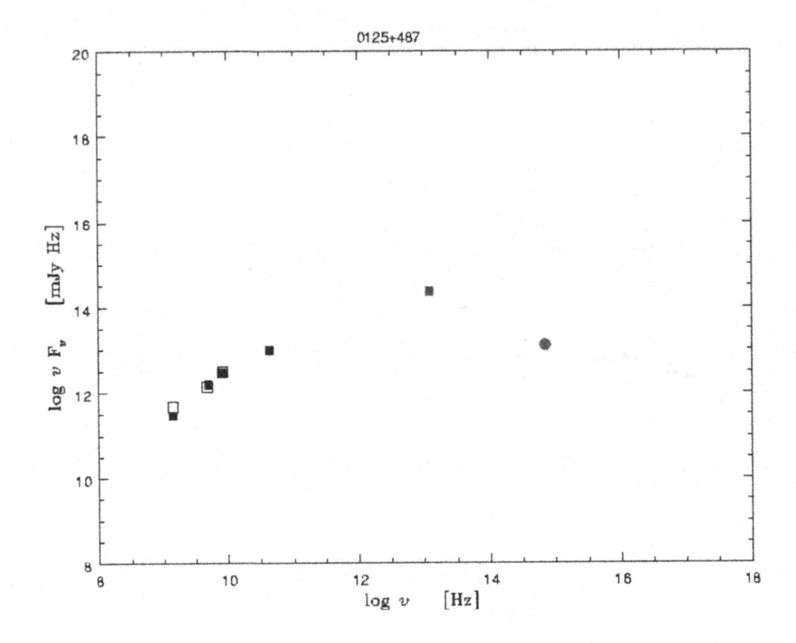

Figure 4. SED of the low radio-luminosity blazar B0125+487 from Anton (2002).

those selected from X-ray surveys. Morris et al. (1992) and Wolter et al. (1991) found from their analysis of BL Lacs from the EMSS survey that these objects appeared to have a higher space density at the present epoch than in the past – negative cosmological evolution – unlike any other AGN population. On the other hand, radio selected BL Lacs show no or mildly positive evolution (Stickel et al., 1991). If the radio-selected and X-ray selected populations do evolve differently, this is an argument against them being two extremes of a single BL Lac population which is the assertion of the BL Lac/FR1 unified scheme. Some of the negative evolution can be explained away by quantifyable selection effects (Marchã & Browne, 1995). However, there remains a worrying difference which may or may not be significant. Clearly larger samples are required.

7. Greater unification

The observational evidence suggests that *all* radio-loud objects have jets which start out highly relativistic and it is from the innermost part of the jet that the non-thermal emission arises and which leads us to classify some radio loud objects as blazars. As discussed above, there is a high degree of similarity between the SEDs of all those objects that are recognized as blazars. Also Heino Falcke (Chapter??) argues that all jets have a common physical mechanism for their emission. It is therefore a reasonable assumption that the radio cores of all radio-loud AGN are intrinsically similar. In fact, we should probably take another step and make a not unreasonable assertion that all radio-loud AGN would be classified as blazars if only we could get an unimpeded view of their non-thermal emission. So, since what interests us is the astrophysics of the non-thermal emission, why not call all radio loud objects blazars? Better still, let's forget about the term blazar altogether and focus only on the astrophysics of the emission!

References

Anton,S., (2002), *PhD thesis, University of Manchester*
Bade, N., et al., (1998) *A&A*, **334**, 459
Browne, I.W.A. (1971), *Nature.*, **231**, 515
Caccianiga, A. et al., (1999) *ApJ.*, **513**, 51
Caccianiga, A. et al., (2002), *MNRAS*, **329**, 877
Donato, D., et al., (2001) *A&A.*, **375**, 379
Fossati, G., et al., (1998) *MNRAS.*, **299**, 433
Ghisellini, G., (1999) *BL Lac Phenomenon, ASP conference series, eds Takalo, L.O. & Silanapaa, A* , **159**, 311
Hoffmeister, C., (1929) *AN*, **236**, 233
Kellermann,K.I., et al., (1968), *Astrophysical Letters*, **2**, 105
Landt, H., et al, (2001) *MNRAS*, **323**, 757
Marchã, M.J.M. & Browne, I.W.A., (1995), *MNRAS*, **275**, 951
Marchã, M.J.M., et al., (1996) *MNRAS*, **282**, 425
Marchã, M.J.M., (2001), *MNRAS*, **326**, 1455
Morris, S. et al., (1992) *ApJ*, **380**, 49
Oke, B. & Gunn, J., (1974), *ApJ.*, **189**, L5
Perlman, E. et al., (1996) *ApJS*, **104**, 251
Stickel, M., et al., (1991), *ApJ*, **374** 431
Stickel, M. et al., (1993), *A&AS*,**98**, 393
Stocke, J. et al., (1991) *ApJS*, **76**, 81
Stritmatter, P., et al., (1972), *ApJ.*, **175**, L5
Wolter, A. et al., (1991), *ApJ*, **369**, 31

OBSERVING BLAZARS WITH VLBI

Some simple considerations

R.W. PORCAS
Max-Planck-Institut fuer Radioastronomie
Auf dem Huegel 69, D53121 Bonn, Germany

Abstract.
This second lecture on blazars is not primarily about their properties but, rather, about various ways in which the VLBI technique can be used to observe them. It covers a number of general points to be considered when making VLBI observations of compact objects, and describes, with examples, various observing modes that can be used to investigate blazar properties. These include studies of structural variability and superluminal motion, multi-frequency and polarization-sensitive VLBI, mm-VLBI and Space-VLBI.

1. Blazars - what are they ?

BLAZAR: The most variable class of active galaxy. This class of active galaxy includes BL Lac objects and some of the most violently variable quasars, the name "blazar" derving from an amalgamation of "BL Lac" and "quasar".

This *definition* is from the Encyclopaedia of Astronomy and Astrophysics (ed. Murdin, 2001). The class includes such well-known quasars as 3C 279, 3C 345, 3C 446 and 3C 454.3, as well as BL Lac itself and OJ 287, another object of its type. An important *property* of blazars is that, although defined by their optical characteristics, they are essentially all strong, flat-spectrum, variable radio sources, with structures dominated by a compact core. Their extended structures are amorphous, with only faint (if any) jets, and they almost never exhibit the "classical double" structures seen in other radio quasars and galaxies. Historically, blazars were amongst the first sources ever detected using the VLBI technique (Broten et al. 1967), and with *real-time* VLBI using a communications

F. Mantovani and A. Kus (eds.), The Role of VLBI in Astrophysics, Astrometry and Geodesy, 93–106.

satellite (Yen et al. 1977). The current *interpretation* of blazars is that they are active galaxies with jets of ejected plasma oriented very close to the observer's line of sight.

The strong and compact nature of their radio emission makes blazars natural and easy targets for VLBI. First, an obvious point: sources with no compact structure are **invisible** to VLBI because the synthesised telescope aperture has a big hole in the middle ! There is no response from emission with spatial frequencies smaller than those measured by the shortest baseline. Equally obviously, strong sources are easier to detect than weak sources, and this is especially true for VLBI where there are detection thresholds. For example, we can think of the *fringe-fitting* process as a necessary fine-tuning of the interferometer pointing ("peaking up") performed after correlation. Blazars are good for this since they generally give a strong enough signal within the interferometer coherence time, even on the longest baselines, so that we can "point" on them directly. A similar detection threshold exists in the *phase self-calibration* process during the *hybrid mapping* procedure; again, blazars are usually strong enough for this, so that *phase-referencing* on a phase-calibration source is not required.

A useful concept is that of the *brightness temperature*, T_b, of a radio emitting region; it is the temperature at which a black body would give the same radio emission per unit area: $T_b = S\lambda^2/2k\Omega$

where S is the source flux density, λ is the observing wavelength, k is Boltzmann's constant and Ω is the observed solid angle of the emission. Since an interferometer has both minimum flux-density and minimum compactness limits we can define a further quantity, the *brightness temperature sensitivity limit* of an interferometer which combines these limits. For baselines of $\sim$ 9000 km this is around 10^9 K with present-day telescopes and systems - clearly unsuitable for *thermal* emission processes. Blazars, however, emit by the *synchrotron* process and usually have maximum T_b in the range 10^{10} to 10^{12} K.

2. A Blazar Archetype - 3C 345

The OVV quasar 3C 345 (redshift = 0.595) has been studied intensely with VLBI. It is strong, variable and conveniently located at declination 39° so that there is long mutual visibility between the telescopes of VLBI arrays in the northern hemisphere, creating relatively full tracks in the u,v-plane, and resulting in high-fidelity hybrid maps of the mas-scale structure. A high-resolution 5 GHz VLBI map is shown in Figure 1. 3C345 exhibits typical blazar features: a dominant, compact "core", apparently unresolved on the eastern edge, and a one-sided, collimated jet pointing in the NW direction. There are regions of enhanced emission in the jet (usually referred to as

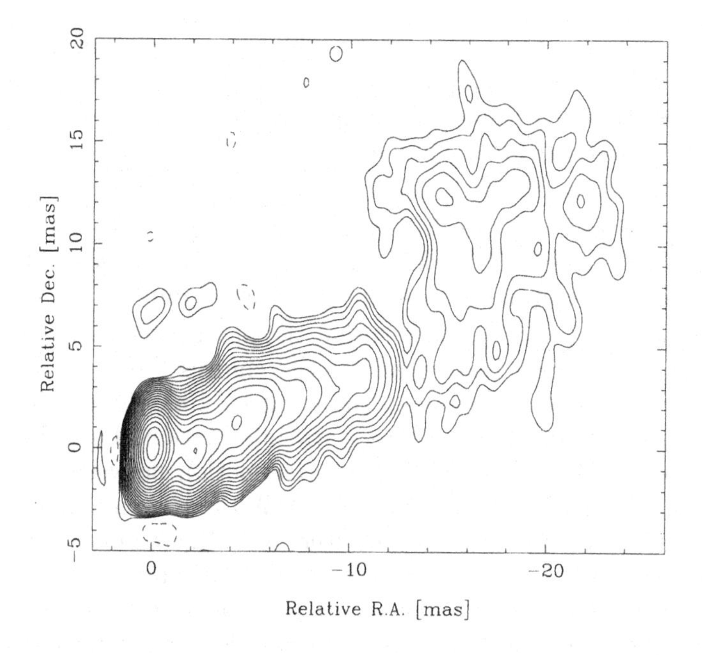

Figure 1. **5 GHz Global VLBI map of 3C 345** (from Lobanov & Zensus 1999)

knots or "blobs") and further down the jet there are more extended, off-axis features. The brightness temperature contrast between these and the core is very high so they can only be revealed by *high dynamic range* imaging. There are changes in the jet direction as it emerges from the core, and a large mis-alignment between the inner jet direction and that of the faint jet seen on arc-second scales. Blazars are not "one map objects", however, and one should, and can study them with many VLBI observations: multiple epochs, multiple frequencies, and multiple objects.

3. Multiple-Epoch Observations

Since blazars are highly variable at radio wavelengths it is natural to make VLBI observations at many epochs to investigate any possible structural changes. Early investigations of this sort revealed the phenomenon of *superluminal motion* in the blazar 3C 279 (Whitney et al. 1971). By today's standards these observations were simple (only a single baseline was used and source structure was deduced from model-fitting the visibility amplitude only !) but they established the basic facts of the phenomenon: apparent faster-than-light expansion between subcomponents in the mas-scale structure. Since then this phenomenon has been observed in many compact radio sources (e.g. 3C 345 - and not only those satisfying the strict blazar definition) from VLBI array monitoring efforts which result in sequences of hybrid maps at a number of observing epochs. The formula relating the apparent faster-than-light velocity, β_{app} (measured in light speed units), redshift z, dimensionless Hubble constant, h (in units of 100 km/s/Mpc), deceleration parameter, q_0 and measured angular proper motion μ (in mas/yr) is:

$$\beta_{app} = 47.4\ \mu\ (q_0 z + (q_0 - 1)(\sqrt{1 + 2q_0 z} - 1))/hq_0^2(1 + z)$$

Note that for objects at cosmological distances (e.g. $z \sim 1$) a superluminal velocity of 10 c corresponds to only a few tenths of a mas/year. Many years of monitoring are required to measure velocities of only 1 c or less.

It is important to note, however, that the phase self-calibration process used for making hybrid maps (and which removes the need to calibrate the real interferometer phase when the number of baselines is larger than the number of antennas) also destroys any information about the absolute source position. Thus it is only the *relative* motion between two or more distinct components which can be observed. Without further information it is not possible to say whether one, the other or both are moving. To proceed further one must find ways to make a correct *registration* of the maps at the different epochs. This can be achieved using the phase-referencing technique (Alef, 1989), whereby the interferometer phase **is** calibrated using a nearby compact reference source (whose structure itself is hopefully stable !). Such observations were first described by Bartel et al. (1986); they established that the compact core of 3C 345 is (more-or-less) stationary and that it is the knots in the jet which move superluminally. Most blazars exhibit an identifiable core component (compact, at the end of a jet-like structure, and unresolved on its outer edge) and there are good theoretical reasons for assuming that this is closely tied to the nucleus of the source; thus it is usually assumed (without measuring) that the core is stationary. In some cases, however, identification of the core may be unclear, especially if there appear to be more than one "stationary" component. Ambiguities in 3C 454.3 (Pauliny-Toth et al. 1987) have recently been investigated by Jorstad et al. (2001) who argue that the easternmost component, identified as the "core", appears to shift upstream after ejection of a component, or else disappears from time to time. Guirado et al. (1995) resolved similar ambiguities in 4C39.25 using the phase-reference technique.

Detailed monitoring of 3C 345 for 20 years has shown that many jet components have been ejected with superluminal speeds (they are referred to as C1, C2, C3,..etc). The velocities of individual components can change (in both magnitude and direction) as they proceed down the jet and the speeds and paths of successive components differ. The different paths of successive components emerging from the core of 3C 345 are shown in Figure 2. The superluminal velocity is usually interpreted as an illusion produced when highly relativistic motion (Lorentz factor $\gamma \sim 10$) occurs at a very small angle to the observer's line of sight. Although the distance

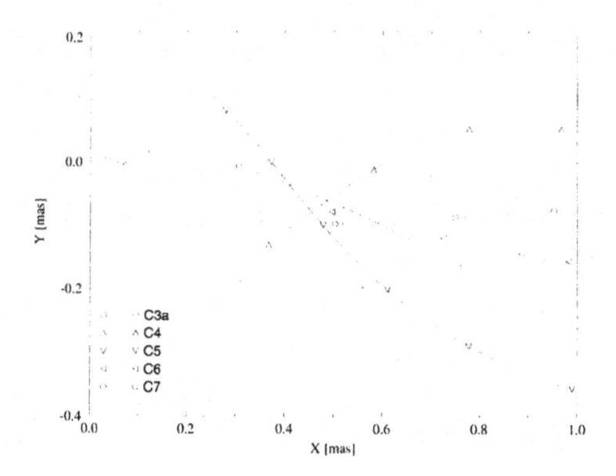

Figure 2. **Component tracks near the core of 3C 345** (from Lobanov 1996)

travelled is reduced in projection by a factor of sin θ the projected motion is speeded up by the Doppler factor $1/(1\text{-}\beta\cos\theta)$ (where θ is the angle to the line of sight and β is the velocity in light units). Apparent velocities as high as γc can be seen if θ is $\sim 1/\gamma$. In this situation, small changes in the true jet angle can be magnified in projection, and they can produce apparent velocity changes. Analysis of the component paths of 3C 345 (Klare, in preparation) suggest a possible precession of the axis of the ejection.

Highly superluminal motion can only occur in those rare circumstances where the source axis is close to the line of sight, and thus the statistics of superluminal motion provide information about the jet orientation in various types of active galaxies. Indeed, the *unified schemes* are driven by the realization that orientation plays a determining role in what radio and optical properties we see. Both Vermeulen (1995) and Ros et al. (2002) present the results of statistical studies of the superluminal velocity distribution in samples of sources. "Slow" superluminal motion is mostly seen in low-luminosity objects, and there are some indications of differences in the mean velocities of BL Lac objects, quasars and radio galaxies, but it is not clear whether, for a given redshift, these differences are statistically significant.

4. Multi-Frequency Observations

In the radio the spectrum can be measured over several decades in wavelength and this is a key quantity for investigating the radio emission process. Multi-frequency VLBI observations thus allow one to study the emission processes in different parts of the mas-scale source structure. The NRAO VLBA is an ideal array for multi-frequency studies because it has the ability to switch between its observing wavelengths (90 to 0.3 cm) in a matter of seconds. The compact "core" at the base of the jets in blazars usually has a flat or inverted (rising with frequency) spectrum whilst further down the jet and in other optically thin, extended regions the radio spectrum shows a power-law decrease with frequency, reflecting the relativistic particle energy distribution. A series of hybrid maps of 3C 345 at 5, 8.4, 15 and 22 GHz is shown in Figure 3. There are some key observational features to remember, however:

a) VLBI observations are *diffraction limited* in that the beamwidth scales with wavelength. Thus observations made with the same array at wavelengths of 6 and 18 cm differ in resolution by a factor of three; two well-separated jet components seen at 6 cm may be merged into a single component at 18 cm.

b) As mentioned before, the hybrid-mapping process loses information about the absolute source position, and it may not always be obvious how to register maps made at different wavelengths - especially as their resolutions may differ. One must either assume that some feature which is recognisable at all wavelengths is *achromatic* (i.e. has no frequency-dependent structure) or one must make careful phase-reference observations (and take into account any possible frequency-dependent structure in the reference source).

c) Although the resolution scales with wavelength, the brightness temperature sensitivity limit does not (because the λ^2 dependency cancels out the beam area change), and simply reflects any differences in receiver or antenna sensitivities. (One cannot, therefore, set higher brightness temperature limits for ultra-compact sources by going to higher observing frequencies - one must go to longer baselines.) However, for optically-thin regions of synchrotron emission the brightness temperature decreases with frequency with a power typically between 2 and 3; thus VLBI observations become dramatically less sensitive to such regions at increasing frequency, and they may, indeed, become invisible to the interferometer if they fall below its brightness temperature sensitivity limit. In such cases the radio spectrum will erroneously be thought to steepen.

Maps of blazars at mas scales are generally dominated by a compact "core", and it is tempting to register maps at different frequencies by assuming that this feature has a frequency-independent position (in the same way that one sometimes assumes it is time-independent). However, both theoretically and observationally it has been shown that this is incorrect. The "core" is not to be understood as an inner nucleus, coincident with some "centre of activity" but, rather, as a point near the base of the jet where radio emission is first seen, at the transition between optically thick and optically thin regions. Theoretical considerations (see e.g. Lobanov 1998) suggest that this point is closer to the jet base at higher frequencies, with a wavelength dependence of roughly λ^{+1}. Figure 4 shows a sketch of the frequency-dependent position of a "core".

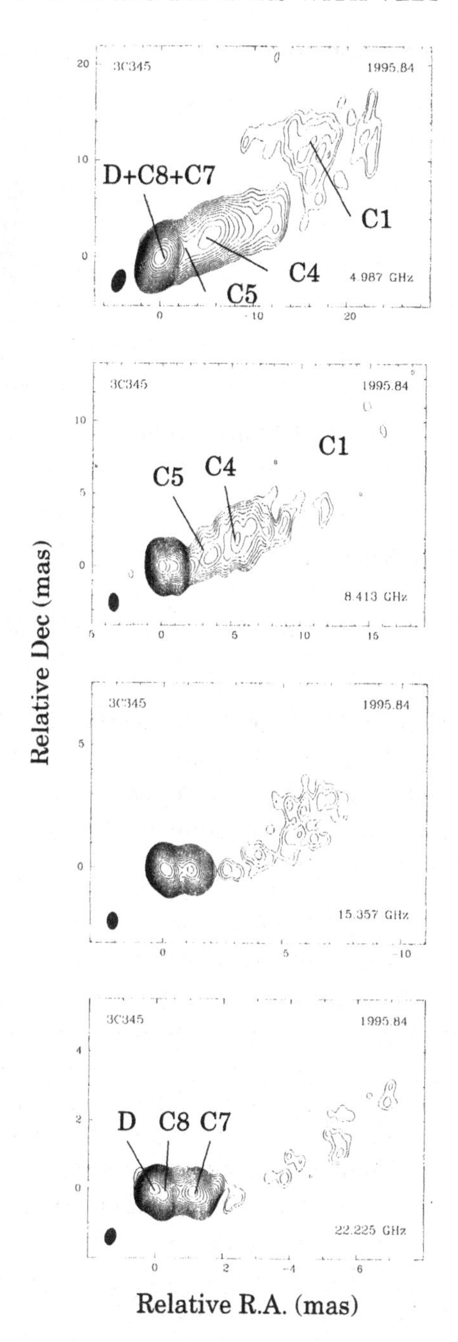

Figure 3. **Multi-frequency VLBI maps of 3C 345. From top: 5.0, 8.4, 15.3, 22.2 GHz** (from Ros et al. 2000)

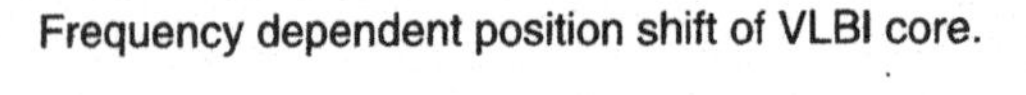

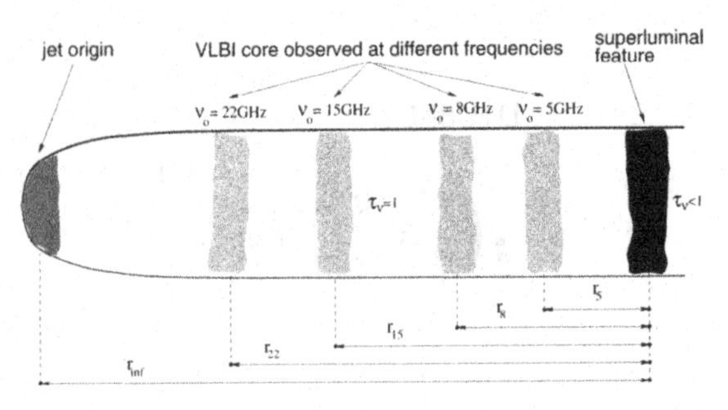

Figure 4. **Sketch of a "core" position at different frequencies** (from Lobanov 1996)

5. Polarization Observations

The polarization characteristics of the radiation from blazars can give additional information about the radio emission mechanism, the local environment and signal propagation. For optically-thin synchrotron radiation the degree of linear polarization indicates the degree of order in the generating magnetic field, and its **E**-vector position angle (the *polarization angle*) emerges perpendicular to the direction of the magnetic field projected in the plane of the sky. (Note, however, that for a relativistically moving jet this projection angle depends on the velocity, β, due to *relativistic aberration*; see Blandford and Konigl, 1979.) VLBI polarimetry ("VLBP") thus allows one to trace the ordered part of the magnetic field down the jet and across the source structure. Ros et al. (2000) have studied the polarized emission from 3C 345 at a number of frequencies, and Denn et al. (2000) have made multiple epoch VLBP observations of BL Lac.

VLBI observations are normally made using circular polarization modes (right and/or left-hand) since the individual antennas, mostly with azimuth-elevation mounts, do not then need to track the source parallactic angle (which can be very different for different sites). Single-polarization observations involve *parallel-hand correlations* (either **RxR** and/or **LxL**) which produce visibilities for the total intensity, I, providing the source is not circularly polarized. For linear polarization VLBI measurements, the *cross-hand correlations* **RxL** and **LxR** must be made, which produce cross-hand visibilities which can be used to make maps of polarized flux, P, degree of polarization, m, and polarization angle, χ. Note that some antennas (notably the space VLBI element HALCA) can only observe with a single circular polarization mode, so that only one of the cross-polarization correlations can be made. This increases the complexity of the mapping process but can nevertheless be made to produce useful linear polarization information on the source structure.

The degree of polarization in blazar cores and jets can be rather low (just a few percent) and this means that a careful calibration of the *instrumental polarization* is necessary. The circular polarization modes delivered by antenna feeds are rarely pure, and are usually contaminated with the orthogonal mode to a small degree, meaning that cross-polarization correlations usually contain a small percentage of parallel-polarization correlation. If the source polarization is low these terms can be comparable. For this reason it is always necessary to determine and remove the effects of instrumental polarization when making linear polarization measurements. Of course, the parallel-hand correlation is similarly contaminated by cross-hand correlation if the source is significantly polarized. However, this effect is much smaller and is usually ignored when making I maps unless high dynamic range is required. A welcome by-product of a careful polarization calibration is that the I and P maps resulting from the analysis will be correctly registered automatically.

The observed polarization angle may, however, be affected by Faraday rotation in any intervening magnetized plasma, which causes a rotation of the angle by an amount proportional to λ^2. The un-rotated angle can be deduced by observing at multiple wavelengths and extrapolating the observed polarization angle to zero wavelength. The *rotation measure*, which characterises the amount of Faraday rotation at each wavelength, contains useful information on the electron density and magnetic field in the intervening medium. Note that if the plasma is at redshift distance z, then the wavelength at which the rotation actually occurs is given by $\lambda_{observed}/(1+z)$. Thus any Faraday rotation which occurs near to a high redshift blazar is diluted by $(1+z)^2$.

Polarization VLBI maps of 3C 345 at 22 and 5 GHz are shown in Figure 5. They illustrate another feature of polarization mapping, namely that of *beam depolarization.* The left map, made at 22 GHz, shows details of the core region, where it can be seen that the core, and components C8 and C7 are linearly polarized up to 10 % but the polarization angles change dramatically between the components. The same region is shown in the 5 GHz map (right) where, because of the four times poorer resolution, these three components are merged into a single component. Whereas the total intensity, I of this component will reflect the sum of those of the core, C7 and C8, the polarized flux is the vector sum of their polarized responses, and these will tend to cancel to some extent. Although we cannot tell what the polarizations of the individual components are at 5 GHz, at least part of the dramatic decrease in percentage polarization (only 2 %) may result from this beam depolarization effect.

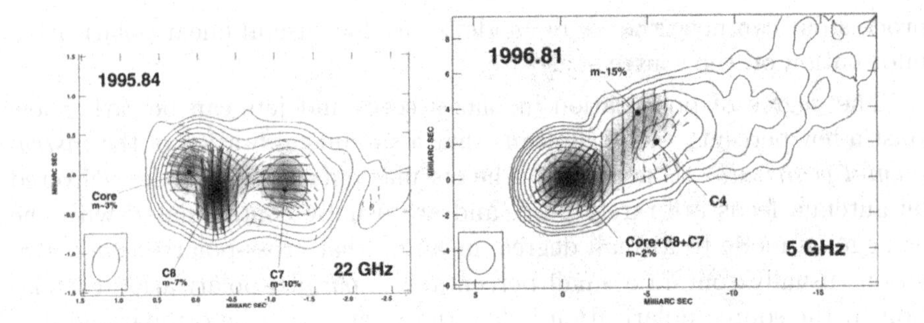

Figure 5. **Polarization VLBI maps of 3C 345 at 5 and 22 GHz** (from Ros et al. 2000). Note the reduction in degree of polarization, m, when the larger (5 GHz) beam blends the three components "core", C8, C7 into one.

Another related effect is the curious "pushing apart" of close components in P maps when their polarisation angles are almost perpendicular, as seen in the case of the blazar OJ 287 (Figure 6). The cancellation which occurs between the two components produces a "null" between them, shifting the component peaks in P away from the null, and they thus appear to straddle the position of the core seen in the I map.

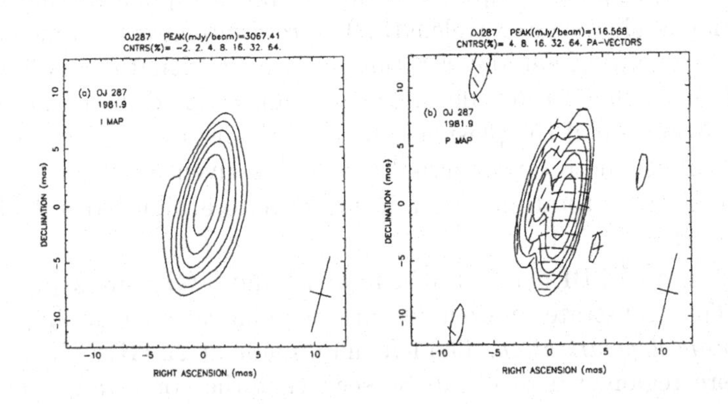

Figure 6. **Total intensity (left) and polarization (right) VLBI maps of OJ 287** (from Roberts et al. 1987). Note how the two polarized peaks with perpendicular polarisation angles "push each other apart".

Cawthorne et al. (1993) made VLBP observations of a sample of compact objects and looked for any relationship between the mas-scale polarization properties and the source structure. They reported a significant difference between two classes of objects; for quasars the polarization angle tends to be perpendicular to the jet angle (i.e. deduced **B** field parallel to the jet) whereas for BL Lac objects the deduced **B** field is perpendicular to the jet.

Radiation can also be *circularly polarized* (i.e. when the source produces an excess of one circular polarization mode over the other). The synchrotron mechanism does not produce this, and it has usually been found that blazars have a negligible (less than 1 %) degree of circular polarization. Nevertheless, VLBP observations of the blazar 3C 279 have recently been made which show a small but significant amount of circular polarization in the jet emission (Figure 7). It is assumed that this is produced by conversion of linear to circular polarization, which requires a low-energy tail in the energy distribution of radiating particles within the jet, which itself suggests the existence of *electron-positron* as well as *electron-proton* plasma in the jet.

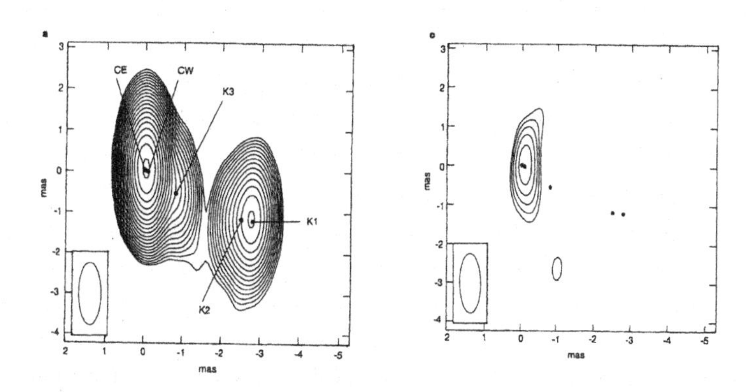

Figure 7. **Total intensity (left) and circular polarization (right) VLBI maps of 3C 279** (from Wardle et al. 1998).

6. Even Higher Resolution: mm-VLBI and Space-VLBI

Since the resolution of an interferometer scales as λ/baseline-length, we can achieve even higher angular resolution by going to shorter wavelengths (mm) or by putting radio telescopes away from the Earth's surface (Earth-Space VLBI). Both require great technical and logistical expertise and a large amount of effort. The prize is resolution down to 50 μas. As usual, blazars are the easiest and most often-observed objects.

At wavelengths of 3 mm (and shorter) the main problems are the poorer performance of most VLBI antennas (usually designed for cm-wavelength operation), less sensitive receivers and, above all, the much greater effect of random pathlength fluctuations in the atmosphere on the interferometer phase. All three of these effects combine to dramatically raise the source detection threshold, typically to $\sim$ 0.5 Jy. In addition, variable absorption in the atmosphere results in amplitude calibration problems, since the effective "antenna" gain depends strongly on the weather and the atmospheric path length. VLBI observations at 3 mm wavelength take place regularly, either with the VLBA (using "dynamic scheduling" to take advantage of favourable weather conditions) or at set time periods together with some of the large mm radio telescopes (such as the IRAM telescopes in France and Spain) when their superior performance can greatly enhance the results. A 3 mm-VLBI map of the blazar 3C 454.3 is shown in Figure 8. VLBI observations at even shorter wavelengths (2 and 1 mm) are still at

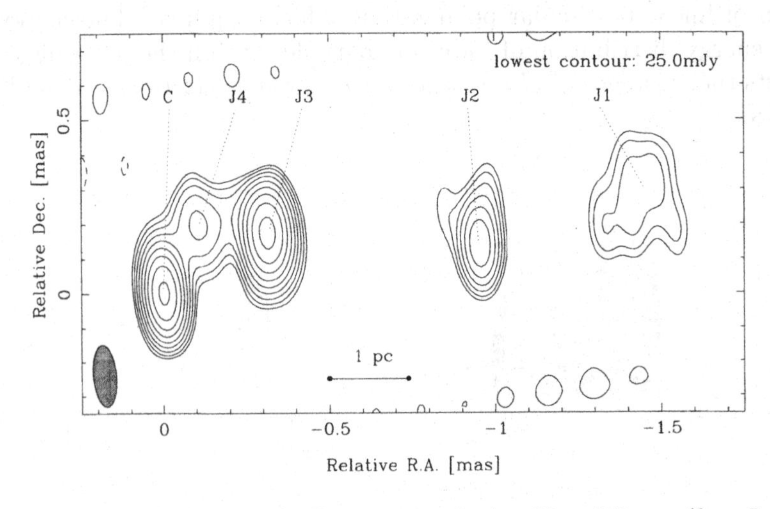

Figure 8. **3 mm-VLBI map of 3C 454.3; resolution 50 x 150 μas** (from Lobanov et al. 2001)

an "experimental" stage. Results from observations at 215 GHz (1.4 mm) are reported by Krichbaum et al. (1997) as a list of source detections - many of them blazars, of course.

Higher resolution at cm wavelengths has been achieved by use of Space-ground VLBI in which a small antenna attached to an Earth-orbiting satellite is used as one end of the baseline. Following a successful demonstration using an existing antenna (Levy et al. 1986), a dedicated satellite named HALCA was launched by the Japanese space agency in 1997, carrying an 8 m dish. A collaborative international program named VSOP has subsequently supported VLBI observations between HALCA and VLBI arrays and antennas on the ground (Hirabayashi et al. 1998). The sensitivity of VSOP observations is necessarily worse then ground-based VLBI due to the poor sensitivity of the HALCA antenna (only 8 m and uncooled receivers) but maps with resolutions three to four times better than ground VLBI have been obtained for many blazars. Another problem which arises is the poor sampling of the u,v-plane since, for many sources, there is a large gap in resolution between the ground-ground and ground-space baselines. Both of these features make a consideration of a suitable *weighting* of the u,v-data extremely important when mapping VSOP data.

Two important aspects of weighting relate to the sampling and the sensitivity of the visibility function. We can either consider that every sample is to be given equal weight, regardless of whether there is another sample very nearby in the u,v-plane, or we can down-weight individual samples if they are bunched close togther, and hence likely to convey nearly the same structural information. These are called *natural* and *uniform* weighting, respectively. Natural weighting usually results in poorer map resolution, since shorter baselines produce a higher density of points in the u,v-plane and these are therefore enhanced in the resulting image. HALCA baselines are not only very long, but the sample density is even lower since HALCA orbits the Earth (making u,v-tracks) four times a day. We can also down-weight visibilities if they have poor sensitivity, and hence larger errors. We refer here to *data weights.* If we use data weights reflecting baseline sensitivities, then HALCA baselines will be further down-weighted; a combination of such weights with natural weighting (which will produce the lowest map noise level) may result in maps in which the (expensive !) HALCA data is hardly used at all. A combination of uniform weighting and equal data weights is therefore most suitable for emphasising the high resolution features which VSOP can measure.

7. Acknowledgements

I thank Alan Roy, Rupal Mittal and Bong Won Sohn for comments on this text, Andrei Lobanov and Eduardo Ros for supplying some of the figures and Franco Mantovani for his infinite patience.

8. References

Alef, W., 1989, in "Very Long Baseline Interferometry", ed. Felli and Spencer, NATO ASI series, 283, 261
Bartel, N. et al. 1986, Nature, 319, 733
Blandford, R.D. & Konigl, A., 1979, ApJ, 232, 34
Broten, N.W. et al. 1967, Nature, 215, 38
Cawthorne, T.V. et al. 1993, Ap. J., 416, 519
Denn, G.R. et al. 2000, Ap.J.Supp., 129, 61
Guirado, J.C. et al. 1995, AJ, 110, 2586
Hirabayashi, H. et al. 1998, Science, 281, 1825.
Jorstad, S. et al. 2001, Ap.J.Supp.,134,181
Krichbaum, T.P. et al. 1997, A+A, 323, L17
Levy, G.S. et al. 1986, Science, 234, 187

Lobanov, A.P., 1996, Ph.D thesis, NMIMT, New Mexico, USA
Lobanov, A.P., 1998, A+A, 330, 79
Lobanov, A.P. & Zensus, J.A., 1999, ApJ, 521, 509
Lobanov, A.P. et al. 2001, A+A, 364, 391
Murdin, P. (ed), 2001, Encyclopaedia of Astronomy and Astrophysics, Inst. of Physics Publishing (Bristol).
Pauliny-Toth, I.I.K. et al. 1987, Nature, 328, 778
Roberts, D.H. et al. 1987, Ap.J., 323, 536
Ros, E. et al. 2000, A+A, 354, 55
Ros, E. et al. 2002, Proc. 6th EVN Symp., 105
Vermeulen, R.C. 1995, Proc. Natl. Acad. Sci. USA, 92, 11385
Wardle, J.F.C. et al. 1998, Nature, 395, 457
Whitney, A. et al. 1971, Science, 173, 225.
Yen, J.L. et al. 1977, Science, 198, 289

THEORY OF RELATIVISTIC JETS: BASIC CONCEPTS

E. KÖRDING AND H. FALCKE
Max-Planck Institut für Radioastronomie
Auf dem Hügel 69, 53115 Bonn, Germany

Abstract. In these lecture notes, we present some basic concepts of relativistic jets. Relativistic effects such as the Doppler shift, superluminal motion and relativistic beaming are discussed. Using simple assumptions for the structure of the jet we derive the radio-spectrum of a freely expanding relativistic jet and compare these results to observations.

1. Introduction

A large number of jets in numerous classes of astrophysical objects have been found using interferometric radio observations. The most prominent sources of jets are active galactic nuclei (AGN), where the jet emission contributes significantly to the spectrum of the source. As an example, the well known FRII radio galaxie Cygnus A is shown in Fig. 1. On small scales one observes a well collimated jet with a flat radio-spectrum, while on large scales the emission is dominated by the hot-spots and has a steep spectrum. Jets have now been found in almost every compact accreting type of objects: Seyfert galaxies, low-luminosity AGN, stellar mass black holes, and neutron-stars. They are also seen in young stellar objects and cataclysmic variables. Many of the different classes of the zoo of active galaxies can be unified if one considers the orientation of the 'central engine' - the disk/jet combination and it surrounding material. Relativistic boosting for example strongly changes the appearance of the jet with aspect angle. A clear sign that these jets are indeed relativistic can be found with VLBI observations: in multiple epoch observations bright spots in the jet can be tracked, which move with an apparent velocity faster than light. Jets also play an important role in X-ray binaries. The central engine of these sources is assumed to be similar to that of Quasars scaled down eight orders of magnitude.

F. Mantovani and A. Kus (eds.), The Role of VLBI in Astrophysics, Astrometry and Geodesy, 107–127.

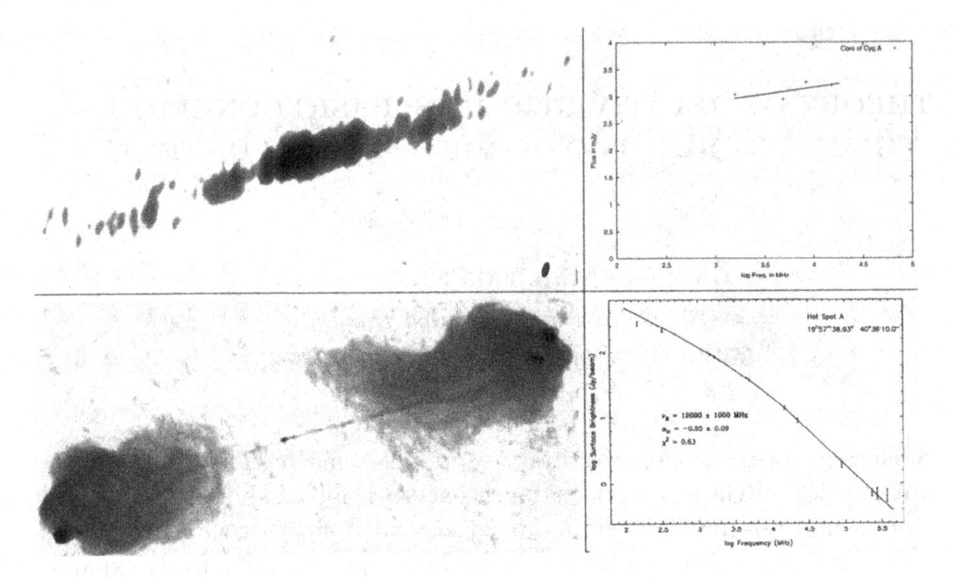

Figure 1. . Top: VLBI-Map of the core of Cygnus-A together with the spectrum of the core component.(Kindly provided by Bach *et. al.*) Bottom: VLA-Map of Cygnus-A and spectrum of one hot-spot. The hot-spots dominate the overall spectrum of the source. Taken from Perley *et. al.* (1984) and Carilli *et. al.* (1991)

It seems that the creation of jets is intimately linked to accretion in many systems, it is therefore important to study these jets and their spectra. Here we discuss some basic theoretical concepts important for relativistic jets. In section two we discuss some effects of relativistic speed: the Doppler effect, relativistic boosting, jet sidedness, and superluminal motion. The third section deals with physical properties of the jet itself - its geometry, basic relativistic hydrodynamics, and spectrum.

2. Effects of relativistic speeds

The exact speed of a jet of an AGN or an XRB is difficult to measure. The jet consists of non-thermal plasma where no spectral lines are visible, so the Doppler-shift of lines cannot be used to derive jet velocities. However, once the jet is relativistic, it is possible to measure the approximate jet speed using relativistic effects. In this chapter we will discuss these effects and their implication for observations.

2.1. RELATIVISTIC DOPPLER EFFECT

Consider a relativistic jet with a plasma blob moving towards the observer. Similar to the classic Doppler-shift we expect that the observed spectrum will be shifted with respect to the emitted spectrum. However, in addition to the geometric effects we have to take time dilation into account. Assume an emitting source moving at a speed $v \lesssim c$ at an angle θ with respect to an observer at rest. Time-dilation tells us that a time difference Δt in the observers rest frame is related to a time difference $\Delta t'$ in the co-moving (primed) frame via

$$\Delta t = \gamma \Delta t', \tag{1}$$

where γ denotes the Lorentz factor $\gamma = \sqrt{1 - \frac{v^2}{c^2}}$. Let us assume that the source emits photon pulses with frequency ν' and period $\Delta t' = \frac{1}{\nu'}$ in its rest frame. In the time Δt the source travels a distance $s = v\Delta t \cos\theta$ towards the observer. Therefore the time difference of the arriving photons is reduced by s/c, i.e.

$$\Delta t_{\mathrm{A}} = \Delta t \left(1 - \frac{v}{c}\cos\theta\right) \tag{2}$$

and the observed frequency is

$$\nu = \frac{1}{\Delta t_{\mathrm{A}}} = \frac{\nu'}{\gamma\left(1 - \frac{v}{c}\cos\theta\right)}. \tag{3}$$

This relation is used to define the Doppler factor

$$\mathcal{D} = \frac{1}{\gamma\left(1 - \frac{v}{c}\cos\theta\right)} = \frac{\sqrt{1-\beta^2}}{1 - \beta\cos\theta}, \tag{4}$$

where we define $\beta = \frac{v}{c}$. We note that the difference to the classical Doppler factor is only the additional Lorentz factor γ from the time dilation. As a consequence we note that all frequencies emitted in the rest frame of the jet will be shifted by the Doppler factor for an observer at rest ($\nu_{\mathrm{obs}} = \mathcal{D}\nu_{\mathrm{jet}}$). The dependence of the Doppler factor on aspect angle and velocity is shown in Fig. 2.

2.2. RELATIVISTIC BOOSTING

To study the effect of relativistic speeds for the emitted flux density S_ν (energy per time, area, and frequency) it is helpful to look at Lorentz invariant quantities. Using the Lorentz invariance of a volume in phase space ($\underset{\cdot}{x}^3 \underset{\cdot}{p}^3$) and the invariance of the number of photons within this volume, it is possible to show that the ratio

$$\frac{S_\nu}{\nu^3} \tag{5}$$

is invariant under Lorentz transformation (see Rybicki & Lightman , 1979, chap. 4.9).

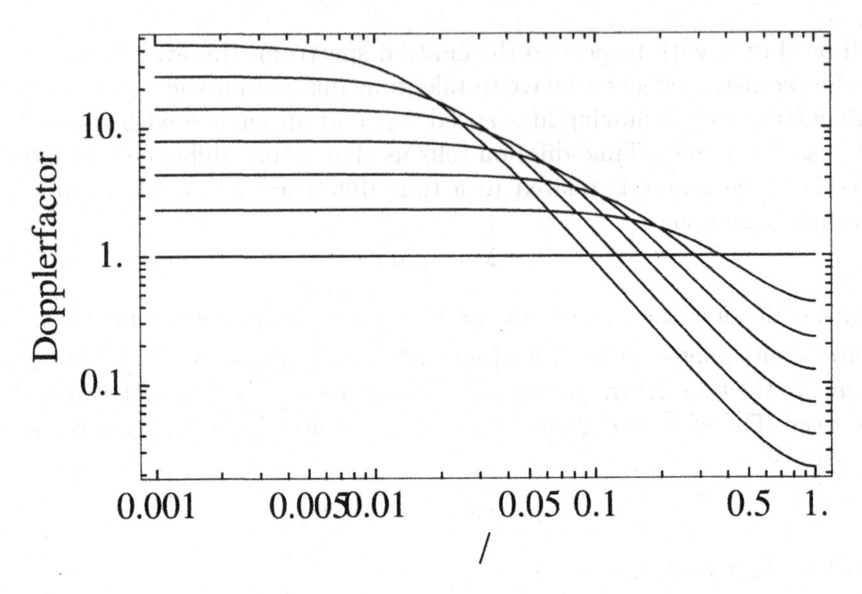

Figure 2. Doppler factor dependence on β and aspect angle. From bottom to top at the left side are $\beta = 0, 0.68, 0.9, 0.96, 0.99, 0.996838, 0.999$

Since the observed frequency is $\nu = \mathcal{D}\nu'$ (Eq. 4) we find that the observed flux has to be (S'_ν = flux density in co-moving frame)

$$S_\nu = \mathcal{D}^3 S'_{\nu'}. \tag{6}$$

The Doppler factor is a strong function of the aspect angle and can become very large for $v \to c$ as plotted in Fig. 2. As the flux density depends on the Doppler factor cubed the effect will be even more drastic. For relatively modest relativistic velocities of $0.97c$ ($\gamma \simeq 4$), for example, the flux in the forward direction can be boosted by a factor 1000, while it is reduced by a factor 1000 in the backward direction!

Note: The dependence on the Doppler factor will be reduced to $\mathcal{D}^2$ when emission from a steady jet is considered rather than from a single blob. This is due to the fact that there is no contraction of the source size if one has a steady jet. However, since boosting will also increase the observed frequency ($\nu = \mathcal{D}\nu'$) the spectral index α ($S_\nu \propto \nu^\alpha$) of the radio spectrum will also enter in the equation, i.e. making it $\mathcal{D}^{2-\alpha}$ ($0.3 \gtrsim \alpha \gtrsim -1$, see Lind & Blandford, 1985) .

The angular effect of Doppler boosting ($\mathcal{D}^3$) is best shown in a polar radiation diagram for a blob moving with speed β (Fig. 3). For a non-moving source the angular emission pattern is a simple circle while it gets highly asymmetric already for $\beta = 0.9$. The transformation from a spherical to an elliptical polar diagram shows that angles are also transformed by relativistic effects. The so-called relativistic aberration (see Rybicki & Lightman ,

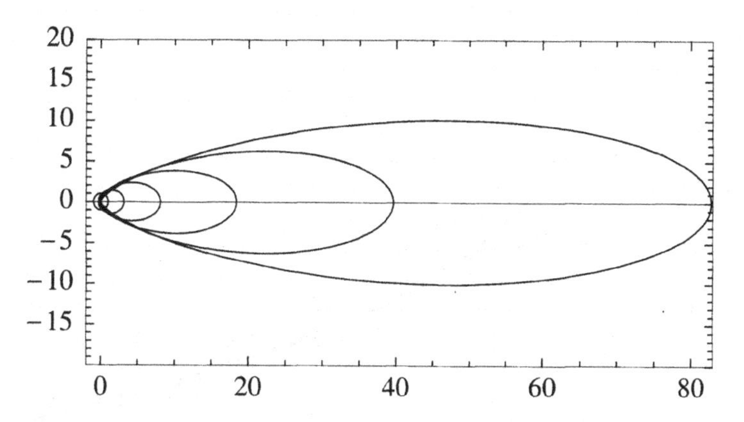

Figure 3. Parametric plot of the boosting cone for $\beta = 0.9, 0.85, 0.75, 0.6, 0.37, 0$. While a non-moving source ($\beta = 0$) radiate isotropic the radiation for a fast source $\beta = 0.9$ is highly beamed.

1979, chap. 4.1) is given by:

$$\tan\theta = \frac{\sin\theta'}{\gamma(\cos\theta' + \beta)}. \tag{7}$$

In the rest frame of the source, half of the radiation will be emitted from $-\pi/2$ to $\pi/2$, hence setting $\theta' = \pi/2$ will give

$$\tan\theta = \frac{1}{\gamma\beta}, \tag{8}$$

thus for $\gamma \gg 1$ half of the radiation will be emitted in a cone (the boosting cone) with half-opening angle

$$\theta \simeq \frac{1}{\gamma}. \tag{9}$$

Consequently appearance of the inner region of the jet will depend strongly on aspect angle. In blazars the jet axis is closely alligned with the line of sight amplifying the radio core. As the jet emission is strongly boosted one also observes strong X-ray and even TeV and γ-ray emission. In radio galaxies the jet is seen under a larger aspect angle, i.e. edge-on, resulting in a fainter core. This can be used to unify different types of AGN (Padovani & Urry, 1992). The same idea has also been applied to unify XRBs and ultra-luminous X-ray sources in galaxies (Körding *et. al.*, 2002).

2.3. JET-SIDEDNESS

For almost all sources the jet on one side of the source is much more luminous than the other. Often the jet on the so-called counterjet side is not

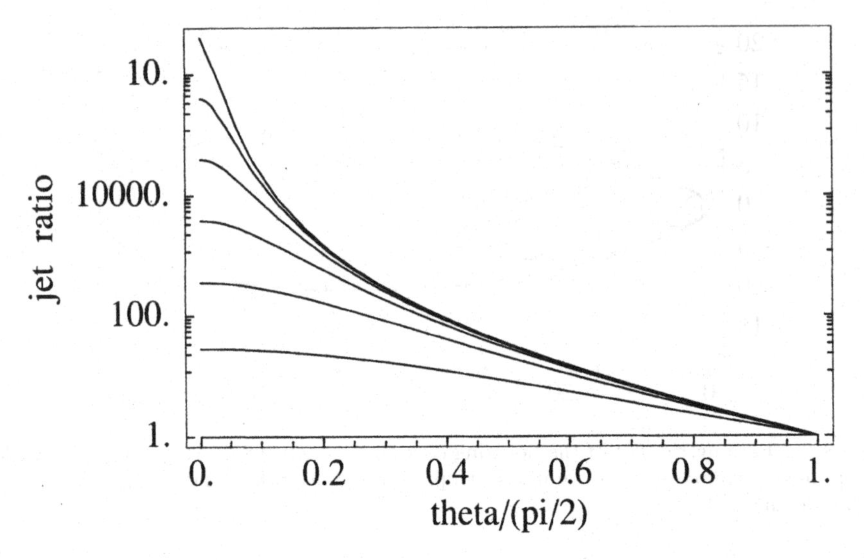

Figure 4. Jet sidedness as a function of aspect angle and Lorentz factor γ. From γ values (from top to bottom) are: $\gamma = 22, 12, 7, 4, 2.2, 1.3$ and 1.

even visible. This can be easily explained with relativistic boosting. Since we expect jets to be intrinsically two-sided, we always have two angles under which the emission is seen by an observer: θ and $\theta + \pi$. We can now calculate the flux ratio R between jet and counter-jet under the assumption of intrinsically symmetric steady jets with spectral index α:

$$R = \frac{\mathcal{D}^{(2-\alpha)}(\beta, \theta)}{\mathcal{D}^{(2-\alpha)}(\beta, \theta + \pi)} = \left(\frac{1 + \beta \cos\theta}{1 - \beta \cos\theta}\right)^{(2-\alpha)} . \tag{10}$$

Even for mildly relativistic jets one side will always be significantly brighter than the other. This behavior is plotted in Fig. 4. This is confirmed by observations:

- *pc-scale:* Most of the strong, compact radio cores seem to come from sources where the angle to the line of sight is small, these jets are always one-sided.
- *kpc-scale:* Even most of the large scale jets appear to be one-sided, even though two extended lobes are seen indicating that really two jets are present (see Fig. 1). Recently Chandra and XMM-Newton detected many X-ray jets on kpc-scale. This is also used as a strong hint that these jets are relativistic on these scales.

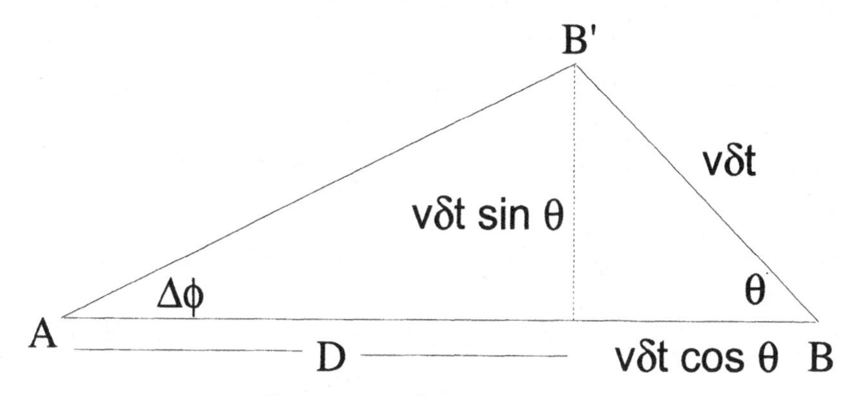

Figure 5. Geometry used for superluminal motion.

2.4. APPARENT SUPERLUMINAL MOTION

One intriguing phenomenon can be observed in many AGN using VLBI techniques: Bright areas - jet components - are moving with an apparent velocity faster than the speed of light. One assumes that these jet components are connected to over-densities in the jet, as one observes in monitoring campaigns that a source first flares and then a new component is ejected. Superluminal motion can be explained as an 'optical illusion' due to relativistic effects.

Let us assume a component in a jet is initially at the position B and moves in time δt to position B' (for the geometry see Fig. 5). The angular separation between the old and the new position of the component for an observer A at a distance D is

$$\Delta\phi = \frac{v\delta t\cos\theta}{D}. \tag{11}$$

Similar to the Doppler effect one has to take the light travel time into account. The measured time interval Δt is therefore

$$\Delta t = \delta t - \beta\delta t\cos\theta = \delta t(1-\beta\cos\theta). \tag{12}$$

Contrary to the relativistic Doppler effect there appears no Lorentz factor as all quantities are measured in the observers frame.

The apparent transverse velocity $v_{\rm app} = D\frac{\Delta\phi}{\Delta t}$ inferred by the observer at position A is

$$\beta_{\rm app} = \frac{v_{\rm app}}{c} = \frac{1}{c}\frac{D\Delta\phi}{\Delta t} = \frac{1}{c}\frac{v\delta t\sin\theta}{\delta t(1-\beta\cos\theta)} = \frac{\beta\sin\theta}{1-\beta\cos\theta}. \tag{13}$$

The function $\beta_{\rm app} = \frac{\beta\sin\theta}{1-\beta\cos\theta}$ is a strong function of β and θ if $\beta \to 1$ as plotted in Figure 6. For Lorentz factors $\gamma \gtrsim 3$ the apparent speed is for almost all angles $v_{\rm app} > c$.

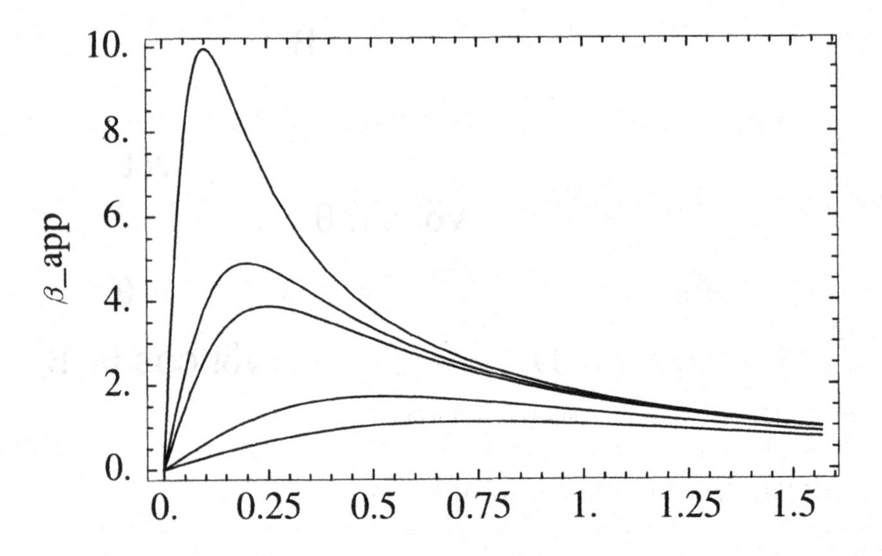

Figure 6. Apparent superluminal motion for $\gamma = 1.5, 2, 4, 5, 10$ (from bottom to top).

The maximum of this function is found by solving

$$\frac{\delta \beta_{\mathrm{app}}}{\delta \theta} = 0, \tag{14}$$

which yields $\cos\theta_{\mathrm{max}} = \beta$ and $\beta_{\mathrm{app,max}} = \beta\gamma$. If one observes a source with $\beta_{\mathrm{app}} = 10$ the jet speed must be at least $\gamma = 10$. However, if the measured β_{app} is below $\beta_{\mathrm{app,max}}$ there are always two values for θ possible. Speeds quoted in the literature from superluminal motion are often derived by assuming that the jet is at it $1/\gamma$ angle, and are therefore lower limits. In Fig. 7 maps of 3C345 are shown - the proper motion of the jet is clearly seen.

3. Compact Radio Cores in AGN and XRB

If one observes an AGN like Cygnus-A with a single dish telescope, one finds a steep spectrum as expected for synchrotron radiation (see Fig. 1). However the core shows a different behavior: The spectrum between 1 and 16 GHz is flat or even slightly inverted. This implies that there is some self-absorption present in this system. In the following sections we will explain the flat spectrum with a simple theory of relativistic jets. A detailed analysis can be found in (Falcke & Biermann, 1995) and (Blandford & Königl, 1979). We will first derive basic properties of the jet and its geometry, which will enable us to derive the spectrum of a freely expanding conical outflow. We will show that a flat spectrum is the most natural result for such a jet.

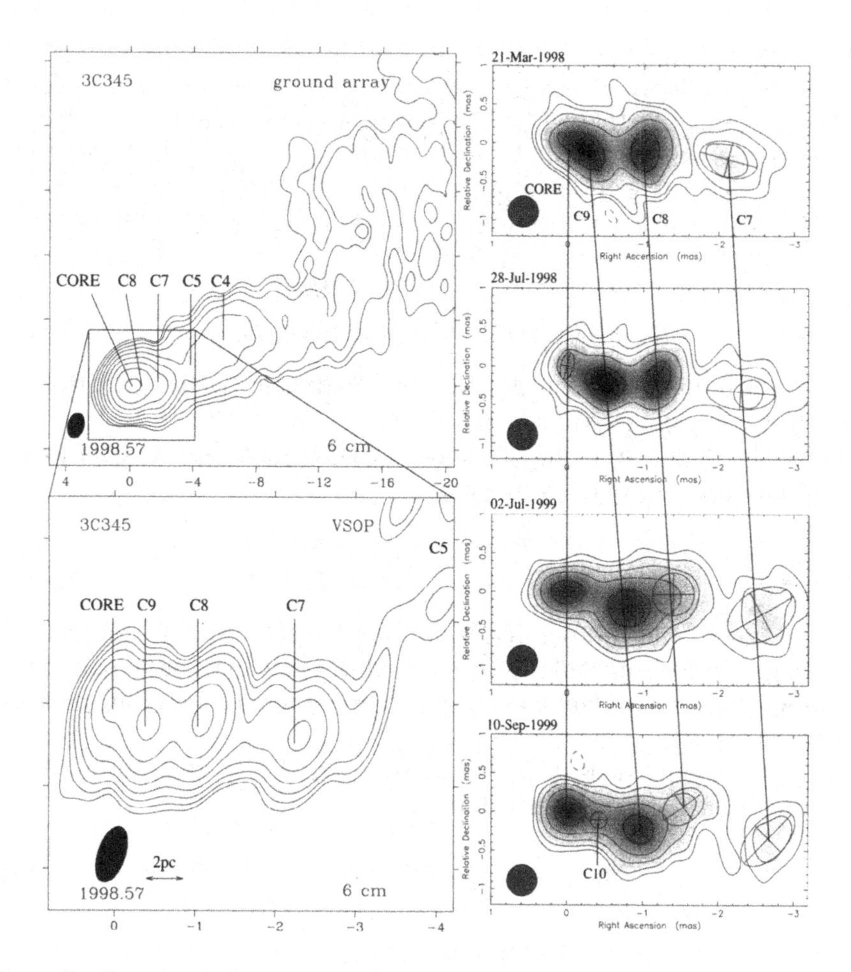

Figure 7. Superluminal motion in 3C345. Apparent velovities around 5 c are found in these 6 cm VSOP observations (Space VLBI). Plot taken from (Klare *et. al.*, 2001) .

3.1. BASIC SETTINGS

As we have no detailed knowledge of the processes in the vicinity of the black hole one has to make some basic assumptions to describe the jet. As usual we assume that the accreting matter forms an accretion disk around the black hole. The jet is fueled by this accretion disk via an unknown link - for example magneto-hydrodynamical processes. For our calculations it is, however, sufficient to assume that a certain fraction $q_m = \frac{\dot{M}_{\rm jet}}{\dot{M}_{disk}}$ of the accreting material is injected into the jet. This parameter can range from zero (no jet) to infinity (no accretion disk and pair creation). Sensible values for most known jets seem to be around $q \approx 10^{-2}$ (Falcke *et. al.*, 1995).

In the jet the dominant contributions to the energy density come from the magnetic fields $U_{\rm B} = B^2/8\pi$, the turbulent kinetic energy density of the plasma $U_{\rm turb}$ and the energy contained in a population of relativistic particles $U_{\rm e+p}$. The latter population is probably created by Fermi acceleration and is the source of the radio-emission we are calculating. We assume that these three energy contributions are approximately in equipartition:

$$U_{\rm turb} \approx U_{\rm B} \quad \& \quad U_{\rm e+p} = kU_{\rm B} = kB^2/8\pi, \;\; k \lesssim 1 \tag{15}$$

$$U_{\rm j} = U_{\rm B} + U_{\rm turb} + U_{\rm e+p} = (2+k)U_{\rm B} \tag{16}$$

where we introduce the factor k for the relativistic particles to discuss derivations from equipartition.

As usually observed in quasars, the relativistic particles are assumed to be distributed as a power law in energy,

$$\frac{dN(\gamma)}{d\gamma} = K\gamma^{-p}, \tag{17}$$

where we ignore the protons for simplicity. This is usually justified for calculating synchrotron radiation ($m_{\rm p} >> m_{\rm e}$), but may be wrong when calculating high-energy emission (e.g., hadronic interactions can become important).

The total energy density in relativistic particles is then obtained by integrating over the energy distribution,

$$U_{\rm e+p} = K \int_{\gamma_1}^{\gamma_2} \gamma^{-p} \cdot \gamma m c^2 {\rm d}\gamma. \tag{18}$$

This integral can be evaluated and using $U_{\rm e+p} = kB^2/8\pi$ we find

$$K = kB^2/(8\pi\Lambda m_{\rm e}c^2) \tag{19}$$

where

$$\Lambda = \left[\begin{array}{c} (\gamma_1^{2-p} - \gamma_2^{2-p})/(p-2) \;\; \text{for } p \neq 2 \\ \ln(\gamma_2/\gamma_1) \;\; \text{for } p = 2. \end{array} \right. \tag{20}$$

The case $p \neq 2$ can be simplified further:

$$(\gamma_1^{2-p} - \gamma_2^{2-p})/(p-2) \propto \left[\begin{array}{l} \gamma_1^{2-p} \;\; \text{for } p > 2 \\ \gamma_2^{2-p} \;\; \text{for } p < 2. \end{array} \right. \tag{21}$$

The equipartition assumption provides us with the proper normalization constant when integrating over the entire energy range,

$$n_{\rm e} = K \int_{\gamma_1}^{\gamma_2} \gamma^{-p} d\gamma \simeq -K\gamma_1^{1-p}/(1-p) \;\; (\gamma_2 \gg \gamma_1) \tag{22}$$

$$\Rightarrow n_\mathrm{e} = \frac{kB^2}{8\pi\gamma_1^{p-1}\Lambda m_\mathrm{e}c^2(p-1)}. \tag{23}$$

Not all the particles are necessarily relativistic and contribute to visible (synchrotron) radiation. We suppose that the *total* particle density n is given by:

$$n = n_\mathrm{e}/x_\mathrm{e} \text{ and } x_\mathrm{e} \lesssim 1. \tag{24}$$

However, we know the total particle density by specifying the mass loss rate of the jet. In this case the particle density is determined by the bulk velocity v, the radius of the outflow cylinder/cone r, and the mass outflow rate $\dot{M}_\mathrm{j}$ of the jet. We have already connected this mass outflow with the accretion rate of the object via the dimensionless parameter q_m. From mass conservation we get

$$\dot{M}_\mathrm{j} = m_\mathrm{p} n \pi r^2 v. \tag{25}$$

Combining the two equations for the density, yields a relation between the outflow rate and the magnetic field as a function of the scaling parameters

$$\dot{M}_\mathrm{j} = m_\mathrm{p}\pi r^2 v \frac{kB^2}{8\pi x_\mathrm{e}\gamma_1^{p-1}\Lambda m_\mathrm{e}c^2(p-1)}. \tag{26}$$

This allows one to express the magnetic field as

$$B = \sqrt{\frac{8\pi x_\mathrm{e}\gamma_1^{p-1}\Lambda m_\mathrm{e}c^2(p-1)\dot{M}_\mathrm{j}}{k m_\mathrm{p}\pi v}} \cdot r^{-1} \tag{27}$$

and we get for a power law index of $p \neq 2$

$$B = 0.3\,\mathrm{G}\sqrt{\frac{x_\mathrm{e}}{\beta k}\left(\frac{p-1}{p-2}\right)\left(\frac{\gamma}{100}\right)\left(\frac{\dot{M}_\mathrm{j}}{M_\odot/\mathrm{yr}}\right)}\left(\frac{r}{\mathrm{pc}}\right)^{-1}. \tag{28}$$

This is scaled to the situation of a jet in a luminous quasar and is roughly of the right order for the B fields derived in their cores.

For an alternative example, consider the Galactic Center, which has an inflow rate determined to be $\lesssim 10^{-7} M_\odot/\mathrm{yr}$ and if we choose the mass ejection parameter $q_\mathrm{m} = 1\%$, we are left with only $\dot{M}_\mathrm{j} = 10^{-9} M_\odot/\mathrm{yr}$ and at 10 $R_\mathrm{g} \sim 3.8 \cdot 10^{12}$ cm we obtain (1 pc $= 3 \cdot 10^{18}$ cm)

$$B = 10\,\mathrm{G}\sqrt{\frac{x_\mathrm{e}}{\beta k}\left(\frac{p-1}{p-2}\right)\left(\frac{\gamma}{100}\right)\left(\frac{\dot{M}_\mathrm{j}}{10^{-9}M_\odot/\mathrm{yr}}\right)}\left(\frac{r}{10R_\mathrm{g}}\right)^{-1} \tag{29}$$

which is a typical value infered for Sgr A^* see (Melia & Falcke, 2001).

3.2. HYDRODYNAMICS

To describe the geometry of the jet we need to derive the sound speed in the jet, as it will describe the sideways expansion of the jet. Therefore we need to study its hydrodynamics as given in Königl (1980). In an perfect gas in thermodynamic equilibrium pressure (P) and energy density (U) are related through the adiabatic index (Γ). Even tough there are shocks present in jets this is a good approximation for most parts of the jet,

$$P = (\Gamma - 1)U. \tag{30}$$

We can treat the mix of relativistic particles and isotropic turbulent magnetic fields as a "photon gas" with adiabatic index $\Gamma = 4/3$. With increasing order of magnetic fields the adiabatic index goes towards towards $\Gamma \to 2$.

The enthalpy density of the jet is given by

$$\omega_{\mathrm{j}} = \frac{\Gamma P}{\Gamma - 1} + m_{\mathrm{p}} n_{tot} c^2. \tag{31}$$

So there are two contributions to the enthalpy. The first term depends on the internal energy (see Eq. 30), the second term represents the rest mass of the particles dominated by the proton mass (the electron mass is negligible). For a perfect gas the sound speed is connected with the enthalpy via

$$\beta_{\mathrm{s}} = \sqrt{\frac{\Gamma P}{\omega_j}}, \tag{32}$$

and we find

$$\beta_{\mathrm{s}}^{-2} = \frac{\omega_j}{\Gamma P} = \frac{1}{\Gamma - 1} + \frac{m_{\mathrm{p}} n_{tot} c^2}{\Gamma P}. \tag{33}$$

If the contribution of the internal energy dominates the enthalpy (a photon plasma) the first term will dominate and the sound speed will reach its maximal value of

$$\beta_{\mathrm{s,max}} = \sqrt{\Gamma - 1}. \tag{34}$$

We find that the maximal sound speed for a photon gas ($\Gamma = 4/3$) is $\beta_{\mathrm{s,max}} = \frac{1}{\sqrt{3}}$ while we have a higher sound speed for higher adiabatic indeces, e.g., $\Gamma = 5/3$ yields $\beta_{\mathrm{s,max}} = \sqrt{\frac{2}{3}}$. However, the rest mass normally contributes (in Eq. 33) as well, resulting in lower sound speeds for heavy jets. We will fist look at the other extreme case, that the rest mass dominates in Eq. (33). We will check when this approximation breaks down to find the parameters when the sound speed comes near its maximal value. In the case that the rest dominates the entropy we find:

$$\beta_{\mathrm{s}} = \frac{2}{3}\sqrt{\left(m_{\mathrm{p}} c^2 n / x_{\mathrm{e}}\right)^{-1} (2 + k) \frac{B^2}{8\pi}}. \tag{35}$$

The relativistic particle density is again related to the magnetic field through equipartition.

$$\beta_s = \frac{2}{3}\sqrt{x_e(2+k)\left(\frac{m_p k}{8\pi\gamma_1^{p-1}\Lambda m_e(p-1)}\right)^{-1}\frac{1}{8\pi}} \quad (36)$$

$$\beta_s = \frac{2}{3}\sqrt{\frac{\gamma_1 m_e}{m_p}\left(\frac{p-1}{p-2}\right)\left(\frac{2+k}{k}\right)x_e}, \quad (37)$$

where we assumed $p \neq 2$. For example, for $p = 3$, $k = 1$ and $x_e = 1$ this yields sub-relativistic velocities, $\beta_s \approx 0.035\sqrt{\gamma_1}$. To check when the internal energy density is high enough that the sound speed approaches the maximal value we evaluate

$$\beta_s < \beta_{s,\max} \Rightarrow x_e\gamma_1 \lesssim \left(\frac{1}{3}\right)\left(\frac{9}{4}\right)\left(\frac{m_p}{m_e}\right)\left(\frac{p-2}{p-1}\right)\left(\frac{k}{2+k}\right). \quad (38)$$

If we plug in reasonable numbers we find

$$x_e\gamma_1 \lesssim 344 \text{ for } p = 3 \text{ and } k \lesssim 1. \quad (39)$$

This means that a *maximal jet* with $c_s = c/\sqrt{3}$, requires either $\gamma_1 \sim 300$ or $x_e \sim 300$. The latter is only possible if one has more electrons than protons (i.e. in a pair-plasma dominated jet), or one has an electron distribution with a low-energy cut-off in the electron distribution around several hundreds ($\gamma_e \sim 100$, $T_e \gtrsim 10^{11}$ K, which is close to typical values for the birghtness temperature T_B observed in AGN).

So we conclude that efficient jets are always "light" (matter is photon-like, that does, however, not exclude the presence of some relativistic protons)!

3.3. SPECTRUM

With the help of the previous equations we can express the particle density, magnetic field, and electron Lorentz factor as a function of dimensionless parameters and the total energy of the system. We can use this to calculate the jet spectrum.

We describe the jet as a freely expanding flow, beyond its collimation point (nozzle). The geometry of the jet is then given by the forward velocity $\gamma_j\beta_j c$ and the side-ways velocity, i.e. the sound-speed $\gamma_s\beta_s c$. An isothermal, non-accelerating jet will then simply take the shape of a cone. The half-opening angle will be given by the Mach cone,

$$\sin\phi \sim \phi = \frac{1}{\mathcal{M}} = \frac{\gamma_s\beta_s}{\gamma_j\beta_j}. \quad (40)$$

The diameter of the jet as a function of the longitudinal coordinate is

$$\Rightarrow r = \sin\phi z \sim \frac{z}{\mathcal{M}} \tag{41}$$

and for a relativistic, maximal jet we have

$$\frac{1}{\mathcal{M}} = \frac{\sqrt{3}}{\gamma_{\rm j}} \sim \frac{1}{6}\left(\frac{10}{\gamma_{\rm j}}\right) \simeq 6^\circ. \tag{42}$$

A high collimation for the jet is therefore an inherent feature of any relativistic hydrodynamic jet, given the flow was initially collimated in a nozzle.

For the particle density and the magnetic field, particle and energy conservation determine the scaling (which we also see from the previous equations 27 & 25)

$$n \propto z^{-2} \ \& \ B \propto z^{-1}. \tag{43}$$

This determines the basic parameters and the geometry of the jet at each point beyond the nozzle and we can calculate an emission spectrum. The dominating process in the radio is synchrotron. The absorption coefficient (see Huege & Flacke, 2003 this volume, Eq. 60) is given for $p = 2$ as

$$\alpha = 4.5 \cdot 10^{-12} {\rm cm}^{-1} \frac{k}{\ln(\gamma_2/\gamma_1)} \left(\frac{B}{\rm G}\right)^4 \left(\frac{\nu}{\rm GHz}\right)^{-3}, \tag{44}$$

assuming a mean pitch angle of $\alpha \sim 54^\circ$ corresponding to an isotropic distribution of pitch angles. For a detailed study see Rybicki & Lightman (1979).

In order to find short equations for the spectrum we rewrite this in a more compact form (with $B \propto z^{-1}$)

$$\alpha = k_1 b_1^4 z^{-4} \nu^{-3}, \tag{45}$$

$$k_1 = 4.5 \cdot 10^{-12} {\rm cm}^{-1} \frac{k}{\ln(\gamma_2/\gamma_1)} {\rm GHz}^{-3}, \tag{46}$$

$$k_1 = 13.9 \cdot 10^6 {\rm pc}^{-1} \frac{k}{\ln(\gamma_2/\gamma_1)} {\rm GHz}^{-3}, \tag{47}$$

$$b_1 = 0.3\,{\rm G} \sqrt{\frac{x_{\rm e} \ln(\gamma_2/\gamma_1)}{\beta k} (p-1) \left(\frac{\gamma}{100}\right) \left(\frac{\dot{M}_{\rm j}}{M_\odot/{\rm yr}}\right)}. \tag{48}$$

In an 'atmosphere' most of the radiation is usually emitted at the frequency and the location where the optical depth becomes unity (e.g., at the peak of the black body).

If the jet has the radius r at the longitudinal coordinate z, the optical depth is given by the absorption coefficient to be

$$\tau = 2r\alpha/\sin i. \tag{49}$$

The $\tau = 1$ surface (i.e., the visual surface) can thus be determined using equation (45):

$$\tau = 1 = 2\frac{z}{\mathcal{M}}k_1 b_1^4 z^{-4}\nu^{-3}/\sin i \tag{50}$$

$$\Rightarrow z_{\tau=1} = \left(2\frac{1}{\sin i\mathcal{M}}k_1\right)^{1/3} b_1^{4/3}\nu^{-1} \tag{51}$$

$$\Rightarrow z_{\tau=1} = 50\, pc \left(\frac{1}{\beta}\frac{k_1}{13.9\cdot 10^6}\right)^{1/3} \frac{b_1}{0.3}^{4/3} \nu^{-1}. \tag{52}$$

This means that frequency and size are inversely related to each other. In a jet one never looks at the same position at different frequencies!

We note that the coordinate of the $\tau = 1$ surface scales with $z_{\tau=1} \propto \nu^{-1}\dot{M}^{2/3}$. If we assume that $\dot{M} \sim 0.03$ and we observe at $\nu = 10$ GHz the $\tau = 1$ surface is at roughly $z \sim 0.5$ pc. At a distance of d=1.5 Gpc 0.5 pc correspond to 0.5 mas, which is just the limit for the resolution of VLBI experiments and explains the existence of unresolved cores in VLBI maps.

The last ingredient needed to calculate the spectrum is the synchrotron emissivity (see Huege & Flacke, 2003 this volume, Eq. 54), given as

$$\epsilon = 5.5\cdot 10^{-19}\frac{\mathrm{erg}}{\mathrm{s\,cm^3\,Hz}}\left(\frac{k}{\ln(\gamma_2/\gamma_1)}\right)\left(\frac{B}{\mathrm{G}}\right)^{3.5}\left(\frac{\nu}{\mathrm{GHz}}\right)^{-0.5}, \tag{53}$$

where we use a mean pitch angle of $\alpha \sim 54°$. As before we rewrite this in a compact form:

$$\epsilon = k_2 b_1^{3.5} r^{-3.5}\nu^{-0.5}, \tag{54}$$

$$k_2 = 5.5\cdot 10^{-19}\frac{\mathrm{erg}}{\mathrm{s\,cm^3\,Hz}}\left(\frac{k}{\ln(\gamma_2/\gamma_1)}\right)\mathrm{GHz}^{0.5} \tag{55}$$

or given in pc:

$$k_2 = 1.6\cdot 10^{37}\frac{\mathrm{erg}}{\mathrm{s\,pc^3\,Hz}}\left(\frac{k}{\ln(\gamma_2/\gamma_1)}\right)\mathrm{GHz}^{0.5}. \tag{56}$$

To get the total emission spectrum in the comoving frame we integrate the emissivity at a given frequency ν in slices along the entire jet, starting at $z_{\tau=1}$:

$$L_\nu = 4\pi\int_{z_{\tau=1}}^{\infty}\epsilon(z)\pi r^2(z)/\sin(i)\mathrm{d}z. \tag{57}$$

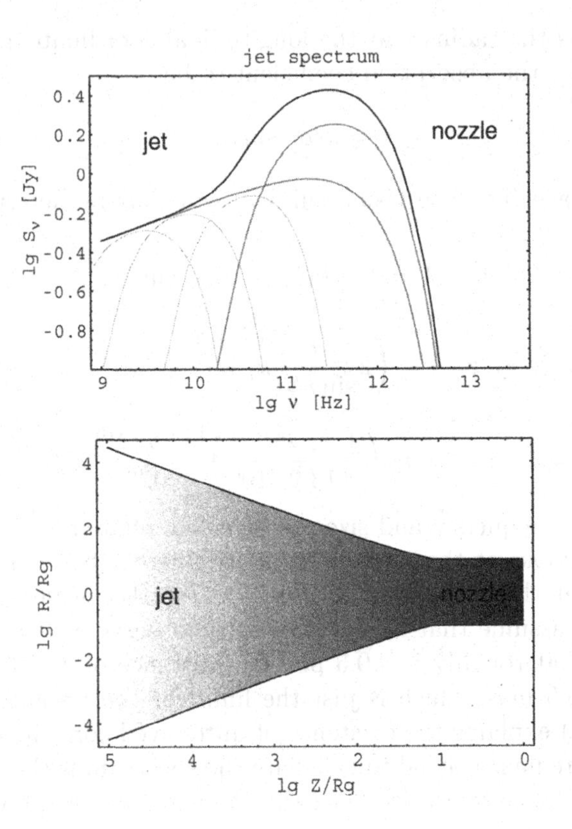

Figure 8. Schematic geometry and spectrum of a jet (here for Sgr A^*) .

We insert equation 54 and find

$$L_\nu = 4\pi^2 \mathcal{M}^{-2} \sin^{-1}(i) k_2 b_1^{3.5} \mathcal{M}^{3.5} \nu^{-0.5} \int_{z_{\tau=1}}^{\infty} z^{-3.5} z^2 \mathrm{d}z, \tag{58}$$

which can be evaluated and yields

$$L_\nu = 4\pi^2 \sin^{-1}(i) k_2 b_1^{3.5} \mathcal{M}^{1.5} \nu^{-0.5} z_{\tau=1}^{-0.5}. \tag{59}$$

Since $z_{\tau=1} \propto \nu^{-1}$ the ν dependence cancels out and we obtain a 'flat' spectrum $L_\nu \propto \nu^0$. What we have essentially done is to divide the jet into many slices along the z-direction, each contributing an absorbed synchrotron spectrum. For a free jet the numbers 'conspire' to give a continous flux level for all slices at $\tau = 1$. We conclude that a flat radio spectrum is a natural signature of a freely expanding outflow. In Fig. 8 we have illustrated this behavior, if we increase the frequency we look deeper into the

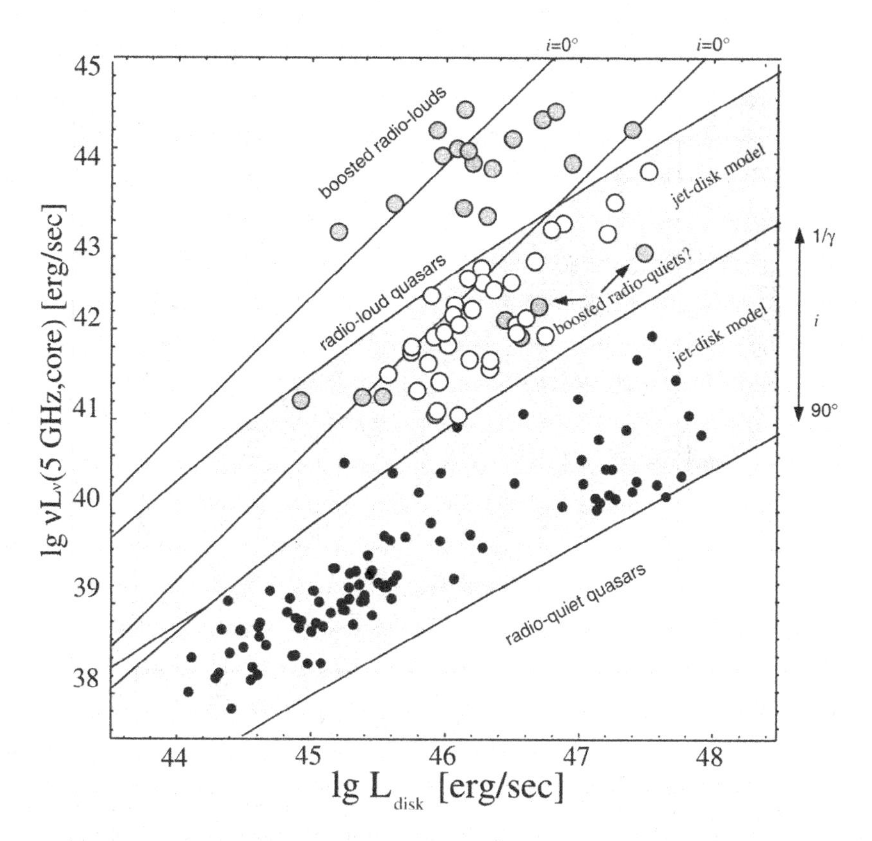

Figure 9. Radio core luminosity versus UV emission for quasars. The big, white circles represent radio loud, steep-spectrum quasars with FR II morphology; big, gray circles are flat-spectrum, variable, core dominated quasars (blazars). The lines represent the jet model. (Updated from Falcke *et. al.* 1995)

jet and the superposition of self-absorbed synchrotron spectra yields the flat spectrum. If one includes relativistic effects we have to consider the relativistic Doppler effect and relativistic aberration. Together this yields an additional $\mathcal{D}^{\frac{13}{6}}$ factor (Falcke & Biermann, 1995).

To obtain the dependence of the absolute flux level on the accretion rate consider that

$$b_1 \propto \sqrt{\dot{M}} \quad \text{and} \quad z_{\tau=1} \propto \dot{M}^{2/3}, \tag{60}$$

while the other parameters do not depend on $\dot{M}$. So we conclude

$$S_\nu \propto \dot{M}^{3.5/2} \dot{M}^{-1/2 \cdot 2/3} \propto \dot{M}^{1.75-0.33} \propto \dot{M}^{1.42}. \tag{61}$$

This non-linear scaling of the radio flux with accretion rate/power remains true until the radiation becomes a significant source of cooling (which we have ignored so far). From this point on, an increase in power can at best

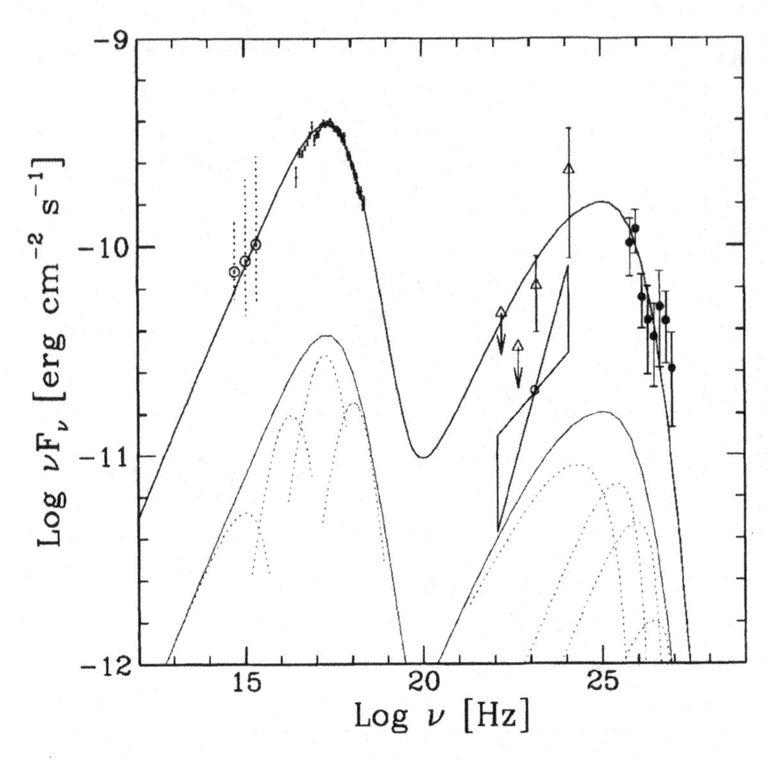

Figure 10. In Mkn 421 the effect of synchrotron and inverse Compton emission is clearly visible. While the first 'bumb' at lower frequencies is created by synchrotron the second is created by inverse Compton scattering (taken from Maraschi *et. al.* (1999)).

lead to a linear increase in jet luminosity. That scaling has been invoked to explain the radio/optical correlation of radio cores (see Fig. 9 and Falcke *et. al.*, 1995). Here one assumes that the optical/UV luminosity of quasars is 'relatively' isotropic and reflects the true engine power. One can plot the radio core luminosity of quasars versus UV-luminosity and see the effects of relativistic boosting and the above scaling law. The lines represent the model where the jet power scales with optical (accretion disk) luminosity plus boosting ($\gamma \sim 3 - 10$) and random orientation. The Blazars cluster at higher radio powers compared to steep-spectrum quasars. This is simply explained by small inclination angles and boosting.

3.4. THE EFFECT OF SYNCHROTRON SELF COMPTON PROCESSES

Up to now we have only considered synchrotron emission from the jet. However, as there is a high relativistic electron population in the jet theses electrons give also rise to inverse Compton processes. Here a high energetic electron will scatter with a seed photon and transfer some of its energy to the photon (for a detailed study see Rybicki & Lightman , 1979, Chapter 7). This process is called synchrotron self-Compton (SSC), if the seed photons are created by synchrotron emission from the same electrons involved in the inverse Compton process. This process will mirror the spectrum of the seed photons to higher energies and stretch it. For example this effect might be seen in Mkn 421 (see Fig. 10), for this TeV blazar the synchrotron component goes up to the X-rays while SSC can create even TeV gamma rays. However, there are also other explainations for the high energy photons, for example, involving hadronic processes.

3.5. MHD-SIMULATIONS OF JETS

Up to now we have calculated the spectrum of a freely expanding relativistic jet use equipartition arguments. However, we did not consider the physical processes creating the jet - the disk/jet link or the particle acceleration mechanism creating the population of highly relativistic particles. Furthermore, we only consider a very simple jet - no fine structure or shocks are present. A detailed analysis of jets can only be achieved via full magnetohydrodynamic simulations. Theses simulations are extremely complex and time consuming. An example for the complex structure arising from such simulations is presented in Fig. 11 which was taken from Krause & Camenzind (2002).

In this article we have presented some of the basic concepts of relativistic jets that are relatively well established. However, jets are far from fully understood, nor is the link between the accretion flow and the jet. The 'central engine' of AGN and XRBs will therefore remain a wide and active research field for the years to come.

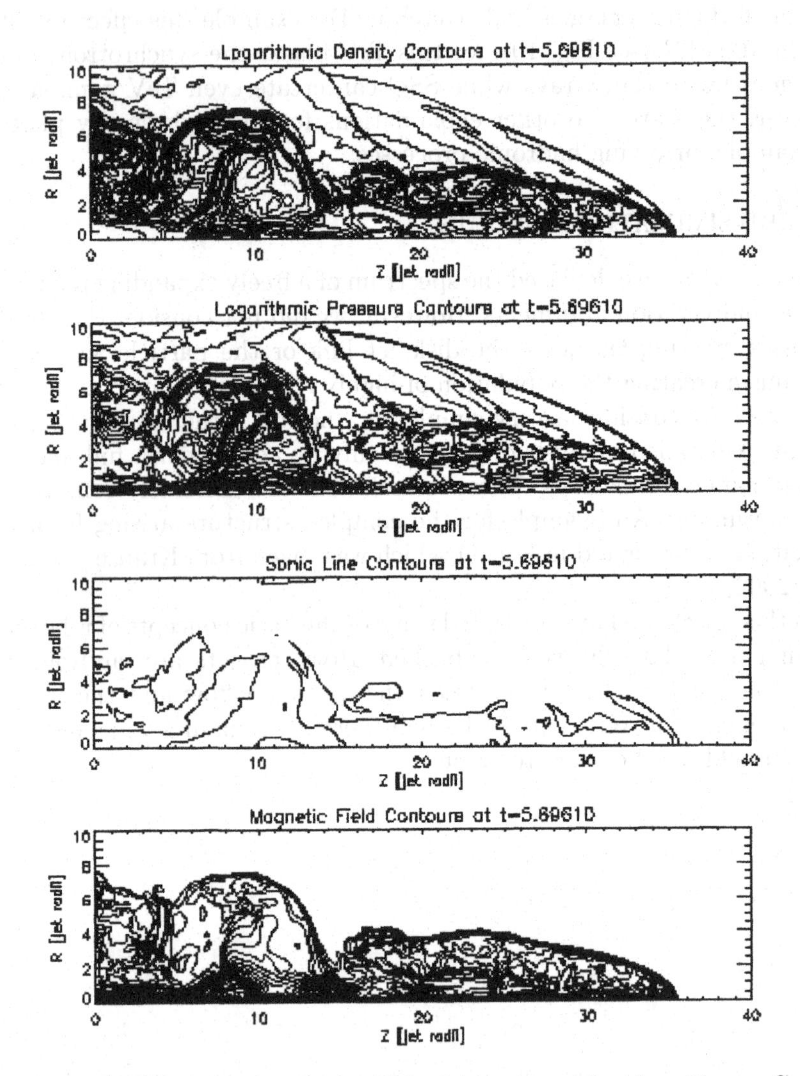

Figure 11. 2d-MHD simulation of a jet. This plot has been taken from Krause, Camenzind (2002)

REFERENCES

Bach, U., Krichbaum, T. et al. in prep
Blandford, R. and Königl, A. (1979), *ApJ*, 232, 34
Carilli, C., Perley, R., Dreher, J., Leahy, J. (1991), *ApJ*, 383, 554
Falcke, H. and Biermann, P. (1995),*A&A* , 293, 665
Falcke, H., Malkan M. and Biermann, P. (1995),*A&A* , 298, 375
Huege, T. and Falcke, H., this volume
Klare, J., Zensus, A., Krichbaum, T. et al. (2001), IAU Symposium, 205, 130
Königl, A. (1980), *Phys. Fluids*, 23, 1083
Körding, E., Falcke, H. and Markoff, S. (2002), *A&A*, 382, L13
Krause, M. and Camenzind, M. (2002),*A&A*, 380, 789
Lind, K., Blandford, R. (1985), *ApJ*, 295, 358
Maraschi M., Fossati, G., Tavecchio, F. et al (1999),*ApJ*, 526, 81
Melia, F. and Falcke, H. (2001), *ARA&A*, 39, 309
Padovani, P. and Urry, C. (1992), *ApJ*, 387, 449
Perley, R., Dreher, J., & Cowan, J. (1984), *ApJ*, 285, L35
Rybicki, G. and Lightman, A. (1979), Radiative Processes in Astrophyscics, *John Wiley & Sons*

PARSEC–SCALE MORPHOLOGIES AND PROPERTIES OF RADIO SOURCES

T. VENTURI
Istituto di Radioastronomia, CNR
Via Gobetti 101
I–40129 Bologna

1. Introduction

In the light of the purpose of this School, in this lecture I will concentrate on one astrophysical issue, where very high resolution radio astronomy turned out to play a crucial role. In particular, the relation between compact flat spectrum radio sources associated with quasars and BL–Lacs, and extended high and low power radio galaxies, respectively FRII and FRI.
I will summarise the kiloparsec– and parsec–scale properties of compact radio sources and of extended radio galaxies. I will show that compact radio sources, including BL–Lacs and quasars, can be understood in terms of Doppler boosting of a relativistic jet aligned at a small angle to the line of sight. The nuclear properties of extended radio galaxies (FRI and FRII sources) are consistent with the idea that they are characterised by relativistic radio jets viewed at large angles to the line of sight. These observational results lead support to the "unified schemes", where compact flat spectrum quasars and BL–Lacs are expected to be the parent population of high power extended radio galaxies and quasars (FRII) and of low power radio galaxies (FRI) respectively. Finally, I will show that FRI and FRII radio galaxies are indistinguishable on the parsec scale.
I will use $H_o = 100$ km $s^{-1}Mpc^{-1}$.

2. Observational properties of compact radio sources

Compact extragalactic flat spectrum sources (CFSRS) are typically associated with quasars and BL–Lacs. The quasar–class includes a variety of objects, such as for example highly polarised quasars (HPQ), optically violent variables (OVV), but also quasars without extreme properties. The

F. Mantovani and A. Kus (eds.), The Role of VLBI in Astrophysics, Astrometry and Geodesy, 129–141.

BL–Lac class includes objects with featureless spectra, or only weak narrow lines. The properties of such objects in the optical band were reviewed by Browne (these proceedings).

2.1. KILOPARSEC–SCALE PROPERTIES

At radio wavelengths, these objects are usually characterised by a strong and compact component on the arcsecond (kiloparsec) scale. Interferometric high sensitivity images for a wealth of compact flat spectrum quasars and BL–Lacs, mainly obtained with the Very Large Array (Socorro, New Mexico), are available in the literature (see for example Antonucci & Ulvestad 1985; Murphy et al. 1983; Cassaro et al. 1999). Together with doubles and core–jet sources, a relevant fraction of BL–Lacs shows extended low brightness emission surrounding the compact component (core–halo morphologies). An example is given in Figure 1. Such extended emission contains a considerable fraction of the total flux density, and Antonucci & Ulvestad (1985) noted that most compact flat spectrum 3C sources would belong to the 3C catalogue even without the flux density contribution of the compact component. The projected linear sizes of these objects are usually small. There seems to be no difference in the large scale radio properties among the various classes of quasars and BL–Lacs.

2.2. FLUX DENSITY VARIABILITY

Flux density variability is another common property of CFSRS. Variability is usually observed over a wide range of bands, from the radio, to the millimetre and optical, up to the X– and γ–ray regimes. The most complete databases of variable sources are found in the optical and radio bands, thanks to the large amount of observing time devoted to this study over the time. An overview of the data available in the literature can be found in Aller et al. (1999, monitoring at cm wavelengths carried out with the 26–m University of Michigan radio telescope, UMRAO), Lähteenmäki & Valtaoja (1999, monitoring at 22 GHz and 37 GHz with the 14–m Metsähovi radio telescope), Venturi et al. (2001, monitoring at 5 GHz and 8.4 GHz with the 32–m Medicina radio telescope).

The radio variability can take the form of flares (or bursts), or continuous trend with modulations. Flux density flares usually start at high frequencies, and propagate to lower frequencies with decreasing amplitude, as expected in the case of an expanding relativistic component. However, more complex behaviours are also common, such as for example flares appearing simultaneously at different frequencies, or overlap of different types of variability.

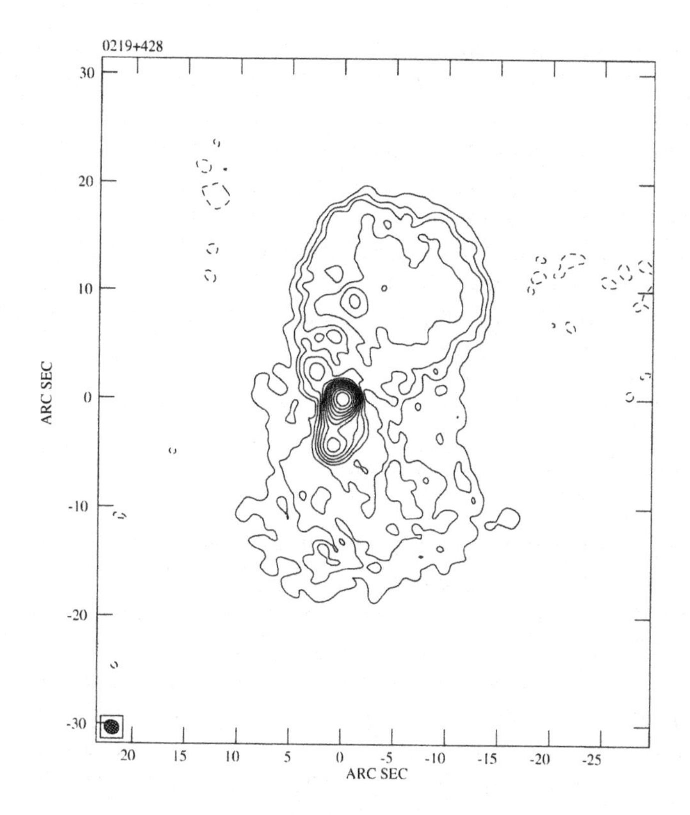

Figure 1. VLA 1.4 GHz image of the BL–Lac 0219+428. The peak in the image is 93 mJy/beam and the lowest contour 0.75 mJy/beam (Cassaro et al. in preparation).

2.3. PARSEC–SCALE PROPERTIES

2.3.1. *Morphology*

Starting from the mid–80s a number of imaging surveys of large samples of CFSRS have been carried out with the Very Long Baseline Interferometry (VLBI) at various frequencies, i.e. 1.6 GHz, 5 GHz, 8.4 GHz, 15 GHz. Here I wish to mention the historical surveys, such as for example the Pearson and Readhead (PR) survey (Pearson and Readhead 1981 and 1988); the Caltech–Jodrell CJ1 and CJ2 surveys (see for instance Taylor et al. 1994, Polatidis et al. 1995, Xu et al. 1995a, Henstock et al. 1995); the variable source survey (Wehrle et al. 1992); the 15 GHz VLBA survey (Kellermann et al. 1998). Hundreds of CFSRS have been imaged through these surveys, which revealed that the overwhelming majority of them ($\geq$ 80%) are asymmetric core–jets on the parsec–scale, with a wide range of

core/jet dominance, jet lengths and jet bending. These observational results suggest that we are selecting a population of objects strongly affected by boosting.
Another important result from this extensive surveying, was the discovery of a consistent fraction ($\sim 10\%$) of compact symmetric objects (CSO). The importance of CSOs will not be dealt with in this paper, however I wish to mention their importance in the evolutionary models developed after the mid–90s, which propose that CSOs are the young precursors of the powerful extended radio galaxies and quasars (Fanti et al. 1995, Readhead et al. 1996, Murgia 2000).

2.3.2. *Jet bending and misalignment*

Jet bending and wiggling is a common feature in parsec–scale jets of CFSRS, associated both with quasars and BL–Lacs. Just to show some examples, high resolution and high sensitivity imaging of the two BL–Lacs MKN 421 (Giovannini et al. 1999a) and MKN 501 (Edwards et al. 2000, Conway & Wrobel 1995) show that the parsec–scale jet is straight and well collimated within the first few mas from the core of the radio emission, and undergoes major bending after $\sim 10 - 15$ mas. In the case of MKN 501 (see Figure 2) the bending angle is $\sim 90°$, and Conway & Wrobel (1995) modelled the jet structure with helical motion generated by an orbiting binary black hole system. They also placed stringent limits on the viewing angle of the jet and on the Lorentz factor, and concluded that the highly relativistic jet in the source is viewed under a very small angle to the line of sight ($\theta \leq 10°$).
An interesting example of a wiggling jet is provided by space– and ground–VLBI images of the BL–Lac 0836+710 (Lobanov et al. 1998), which show that the brightness ridge line in the parsec–scale jet does not follow a straight path. The parsec–scale structure was modelled with a highly relativistic jet undergoing Kelvin–Helmholtz instabilities and variations of the viewing angle.
The statistical relevance of the misalignment between the position angle of the parsec– and the kiloparsec–scale jets (ΔPA) in CFSRS was first pointed out by Pearson and Readhead (1988) for sources in the PR sample, and further strengthened after analysis of the CJ1 survey. The distribution of ΔPA shows two peaks, at $0°$ and $90°$ respectively. Xu (1995b) pointed out that the secondary peak around $90°$ strongly depends on the fraction of flux density contained in the sub–mas core, i.e. it is strongly connected to the degree of relativistic beaming in the sources. In particular, more beamed sources have larger misalignment angles.

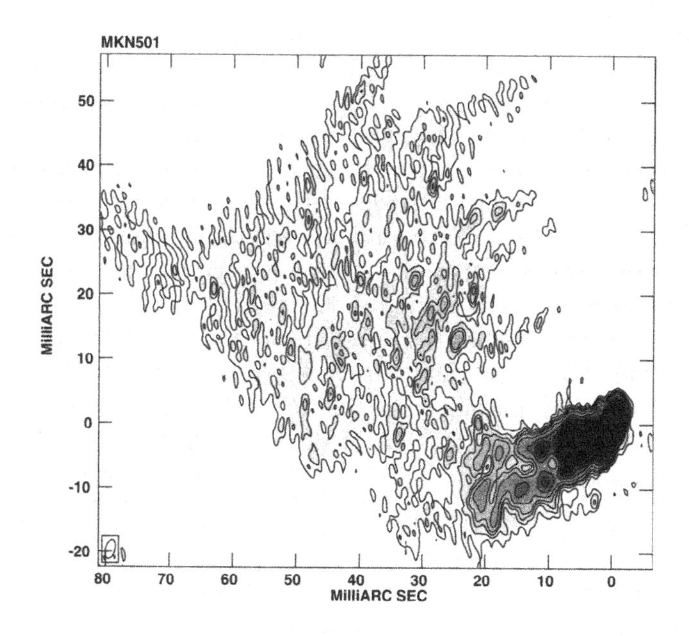

Figure 2. 1.6 GHz VLBA image of MKN 501. Contours superposed to the grey scale image of the total intensity. The resolution is 3×1.5 mas. (Giovannini et al. in prep.)

2.3.3. *Structural variability and superluminal motion*

Multiepoch imaging is now available for a large number of CFSRS, and it is clear that morphological changes and structural variability are a common feature for the sources in this class. The "knotty" appearance of the parsec–scale jets usually allows the identification of jet brightness peaks from epoch to epoch, and proper motion of features along the jets has been found since the late 1970s. In most cases the proper motions observed correspond to superluminal speeds. It is important to point out that only expansion is detected, i.e. the separation between the brightness peaks always increases, and there is no known case of superluminal contraction. A wealth of information on superluminal sources can be found in many conference proceedings, as for example *Superluminal Radio Sources* (Eds. Zensus & Pearson 1987), *Parsec–scale Radio Jets* (Eds. Zensus & Pearson 1990), *Sub–arcsecond Radio Astronomy* (Eds. Davis & Booth, 1993), *Compact Extragalactic Radio Sources* (Eds. Zensus & Kellermann 1994), *Quasars and Active Galactic Nuclei: High Resolution Radio Imaging* (Eds. Cohen & Kellermann 1995), *Astrophysical Phenomena Revealed by Space VLBI* (Eds. Hirabayashi, Edwards & Murphy 2000).

It has long been accepted that superluminal motion is the result of a relativistic jet expanding very close to the line of sight (see also the Lectures given by R. Porcas and H. Falke, these proceedings). The assumption that twin relativistic jets are ejected symmetrically from the source central engine and viewed at a small angle to the line of sight (see for instance Blandford & Konigl 1979), has some observational consequences, such as for example Doppler boosting of the flux density for the approaching jet, and apparent superluminal motion in the same jet.
The observed flux density for the boosted component is a strong function of the Doppler factor δ, defined as $\delta = [\gamma\,(1-\beta_i cos\theta)]^{-1}$, where γ is the Lorentz factor of the emitting plasma, $\gamma = (1-\beta_i^2)^{-1/2}$, β_i is the intrinsic plasma speed and θ is the angle to the line of sight. In particular, $S_{obs} \propto \delta^{3+\alpha}$.
The apparent speed, usually referred to as β_{app}, is a function of β_i and of the viewing angle, according to the formula: $\beta_{app} = \frac{\beta_i\ sin\theta}{1-\beta_i\ cos\theta}$ (Pearson & Zensus 1987). Once the proper motion has been obtained by the observations, this relation allows to define limits to the intrinsic jet speed and to the viewing angles.

2.3.4. *Statistics of superluminal motions*

Thanks to the large number of CFSRS with measured proper motion it is possible to carry out a statistical analysis on the superluminal motion. This is important to test the possible existence of different populations of sources, such as for example quasars and BL–Lacs, through the distribution of their intrinsic Lorentz factors. This study was first carried out by Vermeulen & Cohen (1994) using all the literature data available at the time, and was continued by Kellermann et al. (1999), on the basis of larger datasets. Both studies led to the conclusion that the distribution of Lorentz factors is very similar for radio galaxies, quasars and BL–Lacs, with average apparent speeds in the range 2.1 – 3.2 $\beta_{app}h^{-1}$. Very fast superluminal motions, with apparent speed in excess of 10, seem to be an exception. A corresponding plausible distribution of Lorentz factors γ could peak around 4.
Kellermann et al. (1999) compared the distribution of the Lorentz factors for the γ–ray loud and γ–ray quiet sources, and found a significant difference in the mean values. In particular, the γ–ray loud sources have higher Lorentz factors, which suggests that these sources could be examples of extreme beaming. Similar conclusions were reached by Jorstad et al. (2001), on the basis of high frequency multiepoch imaging and analysis of a sample of γ–ray loud radio sources.

3. The case for unification

A natural question is what would compact flat spectrum radio sources look like, were they viewed under large angles. Orr & Browne (1982) were the first to propose that compact flat spectrum radio quasars and extended radio quasars are the same type of objects seen under different angles. Barthel (1989) proposed a similar unification between radio loud quasars and powerful extended radio galaxies FRII (Fanaroff & Riley 1974). Browne (1984) and, more recently, Urry et al. (1991) and Urry & Padovani (1995) proposed that low luminosity BL–Lacs are the parent population of the extended low luminosity FRI radio galaxies. These basic ideas are known as "unified schemes". Barthel (1989) proposed that the viewing angle separating quasars and FRII radio galaxies is 44.5°, while lower angles, i.e. $\theta \sim 25^\circ \div 30^\circ$, can be considered acceptable limits for BL–Lacs and FRI. Here I will concentrate on those aspects of the unified schemes strictly related to the radio emission. A more general overview of this issue can be found in the Lecture given by Browne (these proceedings).
Assuming that unification holds, we expect that those properties independent of orientation do not differ in the beamed population and in the parent population, in particular the total radio power and the intrinsic jet speed should be similar. On the other hand, those affected by orientation (such as for example the dominance of the extended emission and the parsec–scale morphology) should differ.
In the previous sections I reported on the convincing observational evidence for relativistic bulk flow in the jets of quasars and BL–Lacs. In the following I will show that the parsec–scale properties of the high and low luminosity radio galaxies (FRII and FRI) are consistent with the idea that these objects are the parent population of quasars and BL–Lacs respectively.

4. Observational properties of extended radio galaxies

4.1. KILOPARSEC–SCALE PROPERTIES

I will briefly summarize the most important large scale properties of extended radio galaxies. The up–to–date issues and problems related to these sources were presented and discussed during the conference held in Oxford, in August 2000, and can be found in those conference proceedings, *Particles and Fields in Radio Galaxies* (Eds. Laing & Blundell, in press).
Extended radio galaxies were historically classified into high and low power radio sources, the break between the two classes being approximately at $\log P_{1.4GHz} = 24.5$ W Hz^{-1}. This luminosity classification reflects in a clear morphological difference (Fanaroff & Riley 1974): low power radio galaxies (referred to as FRI) are characterised by a central radio core, coincident

with the nucleus of the optical counterpart, and two twin jets which usually lose collimation at some distance from the nucleus, and "flare", to form to extended lobes of emission. The degree of jet symmetry varies from case to case, and more complicated morphologies can be found. Examples of FRI radio galaxies can be found in Fanti et al. (1987).
High power radio galaxies (FRII) usually have a weak nucleus, coincident with the optical counterpart, very asymmetric (sometimes even invisible) jets, two opposite hot–spots, considered to be the working surface of the underlying jets, and two extended lobes, formed by backflow material left behind by the advancing hot–spots. The prototype of FRII is Cygnus A (Carilli et al. 1991).
The power break between FRI and FRII is not unique, and Owen & Laing (1989) first showed that it is a function of the optical magnitude of the associated galaxy, usually a bright elliptical.

4.2. PARSEC–SCALE PROPERTIES

The results presented in the following sub–sections are mainly based on a long term study of a complete sample of radio galaxies, carried out at VLBI (parsec–scale) resolution over a wide range of frequencies by Giovannini et al. (2001 and references therein). We will refer to this sample as the "Bologna sample", which includes a total of 27 objects (13 FRI, 7 FRII, 5 low luminosity compact sources and 2 BL–Lacs).

4.2.1. *Morphology*

High sensitivity parsec–scale observations of a sample of radio galaxies (Giovannini et al. 2001, and references therein), as well as observations of individual objects, as for example M87 (Junor et al. 1999), Centaurus A (Tingay et al. 2001)and Cygnus A (Krichbaum et al. 1998), show that the nuclear regions in the vast majority of FRI and FRII radio galaxies are asymmetric, i.e. core–jet morphologies are most frequently found. In a few cases, a short counter–jet is also detected. Two radio sources with symmetric morphology were found in the "Bologna sample": 3C 338 (FRI) and 3C 452 (narrow line FRII). Evidence of jet limb–brightening has been found in B2 1144+35 (Giovannini et al. 1999b), a low luminosity radio galaxy showing signs of recently restarted activity. Examples of parsec–scale morphologies in FRIs are givenin Figure 3.

4.2.2. *Proper motions*

Monitoring of the most promising objects in the "Bologna sample" (i.e. radio galaxies showing nuclear variability) revealed proper motion for three radio galaxies. Considering that two more sources in the sample are well

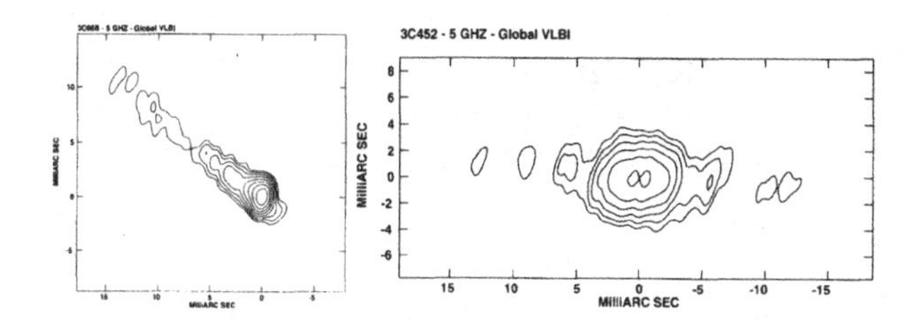

Figure 3. 5 GHz global VLBI images of the two low luminosity radio galaxies 3C66B (left panel) and 3C452 (right panel) at the resolution of $\sim$ 2 $\times$ 1 mas.

known superluminal sources (the FRI radio galaxy M87, Biretta et al. 1995, and the FRII 3C 390.3, Alef et al. 1996), the total number amounts to 5. In one case, 3C 338, the motion is subluminal, while in the other case apparent speeds in the range $\beta_{app} 1 \div 6$ have been found (Giovannini et al. 2001; Venturi et al. 2000).
Thanks to these results, it is now accepted that proper motion can be found in the nuclear regions of FRI as well as in FRII radio galaxies. This is an extremely interesting result, which has completely changed our view of the energy transport in these objects.

4.2.3. *Lorentz factors and viewing angles*

If we make the reasonable assumption that the observed properties of radio galaxies are due to Doppler boosting, we can use the observations to derive some information on the viewing angle θ and on the intrinsic plasma speed β_i, the two quantities we are interested in, in order to test the validity of unification. In particular, (1) the jet–to–counterjet brightness ratio R, (2) the core dominance $P_c(\theta)$, and (3) the proper motion β_{app}, are functions of β_i and θ according to the following relations (see Giovannini et al. 2001 for details and for a thorough discussion of the results):

(1) $\beta_i cos\theta = (R^{1/(2+\alpha)} - 1)/(R^{1/(2+\alpha)} + 1)$;
(2) $P_c(\theta) = P_c(1 - \beta_i cos\theta)^{-(2+\alpha)}$;
(3) $\beta_{app} = \beta_i sen\theta/(1 - \beta_i cos\theta)$.

The values of θ and β_i (and hence of the Lorentz factor γ) for the extended radio galaxies in the "Bologna sample" are all consistent with the idea that they are viewed at "large" angles to the line of sight, i.e. $\theta > 30°$, and that they have at least mildly relativistic jets, i.e. γ of the order of 1.2 $\div$ 1.7, but also up to 7. The only exceptions to this rule are B2 1144+35 and M87, whose parsec–scale properties lead to a very small viewing angle. Concerning B2 1144+35, there is now evidence that the galaxy shows BL–Lac type properties, while M87 is a known peculiar radio galaxy.

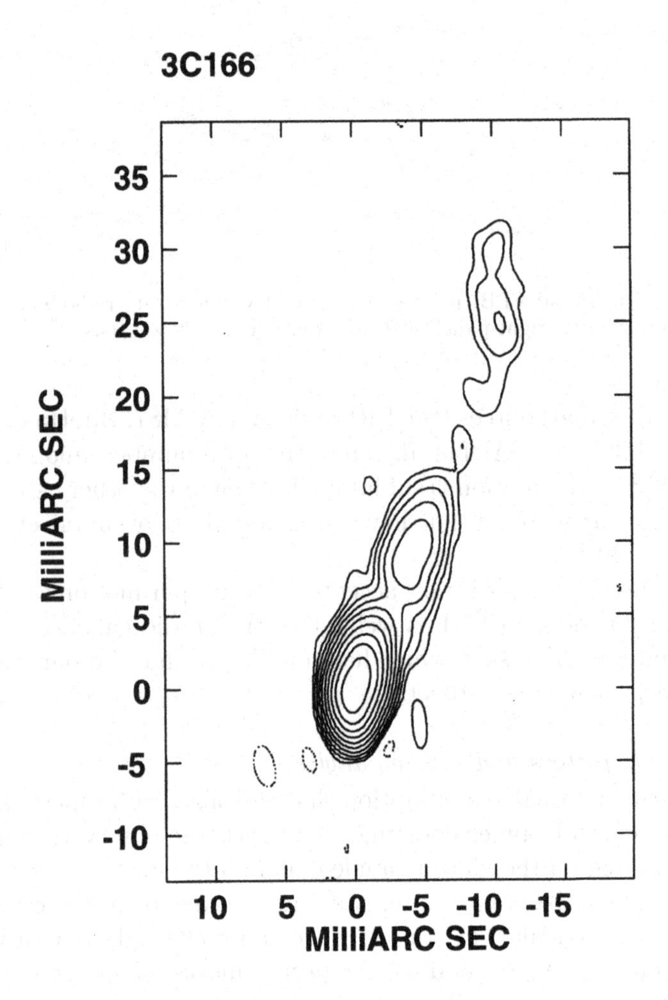

Figure 4. 5 GHz VLBA image of the FRII radio galaxy 3C 166. The restoring beam is 2.94×1.65 mas. The rms in the image is 60 μJy.

4.2.4. *FRI vs FRII*

Another relevant result of the extensive study of FRI and FRII radio galaxies on the parsec–scale, is their astonishing similarity (same morphology and same γ found in both classes) despite their major differences on the kiloparsec–scale (see Figure 4). The explanation of this observational result is challenging, since FRII radio galaxies can maintain relativistic jets out

to the distant hot spots.
One possibility is that the ambient medium plays a crucial role in the propagation of intrinsically identical jets, and/or that the two classes of associated counterparts have different accretion rates. It is expected that joint studies in the radio and optical bands, especially with HST, will prove to be crucial to throw a light on this new puzzle.

5. Conclusions

In this lecture I presented a clear astrophysical issue, where high resolution radio astronomy proved to be crucial in its understanding and development. In particular, the relation between compact flat spectrum radio sources (quasars and BL–Lacs) and extended radio galaxies (FRI and FRII). I showed that the nuclear properties in these classes of objects are consistent with the proposed "unified schemes", which attribute most of the observed differences to orientation effects.

Acknowledgements

I wish to thank Prof. R. Fanti for careful reading of the manuscript and insightful comments. Thanks are due to P. Cassaro and G. Giovannini for providing the images ahead of publication.

References

Alef, W., Wu, S.Y., Preuss,E., Kellermann, K.I., Qiu, Y.H., 1996, 3C 390.3: a lobe-dominated radio galaxy with a possible superluminal nucleus. Results from VLA observations and VLBI monitoring at 5GHz, *Astron. & Astrophys.*, **308**, pp. 376–380

Aller, M.F.,Aller, H.D., Hughes,P.A., Latimer, G.E., 1999, Centimeter–wavelength total flux and linear polarization properties of radio–loud BL Lacertae objects, *Astrophys. J.*, **512**, pp. 601–622

Antonucci, R.J., Ulvestad, J.S., 1985 Extended radio emission and the nature of blazars, *Astrophys. J.*, **294**, pp. 158–182

Barthel, P.D., 1989, Is every quasar beamed?, *Astrophys. J.*, **336**, pp. 606–611

Biretta, J.A., Zhou, F., Owen, F.N., 1995, Detection of proper motions in the M87 jet, *Astrophys.J.* **447**, pp. 582–596

Blandford, R.D., Konigl, A., 1979, Relativistic jets as compact radio sources, *Astrophys. J.*, **232**, pp. 34–48

Browne, I.W.A., 1984, Intermediate scale structure, in *VLBI Compact Radio Sources*, IAU Symp. 110, Eds. Fanti R., K. Kellermann & G. Setti, D. Reidel Publ. Company, Dordrecht, Boston, pp. 1–5

Cassaro, P., Stanghellini C., Dallacasa, D., Bondi, M., Zappalà, R.A., Extended radio emission in BL Lac objects. I. The images, *Astron. & Astrophys. Suppl. Ser.* **139**, pp. 601–616

Carilli, C.L., Perley, R. A., Dreher, J. W., Leahy, J. P., 1991, Multifrequency radio observations of Cygnus A - Spectral aging in powerful radio galaxies, *Astrophys. J.*, **383**, pp. 554–573

Cohen, M.H., Kellermann, K.I., Eds., 1995, *Quasars and Active Galactic Nuclei: High Resolution Radio Imaging*, Proc. Natl. Acad. Sci. USA, Vol. **92**

Conway, J.E., Wrobel, J.M., 1995, A helical jet in the orthogonally misaligned BL Lacertae object Markarian 501 (B1652+398), *Astrophys. J.*, **439**, pp. 98–112

Davis, R.J., Booth, R.S., Eds., 1993, *Sub-arcsecond Radio Astronomy*, Cambridge University Press

Edwards, P.G., Cotton, W.D., Giovannini, G., et al., 2000, A spectral index map from VSOP observations of Mkn 501, *PASJ*, **52**, pp. 1015–1019

Fanaroff, B.L., Riley, J.M., 1974, The morphology of extragalactic radio sources of high and low luminosity, *Mon. Not. R. Astron. Soc.*, **167**, pp. 31–36

Fanti, C., Fanti, R., Dallacasa, D., Schilizzi, R.T., Spencer, R.E., Stanghellini, C., 1995, Are compact steep spectrum young?, *Astron. & Astrophys.*, **302**, pp. 317–326

Fanti, C., Fanti, R., de Ruiter, H.R., Parma, P., 1987, VLA observations of low luminosity radio galaxies. IV - The B2 sample revisited, *Astron. & Astrophys. Suppl. Ser.*, **69**, pp. 57–76

Giovannini, G., Feretti, L., Venturi, T., Cotton, W.D., Lara, L., 1999a, VLBI Observations of Mkn 421 and Mkn 501, in *The BL–Lac phenomenon*, Eds. L.O. Takalo & A. Sillanpaa, *ASP. Conf. Ser.*, Vol. **159**, pp. 439–442

Giovannini, G., Taylor, G.B., Arbizzani, E., Bondi, M., Cotton, W.D., Feretti, L., Lara, L., Venturi, T., 1999b, B2 1144+35: A Giant Low-Power Radio Galaxy with Superluminal Motion, *Astrophys. J.*, **522**, pp. 101–112

Giovannini, G., Cotton, W.D., Feretti, L., Lara, L., Venturi, T., 2001, VLBI observations of a complete sample of radio galaxies: 10 years later, *Astrophys. J.*, **552**, pp. 508–526

Henstock, D.R., Browne, I.W.A., Wilkinson, P.N., et al., 1995, The Second Caltech–Jodrell Bank VLBI Survey. II. Observations of 102 of 193 Sources, *Astrophys. J. Suppl.*, **95**, pp. 1–36

Hirabayashi, H., Edwards, P.G., Murphy D.W., Eds., 2000, *Astrophysical Phenomena Revealed by Space VLBI*, published by the Institute of Space and Aeronautical Science, Sagamihara, Japan

Jorstad, S.G., Marscher, A.P., Mattox, J.R., Wehrle, A.E., Bloom, S.D., Yurchenko, A.V., 2001, Multiepoch Very Long Baseline Array Observations of EGRET-detected Quasars and BL Lacertae Objects: Superluminal Motion of Gamma-Ray Bright Blazars, *The Astrophys. J. Suppl.*, **134**, pp. 181–240

Junor, W., Biretta, J.A., Livio, M., 1999, Formation of the radio jet in M87 at 100 Schwarzschild radii from the central black hole, *Nature*, **401**, pp. 891–892

Kellermann, K.I., Vermeulen, R.C., Zensus, A.J., Cohen, M.H., 1998, Sub-Milliarcsecond Imaging of Quasars and Active Galactic Nuclei, *The Astron. J.*, **115**, pp. 1295–1318

Kellermann, K.I., Vermeulen, R.C., Zensus, A.J., Cohen, M.H., West, A., 1999, Kinematics of quasars and AGN, *New Astr. Rev.*, **43**, pp. 757–760

Krichbaum, T.P., Alef, W., Witzel, A., Zensus, J.A., Booth, R.S., Greve, A., Rogers, A.E.E., 1998, VLBI observations of CygnusA with sub-milliarcsecond resolution *Astron. & Astrophys.*, **329**, pp. 873–894

Lähteenmäki, A., Valtaojia, E., 1999, Total flux density variations in extragalactic radio sources, *Astrophys. J.*, **521**, pp. 493–501

Laing, R.A., Blundell, K.M., Eds., 2002, *Particles and Fileds in Radio Galaxies, ASP Conf. Ser.*, in press

Lobanov, A.P., Krichbaum, T.P., Witzel, A., et al., 1998, VSOP imaging of S5 0836+710: a close–up on plasma instabilities in the jet, *Astron. & Astrophys.*, **340**, pp. L60–L64

Murgia, M., 2000, On the ages of extra galactic radio sources, Ph.D. Thesis, University of Bologna

Murphy, D.W., Browne, I.W.A., Perley, R.A., 1993, VLA observations of a complete sample of core–dominated radio sources, *Mon. Not. R. Astron. Soc.*, **264**, pp. 298–318

Orr, M.J.L., Browne, I.A.W., 1982, Relativistic beaming and quasar statistics, *Mon. Not. R. Astron. Soc.*, **200**, pp. 1067–1080

Owen, F.N., Laing, R.A., 1989, CCD surface photometry of radio galaxies. I - FR class I and II sources, *Mont. Not.R. astron. Soc.*, **238**, pp. 357-378

Pearson, T. J., Readhead, A. C. S., 1981, The milli-arcsecond structure of a complete sample of radio sources. I - VLBI maps of seven sources, *Astrophys. J.*,**248**, pp. 61–81

Pearson, T. J., Readhead, A. C. S., 1988, The milli-arcsecond structure of a complete sample of radio sources. II - First–epoch maps at 5 GHz, *Astrophys. J.*,**328**, pp. 114–142

Pearson T.J., Zensus, A.J., 1987, Superluminal radio sources: Introduction, in *Superluminal Radio Sources*, Eds. J.A. Zensus & T.J. Pearson, Cambridge Univ. Press, pp. 1–11

Polatidis, A., G., Wilkinson, P.N., Xu, W., et al., 1995, The First Caltech–Jodrell Bank VLBI Survey. II. lambda = 18 Centimeter Observations of 25 Sources, *Astrophys. J. Suppl.*, **98**, pp. 1–32

Readhead, A.C.S., Taylor, G.B., Pearson, T.J., Wilkinson, P.N., 1996, Compact Symmetric Objects and the Evolution of Powerful Extragalactic Radio Sources, *Astrophys. J.*, **460**, pp. 634–643

Taylor, G.B., Vermeulen, R.C., Pearson, T.J., et al., 1994, The second Caltech-Jodrell Bank VLBI survey. 1: Observations of 91 of 193 sources, *Astrophys. J. Suppl.*, **95**, pp. 345–370

Tingay, S.J., Preston, R.A.,Jauncey, D.L., 2001, The Subparsec-Scale Structure and Evolution of Centaurus A. II. Continued Very Long Baseline Array Monitoring, *The Astron. J.*, **122**, pp. 1697–1706.

Urry, M.C., Padovani, P., Stickel, C.M., 1991, Fanaroff-Riley I galaxies as the parent population of BL Lacertae objects. III - Radio constraints, *Astrophys. J.*, **382**, pp. 501–507

Urry, M.C., Padovani, P., 1995, Unified Schemes for Radio-Loud Active Galactic Nuclei, *Publ. Astr. Soc. Pac.*, **107**, pp. 803–845

Venturi, T., Dallacasa, D., Orfei, A., et al., 2001, Radio monitoring of a sample of X– and γ–ray loud blazars, *Astron. & Astrophys.*, **379**, pp. 755–766

Venturi. T., Giovannini, G., Feretti, L., Cotton, W.D., Lara, L., 2000, Parsec–scale properties of a complete sample of radio galaxies, *Proceedings of the 5th European VLBI Network Symposium*, Eds. J.E. Conway, A.G. Polatidis, R.S. Booth & Y.M Pihlström, Publ. Onsala Space Observatory, Chalmers University, Göteborg, Sweden, pp. 15–18

Vermeulen, R.C., Cohen, M.H., 1994, Superluminal motion statistics and cosmology, *Astrophys. J.*, **430**, pp. 467–494

Xu, W., Readhead, A.C.S., Pearson, T.J., et al., 1995a, The First Caltech–Jodrell Bank VLBI Survey. III. VLBI and MERLIN Observations at 5 GHz and VLA Observations at 1.4 GHz, *Astrophys. J. Suppl.*, **99**, pp. 297–348

Xu, W., 1995b, Ph.D. Thesis, California Institute of Technology

Wehrle, A.E., Cohen, M.H., Unwin, S.C., et al. 1992, The milliarcsecond structure of highly variable radio sources, *Astrophys. J.*, **391**, pp. 589–607

Zensus, A.J., Pearson, T.J., Eds., 1987, *Superluminal Radio Sources*, Cambridge Univ. Press

Zensus, A.J., Pearson, T.J., Eds., 1987, 1990, *Parsec–Scale Radio Jets*, Cambridge University Press

Zensus, A.J., Kellermann, K.I., Eds, 1994, *Compact Extragalactic Radio Sources*, Proceedings of a workshop held at Socorro, New Mexico, NRAO

SUPERMASSIVE BLACK HOLES IN THE UNIVERSE

ANDREAS BRUNTHALER AND HEINO FALCKE
Max-Planck-Institut für Radioastronomie
Auf dem Hügel 69, 53121 Bonn, Germany

Abstract. In these lecture notes we will give a very basic introduction to supermassive black holes. We derive the basic luminosity expected for accretion and introduce the Eddington limit. Kinematical evidence for black holes in the centers of galaxies from optical and radio spectroscopy is discussed as well as the link between black holes and host galaxy properties. Finally we consider the evidence for the black hole in the Galactic Center.

1. Introduction

In the early sixties, a number of strong radio sources were detected that were associated with optical point sources. Although they appear stellar on photographic plates, they had very strange optical emission lines which were different from anything seen before in stars. Schmidt (1963) realized for one of this "quasi stellar radio sources" (quasars), 3C 273, that the emission lines can be explained with hydrogen emission at a redshift of $z = 0.158$.

This unusual high redshift lead to another problem. The optical brightness of $m_B = 13.1$ corresponds to a luminosity of $L = 2 \times 10^{46}$ erg s^{-1} or 0.5×10^{13} L$_\odot$. Since the sources show variability on timescales of less than a year, the size of the emitting region is restricted to be smaller than one light year. Although confined into a 10^{-12}th fraction of the volume of a galaxy, the luminosities of these quasars exceed the luminosity of an entire galaxy. The most efficient way to release energy is accretion in a deep, relativistic gravitational potential. Only supermassive black holes have survived today as major paradigm to explain the huge energy output of the central engine of active galactic nuclei.

The next chapter will give a very basic introduction into the accretion of black holes. In chapter 3 we will review some evidence for supermassive

F. Mantovani and A. Kus (eds.), The Role of VLBI in Astrophysics, Astrometry and Geodesy, 143–156.

black holes in external galaxies, while chapter 4 will focus on the black hole in our Galactic Center.

2. Black Holes – Basic Principles

A key feature of general relativity is that the presence of mass distorts space-time. In a black hole the space-time is curved to such an extend that light can no longer escape. The characteristic size of a black hole is given by its Schwarzschild radius. This radius is in classical terms the radius at which the escape velocity is equal to the speed of light:

$$v_{esc} = \sqrt{\frac{2GM}{R}} = c \Rightarrow R_S = \frac{2GM_{BH}}{c^2} = 3\ \mathrm{km} \times \frac{M_{BH}}{\mathrm{M}_\odot}. \quad (1)$$

Any material or radiation that falls into this event horizon is trapped forever in the black hole.

2.1. ACCRETION

The strong gravity field of the black hole will attract gas clouds in the central region of its host galaxy. This gas clouds will collide with each other in the vicinity of the black hole and transform kinetic energy into frictional heating. Usually the gas clouds will have a certain amount of angular momentum which has to be conserved. This will lead to the formation of an accretion disk. The particles in the disk will continue to interact with each other and heat the disk even more. In this process the particles will lose kinetic energy and move inwards while angular momentum is transported outwards, until the gas finally reaches the last stable orbit and falls into the black hole. The heated accretion disk will radiate predominantly in the optical and UV and produce the enormous luminosities observed in quasars.

The energy output is related to the rate of accreted material. A particle with mass m which falls in a gravitational potential of a central mass M from an infinite distance to a distance R from the black hole gains the energy:

$$U = \frac{GMm}{R}. \quad (2)$$

In the case of a black hole, the accreted material can only radiate until it reaches the event horizon R_S. If the energy is converted with an efficiency of η into radiation, the luminosity depends on the accretion rate $\dot{m}$ as:

$$L = \eta \dot{U} = \eta \frac{GM_{BH}}{R_S}\frac{dm}{dt} = \eta \frac{1}{2}\dot{m}c^2 \quad (3)$$

or

$$L = 10^{46}\frac{\mathrm{erg}}{\mathrm{s}}\left(\frac{\eta}{0.1}\right)\left(\frac{\dot{m}}{\mathrm{M}_\odot \mathrm{yr}^{-1}}\right). \quad (4)$$

Most of the luminosity will be radiated in the inner 10 R_S over an area $A = \pi(10R_S)^2$. If we assume the emission is black body radiation, we can use the Stefan-Boltzmann law

$$\sigma T^4 = \frac{L}{A} \Rightarrow T = \left(\frac{L}{A\sigma}\right)^{\frac{1}{4}} = 1.6 \cdot 10^5 \mathrm{K} \left(\frac{\dot{m}}{\mathrm{M}_\odot yr^{-1}}\right)^{\frac{1}{4}} \cdot \left(\frac{M_{BH}}{10^8 \mathrm{M}_\odot}\right)^{-\frac{1}{2}} \quad (5)$$

where σ is the Stefan-Boltzmann constant to estimate a temperature. With a temperature exceeding 10^5K, the central engine radiates mainly in the ultraviolet.

Hence, a quasar with a luminosity of 10^{46} erg s^{-1} and an efficiency of $\eta = 0.1$ would accrete 1 $\mathrm{M}_\odot \mathrm{yr}^{-1}$. If there is $\sim 10^{10}\mathrm{M}_\odot$ of material in the central region of the galaxy, the lifetime of the quasar would be limited to a few times 10^9 years. Hence, the quasar phase can be just a short fraction of the lifetime of a galaxy.

2.2. THE EDDINGTON LIMIT

The central engine is generating a very large amount of radiation. This radiation exerts a force on the accreting material and sets an upper limit on the luminosity. The Eddington Limit is reached if the radiation force is equal to the gravitational force. The radiation interacts mainly with the electrons, while the gravitation affects predominantly the heavy protons,

$$\frac{\sigma_{th} L}{4\pi R^2 c} < \frac{G M_{BH} m_p}{R^2}. \quad (6)$$

Here is σ_{th} the Thomson cross section for electron photon scattering, L the luminosity, and m_p the proton mass. This requirement gives an lower limit of the black hole mass for a given luminosity assuming isotropic emission:

$$M_{BH} > 0.8 \cdot 10^8 \mathrm{M}_\odot \left(\frac{L}{10^{46}\ \mathrm{erg\ s}^{-1}}\right). \quad (7)$$

The Schwarzschild radius of a black hole gives a characteristic size scale and one can calculate an *equivalent* mass density of a black hole:

$$\rho_{BH} = \frac{M_{BH}}{4/3\pi (R_S)^3} = 1.8\ \frac{\mathrm{g}}{\mathrm{cm}^3} \left(\frac{M_{BH}}{10^8 M_\odot}\right) \quad (8)$$

Hence, a black hole with a mass of $5.5 \times 10^7 \mathrm{M}\odot$ would have the same *density* as water and a size slightly larger than 1 AU – the size of the Earth orbit. This *density*, of course, should not be taken too serious.

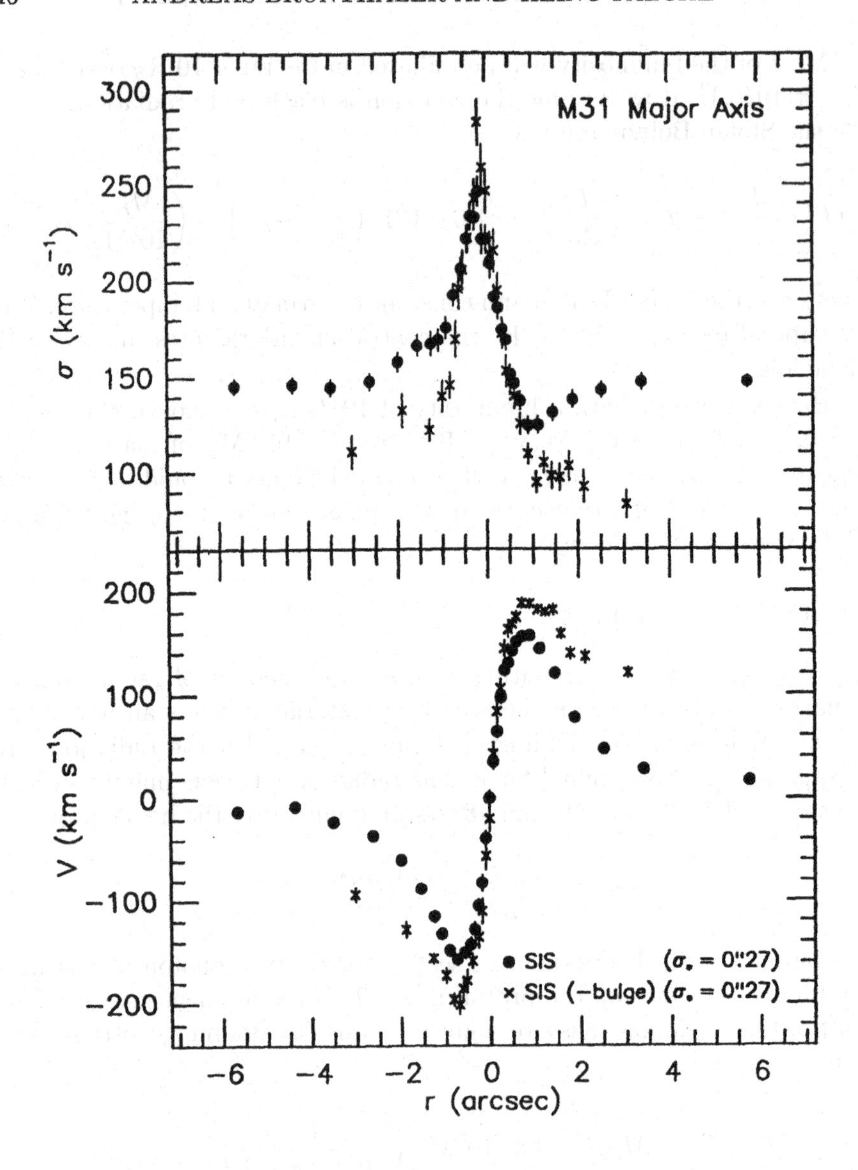

Figure 1. Velocity $V(r)$ (bottom) and velocity dispersion $\sigma(r)$ (top) profiles along the nucleus major axis of M31. (Taken from Kormendy and Richstone 1995).

3. Observational evidence

Black Holes were used to explain the luminous emission of AGN. Quasars are found in many normal galaxies and the AGN-phase is just a short fraction of the lifetime. This leads to the prediction of massive black holes at the center of many galaxies. The fact that accretion onto a black hole

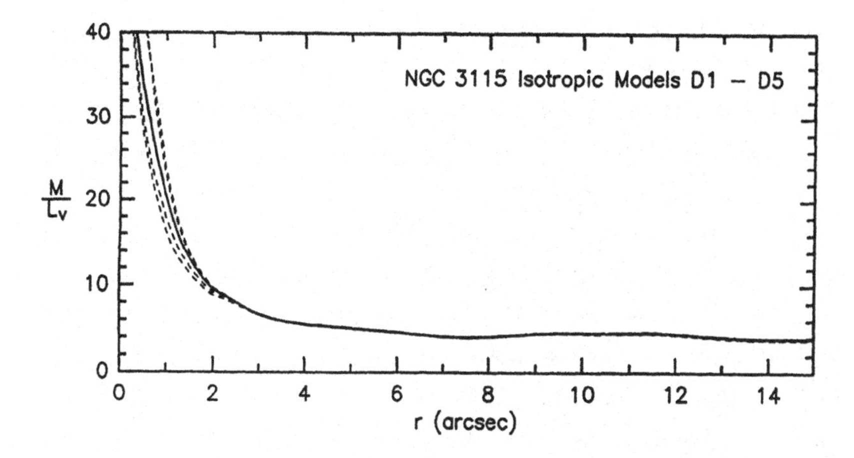

Figure 2. Mass-to-light ratio interior to radius r for NGC 3115. M/L_V increases at $r < 2''$ by a factor of ~ 10 (Taken from Kormendy and Richstone 1995).

can explain the observed luminosities, is no proof for the existence of black holes. To find stronger evidence for supermassive black holes, one has to search for dynamical evidence for *dead quasars* in galaxies. This can be done simplest in nearby galaxies.

3.1. STELLAR VELOCITY DISPERSION

Long-slit spectroscopy of stellar absorption lines in nearby galactic nuclei shows high rotational velocities and a high velocity dispersion towards the center. The velocities are measured from the Doppler shift of the line centroids while the velocity dispersion is inferred from the line width. For an isotropic velocity field with net rotation v and velocity dispersion σ we can derive a central mass of

$$M(r) = \frac{Rv^2}{G} + \frac{R\sigma^2}{G} \tag{9}$$

using essentially the Kepler law and assuming that the stars are virialized.

Figure 1 shows the velocity and velocity dispersion profiles along the nucleus major axis of M31. The velocity dispersion increases towards the center of M31 to ~ 250 km s^{-1} while the velocities increase to ~ 200 km s^{-1} before they drop to 0 km s^{-1} in the center. Kormendy (1988) derived a dark object mass of $\sim 10^7$ M$_\odot$ for this galaxy.

Using formula 9, one can calculate the mass inside a given radius and compare it with the measured stellar light. Figure 2 shows the mass-to-light ratio M/L_V for the galaxy NGC 3115 derived in this way. It stays roughly constant in the outer parts and increases drastically towards the central

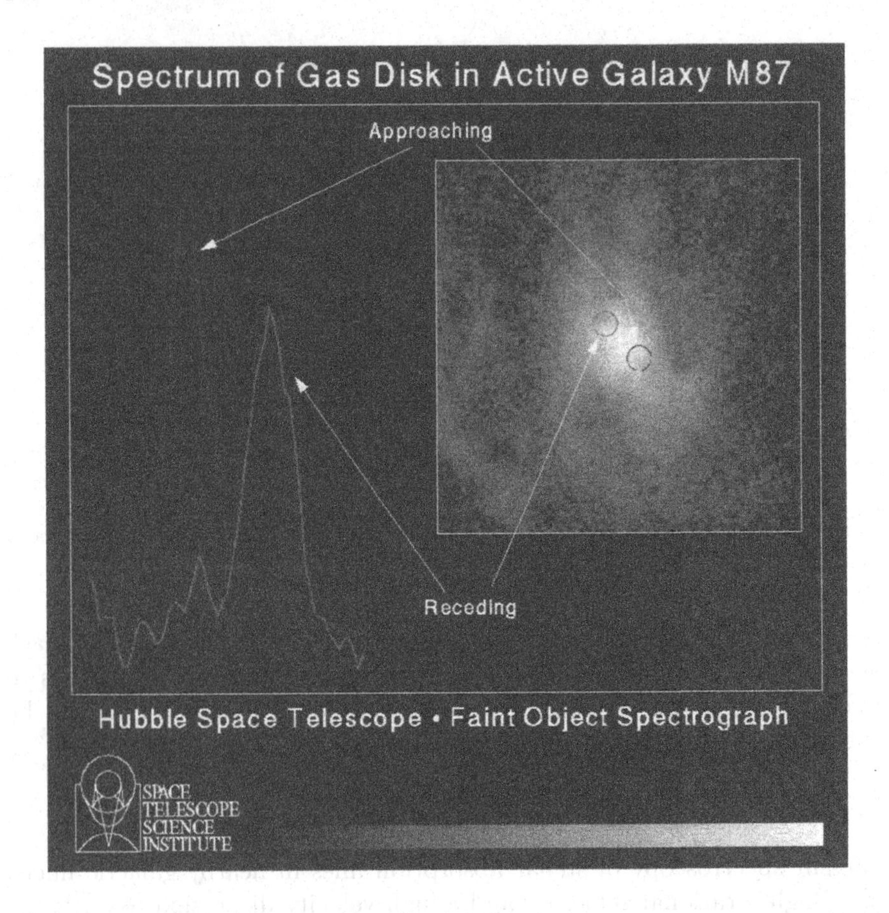

Figure 3. Hubble Space Telescope image of a rotating disk in M87. The emission lines of the receding part are redshifting, while the lines of the approaching part are blue-shifted with respect to the systematic velocity of the galaxy, yielding evidence for a massive dark object – presumably a black hole. (Ford *et al.* 1994)

region of the galaxy. The kinematic data indicates that NGC 3115 harbors a massive dark object with a mass of 10^9 $M_\odot$ (Kormendy and Richstone, 1992) in its center.

3.2. HIGH-RESOLUTION SPECTROSCOPY OF GAS

A further hint for massive matter concentrations comes from the kinematic behavior of the interstellar gas close to the core of galaxies. The Hubble Space Telescope is able to resolve the dynamics of stars and the gas in nearby galaxies. A prime example is the nearby giant elliptical galaxy M87 which contains a hot rotating disk in the center (see Figure 3). Measurements of the radial velocities using spectroscopy of emission lines from

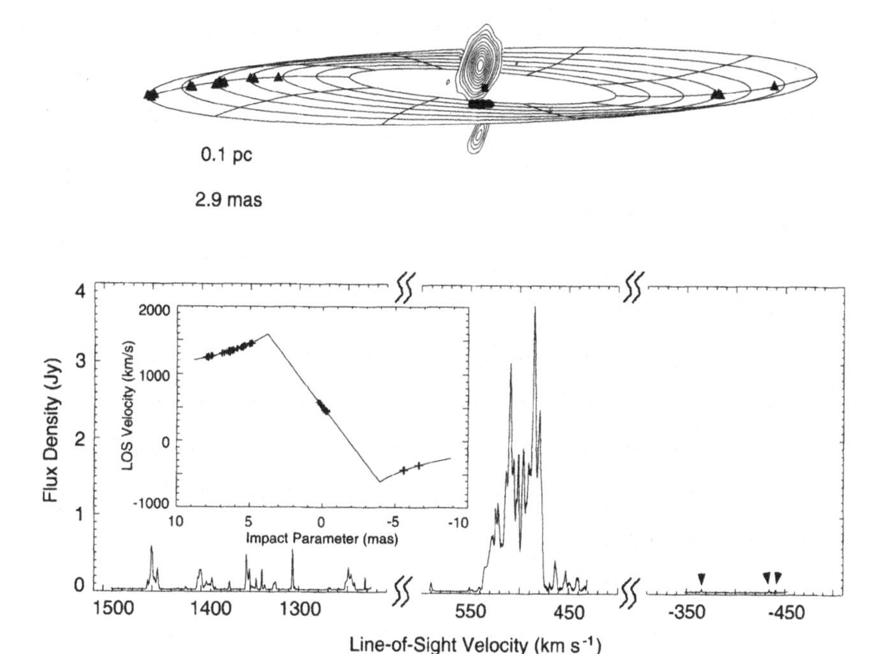

Figure 4. The warped disk model, the maser positions and the 22 GHz continuum emission of a sub-parsec-scale jet from VLBI observations of NGC 4258 (top). Also shown is the total spectrum with masers at the systematic velocity of $\sim$ 470 km s^{-1} and the *high-velocity* masers Doppler shifted by ± 1000 km s^{-1}. The inlay shows the line-of-sight velocity versus the the impact parameter for a Keplerian disk with the maser data superposed. (Taken from Herrnstein *et al.* 1999).

the ionized gas show the gas to be in Keplerian rotation about a mass of $M = 2.4 \times 10^9$ M$_\odot$ within the inner 18 pc of the nucleus (Ford *et al.* 1994; Harms *et al.* 1994).

3.3. WATER MASERS - NGC 4258

Strong evidence for a supermassive black hole comes also from interferometric spectral line observations of water vapor maser emission in the centers of galaxies. The spectrum of the H_2O maser emission in the Seyfert galaxy NGC 4258 consists of maser components at the systematic velocity of the galaxy as well as high-velocity masers which are Doppler shifted by ± 1000 km s^{-1}. High resolution Very Long Baseline Interferometry (VLBI) observations show the maser spots in a thin warped disk around the center. The masers with the systematic velocity appear in front of the nucleus, while the blue- and red-shifted components are on the approaching and receding sides of the disk which is in perfect Keplerian rotation (Figure 4). From the rotation and the distance of the source, one can estimate an

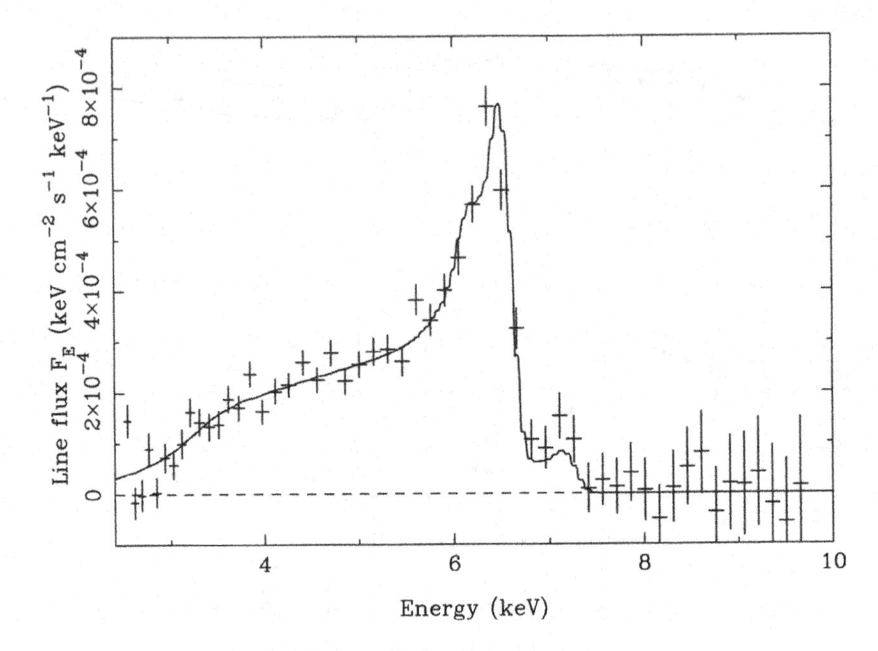

Figure 5. Continuum subtracted relativistic iron line profile from an XMM-Newton observation of MCG-6-30-15 (Taken from Fabian *et al.* 2002).

enclosed mass of $3.6 \times 10^7 M_\odot$ within 0.1 parsecs (Miyoshi *et al.* 1995; Herrnstein *et al.* 1997). This requires a mass density of $> 4 \times 10^9$ $M_\odot pc^{-3}$.

3.4. RELATIVISTICALLY BROADENED IRON Kα-LINE

X-ray spectroscopy of Seyfert galaxies have revealed enormously broadened iron Kα emission lines with line widths of $\sim 100,000$ km s^{-1}. ASCA (Advanced Satellite for Cosmology and Astrophysics) observations by Tanaka *et al.* (1995) found that the iron line profile in MCG-6-30-15 could be explained by emission from an accretion disk around a black hole. Recent XMM-Newton observations (see Figure 5) find also extremely broad and redshifted emission indicating an origin in the central regions of an accretion disk around a rotating black hole (Wilms *et al.* 2002; Fabian *et al.* 2002).

3.5. BLACK HOLES AND THEIR HOST GALAXIES

Many nearby galaxies show evidence for a massive dark object in their center. This raises the question whether the evolution of black holes and galaxies are connected. Hence it is important to search for correlations between properties of the black hole and those of galaxies. Gebhardt *et al.* (2000) and Ferrarese and Merritt (2000) found a strong correlation be-

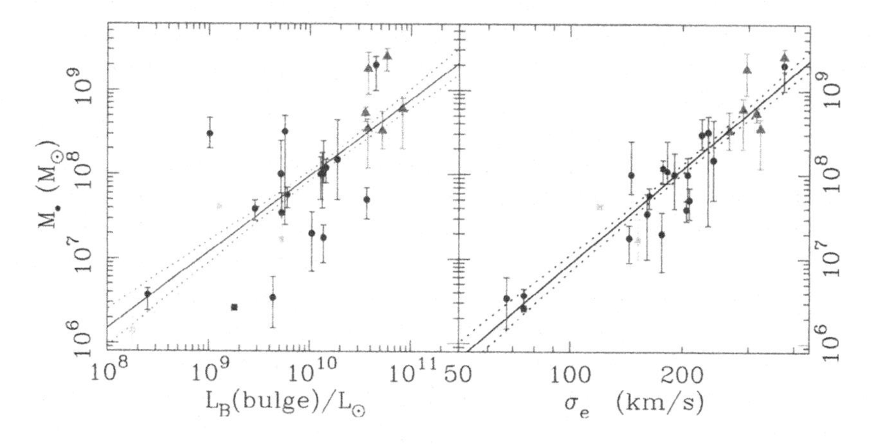

Figure 6. Black hole mass versus bulge luminosity (left) and the velocity dispersion (right). The velocity dispersion shows a tight correlation with the the mass of the cental black hole. Solid and dotted lines are the best fit and their 68% confidence bands. (Taken from Gebhardt et al. 2000).

tween the black hole mass and the velocity dispersion of stars in the bulge of the host galaxy. Figure 6 shows this correlation for 26 galaxies with black hole masses from kinematics of stars, gas and masers.

The velocity dispersion of stars in the bulge depends on the mass of the spheroidal stellar component and this correlation indicates that the more massive the bulge the heavier is the black hole. This suggests that the evolution of black holes and their host galaxies are intimately linked.

There are various processes for bulge and black hole mass evolution. A primordial hydrogen cloud collapses around a small black hole. Infalling gas feeds the black hole and forms stars. Finally the collapse yields a giant elliptical galaxy or bulge and the black hole growth stops. Another scenario is the merger of two spiral galaxies with black holes. The galaxies collide and the merger yields an elliptical galaxy with a larger central black hole. The central black hole could also grow throughout the cosmological history by accretion of ordinary or dark matter through the galactic disk or halo into the center. Hence, while black holes can grow in different ways, the reason why black holes and bulges are so intimately linked remains a big puzzle.

4. The Dark Mass in the Galactic Center

The closest place to look for a supermassive black hole is the center of our own Galaxy. The compact radio source Sgr A* in our Galactic Center is thought to contain a black hole. For a detailed review about the inner parsecs of our Galaxy see Melia and Falcke (2001).

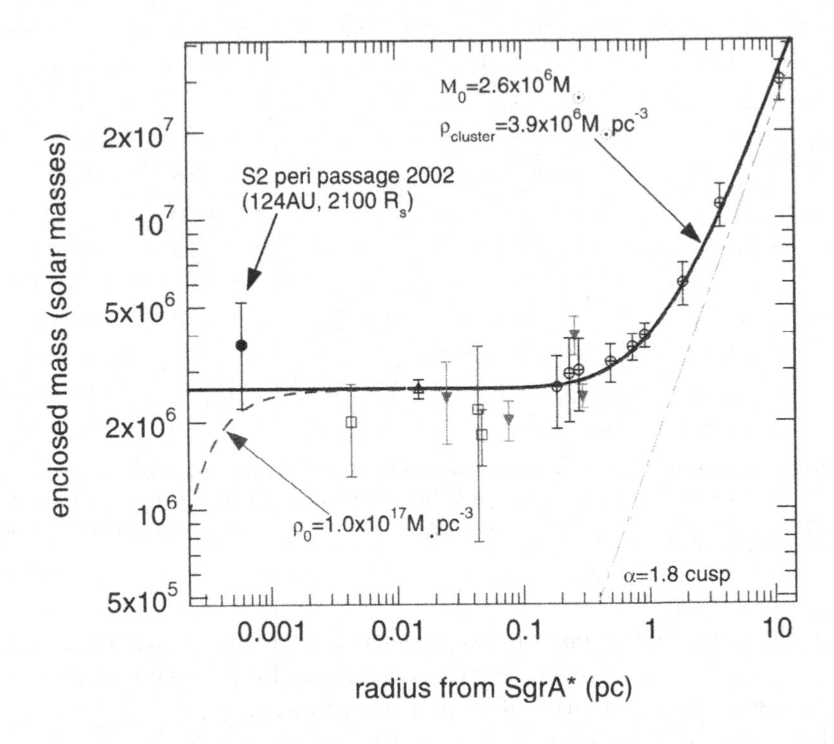

Figure 7. Enclosed mass as a function of radius from Sgr A* in the Galactic Center (Taken from Schödel et al. 2002).

Measurements of stellar proper motions in the vicinity of Sgr A* have revealed a dark mass in the Galactic Center. The center of gravity coincides with Sgr A* within 0.01 light years. Recently Ghez *et al.* (2000) and Eckart *et al.* (2002) detected for the first time acceleration in the proper motions. This allows one to constrain the possible orbits around Sgr A* and locate the center of mass even better.

For one star, both peri- and apo-center passages have been observed recently that show a highly elliptical Keplerian orbit with an orbital period of 15.2 years and a peri-center distance of 17 light hour (Schödel *et al.*, 2002). This orbit requires an enclosed mass of $3.7 \pm 1.5 \times 10^6$ $M_\odot$. Figure 7 shows the enclosed mass as a function of the radius from Sgr A*. The solid curve is the best fit to all data points and represents the sum of a $2.6 \pm 0.2 \times 10^6$ $M_\odot$ point mass and a stellar cluster with central density 3.9×10^6 $M_\odot$ pc^{-3}, core radius 0.34 pc and power-law index $\alpha = 1.8$.

Further evidence about the nature of Sgr A* comes from its proper motion with respect to background quasars that has been measured with VLBI. Sgr A* apparently moves with 219 km s^{-1} along the Galactic Plane (see Figure 8), which entirely reflects the motion of the sun around the

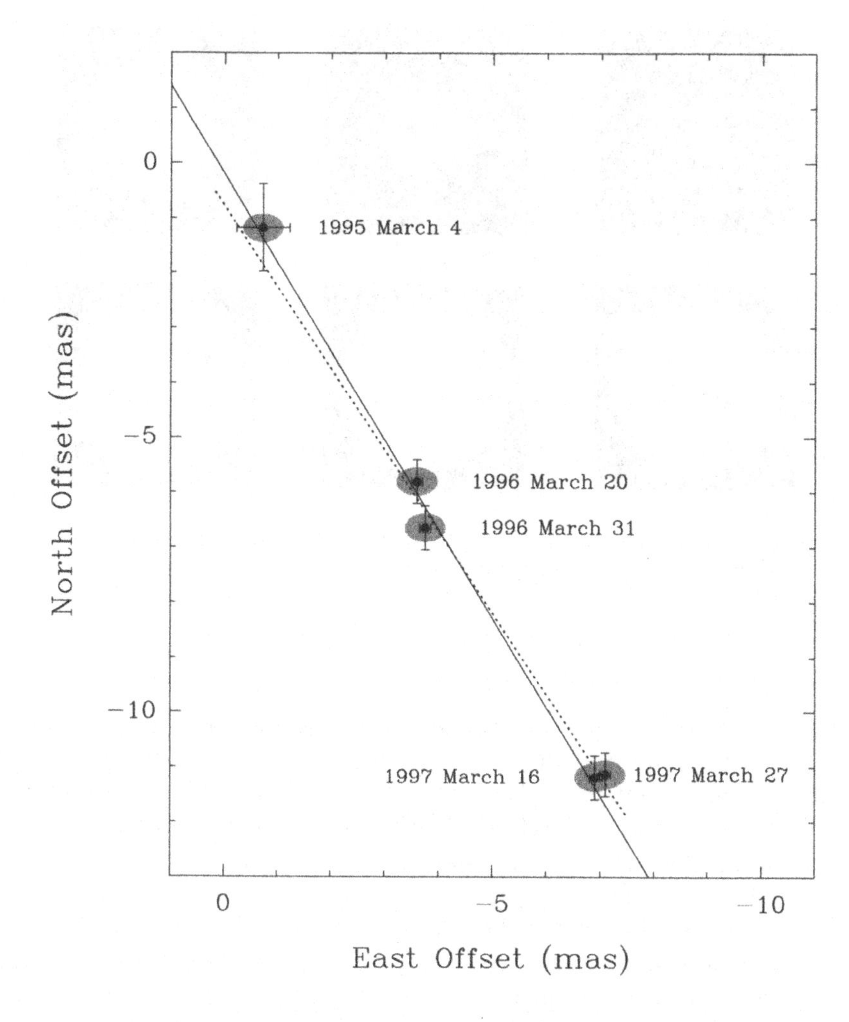

Figure 8. Proper motion of Sgr A* with respect to background quasars measured with VLBI. The solid line gives the orientation of the Galactic plane (Taken from Reid *et al.* 1999).

Galactic Center. Hence the proper motion of Sgr A* itself is consistent with zero (Reid *et al.* 1999; Backer & Sramek 1999). This is in clear contrast to the velocities of stars in the central region which move at speeds that exceed 1000 km s^{-1}. Thus, Sgr A* has to be much more massive than these stars and the upper limit on the speed gives a lower limit on the mass of $\sim 10^3 M_\odot$. Further VLBI observations (Reid *et al.*, 2003) and improved theoretical models (Chatterjee, Hernquist and Loeb, 2002) increase the lower limit to $\sim 10^5 M_\odot$.

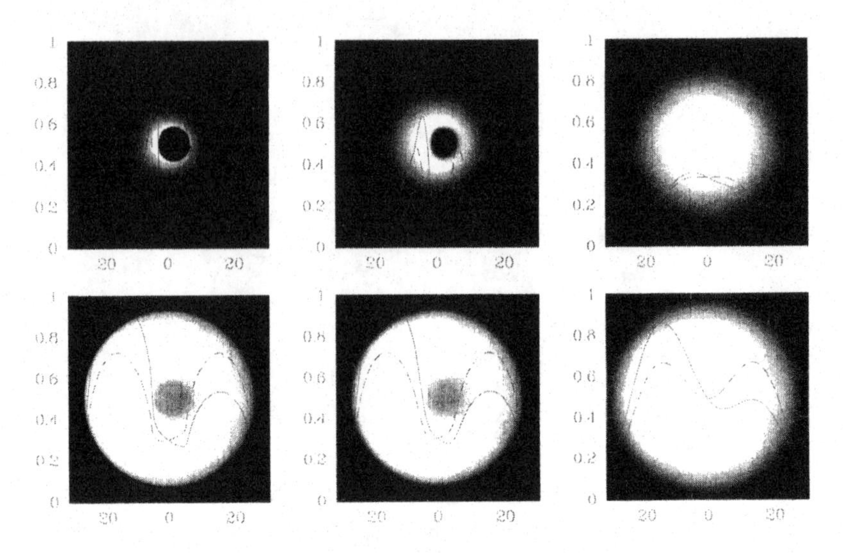

Figure 9. Shadow of a black hole. For a description see text. (Taken from Falcke, Melia and Agol 2000).

Black holes emit strong emission at scales that are affected by General Relativity. The photon orbits are bent in the vicinity of the black hole and can become circular at distances of $\sim 2-3R_S$. Closer orbits will end in the event horizon and produce a *shadow* in the emitting region around the black hole. Figure 9 shows the model image of an optically thin emission region surrounding a black hole with the characteristics of Sgr A* at the Galactic Center. The black hole is either maximally rotating (upper row) or non rotating (lower row). Images (a,d) show ray-tracing calculations, (b,e) are the images seen by an idealized VLBI array at 0.6 mm wavelength, taking interstellar scattering into account. The images (c,f) are for a wavelength of 1.3 mm. The intensity variations along the x-axis (solid curve) and the y-axis (dashed curve) are overlayed. For Sgr A*, the predicted size of this shadow is ~ 30 μas and approaches the resolution of current VLBI experiments (Falcke, Melia and Agol, 2000).

Recently Sgr A* was detected for the first time with VLBI at 1.4 mm on one baseline (Plateau de Bure - Pico Veleta). Krichbaum *et al.* (1998) derived a source size of 0.11 ± 0.06 mas or 17 ± 9 Schwarzschild radii for a 2.6×10^6 $M_\odot$ black hole. Figure 10 shows the source size of Sgr A* versus wavelength. The data points follow a λ^2 behavior which is expected from scatter broadening by the interstellar medium. The source size at 1.4 mm is significant larger than the scattering size and may be intrinsic. This size is just a factor of 3 away from the shadow.

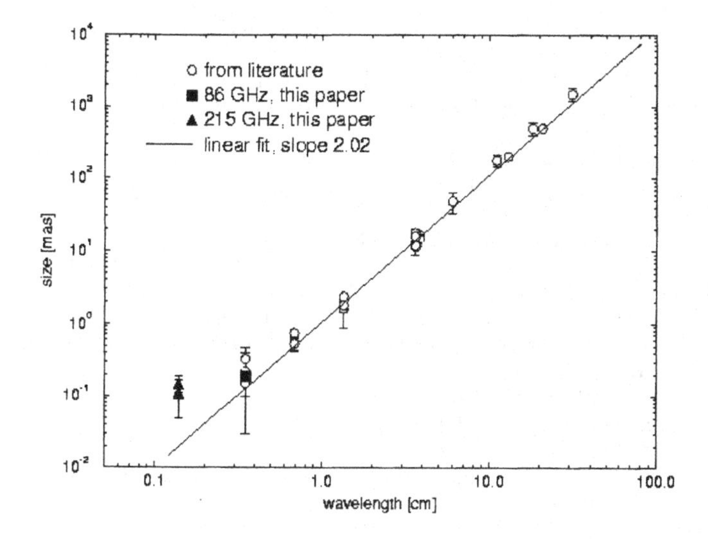

Figure 10. Size of Sgr A* versus observing wavelength. (Taken from Krichbaum et al. 1998).

5. Conclusion

Black holes were introduced into astrophysics as purely theoretical concepts which were needed to explain quasar luminosities. In the last years, strong evidence was found that the nuclei of non-active galaxies harbor large dark point masses. Studies of the properties of the host galaxies show that the evolution of black holes and their host galaxies is closely linked. The best black hole candidates are Sgr A* and NGC 4258 and submm-VLBI will soon approach the event horizon for the black hole in the Galactic Center to finally prove the concept of supermassive black holes.

References

Backer, D.C. and Sramek, R.A. 1999, ApJ 524, 805
Chatterjee, P., Hernquist, L. and Loeb, A. 2002, ApJ 572, 371
Eckart, A. and Genzel, R. 1996, Nature 383, 415
Eckart, A., Genzel, R., Ott, T., and Schödel, R., 2002, MNRAS 331, 917
Fabian, A.C. et al. 2002, MNRAS 335, L1
Falcke, H., Melia, F. and Agol, E. 2000, ApJ 528, L13
Ferrarese, L. and Merritt, D. 2000, ApJ 539, L9
Ford, H.C. et al. 1994, ApJ 435, L27
Gebhardt, K. et al. 2000, ApJ 539, L13
Ghez, A.M., Morris, M., Becklin, E.E., Tanner, A. and Kremenek, T. 2000, Nature 407, 349
Harms, R.J. et al. 1994, ApJ 435, L35
Hazard, C., Mackey, M.B. and Shimmins, A.J. 1963, Nature 197, 1037

Herrnstein, J.R., Moran, J.M., Greenhill, L.J., Diamond, P.J., Miyoshi, M., Nakai, N. and Inoue, M. 1997, ApJ 475, L17
Herrnstein, J.R., Moran, J.M., Greenhill, L.J., Diamond, P.J., Inoue, M., Nakai, N., Miyoshi, M., Henkel, C. and Riess, A. 1999, Nature 400, 539
Kormendy, J. 1998, ApJ 325, 128
Kormendy, J. and Richstone, D. 1992, ApJ 393, 559
Kormendy, J. and Richstone, D. 1995, ARAA 33, 581
Krichbaum, T.P. et al. 1998, A&A 335, L106
Melia, F. and Falcke, H. 2001, ARAA 39, 309
Miyoshi, M., Moran, J., Herrnstein, J., Greenhill, L., Nakai, N., Diamond, P. and Inoue, M. 1995, Nature 373, 127
Reid, M.J., Readhead, A.C.S., Vermeulen, R.C. and Treuhaft, R.N. 1999, ApJ 524, 816
Reid, M., Menten K.M., Genzel R., Ott T., Schödel R., Brunthaler, A. 2003, Astronomische Nachrichten, 324, 505
Schmidt, M. 1963, Nature 197, 1040
Schödel, R. et al. 2002, Nature 419, 694
Tanaka, Y. et al. 1995, Nature 375, 659
Wilms, J., Reynolds, C.S., Begelman, M.C., Reeves, J., Molendi, S., Staubert, R., and Kendziorra, E. 2001, MNRAS 328, L27

YOUNG POWERFUL RADIO SOURCES AND THEIR EVOLUTION

R. FANTI
Dept. of Physics - Via Irnerio 46 - 40126 Bologna - Italy

1. A brief historical overview

1.1. RADIO SOURCES EVOLUTION

Although the study of the physical evolution of powerful extragalactic radiosources is an old item (see e.g. Shklovskii, 1963), to my knowledge the first discussion on it, in terms of comparison of a physical model with the properties of a statistically reprerentative sample of the radio source population, was given by Baldwin at the IAU 87 Symposium in Albuquerque (Baldwin 1982) and is still largely valid nowdays.

Baldwin pointed out the great importance of the study of the distribution of radiosources in the " Radio Luminosity - Linear size plane " (P_r - LS plane), that he called the "*HR diagram of radioastronomers*" and he showed how it strongly constrains the evolutionary behaviour of radio sources. He modified the seminal model of Scheuer (1974), including in it the more recent information on the properties of the source ambient medium, and gave relations describing the source growth and luminosity change with time. I will refer to it as the *Scheuer-Baldwin model* (*SB*).

These relations allowed him to discuss the implications of the distribution in the "P_r-LS" plane of the radio sources pf the 3CR catalogue (Laing et al., 1983). His main conclusion was that during the source life the radioluminosity decrease as source size increases.

1.2. THE "DISCOVERY" OF THE YOUNG SOURCES POPULATION

In the early eighties Kapahi (1981) and Peackock and Wall (1982) pointed out that, besides the classical populations of "flat compact" and "steep extended" radio sources there appeared to be a third population, the "Compact Steep-spectrum Sources" (CSS). These new objects were characterized by high luminosit, sub-Kpc structure and a more or less pronounced low

F. Mantovani and A. Kus (eds.), The Role of VLBI in Astrophysics, Astrometry and Geodesy, 157–174.

frequency spectral curvature, bearing some similarity with the "flat spectrum" population.

Evidences of this population were found earlier from the spectral studies of 3C radio sources by Conway et al. (1963) and from the first long baseline interferometry observations (e.g. Palmer et al., 1971).

On the basis of the (rather poor) structure information available at that time and of the similarity of the counts, Peacock and Wall, analyzing their northern sky catalogue at 2.7 GHz (PW catalogue), concluded: *Thus there are ground for regarding the compact steep-spectrum sources as being more closely related to extended sources than to compact flat-spectrum sources, but much further study of these objects is required before their relationship to the rest of the radio population is clear.*

Similarly, Kapahy had concluded: *The higher fraction of unresolved sources among normal spectrum population in the present 5 GHz sample can arise due to the larger representation in high frequency surveys of compact sources that have a normal spectrum at high frequencies but show a low frequency turnover in their spectra due to synchrotron self absorption.*

Those CSSs, whose spectrum peaks at about 1 GHz, were named **G**Hz **P**eaked **S**ources (GPS).

In those years a number of authors embarked on studies at high resolution (either VLBI or VLA) of this newly discovered population.

Phillips and Mutel (1982) discussed the properties of a small sub-class of this population, which appeared to have double, < 1 kpc, structure. They pointed out the similarities with the large size powerful radio sources and excluded the presence of beaming effects. They concluded: *The close similarity in morphology and spectral index of compact and extended symmetric sources is further strong evidence that the same underlying mechanism is responsible for both types of sources. Spectra of extended sources differ from those of compact symmetric sources principally in the turnover frequency... .. As the lobes evolve and expand to larger sizes, they would become optically thin and the turnover frequencies move downward.*

Subsequently this has beem called the "**Youth scenario**".

Van Breugel et al. (1984) suggested that the CSS may be embedded in a relatively dense interstellar medium associated to the "Narrow Line Region". They proposed that interaction with this medium would explain several properties of the CSS/GPS, as the low frequency turnover by free-free absorption, bright emission lines (through photon-and/or shock ionization) and asymmetric or distorted morphologies. With time it was assumed that this environment is so dense to inhibit the growth in size of these objects and to maintain them small for their whole life. This has been called "**Frustration scenario**".

For about ten years these competing scenarios have been around, perhaps with some more preference for the "frustration scenario", the problem with the "youth scenario" being that (with the subliminal assumption of constant luminosity during the source life) the number of young sources should be given by the ratio of the life in the young phase to the total life, of the order of the ratio of CSS sizes over the size of the larger source. An expected number exceedingly small, against a fraction of CSS, which is found to be > 15 - 20 % in radio catalogues.

With the advent of high resolution instruments (VLA, Merlin, VLBI) several, more or less complete, samples of such sources have been or are being studied in great detail. See, e.g., Gopal-Krishna et al., 1983; O'Dea et al., 1990; Pearson and Readhead 1988 (referred as PR); Fanti et al., 1995 (3CR&PW); Polatidis et al. 1995 and Xu et al. 1995 (the "First Caltech-Jodrell Bank VLBI Survey", referred as $CJ1$); Stanghellini et al., 1998; Augusto et al., 1998; Snellen et al., 1998a; Marecki et al., 1999; Polatidis et al., 2000; Peck and Taylor, 2000); Fanti et al., 2001.

An extended and comprehensive review on the subject has been presented by O'Dea (1998).

2. Radio properties of CSSs/GPSs

2.1. MORPHOLOGY

Many of the CSS/GPS, at all sizes, especially galaxies, show radio emission on both sides of a weak core. At the appropriate resolution most of them appear as scaled down versions of powerful extended radio sources as Cygnus A.

Terms like CSO, MSO have been introduced to indicate different sub–classes of those **O**bjects displaying the two sided (or **S**ymmetric) emission and are either **C**ompact ($LS_{Kpc} < 1$, $CSOs$) or **M**edium sized ($MSOs$, $1 < LS_{Kpc} < 20$), while $LSOs$ is adopted to indicate the powerful **L**arge–sized **S**ymmetric **O**biects, typically FRII (Readhead et al., 1994; Fanti et al. 1995). To be precise, this nomenclature implies the existence of a central core on both sides of which the radio emission is located. Not in all $CSOs$ or $MSOs$ a core has been detected yet, but in most cases the visible morphology suggests that what is observed are actually two lobes, opposite with respect to a (yet un–seen) centre of activity. This two–sidedness is often referred to as "symmetry" though the characteristics of the two (mini)lobes may be quite different. For istance, the arm ratios and the flux density ratios may indeed be quite far from one.

The ratio of the component trasverse size to the total source size, ranges from ≈ 0.03 to ≈ 0.3, with median value 0.12.

A fraction of CSSs, mostly associated with Quasars, show a jet in between the lobes.

In a minority of cases more complex morphologies, including very distorted and one–sided ones, are found.

Some 10% do show weak arc second extended features (see, e.g., Stanghellini et al. 1998) accounting for $\leq 5\%$ of the total flux density. This is suggestive of recurrent cycles of activity, where the extended weak structures may represent the relic of an earlier phase of activity and the CSS is the young re–born radio source (Baum et al., 1990).

2.2. LOW FREQUENCY SPECTRAL TURNOVER

The turnover in the radio spectrum could be attributed to either Synchrotron Self–Absorption (SSA) or Free–Free Absorption (FFA).

Fanti et al. (1990) and O'Dea and Baum (1997) pointed out to a good anti–correlation between source linear size LS and the rest-frame turnover frequency, well represented by the relationship $\nu_m^{rest} \propto (LS)^{-0.65\pm0.05}$ over four order of magnitudes in LS. This anti-correlation is confirmed by adding the sources of new samples, as Snellen (1998a) and Fanti et al. (2001).

This is what expected from SSA. For a single homogeneous component, a well known relation holds between the observed parameters:

$$\nu_m \propto H^{1/5} S_m^{2/5} \theta^{-4/5} (1+z)^{1/5} \tag{1}$$

where ν_m and S_m are the observed peak frequency and flux density, θ the component angular diameter, z the source redshift and H is the magnetic field. It is convenient to re–write the previous equation as a function of the source power P_R at a fixed reference frequency (ν_R) chosen in the thin part of the radio spectrum, and of linear size (LS). In order to do this, I make two assumptions: *1)* the "average" component linear size d is a function of the source total size i.e. $d = d_0(LS/LS_o)^\xi$ ($\xi = 1$ indicates a self–similar morphological source evolution); *2)* the radio source in roughly in equipartition condition so that H is determined as a function of the other parameters.

In the source rest-frame, for a spectral index $\alpha = 0.75$, the equation becomes:

$$\nu_m^{rest}(GHz) \approx 6 \times 10^{-9-2.25\xi}\, d_{0,Kpc}^{-0.75}\, LS_{Kpc}^{-0.75\xi}\, P_{R,W/Hz}^{0.35}\, \nu_{R,GHz}^{0.26} \tag{2}$$

It can be seen that ν_m^{rest} has a moderate dependence on the source power and a strong dependence on d_0 and LS. If all sources had the same P_R and the same coefficient d_0, their ν_m^{rest} would sit on the line $\propto (LS)^{-0.75\xi}$. For

$\xi \approx 1$ the above equation agrees pretty well with the best–fit line of the $P_R - \nu_m$ relationship although data points from different samples show systematic shift from each other. Furthermore the dispersion is large. Scaling all the ν_m^{rest} according to SSA equation (2) to a fixed P_R the systematic shifts between different samples are largely removed. The dispersion instead is reduced only marginally. What is left is roughly justified by the dispersion in the parameter d_0 which relates the source component and total size.

As pointed out by O'Dea and Baum (1997), the good linear correlation $P_R - \nu_m$ does not necessarily represent the evolutionary track that radio sources follow as they age and grow in linear size.

Given the evidence that these sources lay within the Narrow Line Emission Region (NLR) and the strong radio source depolarization (Sect. 2.3), it could be argued, alternatively, that the gas density is high enough that FFA plays a major role (see, e.g., van Breugel, 1984).

Bicknell et al. (1997) have developed a model in which the jets are propagating through a medium with $n_{ext} \propto (LS)^{-\beta}, \beta \approx 1 - 2$, with a density $10 - 100$ cm^{-3} at 1 kpc. In this model the $\nu_m - LS$ relation is explained by FFA due to the ionized gas produced in radiative shocks surrounding the expanding radio source. If FFA is important in determining the spectral turnover in GPSs, then the unavoidable SSA must occurr at frequencies lower than observed and be hidden by the optically thick part of the spectrum.

However, the high densities of gas required by the model are not consistent with the large observed expansion velocities (Sect. 4.1.1).

In view of these arguments it is felt that SSA is the more likely explanation for the low frequency turnover. Furthermore, we take the above result as an indication that the CSS/GPS are close to equipartition conditions.

2.3. POLARIZATION

Usually CSSs are less polarized and display larger rotation measures than large size radio sources.

Examples of large rotation measures ($\geq 1000\ rad\ m^{-2}$ in the source rest frame) are reported in Taylor et al. (1992). Note, however, that there are also sources with negligible rotation measure (see, e.g., Cotton et al., 1997, for the case of 3C138).

Some of them show strong depolarization asymmetries, recalling the well known Laing–Garrington effect for LSO (Garrington & Akujor, 1996).

The above findings imply that the Faraday dispersion Δ is very high, $\approx 2500\ cm^{-3}\ \mu G$ pc, and $n_e H \approx 1\ cm^{-3} \mu G$, some 30 times larger than in typical $LSOs$. In a plot $\Delta - LS$ they continue the trend established for large size radio sources (Garrington & Akujor, 1996). The above trend is likely to represent the radial decline of the gas density external to the radio radiosource. From the comparison with *LSOs*, Garrington & Akujor find $n_{ext} \propto LS^{(-1.1\pm0.1)}$.

2.4. VARIABILITY

CSSs show little or no flux density variations. This, together with the fact that their radio spectrum does not rise at high frequencies, means that they do not contain extremely compact luminous components, like a boosted radio core.

Variability, however, may be present in very small (GPS) sources. Therefore, in order to select "genuine" samples of GPSs or *HFPs* (Sect. 7), it is necessary to obtain "simultaneous" radio spectra so as to reject flat spectrum sources which, due to occasional flux density variations at some frequencies, may produce "erroneously" peaked spectra.

3. Properties of CSS/GPS at IR, optical and X-ray wavelengths

CSS/GPS sources are identified with both galaxies and quasars. They, however, are not evenly distributed in redshift, quasars being found, on average, at a higher redshift than galaxies. Available median values are: $z_G \sim 0.4$ and z_Q ranging from 0.8 (3CR&PW, CJ1+PR) to 1.3 (Stanghellini et al., 1998) to $z_Q \sim 1.8$ for the *HFP* presented in Sect. 7 (see discussion in Sect. 8).

In the recent years significant progresses have been obtained in understanding the properties of CSS/GPS host galaxies at wavelengths different from the radio band. These observations provide important information on both the nature of the host galaxies and on the properties of the environment. A brief summary of the major results is given here below.

3.1. OPTICAL AND NIR CONTINUUM PROPERTIES

Snellen et al. (1996) and O'Dea et al. (1996a) have studied the R-band Hubble relation of GPS hosts and found a dispersion significantly smaller than for 3CR radio galaxies. In addition GPS host galaxies are fainter than the 3CR ones at $z > 0.5$. This is interpreted as due to GPS galaxies not having the extended optical-UV component, aligned with the radio source, present in the powerful high z 3CR hosts The lack of the "aligned emission" in the continuum is confirmed by HST imaging in bands free of emission lines (Axon et al., 2000).

Additional information comes from observations in the NIR, either ground-based (Snellen et al., 1996, de Vries et al., 1998) or with HST (de Vries et al., 2000). In the K band GPS/CSS hosts have absolute magnitudes similar to the hosts of the large radio size FRII radio galaxies, about one magnitude fainter than the brightest cluster galaxies.

Both the Hubble diagram and the redshift - colour observed relationships are consistent with no–evolution or passive evolution models of elliptical galaxies.

3.2. LINE EMISSION

It is known since some years (Gelderman and Whittle, 1994; Morganti et al., 1997) that CSS/GPS have large Narrow Lines (NL) luminosities. In a plot L_{radio} vs L_{NL} they lay close to the region of the large size radio galaxies of similar radio power. However Geldermann and Whittle noticed that the NL profiles are broader and complex and suggest that this is the result of the source interaction with the external medium.

High spatial resolution narrow band imaging centered on either O[II]λ3727 or O[III]λ5007 line have been obtained with the HST (de Vries et al., 1999; Axon et al. 2000). Most of the line emission appears aligned along the source axis. This morphology is discussed in Axon et al. (2000) in terms of photo-ionization from a central AGN (obscured in galaxies) and in terms of radio jet induced shocks by de Vries et al. (1999).

3.3. FIR EMISSION

Early analysis was performed by Heckmann et al. (1992, 1994) using IRAS data ($12.5\mu \leq \lambda_{obs} \leq 100\mu$) and more recently by Fanti et al., (2000) who observed with ISOPHOT in the interval $60\mu \leq \lambda_{obs} \leq 200\mu$ a sample of CSS/GPS radio galaxies and a comparison sample of large size radio galaxies matched in redshift and radio power. No significant difference was found between the two populations. Both are, on average, strong MFIR emitters ($L_{MFIR} \approx 2-5 \times 10^{44}$ erg sec^{-1}). These high luminosities may be due to the UV/X radiation of a hidden AGN reprocessed by the absorbing material, or may be due to star burst activity or both.

3.4. X-RAY EMISSION

O'Dea et al. (1996b, 2000) have observed two GPS radio galaxies (1345+125 and 2352+495) with ROSAT and with CHANDRA. In both cases $L_X \leq 10^{42}$ erg sec^{-1} in the band 0.2 – 2 keV. These observations constrain the pressure of any hot environment to a range too low to frustrate the radio sources.

For a few other low redshift CSS radio galaxies the published results from the ROSAT ALL SKY SURVEY (Brinkmann et al. 1995) can be used. No detections were found, and this points to similar conclusions.

4. Which scenario?

After several years the "frustration scenario" has been around no compelling evidence has come out for a medium as dense as required by it. On the contrary two pieces of evidence point definitely toward CSOs/MSOs being young objects.

4.1. EVIDENCE FOR YOUTH

An obvious way to test the "youth" model is to determine the source age directly. I will then review the present observational evidence in favour of this.

4.1.1. *Lobe proper motions*

So far lobe proper motions have been measured in a handfull of the smallest *CSOs* by several authors (Owsianik and Conway (1998); Owsianik et al. (1998); Owsianik et al. (2000); Polatidis et al.(2000); Stanghellini et al, (2001a); Taylor et al. (2000); Taylor and Vermeulen (1997); Tschager et al., 2000). Velocities between the edges of the two minilobes up to $0.4c$ have been found.

Measures obtained by different authors sometimes disagree. This reflects the difficulties and uncertainties of the measures, since monitoring of the source structure is required for many years, with mas–sized beams. However, in spite of the uncertainties on the exact values of the separation speed, these data prove that at least *some CSOs* are indeed young.

No measurements of this kind exist so far for larger sources and there are little hopes that such data will become ever available.

From the proper motion detection, based on ram pressure arguments, it is possible to estimate, or set upper limits, to n_{ext}. The values found so far indicate $n_{ext} \leq 1 - 2$ cm^{-3} for $LS \leq 100$ pc (Owsianik & Conway, 1998; Owsianik et al., 1998).

4.1.2. *Radiative ages*

Murgia et al. (1999) have fitted the integrated spectra of the 3CR&PW sample of CSSs with a continuous injection synchrotron model. The model fits the data remarkably well over three orders of magnitude in frequency ($\sim$ $0.1 - 200$ GHz). From the high frequency spectral break ($\nu_{\rm br}$), and assuming the equipartition magnetic field (H_{eq}), they have derived the radiative ages ($\tau_{\rm syn}$) of these radio sources.

These ages do not necessarily represent the source age, but rather the radiative age of the dominant component(s) in the source. Only when the lobes, which have accumulated the electrons produced during the whole source lifetime, dominate the source spectrum, the radiative age is likely to represent the age of the source. If instead the spectrum is dominated by a jet or by hotspots, $\tau_{\rm syn}$ represents the permanence time of the electrons in that structure, and it is likely (much) shorter than the source age. In addition jets and hot spots might have their break frequency up–shifted by Doppler effects.

If the analysis is restricted to the lobe dominated sources in the sample, one finds a good inverse correlation between LS and $\nu_{\rm br}$. The larger is a CSS the longer is its (radiative) age. The estimated ages are in the range $\approx (1 \div 100) \times 10^3$ years and the average expansion speed, derived as $LS/\tau_{\rm syn}$, is $(0.34 \pm 0.07)c$. This value might be slightly higher than found for $CSOs$, but a magnetic field just a factor of two lower would provide a better agreement.

An interesting by product of this analysis is that the source magnetic field seems to be within a factor ≈ 2 of the equipartion value. Particularly interesting is the case of 1943+546, where $\tau_{\rm syn}$, derived from fitting the integrated spectrum on the assumption that $H = H_{\rm eq}$, is remarkably close to the kinematic age obtained by Polatidis et al. (2000) (Murgia, priv. comm.). So, it seems that we can be confident that $H_{\rm eq}$ is a fair estimate for the magnetic field.

4.2. EXPANSION VELOCITIES AND AGES OF LSO

The ages and corresponding expansion velocities of LSO are generally estimated with two independent methods:

a) radio spectral studies across the radio lobes and appliation of "standard synchrotron ageing" studies of the relativistic electrons population;

b) studies of the asymmetries in the lengths of the two lobes, due to light travel time differences betwen the advancing and the receiding lobe.

The first method has been applied to several powerful sources of large size (e.g.: Carilli et al., 1991, for Cyg.A; Leahy et al., 1989, and references therein). The weak point is that one has to know the magnetic field, which is generally assumed to be the equipartition one.

The second method does not allow to obtain information on individual sources, as it involves also the angle of the source axis with the line of sight. It rather gives statistical information of the source population, providing strong upper limits to the maximum expansion speed of the lobes. In a recent paper by Arshakian and Longair (2000), following an earlier approch by Scheuer (1995), this method has been modified adding to the light travel time effect also asymmetries intrinsic to the source or due to inhomogeneities in the external medium. These authors estimate now an average lobe speed $v_{lobe} = (0.11 \pm 0.04)c$.

After Arshakian and Longair, the two methods give consistent results. It is remarkable that the speeds obtained for the LSOs are in the same range as those for the CSOs, so that the two classes of sources would have ages which are in a ratio similar to that of their linear sizes (but beware of the fact that the two classes of sources, according to evolutionary models (see section 6), correspond to different jet power.

5. The source distribution in the "P_R - LS" plane

As pointed out earlier (Baldwin 1982) and applied by several authors later (Fanti et al., 1995; Readhead et al. 1996), the LS distribution of radiosources gives valuable constrains on the evolution of their properties as they age, in terms of expansion velocity and radio power versus size.

If one describes the size distribution as $\frac{dN(LS)}{dLS} \propto LS^{-m}$, and assume that the lobe velocity $v_{lobe} \propto LS^x$ and the radio power $P_R \propto LS^{-h}$, a simple ration holds between m, x and h:

$$m = x + hb \tag{3}$$

where b is a coefficient in the range of 1.5 - 2. Its meaning is related to the exponent of the source counts (log S - log N), if the size distribution is computed for all sources of a flux limited sample (as done in Readhead at al., 1996). It is instead, related to the exponent of the Luminosity Function (integral), if the LS distribution is determined considering, in the flux limited sample, only sources brighter than a given $P_{R,min}$ (as done in Fanti et al. 1995). The above equation has to be taken as a zero order relationship, in view of the laws adopted for N(LS), $v_{lobe}(LS)$ and $P_R(LS)$.

The two ways of counting give essentially the same results. Fanti et al. (1995) found $m \approx 0.7 \pm 0.1$ in the 3CR and in the PW catalogues ($0.3Kpc \leq LS(Kpc) \leq 200$). This relationship is more or less confirmed by the new sample of Fanti et al. (2001) and can be extended further down by inclunding the sources of Stanghellini et al. (1998), as done by O'Dea and Baum (1997).

As pointed out earlier by Baldwin (1982), equation (3) requires either a strong speed up of the souces during their life, or a decrease of their radio luminosity or both.

6. Evolutionary models and the data

Current models relate luminosity and expansion velocity of a source to the *jet power*, F_e, and to the external gas density, described as $n_{ext} \propto r^{-\beta}$.

In very simple terms, following the SB approach, the main steps for a *zero order model* are:

i) the *advance speed* of the head of the lobe, v_h, is regulated by balance between *jet thrust* (F_e/c, if the jet is relativistic) and ram-pressure ($\approx n_{ext} m_p v_h^2$);

ii) the lobe is over-pressured and expands side-ways, regulated by ram-pressure at the lateral expansion speed v_{exp};

iii) the lobe volume is obtained by integration of the lobe expansion;

iv expansion losses only are considered (radiation losses are neglected);

v) the energy carried by the jets is partly spent in doing work in the expansion and partly accumulates in the lobes ($U_{lobe} \propto F_e \times t$);

vi) the radio power P_R is computed from equipartition assumptions ($P_R \propto U^{7/4} Vol^{3/4}$).

Assumptions have to be made on the dependence on LS and F_e of the area, A_h, where the thrust is discharged (the *source head*). Baldwin (1982) assumes $A_h \propto LS^2$ (referred as SB *conical model*), and independent on F_e. Begelman (1996, referred as *Be model*) assumes self-similarity in the source growth.

Under the above assumptions it is possible to integrate the relevant equations (a rather boring piece of work) and get analitic solutions. It is found that both P_R and v_{lobe} are power law functions of LS with exponents -h and x which are related in a simple way to the parameter β of the external density.

In the following table I give the relations between the parameters h and x and β.

The model by Kaiser & Alexander (1997), ($K\&A$) gives relations similar to the *Be* model.

I remark that the luminosity evolution (parameter h) is perhaps more uncertain than the velocity evolution, as it is tied to the assumption of equipartition and, for the hot spots, to more uncertain geometrical parameters. Furthermore radiation losses are not considered. This certainly underestimates the decrement of the high frequencies luminosity at large LS values (see Kaiser et al., 1997).

As one can see, β is the key parameter in determining both the advance velocity (x) and the luminosity (h) as a function of size.

The *Be* model does not differ much from *SB* conical model. Both models converge to similar values for $\beta \approx 2$. Furthermore, it can be shown that the *Be* model is recovered brom the *BS* model taking

$$A_h \propto LS^{\gamma} F_e^{\delta} \qquad for \ \ \gamma = \frac{\beta + 4}{3} \qquad and \qquad \delta = 1/3 \tag{4}$$

In these models the lobe luminosity increases ($h < 0$) with LS until $\beta \leq 1.2$, then it starts decreasing while the hot–spot luminosity is always decreasing with LS, ($h > 0$) although slowly, independent of β.

TABLE 1. Evolution Models

Model	x	h_{lobe}	$h_{h.spot}$
SB (con. jet)	$\frac{\beta-2}{2}$	$\frac{11\beta-14}{16}$	0.5
SB (cil. jet)	$\frac{\beta}{2}$	$\frac{7\beta-4}{16}$	0.0
Be $K\&A$	$\frac{\beta-2}{3}$	$\frac{7\beta-8}{12}$	$\frac{\beta+4}{12}$

$v_l \propto (LS)^x, P \propto (LS)^{-h}$

The lobe speed, instead, is always decreasing, or constant, with LS ($x \leq 0$) in the Be and SB conical model, unless the external gas density has a very steep gradient ($\beta > 2$). In SB cilindrical model ($\gamma = 0$), instead, the lobe speed is always increasing with the source size, for $\beta \geq 0$.

A *King law* is generally assumed for the external density profile, namely $\beta \approx 0$ for $r \leq r_0$ (*core radius*) and $\beta \approx 1.5 \div 2$ for $r \leq r_0$. Under these assumptions the lobe luminosity increases ($h < 0$) with LS within the core radius and v_h decreases. Outside the core radius the luminosity declines, while the deceleration is reduced or absent.

The observed linear size and radio luminosity distribution for the 3CR&PW sample ($LS \geq 0.3\ kpc$, $logP_{2.7} \geq 26.0$) give:

$$m \approx 0.7 \pm 0.1 \quad b \approx 2 \pm 0.2$$

Entering these values into eq. (3) along with the dependence on β of x and h (Table. 1) we estimate

$$\beta \approx 1.8 \pm 0.15$$

for both SB conical jet and Be models. This value would indicate that the lobes of the CSSs in the 3CR&PW sample are already out of the core radius, (which appears then to be ≤ 0.5 kpc) in the region where the gas density rapidly decreases with distance.

The lobes are expected to decrease in luminosity as $P \propto (LS)^{-0.5}$ and their speed is expected to be roughly constant.

7. Towards infant sources

We refer here to finding radio sources peaking at even higher frequencies than the present GPSs. It is only observational limitations that have largely prevented, so far, their selection. Their number and their mas morphologies may place stringent constraints on evolutionary models. In fact, according to the $LS - \nu_m$ relation, these sources are expected to be very small, and hence, if the *youth* model is correct, very recently born.

The first announced discovery of sources peaking at very high frequencies came from Edge et al., (1996) who found two radio sources, associated with the X–ray objects RXJ 0952.7+5151 and RXJ 1459.9+3337, peaking at 20 and 30 GHz respectively. These sources, could be < 100 years old.

Edge et al. started a systematic search for such sources, by cross correlating the 87GB at 4.9 GHz (Gregory et al., 1996) and the NVSS catalogue at 1.4 GHz (Condon et al., 1998). They considered radio sources with $S_{5\mathrm{GHz}} > 50$ mJy and more inverted than $\alpha_{1.4}^{5} = -0.5 \quad (S(\nu) \propto \nu^{-\alpha})$. A total of 180 candidates were selected and observed with the Ryle telescope at 15 GHz. A large multi–frequency VLA follow–up was announced for 55 objects, but no further results have become available, to my knowledge.

Following Edge's report, a similar systematic search for such objects has been undertaken by Dallacasa et al. (2000). I'll use their nomenclature and call these objects **H**igh **F**requency **P**eakers (HFP).

Also Dallacasa et al. cross correlated the 87GB and the NVSS catalogue and selected 103 sources with $S > 300$ mJy at 5 GHz and a possibly inverted spectrum ($\alpha_{1.4}^{5} < -0.5$). In order to evercome problems due to flux density variability, all these HFP candidates were re–observed with the VLA simultaneously in 5 different bands from 1.4 to 22.5 GHz, in order to obtain an "istantaneous" radio spectrum. Out of the original list only 43 turned out to be "genuine" $HFPs$ with $S_{5\mathrm{GHz}} \geq 300$ mJy.

The observed median $< \nu_{peak} >$ of the turnovet frequency is ≈ 6.5 GHz, to be compared with $< \nu_{peak} > \approx 1$ GHz in Stanghellini (1998) and $\nu_{peak} \approx 1.8$ GHz in Snellen (1998a). Fifteen objects ($(35 \pm 10)\%$) peak above 10 GHz, 7 of which ($(16 \pm 7)\%$) above the highest observed frequency (22 GHz).

So far the identified objects include 3 galaxies, 20 quasars, 2 BL Lac objects, and 8 stellar objects which require further investigation. The remaining 10 are empty fields on the Sky Survey. The fraction of QSS is therefore not less than 50%, but their number could even increase when the optical identification will be complete. Their redshift seems even higher than in the Stanghellini et al. (1998) GPS sample, although the difference is not statistically significant. This is one of the situations in which boosting may become important (Sect. 8).

8. Problems

In spite of the success in describing the gross features of the source distribution in the "$P_R - LS$" plane, a number of questions and problems were raised in the last few years. In our opinion the main ones are:

a) why don't we see in the LS distribution any indication of the *core radius*?

b) are there any short lived CSSs/GPSs ?

c) why is the galaxies/quasar ratio so much changing with redshift?

d) why CSSs/GPSs have *narrow line* and *FIR* luminosities so similar to those of LSOs?

We examine briefly each of these questions.

a) In Sect. 6 we found that the LS distribution of CSSs/GPSs points to a $\beta \approx 1.5 \div 2$ down to $LS \approx 20$ pc. For sources within the core radius ($\beta \approx 0$) the models predict an increase with LS of the lobe luminosity ($h \approx 0.75 \div 0.9$) and a decrease of the advance speed ($x \approx -0.66 \div -1.0$). With these values we derive $m \approx -1.2 \div -2$ which implies a strong drop in the number of sources of small diameter, in contrast with what is observed in the Stanghellini et al. (1998) and Snellen et al. (1998a) samples, where, instead, $m \approx 0.7$. One could argue that at small sizes the source luminosity becomes dominated by the hot–spot luminosity, which is always increasing with decreasing size ($h > 0$). However the models predict $m \approx 0$, still in disagreement with what is found.

A way to overcome the problem is to say that the gas density never flattens as we approach the galaxy nucleus. Alternatively one could suggest that at small LS another population of radio sources appears, not related via evolution to the LSO population.

A hint for an intruder population can be seen in some irregularities in the LS distribution. The *B3–VLA CSS* sample of Fanti et al. (2001) shows a marginally significant deficiency of sources of ≈ 1 kpc size. Although at a less significant level, the PW sample, which is independent of the previous one, does show the same effect. We remind that also O'Dea & Baum (1997) pointed out something similar. So it could be that at a few kpc we see the separation of two different populations.

We think that a population of small *frustrated sources* can be excluded on the basis of the fast motions just observed in the smaller sources. More realistic possibilities are: *i)* a population of short lived sources; *ii)* a population of intrinsically faint sources, whose luminosity is boosted by relativistic beaming effects. These two possibilities are related to other questions discussed below.

b) Following the work of Murgia et al. (1999), a spectral study was undertaken (Zappacosta et al. in prep.) also for the *B3–VLA CSS* sample.

Surprisingly a minority of sources ($\leq$ 10%) showed radio spectra typical of a *dying source*, namely with a strong steepening beyond the break frequency. These spectra were well fitted by either the Kardashev (1962) or the Jaffe & Perola (1974) model, implying a long time of inactivity of the central source of energy and suggesting that they may be transient objects (see also Readhead et al. 1994). Actually it is also possible that these sources are not really dead, but are just undergoing a temporary phase of inactivity. Reynolds & Begelman (1997) have discussed this *intermittence scenario* and have shown that it would produce irregularities, as gaps or plateaus, in the LS distribution, similar to what mentioned in point *a)*.

c) CSSs/GPSs are identified with both galaxies and quasars. However GPS quasars dominate at high redshifts ($z \geq 1$), while galaxies are typically at low redshifts ($z \leq 0.7$) (O'Dea, 1990; Snellen et al. 1999). How does this fit in the framework of the unified models by orientation?

Snellen et al. (1999) propose that at high redshifts the GPS galaxies and quasars are not unified by orientation but are two distinct classes of objects, or that there are only *GPS naked quasars.* O'Dea (1998) points out that GPS are selected at centimeter wavelength, where the contribution from Doppler boosting in the jet is larger, thus making them visible over larger distances (in a flux density limited catalogue) as compared to the mis–oriented GPS galaxies.

It might be possible that the lack of evidence for core radius from the LS distribution is apparent and is due to an excess of small size very boosted sources. Actually Stanghellini et al. (2001b) show that in their GPS sample most of the quasar radio emission is jet dominated and, therefore, possibly boosted. The same seems true in the HFPs of Dallacasa et al. (2000).

However, where are at low redshifts the "oriented" galaxies? If boosting is at work, the jet and core luminosity in these objects might overwhelm the lobe luminosity, the source spectrum become wider or flatter, and the source not recognized as GPSs (see Snellen et al. 1998b).

d) A problem for the *youth* model, pointed out by Fanti et al. 2000, (and earlier by O'Dea, 1998) is that if the radio luminosity decreases with increasing source size, CSSs/GPSs and LSOs of similar radio luminosities, but a factor ≈100 different in size, would have "radio engines" of power different by a factor 10 or so (CSSs/GPSs being less powerful). In the past there have been strong suggestions that the jet power, F_e, is proportional to the bolometric luminosity of the AGN (e.g., Rawling & Saunders, 1991) and therefore to the NL and $MFIR$ luminosities, which are a reprocessed fraction of it. On this basis one would expect the CSSs/GPSs to be definitely less powerful in the $MFIR$ and in the NL emission, contrary to what seen in the observations (Fanti et al. 2000; Morganti et al. 1997)

A way to explain this apparent contradiction would be that there is an additional source of ionization and heating arising from conversion of a fraction of the jet power when the jet interacts with the interstellar medium (as, e.g., in the model of Bicknell et al. 1997). This may be more effective at distances close to the nucleus and less so further out as the source ages. Such an effect should roughly compensate for the lower radiation power from the central engine in CSSs/GPSs. Alternatively O'Dea (1998) suggest that in young sources the ambient gas, not yet settled in a disk, has a much larger covering factor and therefore intercepts a larger fraction of the continuum radiation of the AGN.

9. Conclusions

At present there seem to be little doubts that the small (CSS/GPS) and the large (LSO) sources are related via an evolutionary scenario. First order evolutionary models seem to justify well the gross properties of the radio source distribution in the "$P_R - LS$" plane. The space distribution of the ambient gas can be traced from the source LS distribution.

However we still require improvements on the information we have.

The possibility that the sources selected at small sizes (≤ 1 kpc) contain a significant fraction of boosted objects has to be further investigated.

The *dying sources* or the *intermittence* are other interesting scenarios to be examined.

Finally one should perhaps try to adapt the Bicknell et al. (1997) model to the physical conditions of high advance velocities of the radio lobes, in order to justify the radio power – line and FIR luminosity, which appears to be independent of source size.

References

Arshakian, T. G., & Longair, M. S. 2000, *Mon. Not. R. Astron. Soc*, 311, 846

Augusto, P., Wilkinson P.N., Browne I.W.A. 1998, *Mon. Not. R. Astron. Soc*, 299, 1159

Axon, D. J., Capetti, A., Fanti, R., Morganti, R., Robinson, A., & Spencer, R. E. 2000, *Astron. J.*, 120, 2284

Baldwin, J. 1982, in "Extragalactic Radio Sources", in IAU Symp. 97, eds D.S. Heeschen & C.M. Wade (Dordrecht: Reidel), 21

Baum, S.A., O'Dea, C.P., Murphy, D.W., de Bruyn, A.G. 1990, *Astron &Astrophys*, 232, 19

Begelman, M. C., 1996, in " CyA: Study of a Radio Galaxy" eds C. Carilli & D. Harris (Camb.Univ. press), 209

Bicknell, G., Dopita, M. A., & O'Dea, C. P. 1997, *Astrophys. J*, 485, 112

Brinkmann, W., Siebert, J., Reich, W., et al. 1995, *Astron &Astrophys*, 109, 147

Carilli C. L., Perley R. A., Dreher J. W., Leahy J. P. 1991, *Astrophys. J*, 383, 554

Condon, J. J., Cotton, W. D., Greisen, E. W., Yin, Q. F., Perley, R. A., Taylor, G. B., & Broderick, J. J. 1998, *Astron. J.*, 115, 1693

Conway R.G., Kellermann K.I., & Long R.Y., 1963, *Mon. Not. R. Astron. Soc*, 125, 313

Cotton W.D., Dallacasa D., Fanti C., Fanti R., Fooley A.R., Schilizzi R.T., Spencer R.E., 1997, *Astron. Astrophys*, 325, 493
Dallacasa, D., Stanghellini, C., Centonza, M., & Fanti, R. 2000, *Astron &Astrophys*, 363, 887
de Vries, W.H., O.Dea, C.P., Baum, S.A., et al. 1998, *Astrophys. J*, 503, 156
de Vries, W .H., O'Dea, C. P., & Baum, S. A. 1999, *Astrophys. J*, 526, 27
de Vries, W. H., O'Dea, C.P., Barthel, P. D., Fanti, C., Fanti, R., & Lehnert, M. D. 2000, *Astron. J.*, 120, 2300
Edge, A.C., Jones, M., Saunders, R., et al 1996, The 2nd workshop on GPS and CSS radio sources - Leiden 30 Sept.– 2 Oct. 1996, I.A.G. Snellen et al eds., p. 43
Fanti, C., Fanti, R., Schilizzi, R. T., Spencer, R. E., Rendong, N., Parma, P., van Breugel, W. J., & Venturi, T., 1990, *Astron &Astrophys*, 231, 346
Fanti, C., Fanti, R., Dallacasa, D., Schilizzi, R. T., Spencer, R. E., & Stanghellini, C. 1995, *Astron &Astrophys*, 302, 317
Fanti, C., Pozzi, F., Fanti, R., Baum, S. A., O'Dea, C. P., Bremer, M., Dallacasa, D., Falcke, H., de Graauw, T., Marecki, A., Miley, G., Röttgering, H., Schilizzi, R. T., Spencer, R. E., & Stanghellini, C. 2000, *Astron &Astrophys*, 358, 499
Fanti, C., Pozzi, F., Dallacasa, D. Fanti, R., Gregorini, L., Stanghellini, C., & Vigotti, M. 2001, *Astron &Astrophys*, 369, 380
Garrington, S.T., Akujor, C.E. 1996, in "Extragalactic Radio Sources", IAU Symp. 175, R. Ekers et al., eds., p.77
Gelderman, R., Whittle, M. 1994, *Astrophys. J. Suppl.* 91, 491
Gopal–Krishna, Patnaik, A. R., & Steppe, H., 1983, *Astron &Astrophys*, 123,107
Gregory, P. C., Scott, W. K., Douglas, K., & Condon, J. J. 1996, *Astrophys. J. Suppl.*, 103, 427
Heckman, T.M., Chambers, K.C., Postman, M. 1992, *Astrophys. J*, 391, 39
Heckman, T.M., O'Dea, C.P., Baum, S.A., et al. 1994, *Astrophys. J*, 428, 65
Jaffe, W. J., & Perola, G. C., 1974 *Astron &Astrophys*, 26, 423
Kaiser, C. R., & Alexander, P. 1997, *Mon. Not. R. Astron. Soc*, 286, 215
Kaiser, C. R., Dennet-Thorpe J., & Alexander, P. 1997, *Mon. Not. R. Astron. Soc*, 292, 723
Kapahi, J. K. 1981, *Astron &Astrophys*, 43, 181
Kardashev N. S., 1962, Sov. Astron., 6, 317
Laing, R. A., Riley, J. M, & Longair, M. S., 1983, *Mon. Not. R. Astron. Soc*, 204, 151
Leahy, J. P., Muxlow, T. W. B., & Stephens, P. W. 1989, *Mon. Not. R. Astron. Soc*, 239, 401
Marecki, A., Falcke, H., Niezgoda, J., Garrington, S. T., & Patnaik, A. R. 1999, *Astron &Astrophys*, 135, 273
Morganti, R., Tadhunter, C. N., Dickson, R. & Shaw, M., 1997, *Astron &Astrophys*, 326, 130
Murgia, M., Fanti, C., Fanti, R., Gregorini, L., Klein, U., Mack, K.-H., & Vigotti, M. 1999, *Astron &Astrophys*, 345, 769
O'Dea, C. P., Baum S.A., Stanghellini C., 1990, *Astrophys. J.*, 380, 66
O'Dea C. P, Stanghellini C., Baum S.A., Charlot S., 1996a, *Astrophys. J*, 470, 806
O'Dea, C. P., Worrall, D. M., Baum, S. A., Stanghellini C., 1996b, *Astron. J.*, 111, 92
O'Dea, C. P., & Baum, S. A. 1997, *Astron. J.*, 113, 148
O'Dea, C. P. 1998, *Pubbl. Astron. Soc. Pac.*, 110, 493
O'Dea, C. P., de Vries, W. H., Worrall, D. M., Baum, S. A., & Koekemoer, A. 2000, *Astron. J.*, 119, 478
Owsianik, I., Conway, J. E., & Polatidis, A. G. 1998, *Astron &Astrophys*, 336, L37
Owsianik, I., & Conway, J. E. 1998, *Astron &Astrophys*, 337,69
Owsianik, I., Conway, J. E., & Polatidis, A. G. 2000, IV EVN/JIVE Symp., eds. M.A. Garrett, R.M. Campbell, L.I. Gurvits, *New Astron. Review*, p. 669
Palmer H.P., Rowson B., Anderson B., Donaldson W., Miley G.K., 1971 *Nature*, 213, 789
Peacock, J. A., & Wall J. V. 1982, *Mon. Not. R. Astron. Soc*, 198, 843

Pearson, T. J., & Readhead, A. C. S. 1988, *Astrophys. J*, 328, 114
Peck A.B., Taylor G.B., 2000, *Astrophys. J.*, 534, 90
Phillips, R. B., & Mutel, R. L. 1982, *Astron &Astrophys*, 106, 21
Polatidis, A. G., Wilkinson, P. N., Xu, W., Readhead, A. C. S., Pearson, T. J., Taylor, G. B., & Vermeulen, R. C. 1995, *Astrophys. J. Suppl.*, 98, 1
Polatidis, A., Wilkinson, P. N., Xu, W., Radhead, A. C. S., Pearson, T. J., Taylor, G. B., & Vermeulen, R. C., 2000 IV EVN/JIVE Symp., eds. M.A. Garrett, R.M. Campbell, L.I. Gurvits, New Astron. Review, p. 657
Rawling, S., & Saunders, R. 1991, Nature, 349, 438
Readhead, A. C. S., Xu, W., Pearson, T. J., Wilkinson, P. N., & Polatidis, A. G., 1994, in Compact Extragalactic Radio Sources, Zensus A., Kellermann K.I. eds, NRAO
Readhead, A. C. S., Taylor, G. B., Xu, W., Wilkinson, P. N., & Polatidis, A. G. 1996, *Astrophys. J*, 460, 612
Reynolds, C. S., & Begelman, M. C. 1997, *Astrophys. J*, 487, L135
Scheuer, P. A. G. 1974, *Mon. Not. R. Astron. Soc*, 166, 513
Scheuer, P. A. G., 1995, *Mon. Not. R. Astron. Soc*, 277, 331
Shklovski I.S., (1963), *Sov. Astron*, **6**, pp. 465
Snellen, I.A.G., Bremer, M.N., Schilizzi, R.T., and Miley, G.K. 1996, *Mon. Not. R. Astron. Soc*, 166, 513
Snellen, I. A. G., Schilizzi, R. T., de Bruyn , A. G., Miley, G. K., Rengelink, R. B., Röttgering, H. J., & Bremer, M. N. 1998a, *Astron &Astrophys Suppl. Ser.*, 131, 435
Snellen, I. A. G., Schilizzi, R. T., de Bruyn, A. G., & Miley, G. K. 1998b, *Astron &Astrophys*, 333, 70
Snellen, I. A. G, Schilizzi, R. T., Bremer, M. N., Miley, G. K., de Bruyn, A. G, & Röttgering, H. J. A. 1999, *Mon. Not. R. Astron. Soc*, 307, 149
Stanghellini, C., O'Dea, C. P., Dallacasa, D., Baum, S. A., Fanti, R., & Fanti, C. 1998, *Astron &Astrophys Suppl. Ser.*, 131, 303
Stanghellini, C. Dallacasa, D., Bondi, M., Liu Xiang, 2001a, V EVN/JIVE Symposium, eds. J.E. Conway, A.G. Polatidis, R.S. Booth, Y. Pihlström, p. 99
Stanghellini, C. Dallacasa, D., O'Dea C.P., Baum S.A., Fanti R., Fanti C., 1991b, *Astron. Astrophys.*, 377, 377
Taylor G.B., Inoue M., Tabara H., 1992, *Astron &Astrophys*, 264, 421
Taylor, G. B., & Vermeulen, R. C., 1997 *Astrophys. J*, L485, L9
Taylor, G. B., Marr, J. M., Pearson, T. J., & Readhead, A. C. S. 2000, *Astrophys. J*, 541, 112
Tschager W., Schilizzi R. T., Röttgering H. J. A., Snellen I.A.G., & Miley G. K. 2000, *Astron &Astrophys*, 360, 887
van Breugel, W. J. M., Heckman T., & Miley, G. K. 1984, *Astron. J.*, 89, 5
van Breugel, W. J. M., 1984, IAU Symp.110 "VLBI and Compact Radio Sources", eds. Fanti, Kellermann and Setti, Reidel, p. 59
Xu, W., Readhead, A. C. R., Pearson, T. J., Polatidis, A. G., & Wilkinson, P. N., 1995, *Astrophys. J. Suppl.*, 99, 297

AGN UNIFICATION; PAST, PRESENT AND FUTURE

I.W.A. BROWNE
University of Manchester
Jodrell Bank Observatory,
Macclesfield,
Cheshire, SK11 9DL,
UK

1. Introduction

Unification is the pursuit of simple models which can account for the observed diversity in AGN properties. Historically the first "unified schemes" focussed on explaining things solely in terms of orientation effects. AGN are not spherically symmetric and projection, Doppler boosting and dust extinction can alter drastically the observed properties of an object depending on the direction from which it is viewed.

When an emitting body is moving relativistically the radiation received by the observer is a very strong function of the angle between the line of sight and the direction of motion of the emitter. The observed flux density S_{obs} is related to that in the emitted frame S_{em} by

$$S_{obs} = S_{em}\delta^{(3-\alpha)},$$

where δ is the Doppler factor and α is the spectral index defined S $\propto \nu^{\alpha}$. (Note that, for a continuous jet, the exponent is (2-α) not (3-α)). The Doppler effect changes the energy and frequency of arrival of the photons while relativistic aberration changes their angular distribution. Lorentz factors of in the range 5 to 10 can boost the flux density by factors $\sim$1000.

Observational evidence that some radio-emitting regions have bulk relativistic motions comes from the observation of superluminal motions in the cores of many radio sources (Falcke, this volume). The pronounced jet/anti-jet asymmetries which persist up to $\geq$100 kpc from the core in some high luminosity radio sources suggest that these jets still move relativistically up to the point where they interact with the ISM and form a hotspot.

F. Mantovani and A. Kus (eds.), The Role of VLBI in Astrophysics, Astrometry and Geodesy, 175–189.

In the rest of this chapter I will outline the historical development of AGN unified models, summarize the current state of affairs and speculate on how such models may be refined in the future. Having neglected the study of AGN for the last decade brings its advantages and disadvantages; I approach the subject with a fresh enthusiasm but also with big gaps in my knowledge of the literature for which I apologize in advance.

2. History

2.1. RADIO UNIFICATION

As far as I am aware the first attempt at AGN unification was by Rowan-Robinson (1976) who tried to unify Seyfert galaxies and radio sources. He recognized the importance of dust extinction and infrared emission from the dust but did not include relativistic beaming. It was the observation of superluminal motions in flat spectrum radio sources and its interpretation in terms of bulk relativistic motion that stimulated the first successful unified models. Blandford and Rees (1978) laid the foundation for such models. They concentrated on radio-loud objects only, while Scheuer & Readhead (1979) proposed that radio core-dominated quasars and radio-quiet quasars could be unified – the former being the beamed versions of the latter. Orr & Browne (1982) pointed out that the Scheuer & Readhead scheme would not work because observations with the VLA and MERLIN were showing that core-dominated quasars nearly all had extended (isotropic) radio emission and hence there was no way that un-beamed versions (the parent population) of the core-dominated radio sources could be radio-quiet. Orr & Browne looked for a non-radio-quiet parent population and showed that the statistics were consistent with the idea that core-dominated and lobe-dominated quasars could be one and the same.

The idea of looking for an un-beamed parent population for beamed BL Lacs was also pursued (Browne, 1983, Antonucci & Ulvestad, 1985). FR1 radio-galaxies were identified as the most likely candidates. Similarly the idea that FR2 radio-galaxies and quasars might be related was widely discussed (see for example Scheuer (1987)), culminating in a presentation of a full FR2/quasar unification scheme by Barthel (1989). What distinguishes quasars from radio-galaxies is that the former have strong broad emission lines and a dominant blue nuclear continuum component to their emission. The Barthel proposal was that there existed an optically thick axisymmetric region, possibly a "molecular torus", which hides the broad lines and nuclear continuum from view in the radio-galaxies but allows a un-obstructed view in the quasars. The progression with increasing angle to the line of sight is illustrated in Figure 1. It is now widely accepted that this basic picture approximates to the truth, though there are many

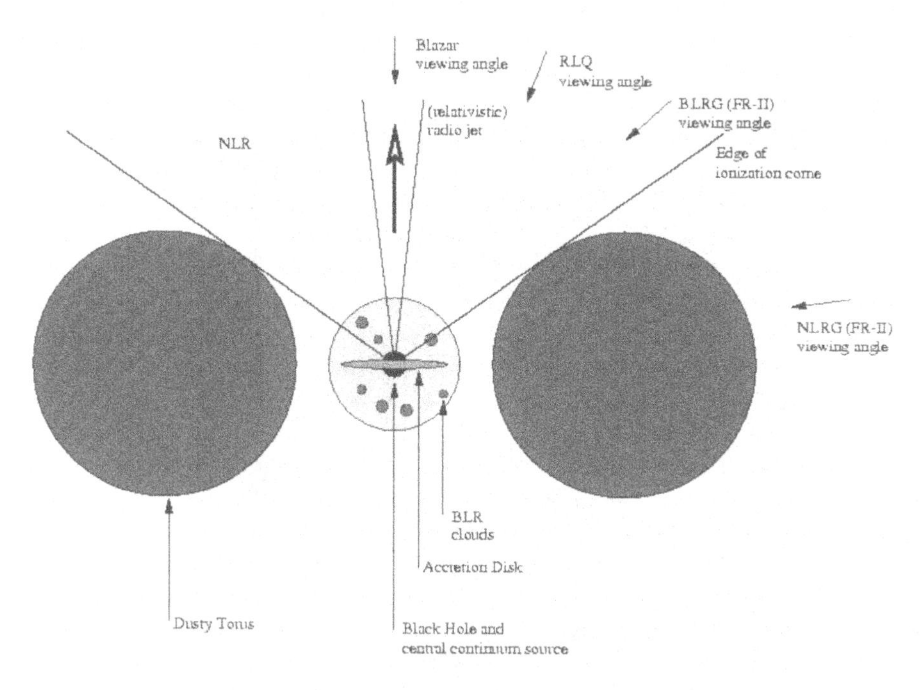

Figure 1. Unification scheme for blazars, radio-loud quasars and FR2 radio galaxies after Chris Reynolds.

detailed refinements required to make a model which is consistent with all the available observational data.

2.2. GENERAL AGN UNIFICATION

The initial stimulus for unified models came from the observation of superluminal motion and the realization that orientation could change drastically the observed properties of an object. But since there was a cross-fertilization between the radio and general AGN developments, I will briefly summarize the AGN unification history. During the late 1980s the idea that orientation was important for all AGN became widely accepted. Lawrence & Elvis (1982) were one of the first to argue for axisymmetric extinction in Seyfert galaxies as an explanation of the systematic differences in X-ray properties between the narrow-line Seyfert 2 type galaxies and the broad-line Seyfert 1s. The breakthrough occurred when Antonucci & and Miller (1985) reported that the Seyfert 2 galaxy NGC 1068 showed broad emission lines when observed in plane polarized light. They interpreted the result in terms of the polarization being a result of scattering off electrons or dust and that what was being observed in the scattered light was a normal broad-line emission (BLR) that was hidden from direct view. Their results

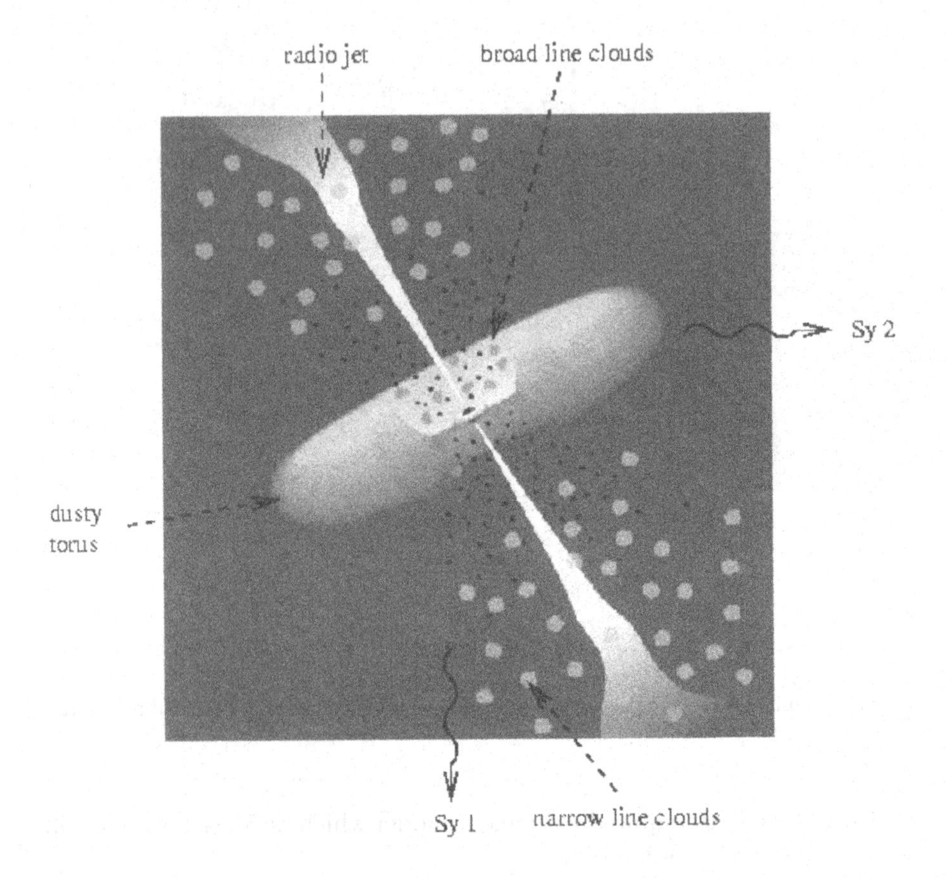

Figure 2. AGN unification diagram from Urry & Padovani (1995) and annotated by M. Voit.

led to proliferation of unification work, culminating in the adoption of a "standard" AGN model as depicted in Figure 2. The key ingredients are an optically thick "torus", a BLR and a source of UV continuum which is hidden from view from certain directions by the torus. The narrow line emission region (NLR) being of larger size than the torus can be seen from all viewing angles. Of course, something has to happen to the energy intercepted by the torus; it raises the temperature of the dust within the torus and this warm dust then re-emits the energy in the infrared part of the spectrum. This nicely accounts for the fact that many AGN have peaks in their SEDs in the infrared.

3. Radio-loud objects; the current situation

The consensus view is that there are two complementary unified models; there is one unifying BL Lac objects with FR1 radio-galaxies and another unifying Quasars and FR2 radio-galaxies. There are several good reviews of the subject (e.g. urry & Padovani, 1995; Padovani, 1999) and I will not try and duplicate all the material contained in them, but simply highlight some of the key features. I will also briefly explore the possibility that FR2s may evolve into FR1s and thus try and introduce a further degree of unification.

3.1. EVIDENCE IN FAVOUR OF BL LAC/FR1 UNIFICATION

Though there has been plenty of evidence for relativistic motion in the jets of BL Lac objects from VLBI detections of superluminal motions (e.g. Mutel et al., 1991), evidence for the corollary– relativistic motion in the FR1 parent population – has been harder to come by. The most convincing results are for the low luminosity nearby radio-galaxy M87. Biretta et al. (1995) have detected superluminally moving components in its jet using multi-epoch VLA 15 GHz observations. More recently (Biretta et al., 1999) have obtained similar results from optical HST observations. Direct observational evidence for bulk relativistic flows in the cores or jets of other FR1s is rare but there is indirect evidence from the asymmetry of small-scale nuclear jets that there is relativistic motion in these too.

Population statistics have been used as supporting evidence for unification. Padovani & Urry (1991) have shown that one can successfully predict the BL Lac (beamed) luminosity function by taking the luminosity function for FR1 radio-galaxies and use the Doppler boosting equation to work out what the luminosity function should be for the beamed core luminosity. Some distribution of Lorentz factors is assumed and it is found that the statistics fit reasonably well for the range factors which is consistent with those deduced from the observed superluminal motions measured in BL Lac objects.

The observation that BL Lac and a couple of other similar objects have weak broad lines (Vermeulen et al, 1995) raises the interesting question of whether or not all FR1 radio-galaxies have BLRs, but hidden by obscuring disk or torus. A recent finding of a correlation between optical nuclear and radio core luminosities (Chiaberge et al., 2001) argues for the beaming hypothesis and against there being significant nuclear extinction (and hence against there being a hidden BLR). The occasional occurrence of BLRs in BL Lacs is still a bit of a mystery.

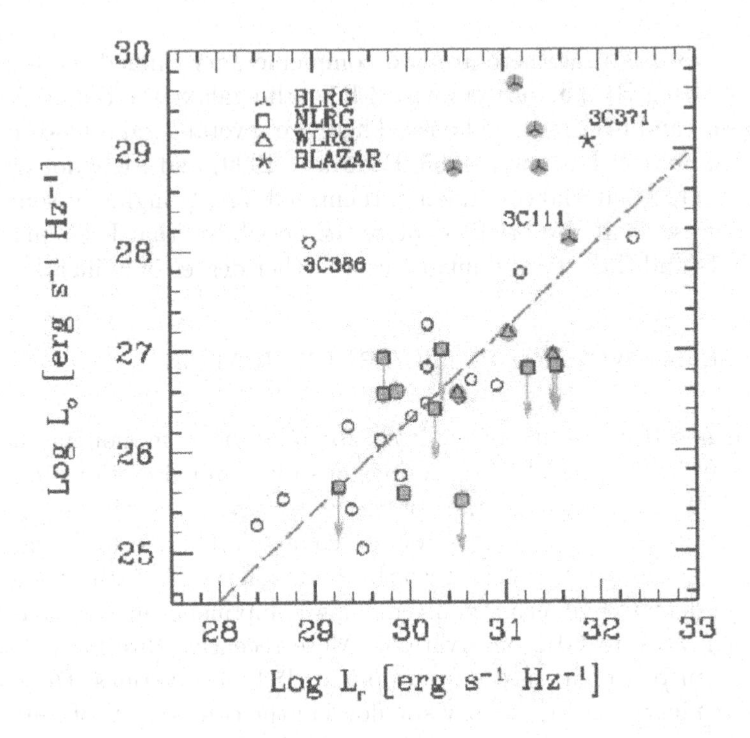

Figure 3. The correlation between radio core luminosity and optical nuclear luminosity, after Chiaberge et al., 2001.

3.2. EVIDENCE FOR RADIO-GALAXY/QUASAR UNIFICATION

The detection of polarized (scattered) broad emission lines in most FR1 narrow-line radio-galaxies would leave no doubt about unification. But this has not been possible. A broad polarized MgII line has been detected in the prototype FR2 radio-galaxy, Cygnus A by Ogle et al. (1997) – which is reassuring. However, detections of such lines in FR2 radio-galaxies is still a rarity.

There are other, more indirect, predictions of the scheme that can be tested. Not all give unqualified support to the scheme:

- The far infrared emission of matched FR2 radio-galaxies and quasars should be the same since this emission should be isotropic. The results are a little ambiguous but are not inconsistent with predictions (van Bemmel et al., 2000).
- The relative numbers of FR2 radio-galaxies and quasars should give the un-obscured cone angle (2:1 $\Rightarrow$ a half-cone angle $\sim 45°$) which in

turn should be related to the average projected linear sizes of objects in the two classes. The results are somewhat inconclusive, but can be said to be consistent with the predictions (see for example Gopal-Krishna et al. (2000)).

- The observed strength of narrow emission lines, which are produced well away from the nuclear obscuring torus, should be the same in all directions; i.e. indistinguishable in radio-galaxies and quasars. Jackson and Browne (1991) tried to test this prediction using the [OIII] luminosity and found that the quasars were more luminous than the matching radio-galaxies and thus concluded that either [OIII] emission was not isotropic or that there were problems with the unification scheme. More recently the test has been repeated (Hes et al., 1993) with [OII] luminosities – the [OII] line being of lower ionization originates further from the nucleus than [OIII] – and in this case the luminosities of the two groups are indistinguishable and things look better for the scheme.
- The cluster environments of quasars and radio-galaxies should be the same. The most recent evidence from a study of low redshift 3CR radio-galaxies and quasars does support unification; in the range $0.15 < z < 0.4$ (Harvanek et al., 2001) the radio-galaxies are found in richer environments than quasars. At higher redshift, however, the environments appear to be the same.

Barthel's original paper concentrated on quasars and radio-galaxies in the redshift range $0.5 < z < 1.5$ and it appears that radio galaxy/quasar unification in its simplest form works well in this range.

3.3. UNIFICATION AS A TOOL

3.3.1. *Using unification to probe the structure of AGN*

There has been a long-standing hope that unified models would become sufficiently well-established that the measurement of some easily observed parameter could be used to give an orientation of the AGN axis of symmetry with respect to the line of sight to the observer. Then any correlations of other AGN properties with orientation would provide information on the distribution of different AGN components with respect to this axis of symmetry. What are suitable orientation indicators in radio-loud AGN? The longest established is the radio-core prominence parameterized as R, the ratio of core to extended radio emission (Orr & Browne, 1982), or the core fraction F_c (Kapahi & Saikia, 1982). The suggestion by Wills & Browne (1986), based on the interpretation of an anti-correlation between R and H-β FWHM, that quasar BLRs were rotating disks is a good illustration of how unification has been used to learn something useful. Disk rotation velocities deduced from line widths are now being used as indicators for black hole masses in quasars (McClure & Dunlop, 2001).

Wills & Brotherton have proposed R_{5000}, the ratio of radio-core flux density to the 5000Å optical continuum flux density, as an alternative orientation indicator for quasars. Empirically, this seems to work as well as, or even better than, R in the sense that R_{5000} correlates with other observables like H-β FWHM just as well as R. The implication is that the 5000Å flux density of a quasar itself does not depend much on orientation. This is a perplexing result because there are several strands of evidence to suggest that the optical continuum is highly anisotropic, even for objects whose emission in the optical is not dominated by non-thermal radiation; i.e. non-blazars.

- Browne & Wright (1984) showed that flat spectrum quasars are systematically brighter than steep spectrum quasars.
- The combination of evidence indicating that narrow line emission is isotropic – e.g. the correlation of jet power deduced from extended radio emission with narrow line luminosity – and a strong anti-correlation of [OIII] and [OII] equivalent widths with R, suggest that optical continuum is a strong function of orientation. (Jackson & Browne, 1991; Baker & Hunstead, 1995).
- The optical continuum is likely to be thermal emission from the AGN accretion disk and theory would suggest that this should be anisotropic.

Over the years there has been an impressive amount of work done trying to use orientation indicators to learn more about the internal structure of AGN. Particular attention has been devoted to measuring different emission line properties. In addition to work quoted above, that by Corbin (1991,1995), Boroson & Green (1992), Baker & Hunstead (1995) and Brotherton (1996) deserves mention. Though he ignores orientations in his analysis, the approach adopted by Boroson (2002) is worth particular attention: he has used Principal Component Analysis on samples of quasars with high quality optical spectra to try and identify key features that differentiate different types of object. Components are identified that, for example, produce a clear separation between radio-loud and radio-quiet objects. Boroson also presents evidence that points to high black hole mass being a necessary condition for radio loudness. Important factors that must be built into more comprehensive unified models have been identified.

However, despite the wealth of data in optical spectra and the best endeavours of a lot of people, progress in understanding the inner workings of AGN has been slow over the last decade. Perhaps the interplay of quite complex astrophysics with a multiplicity of selection effects is the reason for the slowness of progress. Some further thoughts on this matter are presented below.

3.3.2. *Unification and radio source population statistics*
Radio source counts and luminosity functions have long been used as consistency checks for unification models (Orr & Browne, 1982; Padovani and Urry, 1991). In a novel approach, Jackson & Wall (1999) turn the argument around; they use the knowledge unified models to relate the relative numbers of objects appearing in surveys made at different frequencies in order to exploit the source counts at these different frequencies to gain knowledge of the luminosity functions for different types of objects. Refining the parameters of the unified models comes as a bye-product. This holistic approach, combining physical model with population statistics, has a lot of potential and is discussed further below.

4. Unification; future prospects

In this section a broader view of unification will be taken. The aim will be to explore ways in which the whole diversity of AGN properties might be incorporated into a coherent picture. What are the interesting questions one might address?

1. Is it really necessary to have separate FR1 and FR2 unified models?
2. Can one finally exploit unified models to get a better understanding of AGN structure?
3. What is the origin of the large range in peak frequencies in blazar SEDs?
4. Can we relate AGN activity to the formation and evolution of galaxies?
5. What is the origin of the radio-quiet/radio-loud dichotomy? Is there really a dichotomy?

A lot of what follows consists of some of my partly-formed ideas and speculations. The excuse for including it is to encourage people to look at the big picture. The hope and expectation is that they can do better than the author.

4.1. UNIFICATION ACROSS THE FR1/FR2 DIVIDE

There seems to be a real distinction between FR1 radio-galaxies and FR2 radio-galaxies and their respective beamed counterparts, the BL Lacs and the quasars (See Table 1 below). But the host galaxies and the properties of the nuclear engine driving the activity is similar in both. Could there be unification across the divide using time (or equivalently, the age of the AGN) as the unifying variable in the sense that (some) FR2s evolve into FR1s? There are strong arguments that this will happen, at least for a subset of radio sources. I will outline a one simple scheme which illustrates how this could happen but, before doing this, it is appropriate to draw attention to the analysis of Ghisellini & Celotti (2001). They suggests that the FR1/FR2 divide arises when the ratio of jet power to black hole mass reaches a critical limit or, equivalently, at a certain accretion rate for the system relative to the Eddington limit (Let's call this the Eddington ratio). A concern with their proposal is that the outward manifestation of the FR1/2 divide often occurs when the radio-jets have travelled many kpc from the site of jet formation. For the Ghisellini & Celotti idea to work the jet has to carry with it a memory of the conditions in the black hole/disk system that tells it whether to form an FR1 or an FR2. I have some difficulty in imagining what the mechanism for the memory might be. It seems much more natural to believe that whether or not a relativistic jet will go on to form an FR1 or an FR2 is related primarily to how long it manages to remain relativistic in the environment through which it propagates. This is the idea that underpins the proposal below.

The basic assumption is that those objects that are observed as FR2s have jets that remain relativistic until they interact catastrophically with the ISM to produce hotspots; those objects in which the jets slow to sub-relativistic speeds before terminating will be classified as FR1s[1] A secondary assumption, which has observational support and which is more in the spirit of Ghisellini & Celotti's model, is that the jet power determines how far the jet can propagate and still remain relativistic; high power jets can propagate further than low power jets.

Let's look at the life history of sources with low and with high power jets. A very young source with the low power jets will start out as an FR2 because the jets are initially relativistic and do not have time to slow down before interacting with the ISM. But as time progresses and the source grows in linear size the outer ends of the jets slow down. The hotspot prominence will decrease, and finally go on to fade completely, leaving the source to live the rest of its life as an FR1. On the other hand, the high-power jet source will be able to maintain its FR2 morphology throughout the majority, or even all, of its active life.

One might ask how this scheme can account for the observation that the division between sources FR1 and FR2 morphologies occurs at a roughly constant observed total radio luminosity irrespective of source linear size? The answer could be quite simple. It has been argued that the efficiency with which jet power is converted to radio power decreases as the hotspots advance further away from the nucleus (Snellen et al., 2001). If this is in-

[1]It is often argued that it is when the hotspot advance speed slows to sub-sonic speeds that marks the transition between FR2 and FR1 structures, rather than the jet slowing to sub-relativistic speeds. Whichever is the case has little effect on the above argument. But the absence of detectable counter-jets in FR2 lobes argues in favour of the jets remaining relativistic up to the hotspots in FR2s.

deed the case, then it is possible for this jet efficiency vs length dependence, and the jet disruption length vs jet power dependence, to conspire to make all jets disrupt at approximately the same observed radio luminosity, irrespective of their intrinsic jet power. Put simply, low power jets disrupt early when their efficiency is high while high power jets disrupt later when the source is larger and the efficiency is lower.

A further question is whether or not such a scheme is consistent with the different cosmological evolution observed for the FR1 and FR2 populations? As lower luminosity objects will spend more time as FR1s, it is simply required, not unreasonably, that high power AGN evolve more strongly than low power AGN.[2]

TABLE 1. Comparison of FR1 and FR2 properties relevant to unification

Property	FR1	FR2
Radio structure	Relaxed, no hotspots	Well-collimated, prominent hotspots
Radio luminosity	Low luminosity	High Luminosity
Emission line properties	Weak lines, no BLR	Strong lines with BLR
Cosmological evolution	No evolution	Strong evolution

4.2. BLACK HOLES AND DIVERSITY

Unification is about simplification and a most remarkable simplification has become apparent in the last few years. This is the recognition of the tight relationship between bulge mass/velocity dispersion and black hole mass. Every galaxy contains an engine capable of making it an AGN (and it probably was one once). Fuel supply is likely to be the key to the level of activity; it determines the luminosity and perhaps, as suggested by Boroson (2002) and others, the Eddington ratio influences the degree of radio-loudness. There is a fairly widely held view (e.g. Lacy et al., 2001; Ghisellini & Celotti, 2001) that radio-loud objects[3] are powered by the most massive

[2]There is also the effect due to the fact that FR1s evolve into FR2s so, if the life-cycle occupies a significant fraction of the Hubble time, we would expect more FR1s now than in the past. However, all the lifetime evidence suggests that this is probably a negligible effect.

[3]I use the term radio-loud in a loose sense to encompass the bulk of objects found in radio surveys. There is a more rigorous definition in terms of the ratio of S_{5GHz}/S_B where S_B is the optical flux density in B band. This latter definition includes many low luminosity AGN in the radio-loud class.

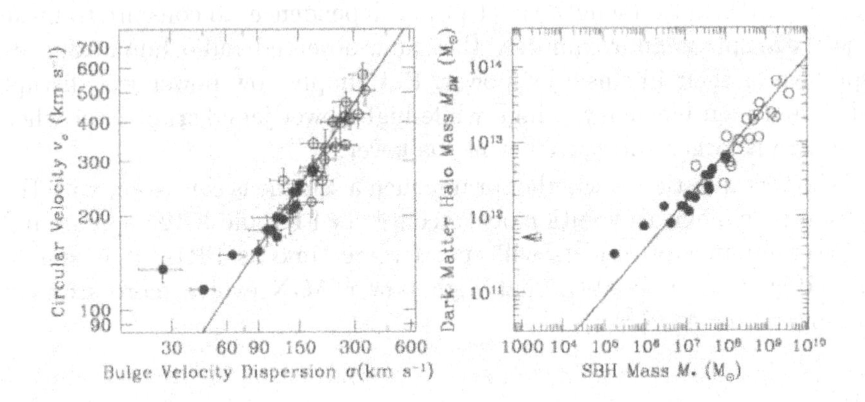

Figure 4. Correlation between black hole mass and bulge luminosity and also for velocity dispersion.

black holes and low Eddington ratios. (But also see Woo & Urry, 2002 for dissenting view.)

A particular question I would like to try and address is what produces the diversity amongst the radio-loud AGN, more specifically what determines the properties of their non-thermal SEDs which we discussed in Chapter??. There is much to suggest that the basic physics of the nuclear jets is very similar except perhaps for the lowest luminosity objects. What governs the jet power produced? What governs the shapes of the non-thermal SEDs, particularly where the synchrotron and inverse Compton peaks occur?

Since it is probable that the majority of radio-loud objects are hosted by massive early-type galaxies, the bulge mass/black hole mass relation suggests that fuelling rate is the only obvious candidate to account for both the range of AGN powers present and the differences in SEDs. In this context, Ghisellini (1999), Fossati et al. (1998) and Donato et el. (2001) argue that, as the photon flux from the accretion disk increases (with fuelling rate), these photons are more likely to be inverse Compton scattered by the relativistic electron in the jet, thus preferentially removing the highest energy electrons from the jet. This seems to provide a natural explanation for the correlation presented by Fossati et al. (1998) and Donato et el. (2001) which suggests that the peak frequencies in the SEDs of blazars systematically increase with decreasing power. I hope this simple picture is true. Being potentially such an important result it deserves careful scrutiny. For example:

– If possible, the correlations should be done with a more direct estimate of the AGN power since the non-thermal power is only a crude indicator of the AGN power (in blazars there is an unknown beaming factor).

- It would be very interesting to see if the correlation improves if it is tried with Eddington ratio rather than power.
- The existing correlation may be subject to strong selection effects. For example most of the low power objects are selected from X-ray surveys and therefore there is a potential bias amongst these towards objects with a high synchrotron turn over frequency. Attempts should be made to model the selection effects.
- There are claims that some low luminosity objects selected from radio surveys have very low synchrotron cut-off frequencies (Anton, 2001). It is important to confirm if such objects really do exist since they would tend to destroy the correlation.

4.3. THE STRUCTURE OF AGN; EXTENDING THE JACKSON & WALL APPROACH

Trying to exploit all the available observations (line, continuum, radio, infrared, optical, X-ray, etc.) of AGN to get a clear understanding of their structures has up to the present been a slow process. Even in the simplest AGN, the blazars whose emission is predominantly non-thermal, selection effects have severely hampered progress[4]. Therefore it may not be too surprising that progress has been slow for objects like quasars and radio-galaxies with richer selection of astrophysics on view. We seem to be reaching the limit of our ability to separate those correlations which are due to selection and those that are telling us about real astrophysics. How can we make progress? For FR2 objects is there, for example, a simple picture which can account for the apparently conflicting correlations of emission line and continuum properties with different possible orientation indicators?

The approach adopted by Boroson(2002), using principal component analysis, looks promising. He uses the results of his analysis to suggest a simple picture that ties together parameters like black hole mass and Eddington ratio for the different types of active galactic nuclei. He also incorporates orientation into his model. This approach gets so far but is still prey to the selection effects appertaining to the choice of samples used in the analysis.

Using the computer to build a "virtual universe" populated with objects obeying to some trial model seems to be a logical next step. (A default cosmological model and luminosity functions would need to be assumed.)

[4]In cynical moments I think that blazar statistics have told as virtually nothing about astrophysics, but lots about selection effects.

The virtual universe can then be sampled using the same set of selection functions that have been used to sample the real universe and then the virtual and the real can be compared. (This is really a generalization of the Jackson & Wall approach.) The key question is how to close the loop? The brute force approach is to try a multi-parameter optimization. More sensible would seem to be to try and mimic the human approach and look at the correlations (perhaps using principal component analyses) between the observables and then manually decide which model parameters to adjust. The advantages of this approach are:

- It simulates the process one tries (and often fails) to do in one's head.
- It sorts out the selection effects from the real astrophysics.
- It can cope with the very large numbers of objects with multi-parameter data which are increasingly found in modern databases.

5. Conclusions

Historically radio observations have been pivotal in the development of unified models. My feeling is that there are exiting possibilities for wider unification, particularly in the area of the relationship between non-thermal SEDs, the AGN fuelling and black hole mass. But we need more awareness of selection-induced correlations and need Urgently to develop better tools to deal with this difficult problem.

References

Anton, S., (2001), *PhD thesis, Victoria University of Manchester*
Antonucci,R.R.J. & Ulvestad,J.S. (1985), *ApJ.*, **294**, 158
Antonucci,R.R.J. & Miller,J.. (1985), *ApJ.*, **297**, 621
Barthel, P.D., (1989), *ApJ.*, **336**, 606
van Bemmel,I.M., Barthel, P.D. & de Graauw, T. (2000), *A&A.*, **359**, 523
Boroson.T.A. & Green, R.F. (1992), *ApJS.*, **80**, 109
Boroson.T.A., (2002), *ApJ.*, **565**, 78
Brotherton,M.S. (1996), *ApJS.*, **102**, 1
Browne, I.W.A. (1983), *MNRAS.*, **204**, 23P
Biretta,J.A., Zou,F., Owen,F.N., (1995), *ApJ.*, **447**, 582
Biretta,J.A., Sparks,W,B., Machetto,F., (1999), *ApJ.*, **520**, 621
Corbin, M.R., (1991) *ApJ.*, **375**, 503
Corbin, M.R., (1997) *ApJS.*, **113**, 245
Donato, D., et al., (2001) *A&A.*, **375**, 379
Fossati, G., et al., (1998) *MNRAS.*, **299**, 433
Garrigton,S. Leahy, J.P., Conway, R.G., Laing, R.A. (1988), *Nature*, **331**, 147
Ghisellini, G., (1999) *BL Lac Phenomenon, ASP conference series, eds Takalo, L.O. & Silanapaa, A* , **159**, 311
Ghisellini,G. & Celotti, A., 2001, *A&A.*, **379**, 1
Gopal-Krishna, Kulkarni,V.K. & Wiita, P.J., (1996) *ApJ.*, **463**, L1
Harvanek, M., Ellington, E. Stocke, J.T. Rhee,G., (2001) *AJ.*, **122**, 2874
Hes, R, Barthel, P.D. & Fosbury,R.A.E. (1983), *Nature*, **362**, 326

Jackson, N.J. & Browne, I.W.A., (1991), *MNRAS*, **250**, 414
Jackson, C.A. & Wall, J.V., (1999), *MNRAS*, **304**, 160
Kapahi,V.K. & Saikia, D.J., (1982) *JApA*, **3**, 465
Lacy, M., et al., (2001) *ApJ*, **551**, L17
Lawrence, A. & Elvis, M., (1985) *ApJ*, **256**, 410
Marchã,M.J.M. & Browne,I.W.A. (1995), *MNRAS*, **275**, 951
McClure,R.J. & Dunlop, J.S., (2001), *MNRAS*, **327**, 199
Mutel,R., et al., (1990), *ApJ.*, **352**, 81
Ogle,P.M., et al., (1997), *ApJ.*, **482**, L37
Orr,M.J.L. & Browne,I.W.A., (1982), *MNRAS.*, **200**, 1067
Padovani, P & Urry,C.M., (1991), *ApJ.*, **368**, 621
Padovani, P., (1999) *BL Lac Phenomenon, ASP conference series, eds Takalo, L.O. & Silanapaa, A* , **159**, 339
Rowan-Robinson, M. (1976), *ApJ.*, **213**, 635
Scheuer, P.A.G. & Readhead, A.C.S. (1979), *Nature*, **277**, 182
Scheuer, P.A.G. (1987), *Superluminal Radio Sources, Cambridge University Press, eds Zensus & Pearson* , p104
Snellen, I. et al., (2001) *MNRAS¿, 319*, 429
Stritmatter, P., et al., (1972), *ApJ.*, **175**, L5
Urry,C.M. & Padovani, P., (1995), *PASP.*, **107**, 803
Wills, B.J. & Browne, I.W.A., (1986), *ApJ.*, **302**, 56
Wills, B.J. & Brotherton,M.s., (1995), *ApJ.*, **448**, 81
Woo, J-H. & Urry, C.M., (2002) astro-ph/0211118

RADIO POLARIMETRY

W. D. COTTON
National Radio Astronomy Observatory
520 Edgemont Road
Charlottesville, VA 22903
USA

1. Introduction

All that we can know about most of the universe is what we can learn from studying the electromagnetic radiation it emits as a function of position on the sky, frequency, and time. Since electromagnetic radiation can be represented as a vector field, its polarization state is one of its fundamental properties. The polarization state of radiation tells much about the source of the radiation and about the medium through which it has traveled. Polarization properties of light are readily measured by radio interferometric techniques. The following sections will first discuss the physics of polarized light, both as it is emitted and then modified in the transmission medium. Following this is a discussion of the application of radio interferometric polarization measurements to astrophysical problems.

2. Physics: What is polarization?

A given photon can be thought of as a finite wave train of oscillating magnetic and electric field. Since it is the electric field that is measured, the magnetic field component of radiation will be ignored in the following, but it is always present. Since the electric field is a vector quantity, it can oscillate in a number of different ways. The manner in which the field vibrates is called its polarization.

The names given to polarization states are derived from the apparent motion of the tip of the electric field vector as seen along the direction of motion. The various possibilities are shown in Figure 1. The two extreme cases are "linear" polarization where the E-vector is confined to a plane;

F. Mantovani and A. Kus (eds.), The Role of VLBI in Astrophysics, Astrometry and Geodesy, 191–204.

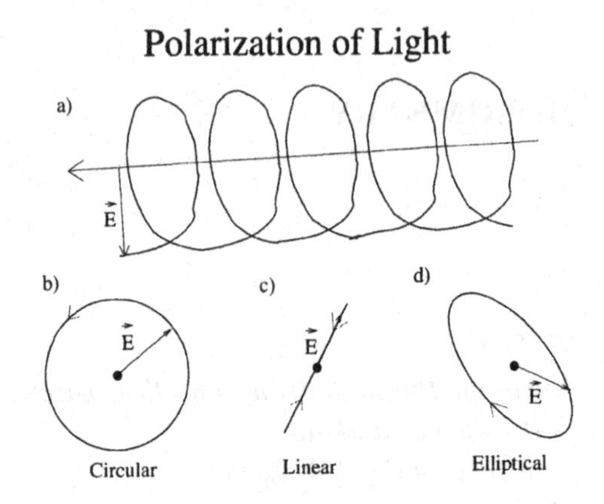

Figure 1. The polarization of light is a description of the oscillations of the E–vector.

"circular" polarization is the case where the E-vector rotates around the direction of motion. In these two cases, the tip of the E-vector traces a line segment or a circle. Intermediate cases are referred to as "elliptical" polarization as the tip of the E-vector traces an ellipse.

A given photon has a well defined polarization state, but if the source of the emission is spatially incoherent, as is almost always the case in astrophysical situations, the photons may not all have the same polarization state. If the photons have random polarization states, the emission is said to be unpolarized.

Each polarization state has an orthogonal state; there are two circular polarizations with opposite senses of rotation. Linear polarizations with a 90° orientation offset are also orthogonal. If the photons in one polarization state are more numerous than in the orthogonal state, the radiation is said to be "partially polarized". For a more complete description of the polarization of light see [1] or [7].

2.1. DESCRIBING THE POLARIZATION STATE

Traditionally the Stokes' parameters [18] are used to describe partially polarized light. There are four Stokes' parameters, I giving the total intensity of the light, Q and U which describe the linear polarization and V which describes the circular polarization. The polarized Stokes' parameters (Q,U,V) give the excess in the favored polarization state over the orthogonal state.

An alternate method of describing the polarization state of light is in terms of the fractional polarization (m_c the fractional circular, and m_l fractional linear) and the orientation (θ) of the E-vector for linear polarization.

The relationship between these and the Stokes' parameters is given by:

$$m_c = \frac{V}{I}$$

$$m_l = \frac{\sqrt{Q^2 + U^2}}{I}$$

$$\theta = \frac{1}{2}\tan^{-1}\frac{U}{Q}$$

2.2. RELATION OF STOKES PARAMETERS TO MEASURABLE QUANTITIES

Measurements of any two orthogonal polarizations are sufficient to completely sample the polarization of an incoming wavefront. With radio interferometers, this is done by having feeds nominally sensitive to orthogonal polarizations, either opposite circular polarization or crossed linear polarization. Thus, for each baseline there are four possible correlations among the combinations of the antenna feeds. These four correlations contain all the polarization information of the incoming wave. The relationship between the various polarizations and the Stokes parameters is given by the following.
Circularly polarized feed correlations:

$$R_{rr} = \frac{1}{2}g_{ip}g^*_{kp}(I + V),$$

$$R_{rl} = \frac{1}{2}g_{ip}g^*_{kq}((D_{ip} - D^*_{kq})I + \ e^{-2j\chi}(Q + jU)),$$

$$R_{lr} = \frac{1}{2}g_{iq}g^*_{kp}((D^*_{kp} - D_{iq})I + \ e^{2j\chi}(Q - jU)),$$

$$R_{ll} = \frac{1}{2}g_{iq}g^*_{kq}(I - V).$$

Linearly polarized feed correlations:

$$R_{pp} = \frac{1}{2}g_{ip}g^*_{kp}(I + Q\ \cos 2\chi + U\ \sin 2\chi),$$

$$R_{pq} = \frac{1}{2}g_{ip}g^*_{kq}((D_{ip} - D^*_{kq})I - Q\ \sin 2\chi + U\ \cos 2\chi + jV),$$

$$R_{qp} = \frac{1}{2}g_{iq}g^*_{kp}((D^*_{kp} - D_{iq})I - Q\ \sin 2\chi + U\ \cos 2\chi - jV),$$

$$R_{qq} = \frac{1}{2}g_{iq}g^*_{kq}(I - Q\ \cos 2\chi - U\ \sin 2\chi)$$

where χ is parallactic angle, gs are instrumental complex gains and Ds are polarization "leakage" terms. This relationship has been simplified by the assumption that the parallactic angle is the same at both antennas. This is not usually the case for VLBI; for a more complete discussion of this case see [19].

The calibration process estimates the g and D terms. Once the calibration is complete, the equations above can be inverted to derive the Stokes' visibilities. The Stokes visibilities can then be used to derive Stokes images by the usual Fourier transform and deconvolution techniques. The one difference between Stokes' I and Q, U or V is that positivity may be used as a constraint in the deconvolution of Stokes I images. Since Stokes Q, U and V can be of either sign, positivity cannot be used as a constraint in the deconvolution.

3. Emission Mechanisms

Electromagnetic radiation is emitted by accelerating charged particles; these are generally electrons (or positrons) as they are the most easily accelerated. The acceleration can be either impulsive as when the particle collides with another or long–term as for an electron spiraling under the influence of a magnetic field. The details of the the emission mechanism determine the preferred polarization state, if any, of the emitted radiation. Several of the more common emission mechanisms giving rise to radio emission from astronomical sources are discussed briefly in the following. In general, thermal emission is unpolarized and nonthermal processes emit polarized emission, although propagation and other effects can depolarize the emission. For more general discussions of astrophysical emission mechanisms see [11] [14] [15].

3.1. THERMAL

In a thermal (nonrelativistic) plasma, the particles will collide and the recoiling electrons will emit; this emission is know as thermal or Bremsstrahlung radiation. The ensemble spectrum from such a source is that of a blackbody at the same temperature as the plasma. The details of the collisions are random so there is no preferred polarization state. Emission from a thermal source is therefore unpolarized.

3.2. SYNCHROTRON EMISSION

Synchrotron emission emission comes from highly relativistic charged particles moving in a magnetic field [16] [14]. The magnetic field causes the particles to spiral, hence to emit. Relativistic aberration causes the emitted radiation to be highly beamed in the current direction of motion of the particle. The observer sees a series of flashes as the beam sweeps past. The frequency of this flashing is the frequency of the radiation. An ensemble of relativistic particles with a distribution of energies and pitch angles to the magnetic field will emit a broad continuum of emission. In optically thin sources, the emission is preferentially linearly polarized with the plane of the E-vector normal to the direction of the magnetic field projected onto the sky. The fractional linear polarization depends on $\frac{\gamma+1}{\gamma+7/3}$ where γ is the electron energy spectral index [16]. Fractional polarization up to 70% can be obtained from sources with highly ordered magnetic field. Disordered magnetic fields, either along the line of sight or across the telescope beam, will reduce the fractional polarization. The situation is more complex if the source is optically thick, in this case the fractional polarization will be less and the E-vectors will align with the magnetic field. Intrinsic circular polarization in synchrotron sources is rather weak [11], but propagation effects inside the source can convert some linear polarization into circular [13].

3.3. GYRO-SYNCHROTRON

Radiation from mildly relativistic charged particles is known as "Gyro-synchrotron" emission [14]. This type of emission can be strongly circularly polarized and is frequently observed from the sun and stars.

3.4. MASER

Maser emission is possible when a meta–stable state in an atom or molecule is out of thermo–dynamical equilibrium. In this case, an incident photon at the wavelength of a transition to a lower energy state can stimulate the emission of a similar photon. This gives the gas a negative absorption at that wavelength; a sufficiently long path length at the same relative radial velocity will give enough gain to produce a powerful beam at that wavelength. A given beam caused by a single incident photon will be coherent and have the same polarization state. However, the source will not be spatially coherent as a large number of incident will be responsible for the total maser emission. Many cosmic masers are observed to have both circular and linear polarization although the reasons for linear polarization are not well understood [9]. Masers are commonly observed in star forming regions and around evolved stars [8].

4. Zeeman Splitting

A magnetic field will interact with the electrons in atoms or molecules causing slight energy shifts between photons emitted in right or left hand circular polarization [10]. This energy shift will give rise to circular polarization. This is known as the "Zeeman" effect and is proportional to $B_{\parallel}$, the component of the magnetic field along the line of sight. The amount of the frequency shift depends on species and transition. Usually, the measurement of the Zeeman effect requires a narrow line and good sensitivity as the shift is small. In the case of OH masers, the shift may be very large.

5. Faraday Rotation

A magnetized plasma is a birefringent medium; right and left circularly polarized electromagnetic waves propagate at different speeds. This causes the angle of linear polarization to rotate, an effect known as "Faraday Rotation" [16]. The rotation of the angle of linear polarization, $\Delta\chi$, is given by:

$$\Delta\chi = \frac{0.93 \times 10^6 \int N\, H_{\parallel}\, ds}{(2\pi\nu)^2}$$

where N is the electron density (cm^{-3}), $H_{\parallel}$ is the component of magnetic field parallel to the direction of propagation, ν is the radio frequency (Hz) and s is distance along the sight (cm). As can be seen from the above, the magnitude of Faraday rotation goes as ν^{-2} or λ^2 and can be a very strong effect at long radio wavelengths. The effect of Faraday rotation is expressed in terms of the rotation measure (RM) where

$$\Delta\chi = RM \times \nu^{-2}$$

In order to unambiguously measure the rotation measure requires observations at multiple frequencies at which the source is optically thin (i.e. has the same intrinsic polarization angle). These measurements can then be used to derive both the intrinsic polarization angle and the rotation measure screen in front of the source. The rotation measure is then the integral along the line of sight of the product of the electron density and one component of the magnetic field, $H_{\parallel}$. The Faraday rotation may arise in the neighborhood of the source, in the interstellar medium in the Galaxy, or at low radio frequencies, in the interplanetary medium or ionosphere.

6. De–polarization

There are numerous effects which can reduce the fractional polarization of a source; this is especially true for linear polarization. Variations in the emitted polarization angle along the line of sight and across the instrumental resolution will reduce the observed fractional polarization. Differences in Faraday rotation can have the same effect. Sources which are polarized at high frequencies and less polarized at low frequencies are referred to as "de–polarized".

7. Astrophysical Applications

Measurements of the polarization state of the emission can be very useful in understanding the physical conditions in astronomical sources and the medium along the line of sight. A number of such applications are discussed in the following.

7.1. SEPARATING THERMAL AND NONTHERMAL SOURCES

Since thermal emission gives rise to random polarization states of the emitted photons, hence is unpolarized, the presence of significant polarization in a source indicates that the emission is nonthermal in origin. An example of the use of polarization to distinguish thermal from nonthermal emission is shown in Figure 2. In this figure, a galactic filament is seen to be partially linearly polarized, hence the emission mechanism is clearly nonthermal. Note that the converse is not necessarily true.

7.2. ORIENTATION AND ORGANIZATION OF MAGNETIC FIELDS

In sources thought to be synchrotron emitters, measurements of the linear polarization give rather directly information about the direction and orderliness of the magnetic fields in the source. For optically thin sources with no intervening Faraday rotation, the observed orientation of the E–vectors on the plane of the sky is orthogonal to the magnetic field. For a region of highly ordered magnetic fields, the fractional polarization will be approximately 70%; disordering of the magnetic fields will reduce this. In the presence of intervening Faraday rotation, measurements at multiple frequencies are needed to separate Faraday rotation and intrinsic polarization angle.

An example is shown in Figure 3 which shows the core and southern lobe of the giant radio galaxy 2146+82 [17]. The core and the jet can be seen in the upper part of the left figure; the magnetic field is basically along the jet (orthogonal to the E-vectors). The field becomes transverse, as well as more disordered, in the middle of the lobe and wraps around its outer edges. The right hand figure shows the polarized intensity and rotation measure. The values of rotation measure are comparable to typical lines of sight through the Galaxy.

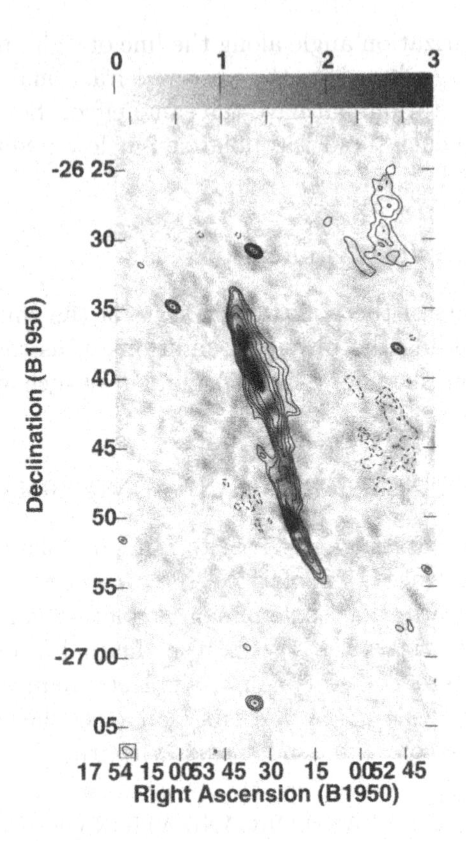

Figure 2. 20 cm measurements of a galactic filament (courtesy F. Yusef-Zadeh). The contours show the total intensity and the gray scale the polarized intensity. The bar at the top shows the gray scale scale; contours are spaced by factors of the square root of two. The filament is seen to be partially linearly polarized, hence a nonthermal emitter.

An example on VLBI scales is shown in Figure 4 [5]. This figure shows the outer jet of the CSS quasar 3C138 with total intensity contours and fractional polarization E-vectors. The Faraday rotation in this source is small, so the E-vectors indicate the normal to the magnetic field direction. In this case, the magnetic field runs parallel to the jet and then wraps around the outer edge.

7.3. MAPPING FARADAY SCREENS IN AGN NUCLEI

Polarized radio sources may be used to probe the intervening material. If there is sufficient magnetized plasma along the line of sight, its presence will be indicated by its Faraday rotation of the background radiation. One application of this technique is the investigation of the emission line regions around active galactic nuclei (AGNs) [20].

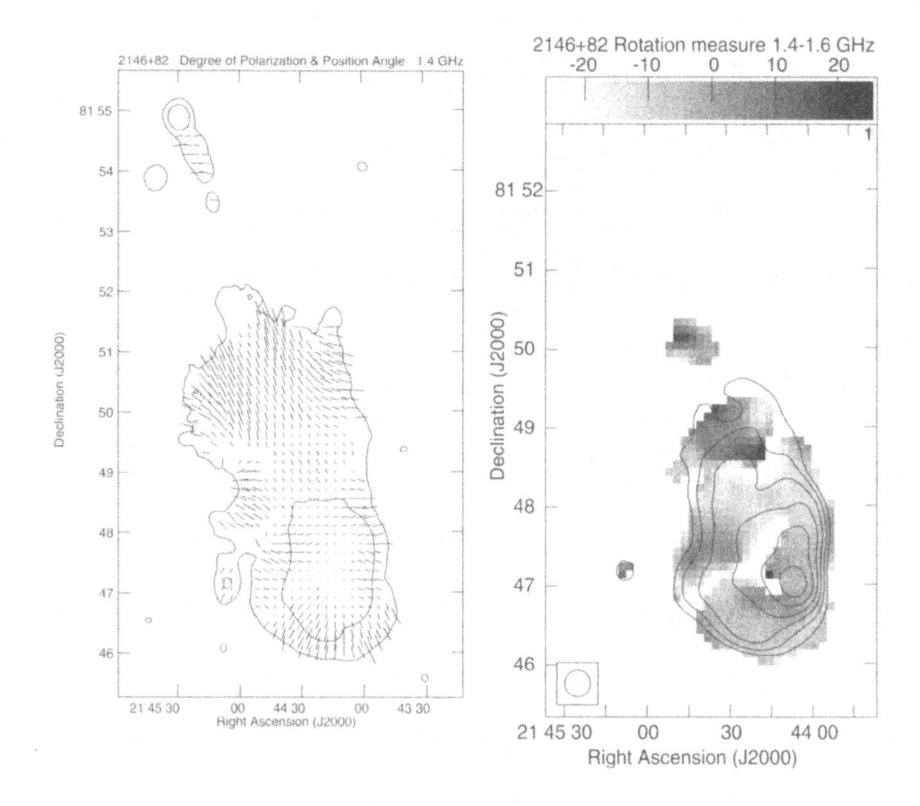

Figure 3. 20 cm observations of the core and southern lobe of 2146+82. Left: Total intensity contours and E-vectors. Right: Polarized intensity contours and rotation measure as gray scale. The bar at the top gives the gray scale in radians m^2. From Palma et al. 2000.

This effect is demonstrated by a polarized, moving knot in the inner jet of 3C138 show in Figure 5. The moving component is the brightest one. The difference in orientation at two frequencies at each epoch shows the effects of Faraday rotation. The difference at the same frequency between the two epochs shows the time variability of the Faraday rotation. These observations have been made at a number of epochs. Preliminary results are shown in Figure 6 which shows the separation of the moving knot from the core and its variations in Faraday rotation.

7.4. MODELING VELOCITY FIELD OF FRI JETS

Another case where polarization measurements play a critical role is in the modeling of relativistic jets. A well known effect of relativistic aberration of radiation from relativistic jets is the "Doppler Beaming" [2] whereby the

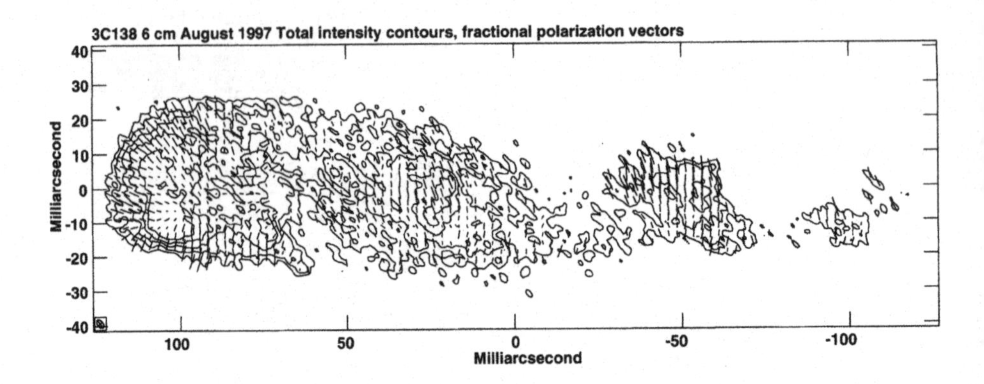

Figure 4. Total intensity contours and fractional polarization E-vectors of the jet of 3C138. From Cotton et al., 1997.

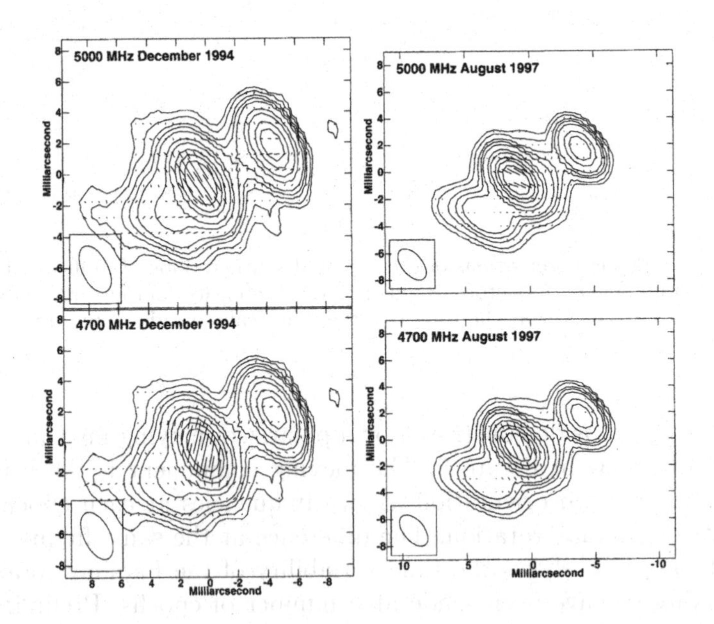

Figure 5. Polarization of the core region of 3C138 at two frequencies and two epochs showing variable Faraday rotation.

radiation is strongly beamed in the direction of motion. The approaching and receding jets will have their brightness strongly enhanced or reduced due to this effect. Relativistic aberration also effects the linear polarization,

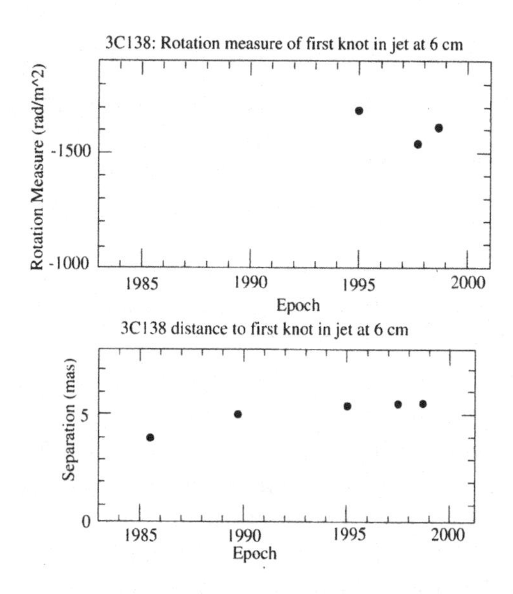

Figure 6. Separation from core and Faraday rotation of knot in 3C138 jet. Uncertainties are smaller that the size of the symbols used.

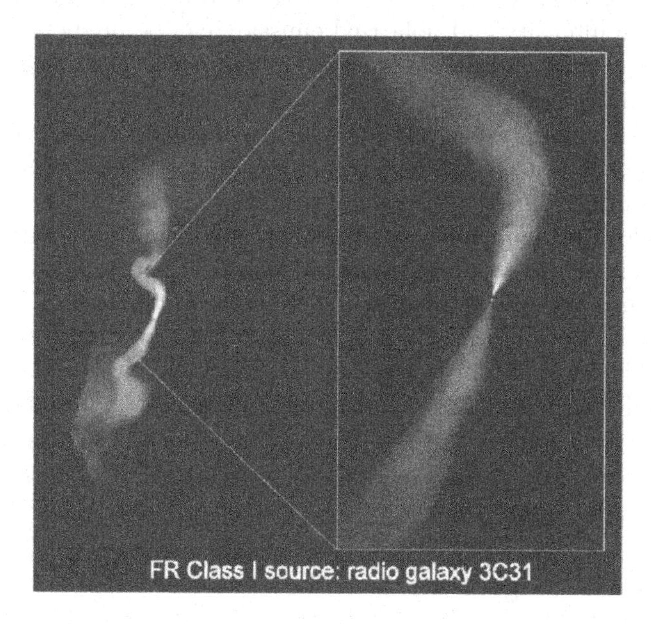

Figure 7. The large and intermediate scale structure of the radio galaxy 3C31 at 3.4 cm wavelength. From Bridle 2000.

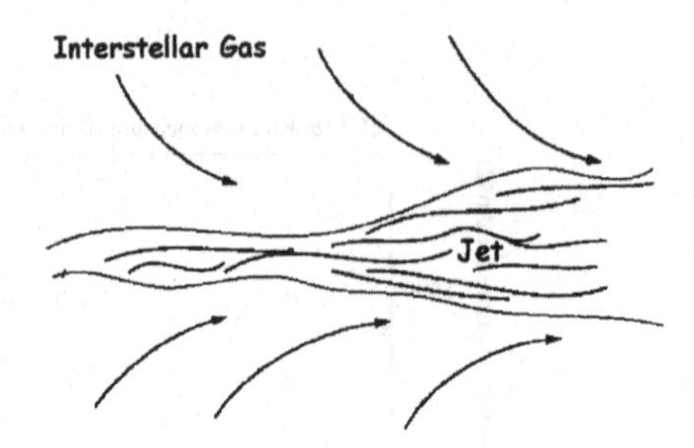

Figure 8. Cartoon illustrating the turbulent entrainment of ISM material along the boundry of the jet. Courtesy A. H. Bridle.

rotating its apparent orientation. Robert Laing and Alan Bridle [3] have been using these effects to model the velocity field in FRI radio sources. In these, relatively low power sources, the jets initially are quite asymmetric and become increasingly symmetric further from the core. This effect is thought to be the result of mass loading of the entrained ISM material slowing the jet from relativistic to sub-relativistic speeds. In order to test this hypothesis, Laing and Bridle have fitted kinematic and magnetic field models to total intensity and linear polarization images of the radio galaxy 3C31. If models of both the magnetic field and kinematics are used, both the total intensity and polarized emission can be predicted. Since the total intensity and polarized emission are independent, this constrains the models much more strongly than total intensity alone.

The large scale structure of 3C31 is shown in Figure 7 together with the inner portion of the jet. The transition from asymmetric to symmetric jets occurs in this inner, straight portion of the jet.

The basic model being fitted has symmetric jets with a highly relativistic spine with a speed that decreases in the transverse direction. Other constrains are optical imaging to get the stellar mass profile for estimates of the stellar density and Chandra X-ray data to estimate the external pressure. The jet is initially slowed by loading from mass loss by stars inside the jet, and then by turbulent entrainment of the ISM through the boundary of the jet, see Figure 8. The final model prediction of the observable quantities are compared with the observations in Figure 9. The derived velocity field is shown in Figure 10.

8. Acknowledgements

The author would like to thank Alan Bridle and Robert Laing for providing prepublication data and discussions on the modeling of FRI radio jets. The author would also like to thank Prisse Könönen for comments.

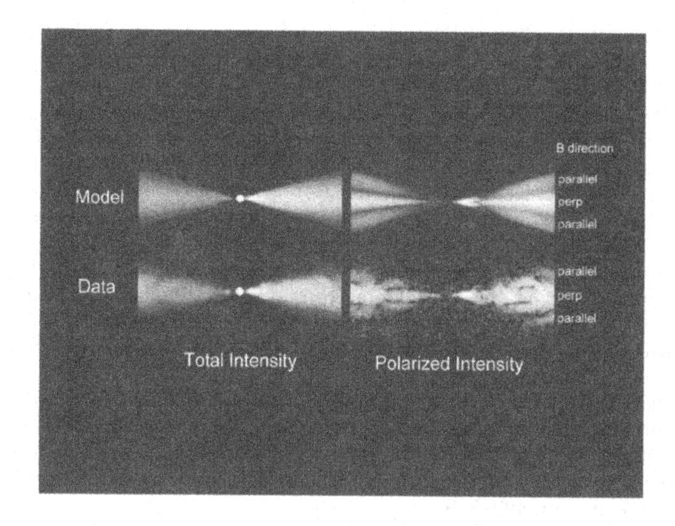

Figure 9. 3C31 inner jet data and model for total and polarized intensity. From Bridle 2000.

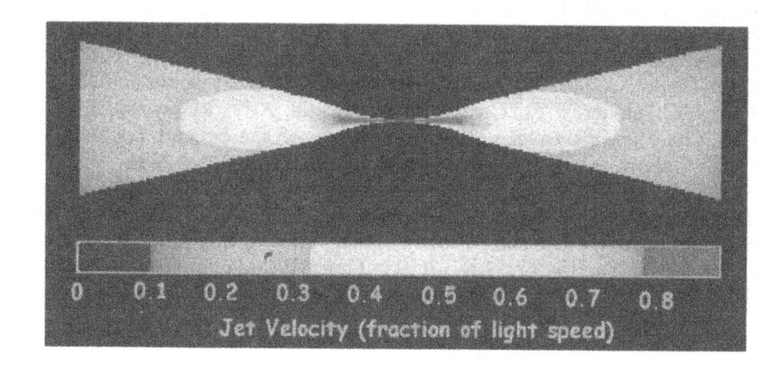

Figure 10. The derived velocity field of the inner jet of 3C31. Courtesy A. H. Bridle.

References

1. Born, M. and Wolf, E, (1975), *Principles of Optics*, Pergamon Press, Oxford
2. Bridle, A.H. and Perley, R.A (1984), Extragalactic Radio Jets, *Ann. Rev. Astron. Astrophys.*, **22**, pp. 319–358
3. Bridle, A.H. (2000), Impact of the VLA: Physics of AGN Jets, in *Radio Interferometry: The Saga and the Science*, Finley, D. G. and Goss, W. M., NRAO Workshop no. 27., Associated Universities Inc., pp. 152–165
4. Cotton, W.D. (1993) Calibration and Imaging Techniques for Polarization Sensitive VLBI Observations, *Astron. J.*, **106**, pp. 1241–1248
5. Cotton, W.D., Dallacasa, D., Fanti, C., Fanti, R., Foley, A.R., Schilizzi, R.T., and Spencer, R.E. (1997) Dual Frequency VLBI polarimetric observations of 3C138, *Astron. Astrophys.*, **1325**, pp. 493–501

6. Cotton, W.D. (1999) Polarization in Interferometry, in Taylor, G., Carilli, C. and Perley, R. eds. (1999) **Synthesys Imaging in Radio Astronomy II**, Astronomical Society of the Pacific, pp. 111–124
7. F. Delplancke (1997), *Conception et réalisation d'un appareil mesurant la diffusion del la lumiére polarisée en vue de la mesure et du contrôle industriel d'états de surfaces et de suspensions de particules*, thesis, Université Libre de Bruxelles, Belgium
8. Diamond, P.J., Kemball, A.J., and Boboltz, D.A. (1996), VLBA Polarization monitoring of SiO masers towards late-type stars, *Vistas in Astronomy*, **41**, no. 2, pp. 175–178
9. Elitzur, M. (1996), Polarization of Astronomical Maser Radiation. III. Arbitrary Zeeman Splitting and Anisotropic Pumping, *Astrophys. J.*, *475*, pp. 415–430.
10. Griem, H.R. (1988), Spectral Line Shapes, in *Astrophysical and Labrotory Spectrosopy*, eds. Brown, R. and Lang, J., Scottish Universities Summer School in Physics, Edinburgh, pp. 105–119
11. Jones, T.W., O'Dell, S.L., and Stein, W.A. (1974), Physics of Compact Nonthermal Sources. I. Theory of radiation processes. . *Astrophys, J.*, **188**, pp. 353-368
12. Kemball, A. (1999) VLBI Polarimetry, in Taylor, G., Carilli, C. and Perley, R. eds. (1999) **Synthesys Imaging in Radio Astronomy II**, Astronomical Society of the Pacific, pp. 499–509
13. Kennett, M. and Melrose, D. (1997) Propagation–induced circular polarization in synchrotron sources, *Publ. Astron. Soc. Aust.*, **15**, part 2, pp. 211-218
14. Melrose, D.B. (1985), Gyromagnetic Emission and Bremsstrahlung, in *Solar Radiophysics*, McLean, D.J., and Labrum, N.R. eds. Cambridge University Press, Cambridge, pp. 177–210
15. Melrose, D.B. (1991), Collective Plasma Radiation Processes, *Ann. Rev. Astron. Astrophys.*, **29**, pp. 31–57
16. Pacholczyk, A.G. (1970) *Radio Astrophysics*, W. H. Freeman and co., San Francisco
17. Palma, C., Bauer, F.E., Cotton, W.D., Bridle, A.H., Majewski, S.R., and Sarazin, C.L. (2000), Multiwavelength Observations of the Second Largest Known FR II Radio Galaxy, NVSS2146+82, *Astron. J.*, **119**, pp. 2068–2084
18. Stokes, G. (1852), *Trans. Cambridge Phil. Soc*, **9**, part 3, pp 399–416
19. Thompson, A.R., Moran, J.M. and Swenson, G.W. Jr. (2001) **Interferometry and Synthesis in Radio Astronomy**, John Wiley & sons, Inc., New York.
20. Zavala, R.T., andTaylor, G.B. (2001), Time Variable Faraday Rotation Measures of 3C273 and 3C279, . *Astrophys, J. Lett.*, **550**, pp. L147–L158

GRAVITATIONAL LENSING WITH A RADIO BIAS

I.W.A. BROWNE[1]
University of Manchester
Jodrell Bank Observatory,
Macclesfield,
Cheshire, SK11 9DL,
UK

1. Introduction

Gravitational lensing is not just a cosmic curiosity; it is a tool that can be used for a surprisingly wide range of astrophysical applications. My aim is to present a basic introduction to gravitational lensing and show some pictures that illustrate typical image configurations. The ways in which one goes about finding lens systems will be reviewed with particular emphasis on radio surveys and how high angular resolution is exploited in order to reject possible lens candidates and find out invaluable information about the real lensed systems. I will draw attention to the fact that the lensing galaxy often leaves an imprint on the radiation as it passes through and this in itself is a unique tool for the study of intermediate redshift galaxies. Finally I will discuss how lens time delays can be measured and how VLBI observations can be crucial in order to convert a time delay into a value for the Hubble constant. For a comprehensive treatment of Gravitational lensing see Scheinder et al.(1992) and for an up-to-date discussion of primarily the radio aspects of lensing see Jackson et al. (2002).

2. The basics of lensing

Radiation is deflected by non-uniform gravitational fields and, just as in conventional optics, images form at extrema in the light travel time surface in accordance with Fermat's principle. In most circumstances there will be just one distorted image, but for radiation passing through a sufficiently deep gravitational potential well, multiple images can be formed. The basic geometry is illustrated in Figure 1.

F. Mantovani and A. Kus (eds.), The Role of VLBI in Astrophysics, Astrometry and Geodesy, 205–217.

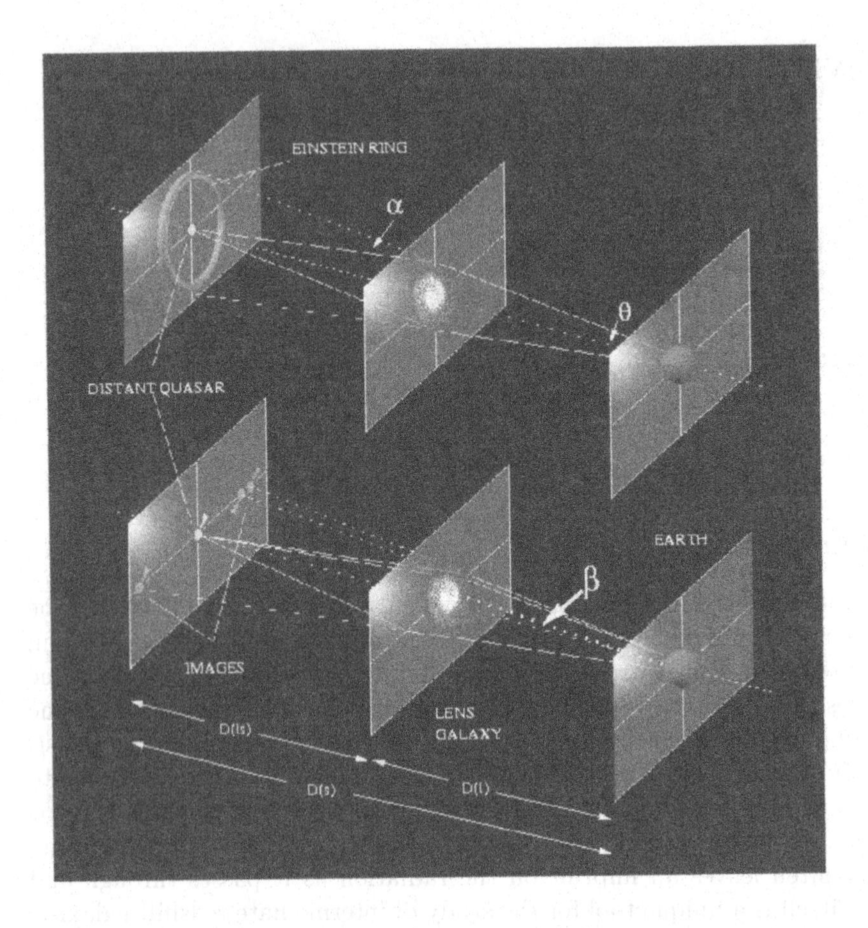

Figure 1. The geometry of gravitational lensing. The top panel shows the condition for the formation of an Einstein ring; i.e. perfect alignment between lens and lensed object. The bottom panel show the situation when the alignment is not perfect and when two distorted images are formed.

For a point mass, M, the deflection angle α is given by

$$\alpha = 4GM/bc^2,$$

where b is the impact parameter. There is a simple geometrical relation between the angles and the distances shown in Figure 1 which must be satisfied if multiple images are to be seen and this is known as the lens equation,

$$\theta - \beta = \alpha D_{ls}/D_s.$$

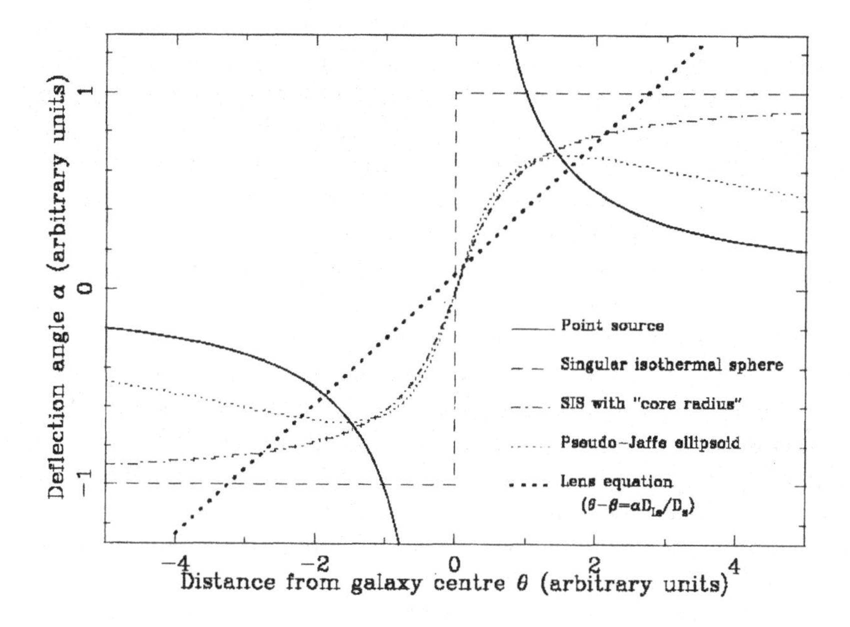

Figure 2. The bend angle diagram. Images are formed where the deflection angle equation and the lens equation are simultaneously satisfied; i.e. where the straight line representing the lens equation intercepts the deflection curves.

For perfect alignment between the lens and the lensed object $\beta = 0$ and

$$\theta_E = \frac{4GMD_{ls}}{bc^2 D_s} = \frac{4GMD_{ls}}{\theta c^2 D_l D_s},$$

since $b = D_l\theta$. Thus θ_E, which is known as the Einstein radius is given by

$$\theta_E = (\frac{4GMD_{ls}}{c^2 D_l D_s})^{1/2}.$$

The Einstein radius gives the characteristic image separation. Note that for non-point masses M is a function of impact parameter so different mass distributions will have different deflection curves. One of the simplest mass distributions which approximates well to real galaxies is the Singular Isothermal Sphere (SIS) which has a constant defection angle as a function of radius. This and other deflection curves are illustrated in Figure 2.

Images are formed at angles where the deflection equation and the lens equation are satisfied simultaneously. This is illustrated in Figure 2, the "bend angle diagram", in which defection curves for a range of different mass profiles are drawn and also a straight line representing the lens equation. Images form where the lens equation line intercepts the deflection curves. For all non-singular mass distributions this happens in three places; three images are expected. Referring back to Figure 1, it can be seen that the path lengths from source to observer are slightly different for the separate images. Thus signals from the source arrive at different times at the observer. As we will see this time delay is the foundation of the method used to determine the Hubble constant from lensing.

It turns out that typical lensing galaxies at typical redshifts produce images separated by $\sim$1 arcsec (Einstein radius $\sim$0.5 arcsec)[1]. In the lensing geometry illustrated in Figure 1, for perfect alignment between observer, lens and source, a ring image (an Einstein ring) is produced. For non-perfect alignment the result is two distorted images of the background object. However, most galaxies do not have circularly symmetric mass distributions and when the symmetry is broken as it is in elliptical galaxies, four images of the background object may sometimes be observed. For even more complex mass distributions, for example when two or more galaxies are involved, higher multiplicities are possible and there is one known example where six images of the same object are seen. Probably the best example so far of an Einstein ring is shown in Figure 3. The top picture shows an near-infrared image taken with the HST. The ring is the image of the host galaxy of the radio source and the lensing galaxy is seen at its centre. Also shown on the same scale is the radio map of the same system. Alignment of the radio emission with the lens is not perfect and so no ring is observed. In addition the imaged radio source has two separate components giving rise to the much more complex radio map. Figure 4 and 5 shows more examples of maps of radio lens systems, all resulting from the CLASS search.

What are the properties of lensed images?

- The images are magnified and distorted.
- There should in theory always be an odd number of images but de-magnification of one image means that almost always an only an even number of images is observed.
- Lensing preserves surface brightness, colour and polarization. But lensing galaxies have an interstellar medium and this can absorb (due to extinction by dust in the optical and perhaps free-free in the radio), change the angular size (in the radio, by multi-path scattering in an irregular ionized medium) and produce Faraday rotation and depolarization. Thus in practice, the observed surface brightnesses, colours and polarizations of lenses images are often not identical. All this is

[1]For stars the Einstein radii are $\sim$1 micro-arcsec, hence the term micro-lensing

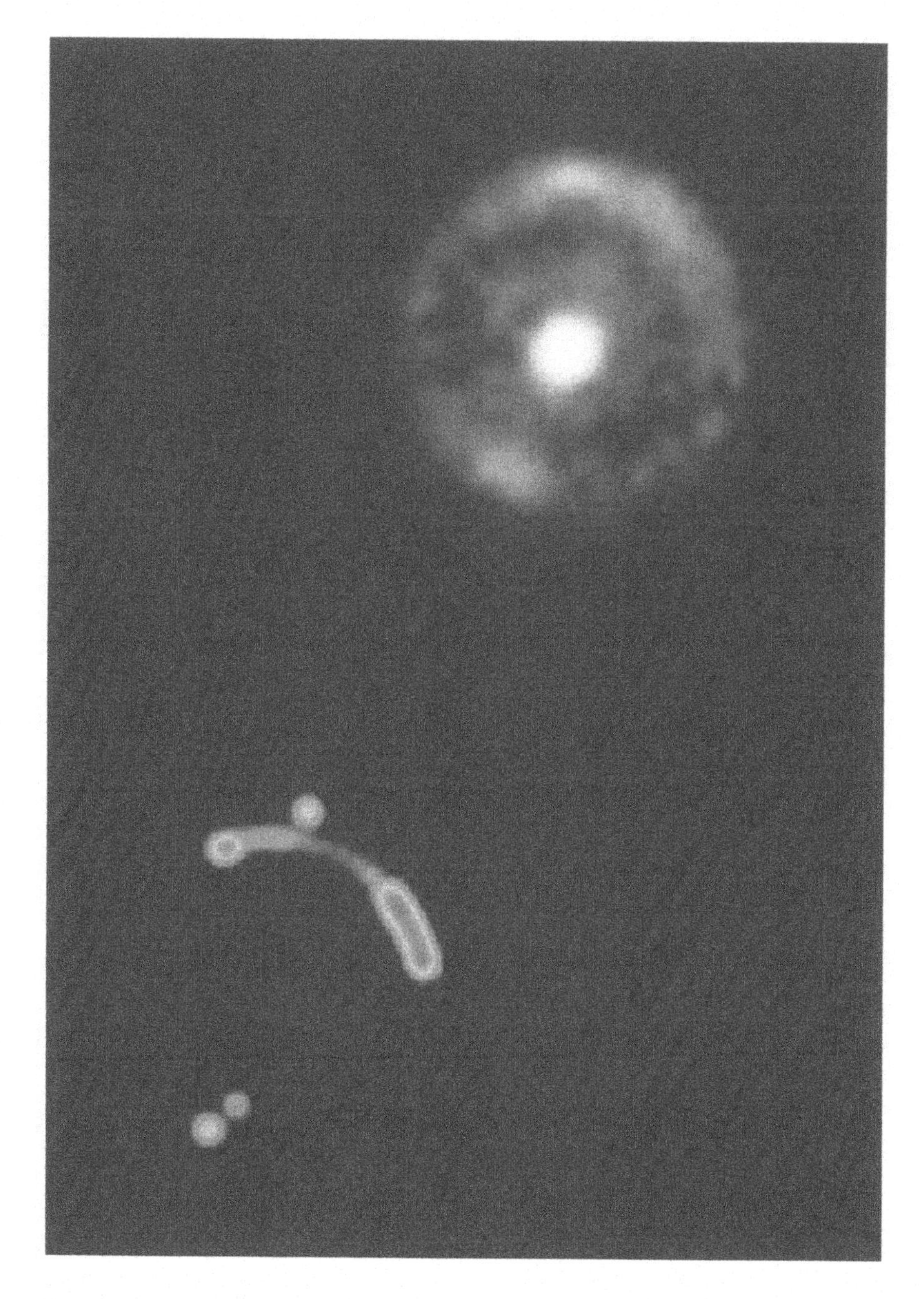

Figure 3. The gravitational lens system B1938+666. The top panel shows the Einstein ring produced when the lensing galaxy, which is visible at the centre of the ring, images the host galaxy of the lensed radio object. The radio image of the same system is shown below. Note that the alignment between the radio components and the lens is not perfect so an arc-image is produced.

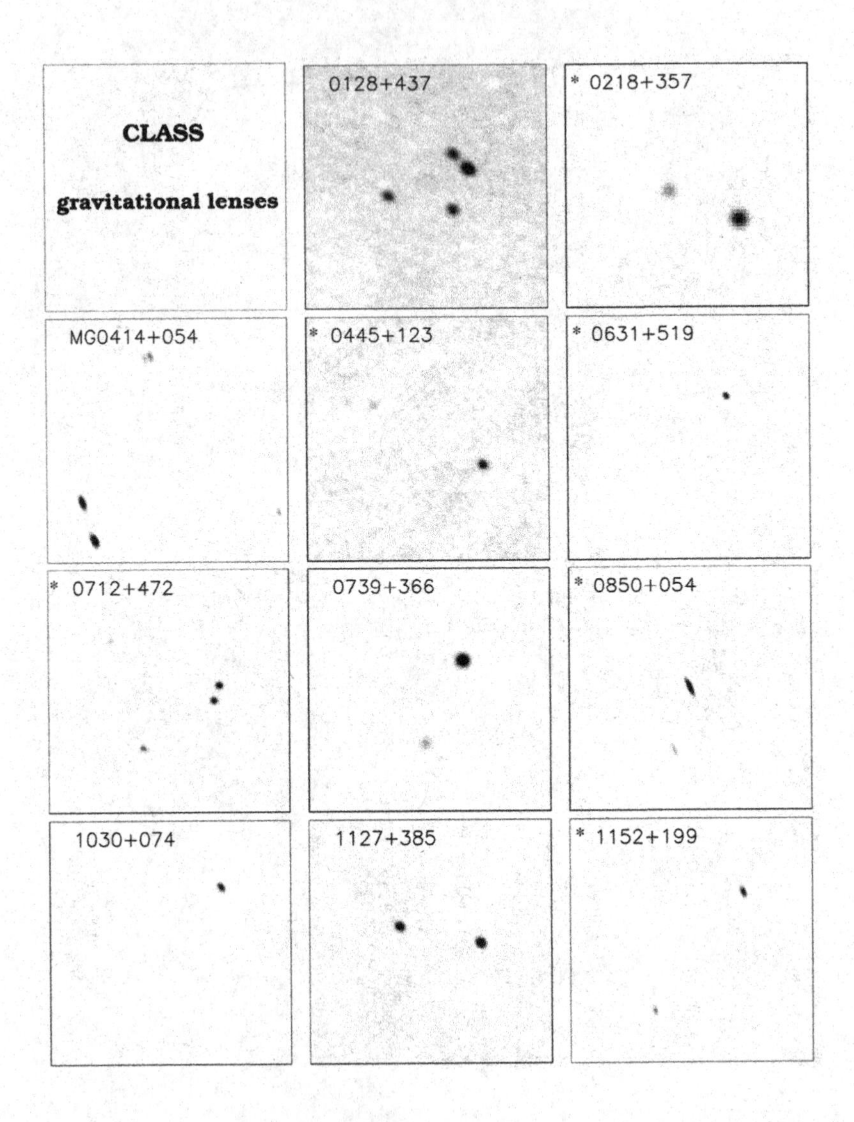

Figure 4. MERLIN maps of CLASS radio lens systems; part 1.

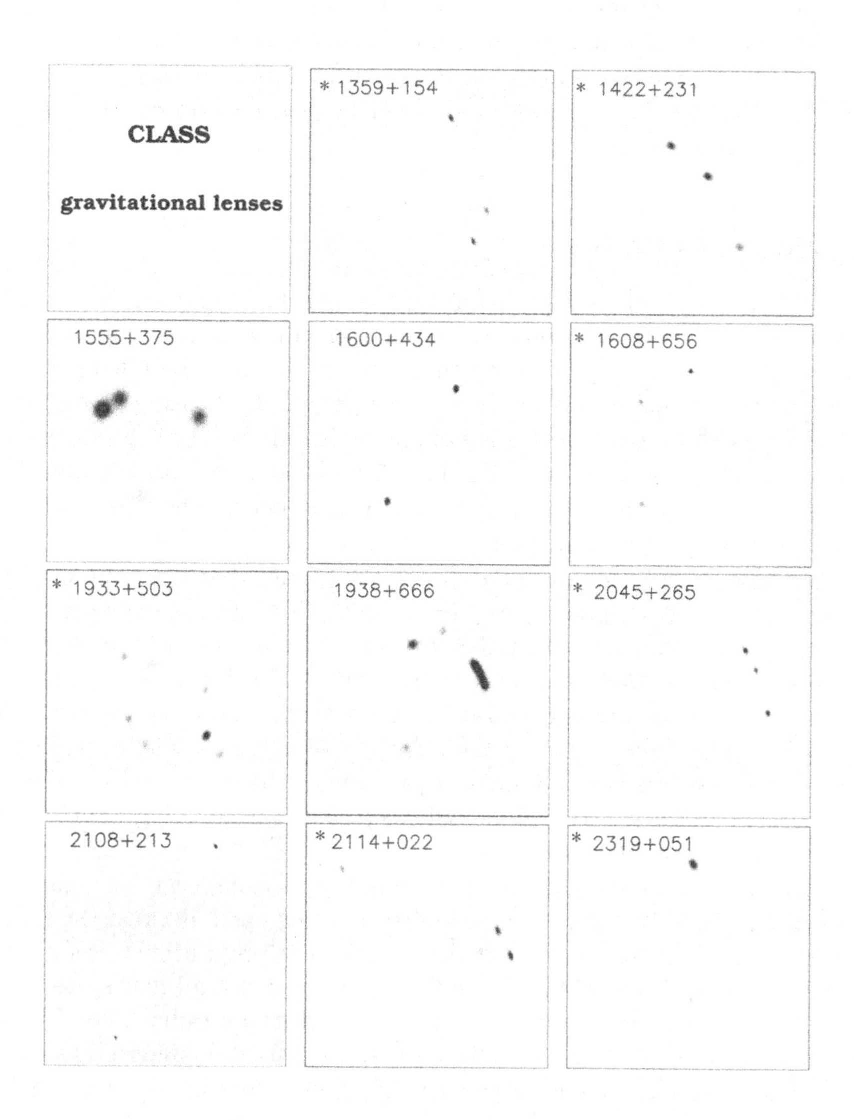

Figure 5. MERLIN maps of CLASS radio lens systems; part 2.

very important to bear in mind when planning lens surveys and when deciding if a candidate lens system is genuine or not.
- If that were not enough, the delays between the different images mean that in variable sources the observed surface brightnesses, polarizations etc. may not be the same, even in the absence of propagation effects in the lens.

3. Finding lens systems

There is a range of astrophysical motivations for lens surveys. Reliable lens statistics can tell us about the cosmological parameters, especially Ω_Λ the "dark energy". Also the search is on for the very best lens systems to monitor for time delays in order to determine H_0. Lensing gives unique information about the distribution of matter in galaxies, both luminous and dark, which can be used to test galaxy formation scenarios. Finally, there are lots of interesting propagation effects that occur when the radiation passes through the lensing galaxy.

For an average compact radio source the chance that it will be multiply imaged by an intervening galaxy is ~1:700. This number depends on the cosmological parameters, the redshift distribution of the background source population and on the slope of the source counts. The latter effect, known as magnification bias, arises because lensing magnifies objects; sources which would not normally be in the flux limited sample, can magnified into it. The size of the reservoir of fainter objects below the survey cut-off depends on the slope of the source counts and steep counts mean a big reservoir and hence a large magnification bias.

Even with some help for magnification bias, lensing events are rare and lens surveys need to cover a large number of targets and, because the typical image separations are ~1 arcsec, they require high resolution observations. Examination of Figures 4 and 5 will indicate that not all lens systems are easy to recognize on the basis of a single picture or radio map. Double-image radio lenses are easily confused with un-lensed double radio sources; at optical wavelengths, especially without colour information, lensed images can get easily lost in, or confused by, the lensing galaxy itself. One particular problem with optical searches is that extinction in the lensing galaxy itself may hide one or more of the lensed images. This is worrysome when trying to produce complete surveys for statistical analysis.

There are many advantages to radio lens searches over optical searches:

- Resolution is no problem, whereas it is for optical searches where the ground-based seeing is of the same order as the typical image separation.

- The radio-sky is relatively sparsely populated so confusion is not an issue.
- Lensing of compact sources is easy to recognize.
- Dust in the lens has no effect on the radio waves.
- Compact radio sources are often variable and thus some of the lens systems will be suitable for time delay monitoring.

4. The CLASS Lens search.

The Very Large Array (VLA) has been the instrument of choice for nearly all radio lens surveys. The pioneers in radio lens surveys were the MIT group led by Bernie Burke (Burke, 1993) who were successful at finding lens systems but had problems in being statistically complete owing to the difficulty they experienced in distinguishing intrinsic doubles from double-image lens systems. This difficulty may be overcome if pre-selection in favour of flat-spectrum sources is used to ensure that the targets have structures dominated by single point-like radio components. This spectral selection of flat-spectrum targets was first used for the JVAS survey (Patnaik et al., Browne et al., Wilkinson et al.) and was subsequently continued with the much larger CLASS project which has subsumed JVAS (Myers et al. 2002; Browne et al., 2002). CLASS (the Cosmic Lens All-Sky Survey) is the largest radio lens survey, and the one I know most about, so I will use it as an illustration of one successful survey technique. CLASS is a collaborative effort involving mainly groups at Jodrell Bank, Caltech, Dwingeloo and NRAO. The basic survey strategy was as follows:

- Sources with $S_{5GHz} \geq 30$ mJy were selected from the GB6 5 GHz survey. The NVSS at 1.4 GHz was used to exclude the steep-spectrum sources.
- They were observed for $\sim$1 min with the VLA in its A-configuration at 8.4 GHz giving a resolution of 0.2 arcsec.
- Candidates lens systems with more than one compact component visible in the VLA maps were selected and followed up at higher resolution, initially with MERLIN and then, if they still looked possible, with the VLBA.
- The criteria used to reject lens candidates on the basis of information in high resolution maps were primarily, large differences in surface brightness and/or a radio structure that was inconsistent with lensing (e.g. a clear core/jet or a CSO-like morphology).

During the CLASS survey $\sim$16,500 sources were observed with the VLA, about 300 sources were followed up with MERLIN and $\sim$50 with the VLBA. Twenty-two lens systems have been found. The survey is complete for image angular separations between 0.3 and 15 arcsec and for image flux density ratios $\leq$10:1. The image separations found range from 0.33 to 4.5 arcsec, with the distribution peaking at around 1 arcsec. The lensing rate is $\sim$1:750.

5. Lensing as an astrophysical tool

5.1. LENSING AND THE HUBBLE CONSTANT; B0218+357

The time delay between signals forming images has both a geometrical and a gravitational component. The geometric contribution is proportional to the angular diameter distances. The measurement of the actual delay gives a real length scale and hence the constant of proportionality. Also required is a good mass model for the lens system, something that VLBI observations can help constrain, so that the gravitational contribution to the delay can be calculated. Knowing the redshifts of the lens and of the lensed object, together with a good mass model and a measured time delay, a value for the Hubble constant can be calculated. The result has a weak dependence on the other cosmological parameters but at the moment this is not the major source of uncertainty, rather it is the accuracy of the mass models. [Check this bit.]

I will use the CLASS lens system B0218+357 (see Figure 6) as an example of the current state of the art. Most of the results I will present come from Biggs et al. (1999, 2001). In some ways measuring the time delay is the easy bit. For this particular system about 50 VLA observations were made over a period of $\sim$3 months. Both total intensity and linear polarization were measured for the two compact images at 8.4 and 15 GHz. Mainly because the image separation is only 0.33 arcsec, the most useful results were obtained at 15 GHz where the resolution is better. Over the period covered by the observations the total intensity, percentage polarization and polarization position angle varied essentially independently, thus giving three separate delay measurements. All gave the same value for time delay of 10.5$\pm$0.4 days at 95% confidence. To show the kind of variations present, the polarization position angle light curves for the two images are shown in Figure 7.

There are several properties of the B0218+357 system that make it a good one for Hubble constant work. There is a good value for the delay and the redshifts of the lens and lensed object are known. Additionally, a particularly useful property is its extensive radio structure. First there is the Einstein ring visible in Figure 6. This probes the gravitational potential at many positions and is good at constraining the position of the lensing galaxy with respect to the lensed images. Being a lens system with such a small image separation this position uncertainty has been one of the major sources of error in the Hubble constant determination until recently. But Wucknitz (2002) has used a variant of the standard CLEAN algorithm,

LensCLEAN, which uses the the fact that there is multiple imaging to reconstruct simultaneously an image of the original lensed object, the lens plane map and to optimize the parameters of the lens model, including the position of the lens.

The images of the compact core of B0218+357 also have quite extensive sub-structure. VLBA observations at 15 GHz by Patnaik et al. (1995) showed that each image contained two components separated by $\sim$1 mas. More recent global VLBI observations by Biggs et al. (2002) reveal that beyond these subcomponents there are jets which extend for 10s of mas. The details of these in turn constrain the radial mass profile of the lensing galaxy.

5.2. LENS STATISTICS AND COSMOLOGY

It is possible to predict the number of lens systems one expects to find and their distributions of image separation, multiplicities and redshifts. One needs to know the properties of the population of the potential lenses as a funtion of redshift and similarly, the properties of the potential lensed objects. In addition, the predictions are sensitive to the values of the cosmological paramaters, particulary Ω_Λ. Thus comparing the predictions with the results of a carefully conducted survey like CLASS is a cosmological test. The most up-to-date and complete analysis using CLASS has been presented by Chae et al. (2002) and Chae (2002). The results are consistent with the "convergence cosmology", $\Omega_M = 0.3, \Omega_{Lambda} = 0.7$. The importance of the results is that they confirm using completely different physics the need for a non-zero cosmological constant. Of course, the bigger the lens sample the more powerful a tool it becomes.

5.3. PROPAGATION EFFECTS

Lensing galaxies are not smooth structureless things; they contain stars, gas in neutral, molecular and ionized form, dust, magnetic fields, dark matter, etc. All can and do imprint some signature on the radiation as it propagates through the lensing galaxy. Many of the effects are seen in just one system, B0218+357. I will simply list some of them:

- Neutral hydrogen absorption has been detected by Carilli et al. (1993).
- Molecular absorption has been detected by Wiklind and Combes (1995).
- Dust extinction affecting the visible light results in the relative brightnesses of the images being different by a factor of about 40 between radio and optical wavelengths.
- Irregularities in the distribution of ionized gas in the lensing galaxy result in scatter-broadenning of the radio images (Biggs et al., 2002).

One manifestation of this is the very strong dependence of the angular sizes of the lensed images on observing frequency; at 1.4 GHz the size of image A is ∼40 mas while at 5 GHz it has shrunk to about 5 mas.
- There is a relative rotation measure of about 800 radians/m^2 between the A and the B image due to differential Faraday radiation occurring in the magnetoionic medium of the lensing galaxy. Another effect of this is the strong depolarization of the A image which is even evident at frequencies as high as 15 GHz. Since the size of the A is ∼1 mas the implied gradient of rotation measure is ≥500 radian/m^2/mas.

Other systems show different effects; B2114+022 shows evidence for free-free absorption (Augusto et al., 2001). Similarly there is strong evidence in B1600+434 for the effects of microlensing by individual stars as exceedingly compact superluminal components move with respect to a microlensing caustic network (Koopmans et al., 2000)

The inability of smooth lens models to explain the observed flux density ratios and radio substructure in lensed images is invoked as evidence for large-scale sub-structure in the matter distributions of the lensing galaxies of the kind predicted by cold dark matter structure formation models (Mao & Schneider, 1998; Dalal & Kochanek, 2002; Metcalf, 2002; Bradac et al., 2002). Finally, lensing tells us about the distribution of all mass; comparing the distribution of luminous mass to that of the total mass obtained from lens modelling should give the profile of the dark matter halo.

6. Future work

It would be great to find and study more lens systems but unfortunately we are reaching the limitations of current radio instruments. To find an order of magnitude more systems than CLASS will mean mapping sources of a few mJy. This is of course possible and the increased sensitivity of upgraded instruments like the EVLA and e-MERLIN will mean that signal-to-noise will no longer be a major problem in a few years time. But observing large numbers of sources (∼100,000) one at a time will still be too time-consuming. Some multiplexing scheme will be required and it is not obvious have this can be achieved.

I do not want to end on a low note. There is still a vast amount of work to be done on the existing lens systems. Using high-resolution, multi-frequency observations to study propagation effects such as scattering, free-free absorption, Faraday rotation occurring in lensing galaxies is a subject only just beginning to be explored. Another almost untouched area is to use VLBI spectral-line observations of absorption systems to study the kinematics and velocity dispersions of lensing galaxies.

For those lens systems that already have time delays (three in CLASS), high-resolution mapping is the key to refining mass models and for making gravitational lensing the method of choice to obtain the most accurate value for the Hubble constant. More and better time delays are required. Monitoring flux densities and polarizations is not the only way to get time delays. Monitoring superluminal motions in images is another possibility but, rather surprisingly, no superluminal lens system has yet been identified. Finally, the extra magnification provided by the lens may be used to increase the effective resolution. This can reveal detail in the host galaxies of lenses objects visible in no other way and may enable us to study the early evolution of compact radio sources, for example CSO, several of which we know to be lensed.

References

Augusto, P., et al., (2001), *MNRAS* **326**, 1007
Biggs, A.D., et al., (1999), *MNRAS*, **304**, 349
Biggs, A.D., et al., (2001) *MNRAS*, **322**, 821
Biggs, A.D., et al., (2002), *astro-ph/0209182*
Bradac, M. et al., (2002), *A&A*, **388**, 373
Browne, I.W.A., et al., (1998), *MNRAS*, **293**, 257
Browne, I.W.A., et al., (2002), *astro-ph/0211069*
Burke, B.F., (1993), *Sub-Arcsecond Radio Astronomy, Eds R.J.Davis, R.S.Booth, CUP*, 123
Carilli, C., (1993), *ApJ*, **412**, L59
Chae, K-H., et al., (2002), *PhRvL*, **89**, 151301
Chae, K-H., (2002), *astro-ph/0209602*
Dalal, N. & Kochanek, C.S., (2002), *ApJ*, **572**, 25
Jackson, N.J., et al., (2002), *The Review of Radio Science 1999-2002, Ed. W.R. Stone, Wiley*, p831
Koopmans, L.V.E., et al., (2000), *A&A*, **358**, 793
Mao,S., & Schneider, P., (1998), *MNRAS*, **295**, 587
Metcalfe, R.B., (2002) *ApJ*, **580**, 696
Myers, S.T., et al., (2002), *astro-ph/0211073*
Patnaik, A.R., et al., (1992), *MNRAS*, **254**, 655
Patnaik, A.R., et al., (1995), *MNRAS*, **274**, 5P
Schneider et al., (1992), *Gravitational lenses, Springer-Verlag*
Wiklind, T., & Combes, F., (1995), *A&A*, **299**, 382
Wilkinson, P.N. et al., (1998), *MNRAS*, **300**, 790
Wucknitz, O., (2002), *PhD thesis, Univerity of Hamburgh*

VLBI OBSERVATIONS OF GRAVITATIONAL LENSES

R.W. PORCAS
Max-Planck-Institut fuer Radioastronomie
Auf dem Huegel 69, D53121 Bonn, Germany

Abstract.
This lecture covers VLBI observations of gravitational lenses, and addresses the particular aspects of gravitational lens studies for which VLBI has proved a useful technique. These include confirming strong gravitational lens "candidates", uncovering the details of complex "merging" images, and providing additional constraints for modelling the lens mass distribution. Examples from a number of known gravitational lens systems are given. Part of the lecture is devoted to considering the special problems of VLBI observing, correlation and analysis when the "source" consists of multiple compact sources in a relatively wide field.

1. Why make VLBI Observations of Gravitational Lenses ?

In *strong* gravitational lensing a mass distribution located between a distant source and the observer produces multiple images of the source. If the source is a sufficiently strong and compact radio emitter we can use VLBI observations to make high-resolution maps of these images, allowing us to examine their mas-scale morphologies and to measure the inter-image separations. VLBI can be used to:

• Confirm lens candidates by demonstrating similar morphologies in the images (e.g. Porcas et al. 1981) or by using the properties that the lens cannot change the spectral index, surface brightness or the degree and angle of polarization of a region in the source (e.g. Patnaik et al. 1995)

• Resolve cases of ambiguous image configurations when lower resolution studies are insufficient (e.g. Garrett et al. 1996a)

• Refine models of the lens mass distribution by providing a *relative magnification matrix* relating the structure of one image to that of another (e.g.

F. Mantovani and A. Kus (eds.), The Role of VLBI in Astrophysics, Astrometry and Geodesy, 219–231.

Garrett et al. 1994a), or determining the values for a parametrized model of the lens potential (e.g. Trotter et al. 2000)

• Provide evidence for deviations from a smooth gravitational potential (mass "granularity') from image distortions (e.g. Garrett et al. 1996b; Biggs & Browne, 2002) or from image flux ratio discrepancies (e.g. Bradac et al. 2002)

• Determine the image relative *time delay* from structural variations (e.g. Campbell et al. 1995)

• Search for "milli-lenses" (10^{6-8} solar masses) by looking for small (1-100 mas) separation images (e.g. Patnaik et al. 1996)

• Investigate the background source using the high magnification of the lens in some images (e.g. Koopmans et al. 2002)

• Investigate the matter around the lens (usually a galaxy) by comparing various propagation effects along the different image paths (e.g. absorption: Carilli et al. 1998; scattering: Jones et al. 1996 & Guirado et al, 1999; and Faraday rotation) and looking for lens proper motion by its effect on the image separations (e.g. Kochanek et al. 1996)

Reviews of VLBI work on gravitational lenses are given in Patnaik and Porcas (1996), Porcas (1998) and Biggs (2003). First, however, we examine some of the peculiar features which arise when making such observations.

2. Special Features of Gravitational Lens VLBI

The main problems with gravitational lens VLBI arise because of the need to have a relatively large field of view (to cover all the images) and because the field may contain multiple compact radio sources of comparable strength. Some of these problems are covered in Porcas (1994) and Garrett et al. (1994b)

2.1. OBSERVING

There are normally no problems in observing the whole field containing the images with a single observation since typical image separations are a few arcseconds and the individual antenna beam widths are ~arcminutes at cm wavelengths. Exceptions are when *phased arrays* are used as sensitive VLBI elements (e.g. VLA and WSRT). It may be necessary to select just a subset of the array (with attendant loss of sensitivity) to create a large enough phased-array beam or, in the case of the VLA, wait for an array configuration where the antennas are sufficiently densely packed.

2.2. CORRELATION

Restriction of the field of view also occurs at correlation. Here the calculated delay and phase difference between the antenna signals from the source are compensated, using an accurate model of the geometry involved (antenna and source positions, sidereal time, orientation of the Earth's axis and station clock offsets). Multiplication of the antenna signals on a pair-wise basis is then made at a number of different delays surrounding the calculated one (to take account of any errors in the model of the delay) and the results are averaged (e.g. for a few seconds) before output. For a lens system the source "position" must be a compromise - some point midway between the images - so the compensation is not exact. This results in a residual delay, phase and phase-rate (*fringerate*) for each image response in the output visibility function. The image response is outside the "correlation field-of-view" if its residual fringerate results in a phase change of more than ~1 radian during the averaging time, or if its residual delay is outside the range of delays correlated. (For an "FX" correlator like that of the VLBA the correlator frequency resolution is equivalent to the delay range. The change of phase across a single frequency channel induced by the residual delay must not exceed ~1 radian.)

The field of view is determined by the baseline lengths, the observing frequency, the output averaging time and the delay range (or frequency resolution). For small image-separation gravitational lenses (~1 arcsec) or moderate baseline lengths (e.g. within Europe) or observing frequencies below 5 GHz the field-of-view of present correlators is usually sufficient. Problems arise when the limits are pushed by observing wide-separation image systems, using high observing frequencies and long baselines.

As an example, the famous double quasar 0957+561 has 2 images separated by 6.1 arcsec. On the baseline Effelsberg to Jodrell Bank at 1.7 GHz each image has a residual fringerate up to 3.3 mHz (after path-length compensation with respect to their mid point) and their residual delays can each reach up to 33 ns. These correspond to phase changes of only 1.2° during a 1 s time average and 6° across a 0.5 MHz frequency channel; thus the images are well within the correlation field-of-view. The same lens observed at 86 GHz between Pico Veleta (Spain) and Owens Valley (California) would cause severe problems, with residual fringerates up to 2.2 Hz for each image (2.2 turns of phase in a 1 s average with a beat frequency between the images of 4.3 Hz) and residual delays of 350 ns (53° across a 0.5 MHz frequency channel). In this case one must use one of the very few advantages of tape-recorded interferometry and correlate the data twice, using the different image positions for the correlator model in the two *passes*. This results in two separate visibility functions which can be analysed individually. Note, however, that the short baselines in each data set will probably still contain partial responses from the "other" image. In addition, since the two images are observed simultaneously with the same antennas, the instrumental phase errors should be identical for the two data sets, and it may be possible to utilise this fact in their data analyses (e.g. by using the *hybrid double mapping* technique of Rioja and Porcas (2000)).

2.3. MAPPING

The main point to realise is that every time we average the visibility function, either in time or frequency, we further restrict the field-of-view (for the reasons explained above). If we use a single-pass correlation and the gravitational lens images only just fit in the field, then we cannot average the data further if we wish to make a single map of the entire field. Such a map will usually contain vast amounts of "empty space" since the size of the compact source structure in each image is much less than the image separation (see Figure 1). In this situation it is sensible to define sub-fields around the known image positions and make the map only in these fields, as can be done with the AIPS task **IMAGR**. Figure 2 shows detailed, high-resolution maps of the 4 image regions of the quad-lens MG 0414+0534, made in this way.

An alternative approach is to *rotate* the (un-averaged) visibility data to new reference positions centred on each of the images in turn (by an appropriate rotation of all the visibility phases), followed by averaging of the visibility function in time and frequency. This shrinks the field-of-view around each image, effectively creating individual visibility functions which can then be mapped as single fields, as if each were a single source. (This approach is the same as when one has to have multiple correlator passes.) The disadvantage of this approach is that there is normally some "contamination" by unwanted images on the short baselines. In addition, some of the images may be too weak for phase self-calibration; if this process is necessary then instrumental phase solutions determined from other, stronger images will have to be transferred and applied.

2.4. SELF-CALIBRATION

If one uses self-calibration processes on the gravitational lens visibilities (e.g. the *fringe-fitting* and *phase self-calibration* steps used in the absence of phase-referencing) then a problem arises because of the need to average in time and frequency in order to achieve enough signal-to-noise ratio to exceed the detection threshold. In the fringe-fitting process the data for some appropriate time interval (the "solution interval") is averaged, using a range of possible fringe-rates and delays, and a search is made to find the

Figure 1. **EVN 1.7 GHz wide-field image of the field of the gravitational lens 0957+561** (from Garrett et al. 1999). Images A and B are the westernmost and southermost features, respectively. The faint emission 1 arcsec to the north of B is presumed to be emission from the lens galaxy. Also visible are compact features in the extended jet/lobe emission to the northeast of A. Note that most of the field is empty !

expected source response at the residual instrumental rate and delay. Different images will normally have responses at different rates and/or delays (e.g. see Figure 3). If the search range is wide enough to include all image responses, the procedure will normally choose the image with the highest response. This process works well if there is one, very dominant, image (e.g. the asymmetric 2-image system 1030+074 with image flux ratio 13:1; Xanthopoulos et al. 1998) but if there are images of comparable strengths the process may pick different images for different baselines and times. This in turn results in inconsistent results for the instrumental residuals.

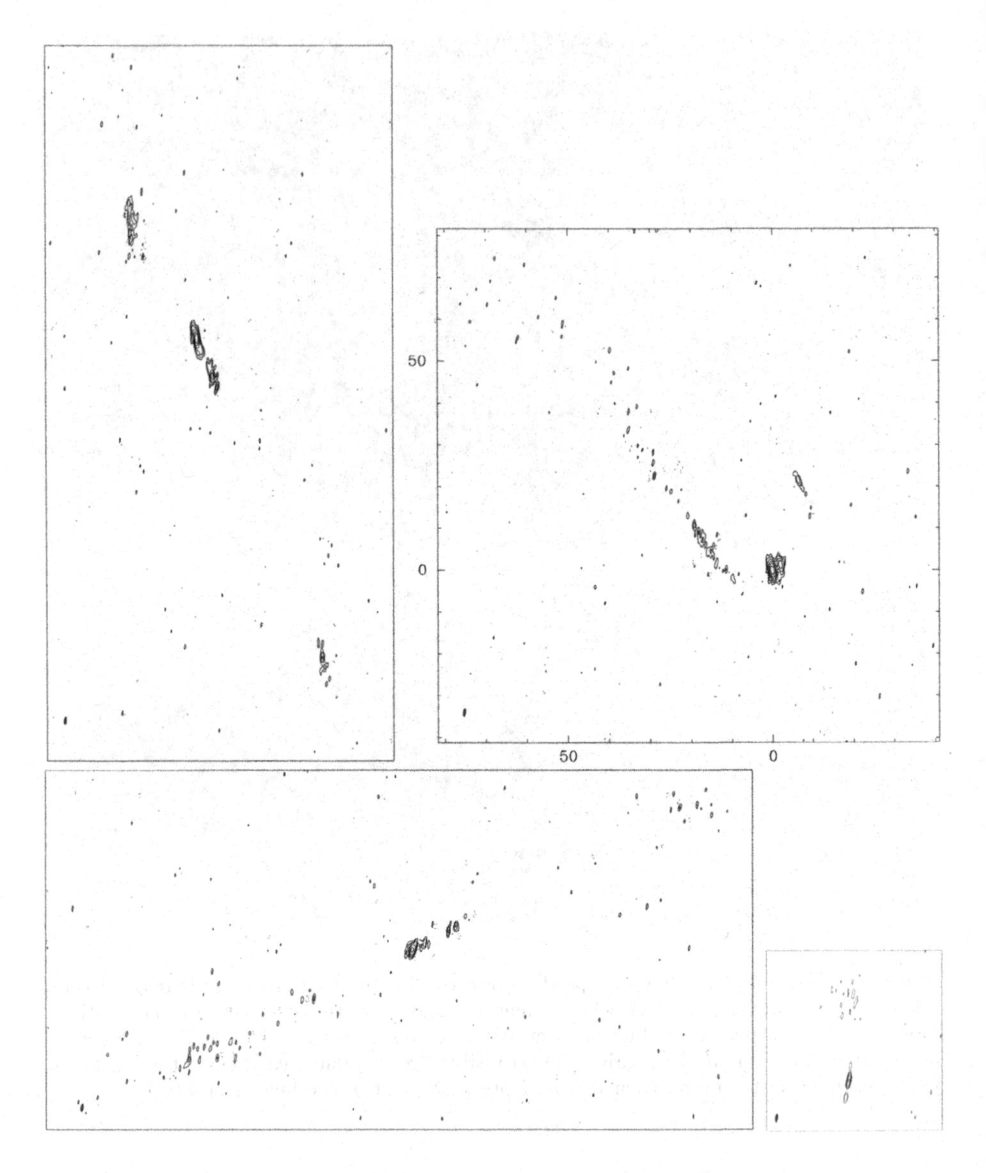

Figure 2. **VLBI 8.4 GHz maps of 4 sub-fields of MG 0414+0534 containing its 4 images** (from Ros et al. 2000). **Top**: images A2 and A1. **Bottom**: images B and C

(The residuals **are** needed, however, for averaging the data in the phase self-calibration step - see below).

One way around this problem is to fringe-fit using a full source model (e.g. provisional maps of all the image fields from a previous mapping iteration); the unaveraged visibility function is divided by the calculated visibil-

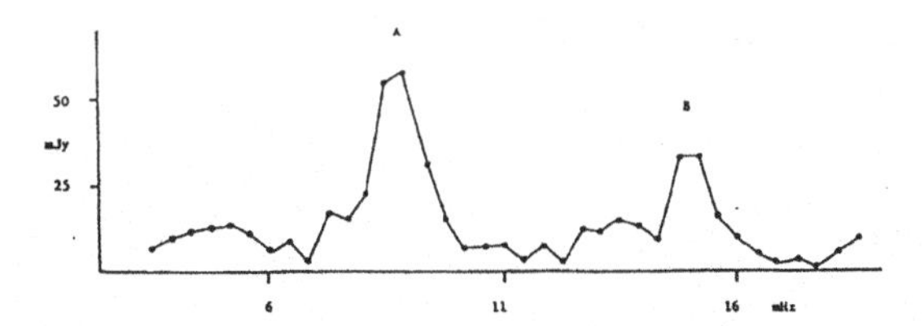

Figure 3. Residual fringe-rate spectrum of 0957+561 from 20 min of data on the Effelsberg-Jodrell baseline at 1.7 GHz (from Porcas et al. 1979). Note the two peaks corresponding to the A and B image responses.

ity function of the model before the data is averaged in fringe-fitting, which moves the image responses to a single rate and delay peak corresponding to the field centre position. Another trick (e.g. for the initial fringe-fit when no model is available) is to use an offset, single point source model to represent just one (e.g. the brightest) image. This is equivalent to rotating the data so that this image is at the search centre, and it can be isolated by restricting the range of rates and delays which are searched.

Averaging the data to achieve a threshold signal-to-noise ratio also occurs in phase self-calibration. The input source model must consist of all the separate sub-fields containing the various images if the structure has been mapped in this way. If one of the images is particularly strong (and compact) one can use a single, offset source model (as described above) for an initial phase self-calibration step.

3. Comparing Images

Having made maps of the various images of a gravitational lens system, the next step is to put them to good use ! One interesting direction is to try and learn more about the details of the mass distribution of the lens by comparing the distortions of the structure of the background object in the various images. This depends upon recognising "corresponding points" in the images, i.e. points in the images which are images of the same point in the background object. This may be problematic, especially if the background object does not have a rich structure, or if the image magnifications are very different. Remember that, since gravitational lensing preserves the background object's surface brightness in all images, a brighter image will have a correspondingly larger surface area, and hence we may see in it features of the background object at much higher resolution than in other, weaker images.

Figure 4 shows high dynamic range, global VLBI maps of the two images of the gravitational lens B 0218+357. The weaker B image (left) exhibits

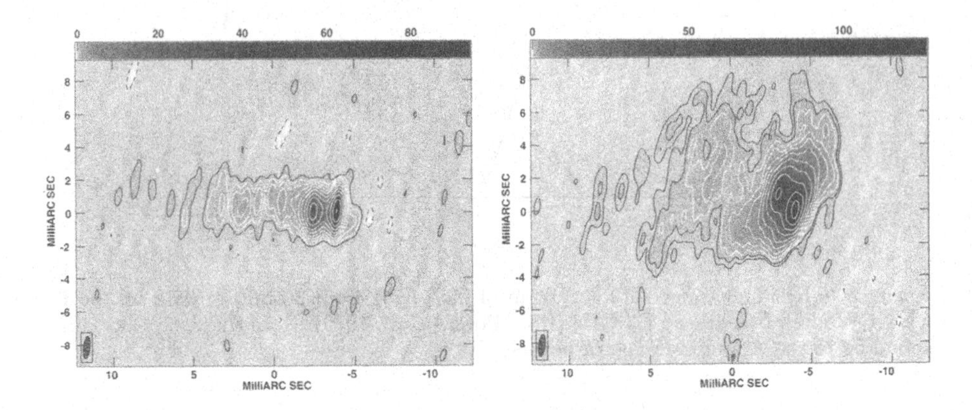

Figure 4. **8.4 GHz Global VLBI maps of the B (left) and A (right) images of the gravitational lens B 0218+357** (from Biggs et al., 2003)

two bright peaks in the west, and an elongated jet-like feature stretching to the east. The same basic morphology can be recognised in the A image (right) where the double peaks appear in the south west, and the jet extends to the north west. Note the relative "stretching" of the A image in position angle $\sim$ -40° and the relative contraction of the jet length, due to the different effects of the lens in the two images. Careful inspection of the bends in the jet path also show that the images have opposite parity (the northward deviation of the jet path in image B corresponds to a southward deviation in image A).

4. Polarization observations

Polarization VLBI imaging of gravitational lenses provides another way of identifying corresponding points in different images, since the gravitational lensing process itself preserves in each image both the degree and angle of polarization of any point in the background object. Recent "VLBP" observations of gravitational lenses are described in Patnaik et al. (1999), Kemball et al. (2001) and Jin et al. (2003).

Polarization VLBI maps of the 4-imaged system B 1422+231, made at 8.4 GHz using the VLBA and the Effelsberg 100 m telescope, are shown in Figure 5. The bright images, A and B, show elongated but relatively featureless structure in total intensity, I, (indicated by contours). By contrast, the polarized emission, indicated by the bars showing the polarization angle (E-field), reveals an asymmetry, with polarized regions in the centres and at only one end of the structures (which must therefore be corresponding ends). The direction of the slight offset of the outer polarization peaks from the source axis defined by the total intensity emission shows that these 2

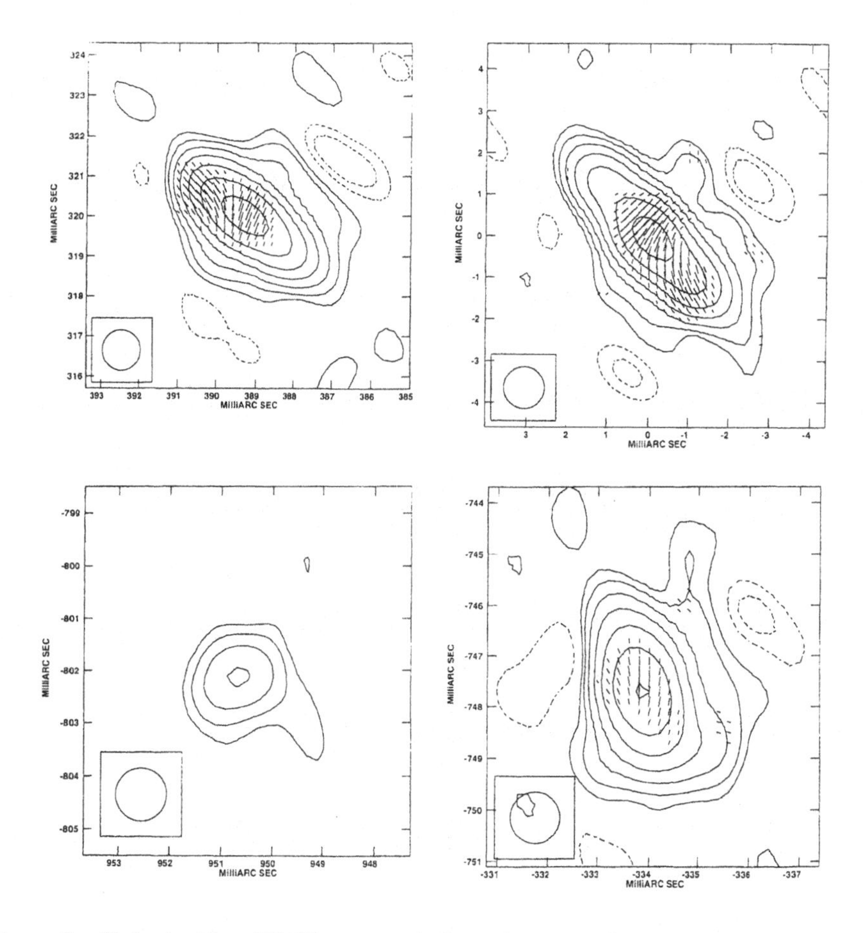

Figure 5. **Polarization VLBI maps of the 4 images of B 1422+231, made at 8.4 GHz** (from Patnaik et al. 1999). Images are **Top:** A (left), B (right); **Bottom:** D (left) and C (right).

images, also, have opposite parity.

Although the polarization angle of the background source emission is preserved by gravitational lensing, the actual measured polarization angle in the different images may, in fact, be different if the different image paths undergo different Faraday rotation. One must therefore take account of this before using the polarization angle to indentify corresponding points. Figure 6 shows VLBP maps of the B (left) and A (right) images of B 0218+357, made at 8.4, 22 and 43 GHz. Note that at 8.4 GHz the compact "core" component (on the right in both A and B maps) does not have the same polarization angle in the two images, although we know these are corresponding points. However, at higher frequencies the effect of the differential Faraday rotation between the two image paths becomes less, and at

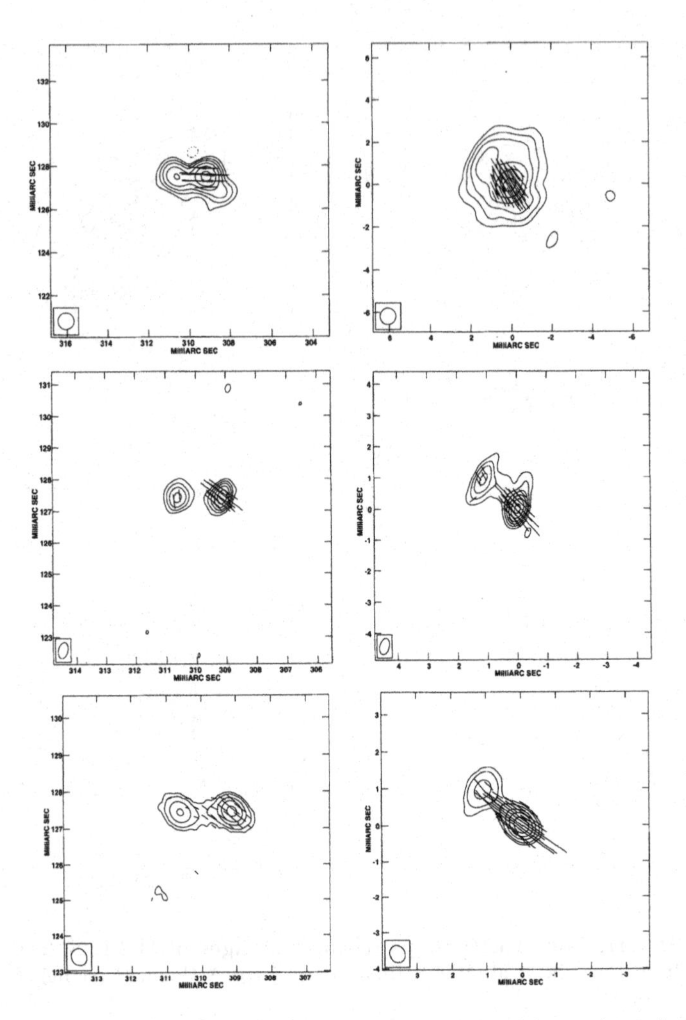

Figure 6. **Polarization VLBI maps of the B (left) and A (right) images of B 0218+357** (from Kemball et al. 2001 and Kemball, Patnaik & Porcas, unpublished). **Top:** 8.4 GHz; **Middle:** 22 GHz; **Bottom:** 43 GHz. Note the equal polarization angles of the bright core emission in the two images at 43 GHz where Faraday rotation is negligible.

43 GHz, where the effect is negligible, the polarization angles are equal. This angle is then the true polarization angle of the background object, although the source structural position angle (which is changed differently in the two images) is not the true one !

5. Relative Magnification Matrix

Having identified the corresponding points in VLBI maps of gravitational lens images it is useful to convert this information into a constraint for modeling the lens mass distribution. This can be done by determining the image *relative magnification matrix* which describes a simple linear transformation of one image into another (for regions where the magnification in the image is constant). In addition to the image flux ratios and positions, this provides 3 further quantities for each image with respect to a reference image. A case where this has been done to good effect is for the double quasar 0957+561 (Garrett et al. 1994a), since both images exhibit the complex jet structure of the background quasar at mas-resolution. Figure 7 shows maps of the A (left) and B (middle) images. Again, the opposite image parity

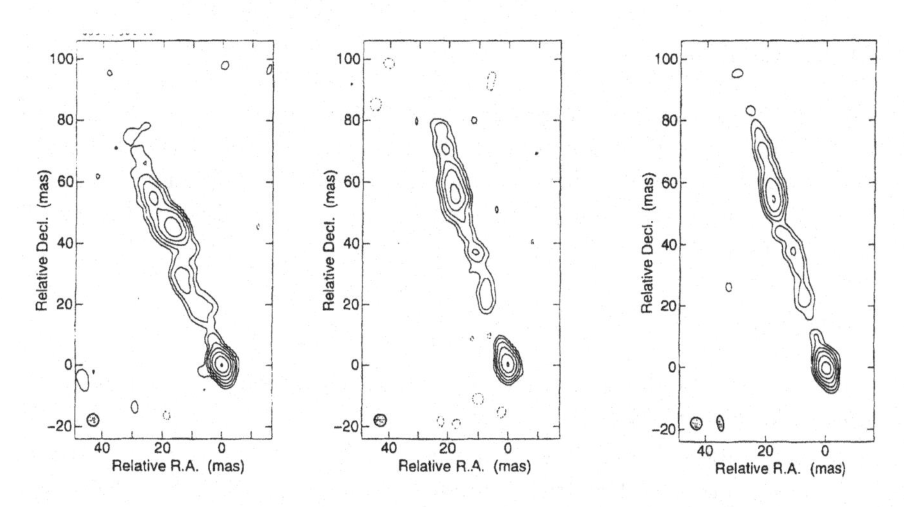

Figure 7. **Global 1.7 GHz VLBI maps of the double quasar 0957+561** (from Garrett et al. 1994a). **Left:** Image A; **Middle:** Image B; **Right:** Image A transformed into image B using the relative magnification matrix.

is revealed by the small kink in the jet. Modelling of the various features along the jet, recognisable in both images, enabled a relative magnification matrix between these two images to be obtained. The map on the right of Figure 7 is the result of transforming the A image into a representation of the B image using this matrix. Note that most of the significant features of the real B image are reproduced by this process. (It is interesting to compare these maps of the A and B images, from observations made in 1989, with those from more recent observations by Haarsma et al. (2001).)

6. Special Analysis Algorithms

Finally, mention should be made of data analysis algorithms which attempt to incorporate the fact that the map being made consists of multiple images of a single source. The LensClean (Kochanek and Narayan, 1992) and LensMEM (Wallington et al. 1996) algorithms attempt to decompose an existing (lens image plane) map into a single (lens object plane) source using a model of the lens, which can be suitably adjusted, whilst the Visibility-LensClean algorithm (Ellithorpe et al. 1996) incorporates this decomposition into the procedure for deriving the image from the interferometer visibilities. These algorithms have rarely (if ever) been used for reconstructing VLBI images, as far as this author is aware. Another, ambitious approach is to attempt to reconstruct the relative magnification matrix for two images from the visibility data alone, using a "shapelet" formalism (Baehren et al. 2002), although this has not yet been tried on real data.

7. Acknowledgements

I thank Eduardo Ros for providing the excellent maps of the images of MG 0414+0534 for inclusion in this lecture and Rupal Mittal for comments on the text. The Monthly Notices of the Royal Astronomical Society, in which Figures 4, 5 and 7 first appeared, are published by Blackwell Publishing, Oxford.

8. References

Baehren, L., Schneider, P. & King, L.J., 2002, Proc. 6th EVN Symposium, ed. Ros et al. (Bonn, MPIfR), 201

Biggs, A.D. 2003, Proc. "Future Directions in High Resolution Astronomy: A Celebration of the 10th Anniversary of the VLBA" ed. Romney, (Socorro, NRAO) *in preparation*

Biggs, A.D. & Browne, I.W.A., 2002, Proc. 6th EVN Symposium, ed. Ros et al. (Bonn, MPIfR), 193

Biggs, A.D., Wucknitz, O., Porcas, R.W., Browne, I.W.A., Jackson, N.J., Mao, S. & Wilkinson, P.N., 2003, MNRAS, 338, 599

Bradac, M., Schneider, P., Steinmetz, M., Lombardi, M., King, L.J. & Porcas, R.W., 2002, A+A, 388, 373

Campbell, R.M., Lehar, J., Corey, B.E., Shapiro, I.I. & Falco, E.E., 1995, AJ, 110, 2566

Carilli, C.L., Menten, K.M., Reid, M.J., Rupen, M. & Claussen, M. 1998, in Proc. IAU Coll. 164, ed. Zensus et al. (Socorro, NRAO), 317

Ellithorpe, J. D., Kochanek, C. S.,& Hewitt, J. N. 1996, ApJ, 464, 556

Garrett, M.A., Calder, R.J., Porcas, R.W., King, L.J., Walsh, D. & Wilkinson, P.N., 1994a, MNRAS, 270, 457

Garrett, M.A., Patnaik, A.R. & Porcas, R.W., 1994b, in Proc. 2nd EVN/JIVE Symposium, ed. Kus et al. (Torun, TRAO), 73

Garrett, M.A., Porcas, R.W., Nair, S. & Patnaik, A.R., 1996a, MNRAS, 279, L7

Garrett, M.A., Nair, S., Porcas, R.W. & Patnaik, A.R., 1996b, Proc. IAU Symp. 173, ed. Kochanek & Hewitt, (Dordrecht, Kluwer), 189
Garrett, M.A., Porcas, R.W., Pedlar, A., Muxlow, T.W.B. & Garrington, S.T. 1999, New Ast. Rev. 43, 519
Guirado,J.C., Jones, D.L., Lara, L., Marcaide, J.M., Preston, R.A., Rao, A.P. & Sherwood, W.A.,. 1999, A+A, 346, 392
Haarsma, D., Lehar, J. & Barkana, R., 2001, Proc. Gravitational Lensing: Recent Progress and Future Goals, ed. Brainerd & Kochanek, (San Fransisco, ASP), 89
Jin, C., Garrett, M.A., Nair, S., Porcas, R.W., Patnaik, A.R. & Nan, R., 2003, MNRAS, 340, 1309
Jones, D.L., Preston, R.A., Murphy, D.W., Jauncey, D.L., Reynolds, J.E., Tzioumis, A.K., King, E.A., McCulloch, P.M., Lovell, J.E.J., Costa, M.E. & van Ommen, T.D., 1996, ApJ L, 470, L23
Kemball, A.J., Patnaik, A.R. & Porcas, R.W., 2001, ApJ, 562, 649
Kochanek, C.S., Kolatt, T.S. & Bartelmann, M., 1996, Ap.J, 473, 610
Kochanek, C. S., & Narayan, R. 1992, ApJ, 401, 461
Koopmans, L.V.E., Garrett, M.A., Blandford, R.D., Lawrence, C.R. Patnaik, A.R. & Porcas, R.W., 2002, MNRAS, 334, 39
Patnaik, A.P. & Porcas, R.W. 1996, IAU Symposium 173, ed. Kochanek & Hewitt, (Dordrecht, Kluwer), 305
Patnaik, A.R., Porcas, R.W. & Browne, I.W.A., 1995, MNRAS, 274, L5
Patnaik, A.R., Garrett, M.A., Polatides, A. & Bagri, D., 1996, Proc. IAU Symp. 173, ed Kochanek & Hewitt, (Dordrecht, Kluwer), 405
Patnaik, A.R. Kemball, A.J., Porcas, R.W. & Garrett, M.A., 1999, MNRAS, 307, L1-L5
Porcas, R.W., Booth, R.S., Browne, I.W.A., Walsh, D. & Wilkinson, P.N., 1979, Nature, 282, 385
Porcas, R.W., Booth, R.S., Browne, I.W.A., Walsh, D. & Wilkinson, P.N., 1981, Nature, 289, 758
Porcas, R.W., 1994, in "Compact Extragalactic Radio Sources", ed. Zensus & Kellermann, (Socorro, NRAO), 125
Porcas, R.W. 1998, IAU Colloquium 164, ed. Zensus et al. (Ast. Soc. Pac. San Francisco), 303
Rioja, M.J. & Porcas, R.W. 2000, A+A, 355, 552
Ros, E., Guirado, J. C., Marcaide, J. M., Prez-Torres, M. A., Falco, E. E., Muoz, J. A., Alberdi, A. & Lara, L. 2000, A+A 362, 845
Trotter C. S., Winn J. N. & Hewitt J. N., 2000, ApJ 535, 671
Wallington, S., Kochanek, C.S. & Narayan, R., 1996, ApJ, 465, 64
Xanthopoulos, E., Browne, I.W.A., King, L.J.,Jackson, N.J., Marlow, D.R., Nair, S., Patnaik, Porcas, R.W. & Wilkinson, P.N., 1998, MNRAS, 300, 649-655

SCINTILLATION IN EXTRAGALACTIC RADIO SOURCES

M. BONDI
Istituto di Radioastronomia del CNR
Via Gobetti 101, I-40129 Bologna, Italy

1. Introduction

Electromagnetic radiation from compact extragalactic radio sources pass through several ionized media: the intergalactic gas, the interstellar medium, the interplanetary medium and the ionosphere. In all these cases, the density fluctuations in the turbulent medium can produce variations of the refractive index and consequently a phase modulation of the wavefront and scattering. As a consequence, a wide variety of phenomena such as intensity scintillation, angular broadening and pulse smearing can be observed. The study of these phenomena provides information on the angular size of the scattered sources and a unique method for the remote analysis of astrophysical plasmas. The various contributions to the scattering can be separated as each medium is sensitive to different angular scales developing scintillation with time-scales ranging from a few seconds to years.

In this contribution I will focus on the intensity scintillation of extragalactic radio sources which is produced by the wave propagation of the extragalactic radiation through the interstellar medium (ISM) of our galaxy. A sketch illustrating the effects of strong scattering on the wave propagation is shown in Figure 1. The plane waves from an extragalactic radio source are scattered passing through the turbulent medium, producing a pattern of dark and bright patches on the observer's plane. The relative motion of the observer through this pattern produces the intensity scintillation. Roughly speaking, the magnitude of the intensity scintillation mainly depends on the source size and the scattering properties of the ISM, while the time-scale on which the scintillation occurs depends on the source size and the distance of the scattering medium from the observer. The reader interested in a more detailed and complete description of interstellar scattering can rely on two conference proceedings [4, 24], and a review paper [18].

F. Mantovani and A. Kus (eds.), The Role of VLBI in Astrophysics, Astrometry and Geodesy, 233–242.

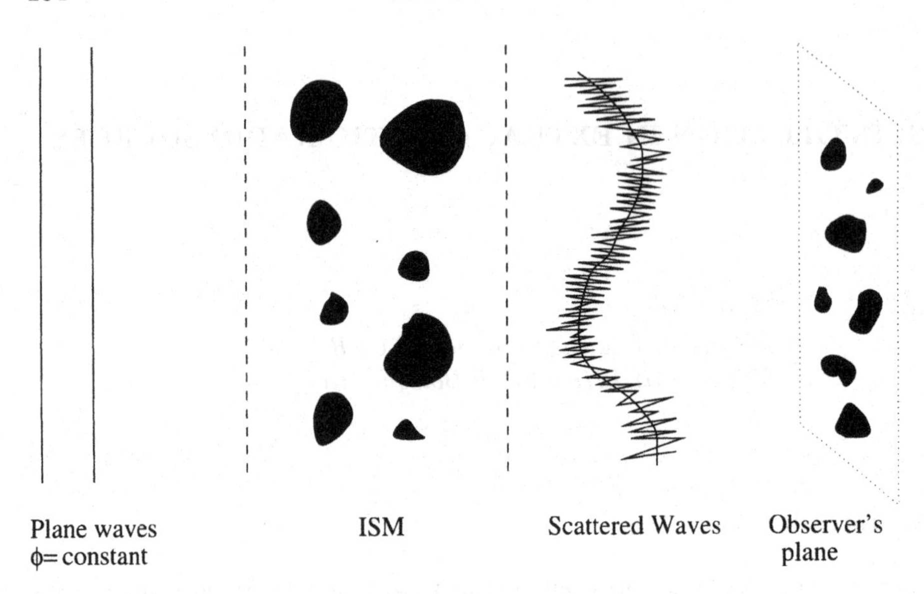

Figure 1. Sketch illustrating the effect of strong scattering on the phase of incident waves. Two component phase variations can be identified: fast diffractive phase fluctuations occurring on short spatial scales, and slower large scale refractive fluctuations.

The outline of this paper is the following. In Section 2, the link between the density fluctuations in the ISM and the intensity fluctuations is explained. Section 3 introduces the basic definitions for the most relevant scattering parameters used throughout the paper. Section 4 gives the different scintillation regimes and the basic relations that can be used to roughly estimate the amount of scintillation at different observing frequencies. Finally, Section 5 reviews three applications of the interstellar scintillation.

2. Density and intensity fluctuations

The density fluctuations in the turbulent medium are described assuming a power-law spectrum:

$$P_{\Delta n}(q) = C_N^2(r)q^{-\beta} \tag{1}$$

where C_N is an amplitude parameter depending on the line of sight and q is the three dimensional wavenumber of the density fluctuations in the plasma. The exponent β is usually assumed to be equal to 11/3, corresponding to the Kolmogorov spectrum for turbulence in neutral gas.

Based on the analysis of Coles [3], in the case of refractive scintillation, the power spectrum of the density fluctuations can be related to the power spectrum of intensity fluctuations [1]:

$$P_{\Delta I}(q) \propto \int_0^L P_{\Delta n}(q)[\sin(q^2 z\lambda/(4\pi))]^2 |V(qz\lambda/(2\pi))|^2 dz \tag{2}$$

where z is the coordinate along the line of sight through the turbulent medium of thickness L, and λ is the observing wavelength. The two-dimensional flux density fluctuation power spectrum is related to the density fluctuation power spectrum through two filter functions: the Fresnel filter, $\sin^2(q^2 z\lambda/4\pi)$, which tends to make scintillation measurements insensitive to large scale fluctuations, and the square modulus of the source visibility, $|V(qz\lambda/2\pi)|^2$, which is determined by the intrinsic structure of the source. Measuring the power spectrum of intensity fluctuations from the monitoring of scintillating sources and the source visibility with VLBI observations, it is possible to derive information on the three-dimensional power spectrum of density fluctuations in the turbulent medium.

3. Scintillation Jargon

In this section I will introduce the definition of the most relevant quantities used in the scintillation theory of extragalactic radio sources.

The scintillation index is defined as the r.m.s. fractional intensity fluctuation, for a point source this is:

$$m_p = \left(\frac{S_{rms}}{< S >}\right)_p \tag{3}$$

Another quantity that can be derived from scintillation observations is the time-scale. This is usually defined as the characteristic time between a minimum and maximum in the light curve. The time-scale will generally depend not only on the source size and the scattering properties of the medium, but also on the relative motion between the medium and the observer and can provide information on the velocity of the irregularities in the ISM.

As we have seen before, the scattering is characterized using a power law spectrum where the amplitude parameter $C_N(r)$ measures the local scattering. The cumulative effect of the ISM inhomogeneities along the line of sight is defined as the scattering measure, $SM = \int C_N^2 dz$.

A relevant quantity is the Fresnel scale, given by (units are cm):

$$R_F = \sqrt{\lambda D/2\pi} \sim 1.2 \times 10^{11} \sqrt{D(kpc)/\nu(GHz)} \tag{4}$$

D is the effective distance through the scattering medium, in kpc, and ν is the observing frequency in GHz. The equivalent angular size of the Fresnel scale is given (in arcseconds) by:

$$\theta_F \sim R_F/D \sim 8 \times 10^{-6}/\sqrt{D(kpc)\nu(GHz)} \quad (5)$$

At about 1 GHz, the angular size of the Fresnel scale is about few microarcsec (μas) for any line of sight across the ISM.

4. Scintillation regimes

Following Narayan (1992) and Walker (1998) it is useful to introduce the concept of scattering strength using the parameter ξ. The scattering strength depends on the properties of the medium, scattering measure and effective screen distance, and on the observing frequency. For a Kolmogorov spectrum of turbulence [26]:

$$\xi \propto SM^{0.6} D^{0.5} \nu^{-1.7} \quad (6)$$

Scintillation is divided into weak and strong according to whether ξ is much smaller or greater than unity. In the strong regime the wavefront is highly corrugated on scales smaller than the Fresnel scale, in the weak regime the phase changes over the Fresnel scale are small. Assuming a model for the distribution of the scattering material [25] it is possible to map the transition frequency ν_0, i.e. the frequency at which $\xi = 1$, and the Fresnel angular scale at the transition frequency, θ_{F0}. At a galactic latitude of 30°, $\nu_0 \simeq 10$ GHz and is roughly constant at higher latitudes, while on the Galactic plane $\nu_0 \simeq 40$ GHz; $\theta_{F0} \simeq 3$ μas at a galactic latitude of 30 degree [27].

4.1. WEAK SCINTILLATION

The spatial scale for weak intensity variations is the Fresnel scale R_F. For sources with angular extent greater than θ_F the scintillation patterns from different parts of the source overlap and smear each other out, eliminating detectable variations:

$$\theta_{\rm weak} = \theta_F = \theta_{F0}\sqrt{\nu_0/\nu} \quad (7)$$

For a point source ($\theta \leq \theta_{\rm weak}$), the scintillation index depends on the observing and transition frequencies only:

$$m_{\rm p,weak} = \xi^{5/6} = (\nu_0/\nu)^{17/12} \quad (8)$$

For a source with $\theta > \theta_{\rm weak}$ the following relation hold:

$$m_{\text{weak}} = m_{\text{p,weak}}(\theta_{\text{weak}}/\theta)^{7/6} \tag{9}$$

As an example we consider a radio source at a galactic latitude of about 30 degree, with a compact core whose size at 22 GHz is $\theta \simeq 0.2$ milliarcsec (mas). For a typical line of sight across the ISM at this galactic latitude the transition frequency is $\nu_0 \simeq 10$ GHz, and the Fresnel angular scale at the transition frequency is $\theta_{F0} \simeq 3$ μarcsec. Therefore, observations at 22 GHz are in the weak scattering regime, and from eq. (7) we have:

$$\theta_{\text{weak}} \simeq 2 \ \mu as$$

and the scintillation index is, from eq. (9):

$$m_{\text{weak}} \simeq 0.2\%$$

We can conclude that weak scintillation can be detected only in strong and extremely compact extragalactic radio sources.

4.2. STRONG SCINTILLATION: DIFFRACTIVE

Diffractive scintillation is an interference effect characterized by fast, narrow-band variations. Smaller scale ($10^8 - 10^{10}$ cm) density inhomogeneities produce fast phase variations occurring on short spatial scales (see Fig. 1). The modulation index is unity for a point source, $m_{\text{p,d}} = 1$, and the interference fringes have a characteristic frequency scale:

$$\Delta\nu/\nu = \xi^{-2} = (\nu/\nu_0)^{17/5} \tag{10}$$

In order to be sensitive to diffractive effects, observations have to be carried out with a frequency resolution of $\Delta\nu$ or better. The angular size on which phase changes of order 1 radian are introduced into the wavefront is named diffractive angular scale:

$$\theta_d = \theta_F \xi^{-1} = \theta_{F0}(\nu/\nu_0)^{6/5} \tag{11}$$

For sources with $\theta > \theta_d$ the modulation index is reduced by a factor θ_d/θ.

Considering our hypothetical radio source, the compact component with size 0.2 mas at 22 GHz is self absorbed and its size increase linearly with the wavelength of observation. So if we have two observing frequencies $\nu_1 = 1.4$ GHz and $\nu_2 = 0.4$ GHz, the size of the compact core are $\theta_1 \simeq 3$ mas and $\theta_2 \simeq 10$ mas, respectively. From eq. (11) the diffractive angular scale at the two frequencies are:

$$\theta_{d,1} = 0.3\mu as \quad \theta_{d,2} = 0.06\mu as$$

The expected scintillation indeces are extremely low due to the finite source size:

$$m_1 \simeq 0.01\% \quad m_2 \simeq 0.0006\%$$

Another problem in detecting diffractive scintillation from extragalactic radio source is the required frequency resolution that in our example is:

$$\Delta\nu_1 \simeq 2 \text{ MHz} \quad \Delta\nu_2 \simeq 7 \text{ kHz}$$

Such narrow bandwidths are not used in continuum observations of extragalactic radio sources, and together with the extremely low values of the scintillation index expected for typical VLBI radio cores readily explain why diffractive scintillation from extragalactic radio sources has not been detected.

4.3. STRONG SCINTILLATION: REFRACTIVE

Refractive scintillation can be understood in terms of ray optics and correspond to lens-like phenomena. It is characterized by slow, broad-band phenomena. Large scale fluctuations ($10^{12} - 10^{14}$ cm) in the ISM produce the focusing and defocusing of the plane waves (see Fig. 1). The refractive scale is given by the scattering disk, much larger than the Fresnel scale, and the time-scale is correspondingly longer:

$$\theta_r = \theta_F \xi = \theta_{F0}(\nu_0/\nu)^{11/5} \tag{12}$$

The point source scintillation index is given by:

$$m_{\text{p,r}} = \xi^{-1/3} = (\nu/\nu_0)^{17/30} \tag{13}$$

Again if $\theta > \theta_r$ the modulation index is reduced by a factor $(\theta_r/\theta)^{7/6}$, while the time-scale increases with θ/θ_r.

As we have already seen our hypothetical radio source has frequency-dependant sizes due to self-absorption, and in particular at $\nu_1 = 1.4$ GHz and $\nu_2 = 0.4$ GHz the size are $\theta_1 \simeq 3$ mas and $\theta_2 \simeq 10$ mas respectively. The refractive angular scales at the two observing frequencies are given by (see eq. 12):

$$\theta_{r,1} \simeq 0.2 \text{ mas} \quad \theta_{r,2} \simeq 3.6 \text{ mas}$$

and, from eq. (13), the scintillation indeces are:

$$m_1 \simeq 1\% \quad m_2 \simeq 5\%$$

Ad odds with what is expected for point like sources where stronger refractive scintillation is expected at higher frequencies, for extended radio sources refractive scintillation increases at lower frequencies.

5. Interstellar scintillation of extragalactic radio sources

5.1. LOW FREQUENCY VARIABILITY

Low frequency variability has been used in the past to indicate flux density variations of the order of 10% on time-scales of months to years observed at radio frequency < 2 GHz. Most of these variations were peculiar, i.e. the lower was the observing frequency the greater was the amplitude of the variability, and they could not be explained in terms of expansion of a synchrotron emitting cloud of plasma. Assuming that the variability is an intrinsic phenomenon, measuring the amplitude and time-scale of the variability it is possible to derive the angular size and brightness temperature of the varying component, and the Doppler factor (δ) required to avoid the Compton catastrophe. The Doppler factors derived from the low frequency bursts were far higher than those measured from proper motions and for this reason the low frequency variability has been a puzzling phenomenon in the '70s and '80s [5]. The solution to the problem was the idea that low frequency variability was not intrinsic to the radio sources but a propagation effect, and refractive scintillation was proposed as the mechanism responsible for the flux density fluctuations [21, 17].

Several observational evidences supporting this hypothesis were found, and the most compelling are:

- dependence of low frequency variability on galactic latitude [2, 10, 22, 9];
- qualitatively and roughly quantitative agreement between the observed scintillation indeces and time-scales and those derived from a "standard model" of the interstellar plasma turbulence [13, 23];
- annual modulation in the radio flux curves of a sample of low frequency variables. This is explained as the signature due to the Earth orbital motion around the Sun on the pattern produced by refractive scintillation [1].

The low frequency radio monitoring and VLBI observations have allowed to derive information on the scattering medium. In particular, it has been found that there is no measurable evidence for a finite propagation speed of the turbulent irregularities responsible for the refractive scintillation. This result can be interpreted in two ways. The first possibility is that the scattering medium is extended along the line of sight. In this case the random velocities of the density irregularities will not produce any net motion. Alternatively, the scattering medium is not uniformly distributed along the line of sight, but it is localized in a thin screen at a certain distance. In this case the velocity of the density irregularities should be low suggesting that they could be associated with the HII regions envelopes, which move at the Alfven speed $V_A \simeq 5$ km/s [1].

5.2. FLICKERING AND INTRA-DAY VARIABILITY

The terms flickering and intra-day variability (IDV) have been used to indicate low amplitude (1% - 5% r.m.s.), short time-scale (few hours to days) flux density variability observed in the frequency range 2-10 GHz in flat spectrum radio sources [11, 14]. In some cases the variations can have substantial amplitude (10-15 %) over few hours (e.g. 0917+624). If intrinsic these variations would imply Lorentz factors of the order of 100. Variations are observed also in polarized flux and position angle. Refractive interstellar scintillation has been claimed to be the cause of these phenomena originally because of a significant trend of increasing amplitude with decreasing galactic latitude [12]. Refractive interstellar scintillation does not produce significant variability in polarized flux by itself, but the assumption of a two component source model with a steady and variable component having nearly orthogonal polarisation angles can explain the observed anticorrelation of the total flux density and polarised flux [15]. A detailed study of the total intensity and polarisation variations of the radio source 0917+624 has been carried out at five frequencies [19]. This paper shows that assuming that the source diameter is linearly dependant on the wavelength, the amplitude and time-scale trends with frequency can be explained in terms of refractive scintillation.

As in the case of the low frequency variability, a confirmation of the extrinsic origin of indra-day variability is the annual modulation detected in two sources, 0917+624 and J1819+3845, [20, 6]. For the last source, a further confirmation that the IDV is due to interstellar scintillation has been recently presented [7]

5.3. SCINTILLATION AS A PROBE OF THE ICM

A new application of the scintillation theory to the study of the intracluster medium (ICM) has been recently suggested [8]. Such a possibility would be extremely interesting as most of the baryons reside in a warm/hot component which is difficult to detect with standard absorption/emission line techniques. In the model proposed by the authors the cluster will act as a foreground screen and the scintillation of a compact quasar behind the cluster could be used to probe the intracluster medium (ICM). A radial profile for the cluster mass density (isothermal β model) is assumed, and the relevant parameters of the model are:

- the mass fraction of the gas in the cluster (in the range $0.04 - 0.2$);
- the distance of the cluster ($z = 0.02$) and of the quasar ($z = 1$), the results are not much sensitive to the distance of the quasar and the distance of the cluster affects mostly the time-scale of the variability;

- the relative velocity between the scintillation pattern and the observer: this is dominated by the velocity of the inhomogeneities assumed of the order of the sound speed in a rich cluster (1000 km/s);
- the size of the quasar;
- the impact parameter, i.e. the distance from the cluster center at which the quasar light intersects the cluster hot gas distribution.

The last two are the most important parameters of the model. In particular, depending on the value of the impact parameter the propagation through the cluster can be in the weak or strong scattering regimes, and the size of the quasar is crucial to determine the scintillation index and time-scale. On the practical side, the scintillation produced by the ICM could be easily detected only for point sources (sources with $\theta < \theta_r$ for line of sight crossing the cluster center, and $\theta < \theta_F$ for larger impact parameters) at high radio frequency (50, 100 GHz). For components larger than about 10 μarcsec the scintillation index will be too low and the time-scale to long to be detected.

References

1. Bondi, M., Padrielli, L., Gregorini, L., Mantovani, F., Shapirovskaya, N.Y., Spangler, S.R. (1994) A&A 287, 390
2. Cawthorne, T.J., Rickett, B.J. (1985) Nature 315, 40
3. Coles, W.A., Frehlich, R.G., Rickett, B.J., Codona, J.L. (1987) AJ 315, 666
4. Cordes, J.M., Rickett, B.J., Baker, D.C. eds. (1988) Proceedings of the American Institute of Physics Conference 174, New York, USA
5. Cotton, W.D., Spangler, S.R. eds. (1982) Proc. NRAO Workshop on Low Frequency Variability of Extragalactic Radio Sources, Green Bank, W. Va
6. Dennet-Thorpe, J., de Bruyn, A.G. (2000) ApJ 529, L65
7. Dennet-Thorpe, J., de Bruyn, A.G. (2002) Nature 415, 57
8. Ferrara, A., Perna, R. (2001) MNRAS 325, 1643
9. Gosh, T., Rao, A.P., (1992) A&A 264, 203
10. Gregorini, L., Ficarra, A., Padrielli, L. (1986) A&A 168, 25
11. Heeschen, D.S. (1984) AJ 89, 1111
12. Heeschen, D.S., Rickett, B.J., 1987 AJ 93, 589
13. Mantovani, F., Padrielli, L., Fanti, R., Gregorini, L., Spangler, S.R. (1990) A&A 233, 535
14. Quirrenbach, A., Witzel, A., Kirchbaum, T.P., Hummel, C.A., Alberdi, A., (1989), Nature 337, 442
15. Quirrenbach, A., Witzel, A., Qian, S.J., Kirchbaum, T.P., Hummel, C.A., Alberdi, A., (1989), A&A 226, L1
16. Quirrenbach, A., Witzel, A., Krichbaum, T.P., Hummel, C.A., Wegner, R., Schalinski, C.J., Ott, M., Alberdi, A., Rioja, M. (1992) A&A 258, 279
17. Rickett, B.J., Coles, W.A., Burgois, G., (1984) A&A 134, 390
18. Rickett, B.J., (1990) ARA&A 28,561
19. Rickett, B.J., Quirrenbach, A., Wegner, R., Krichbaum, T.P., Witzel, A. (1995) A&A 293, 479
20. Rickett, B.J., Witzel, A., Kraus, A., Krichbaum, T.P., Qian, S.J. (2000) ApJ 550, L11
21. Shapirovskaya, N.Y. (1978) Soviet Astronomy 22, 544

22. Spangler, S.R., Fanti, R., Gregorini, L., Padrielli, L., (1989) A&A 209 315
23. Spangler, S.R., Eastman, W.A., Gregorini, L., Mantovani, F., Padrielli, L. (1993) A&A 267, 213
24. Strom, R., Bo, P., Walker, M., Rendong, N. eds. (2001) Proceedings of the IAU Colloquium 182, Astroph. & Space Scie. 278, Vol. 1 & 2
25. Taylor, J.H., Cordes, J.M. (1993) ApJ 411, 674
26. Walker, M.A. (1998) MNRAS 294, 307
27. Walker, M.A. (2001) MNRAS 321, 176

SUPERNOVAE

J.M. MARCAIDE
Depto. de Astronomía y Astrofísica
Universidad de Valencia
Edificio de Investigación
Calle Dr. Moliner, 50
E–46100 Burjassot, Valencia, Spain

1. Introduction

Supernovae are classified by their optical spectra: if a supernova does not show hydrogen lines in its spectrum, then it is a Type I; if it does, it is a Type II. Further, if a Type I does show a deep Si II λ 6200Å absorption line early after explosion near maximum brightness, it is a Type Ia; if it does not show the Si II line but does show He I lines, it is a Type Ib; and finally, if it does not show either Si II or He I lines in the early time spectra, it is a Type Ic. All Type I supernovae show very similar light curves: a short pronounced rise to a peak, followed by a rapid decrease and, after a few weeks, a more gradual decrease. Type II supernovae display a wide variety of light curves that can however be grouped into two: Type II-L, with light curves similar to Type I supernovae, and Type II-P with an extended plateau at the peak level before the start of the rapid decrease [for a detailed account see Filippenko (1997)].

Radio emission has never been detected from Type Ia supernovae, which are believed to be the result of an exploding carbon-oxygen white dwarf in a binary system (Woosley, Taam, and Weaver 1986). Instead, radio emission has been detected in Type Ib, Ic and II supernovae, which are believed to be the result of the gravitational collapse of a massive star. Typically, Type II supernovae are more radio luminous than Type Ib and Ic. Yet, only but a few of those luminous radio supernovae have been accessible to study with the VLBI technique. Distance to supernovae plays an essential role: the flux density falls with the square of the distance and, the angular resolution being constant for a given array, the spatial resolution falls linearly

F. Mantovani and A. Kus (eds.), The Role of VLBI in Astrophysics, Astrometry and Geodesy, 243–256.

with distance. Clearly, even a strong radio supernova like SN1979C at the distance of the Virgo Cluster (peak emission of 8 mJy at 6cm) can hardly be studied in detail with earth-wide arrays.

On the other hand, there are strong supernova remnants (SNR) in our Galaxy that have been studied with great detail with the VLA and other connected element interferometers. Some of them like Cas A and Tycho appear extremely spherical (see Figure 1). Their radio emission is synchrotron, caused by relativistic electrons in tangled magnetic fields. Did some of these SNR have their origin long ago in Type Ia explosions that helped compress the interstellar magnetic fields? Or are they more closely related to the radio supernovae, and hence to supernovae of Type Ib, Ic and II? Whatever the answer is to this question, the present radio supernova model, called the standard interaction model (SIM), and initially proposed by Chevalier (1982a) seems somewhat inspired in the SNR.

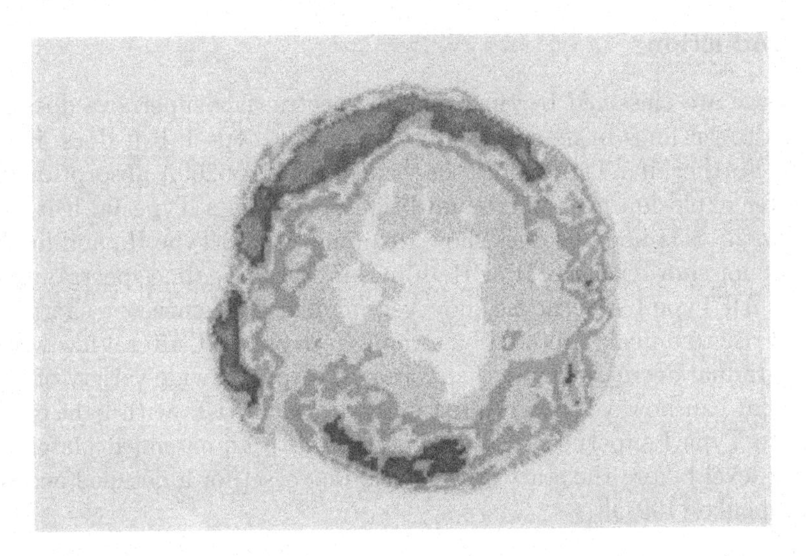

Figure 1. Image of the radio supernova remnant Tycho (Courtesy of NRAO).

2. Standard Interaction Model

In this model, a strong interaction between the expanding supernova ejecta ($\rho_c \propto r^{-n}$) and the circumstellar medium ($\rho_{cs} \propto r^{-s}$) is expected. This interaction causes the formation of a self-similarly expanding ($R_{sh} \propto t^{-m}; m = (n-3)/(n-s)$) shell-like structure from which the observed synchrotron radiation arises. In addition, it is assumed that (1) both magnetic energy density, $\epsilon_B \propto B^2$, and relativistic energy density , $\epsilon_{rel} \propto N_0 E^{2-p}$, evolve with time as the post-shock thermal energy density, $\epsilon_{th} \propto \rho_{sh} v_{sh}^2$; (2) the synchrotron emission is optically thin; and (3) the external absorbing medium has a power-law density profile with $s = 2$. Attempts to fit the radio light curves of some radio supernovae have led several authors to extend the model to incorporate thermal and non-thermal absorbers of the radiation (Weiler et al. 1989, 1990), to consider alternative power-law profiles for the circumstellar medium (Van Dyk et al. 1994, Lundqvist 1994, Fransson et al. 1996, Marcaide et al. 1997), or to include synchrotron self-absorption as an additional absorbing mechanism (Fransson and Björnsson 1998, Pérez-Torres, Alberdi and Marcaide 2001). The latter extension is equivalent to a modification of the assumption (2) of the SIM, in the sense that the synchrotron emission is no longer assumed to be optically thin. Also, these authors have found that radiative losses play an important role.

3. Observations

High angular radio observations of radio supernovae and young supernova remnants are only a few until present. We will describe briefly here most of them.

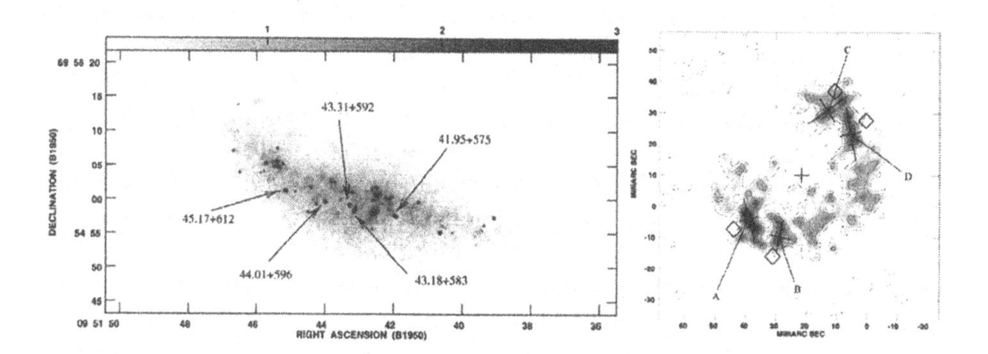

Figure 2. MERLIN+VLA L-band image of M82 showing the positions of the compact young supernova remnant candidate sources detected with VLBI. At right an image of 43.31+592 showing a shell-structure (Pedlar et al. 1999; McDonald et al. (2001). Courtesy of Monthly Notices of the Royal Astronomical Society).

3.1. YOUNG SUPERNOVA REMNANTS IN M82 AND NGC 2146

The nearby starburst galaxy M82 offers a rich population of young SNR candidates that have been intensively studied with high resolution by, e.g.,

Muxlow et al. (1994), Pedlar et al. (1999), and McDonald et al. (2001) using MERLIN, EVN, and global VLBI, respectively. (See Figure 2 for an overview of the galaxy and the compact young SNR candidates.) These authors find that 5 of the SNR candidates are smaller than 1 pc in size, that the candidate 43.31+592 expands at speeds of about 10,000 km s^{-1} with a deceleration parameter $m > 0.73$, and that the strongest candidate, 41.95+575, expands at speeds lower than about 2,000 km s^{-1} casting some doubts that it is a SNR at all. The inset in Figure 2 show clearly the shell-like structure of the source 43.31+592, a young supernova remnant.

Tarchi et al. (2000) have studied the galaxy NGC 2146 with MERLIN and the VLA, have detected 18 compact sources and have shown that only 3 of them have the spectral indices expected for young SNR. There are no VLBI observations of the 3 possible young SNR.

3.2. SN 1987A: A RADIO SUPERNOVA OR A YOUNG SUPERNOVA REMNANT IN THE LMC?

The supernova SN1987A in the Large Magellanic Cloud (LMC), the closest visible supernova in the last 400 years, is a remarkable case. Practically radio quiet since the explosion time for over 4 years, it turned radio loud around day 1400 after explosion (Staveley-Smith et al. 1992, 1993) and evolved louder and louder over the following ten years. Gaensler et al. (1997) have monitored the evolution of its structure with the Australia Telescope Compact Array and have shown that its size has grown almost linearly during the radio loud period at an speed of 2,800±400 km s^{-1} about a tenth the average speed needed during the radio silent period to grow to the initial size of the radio loud period. Figure 3 shows the radio image of SN1987A superposed with the optical ring and the plot of the image radius as a function of time. This case shows pretty clearly that the interaction described by the SIM practically did not take place for over 4 years, probably because the circumstellar density was so low. But, when the free expanding supernova ejecta reached a circumstellar region of high density, corresponding to the final stage of the progenitor as a red supergiant before turning into a blue supergiant, then the conditions were appropriate for strongly slowing down the ejecta, and forming a shock similar to that described by the SIM. Gaensler et al. (1997) show a large asymmetry of the radio remnant of SN1987A and hot spots indicative of directional anisotropies in the ejecta.

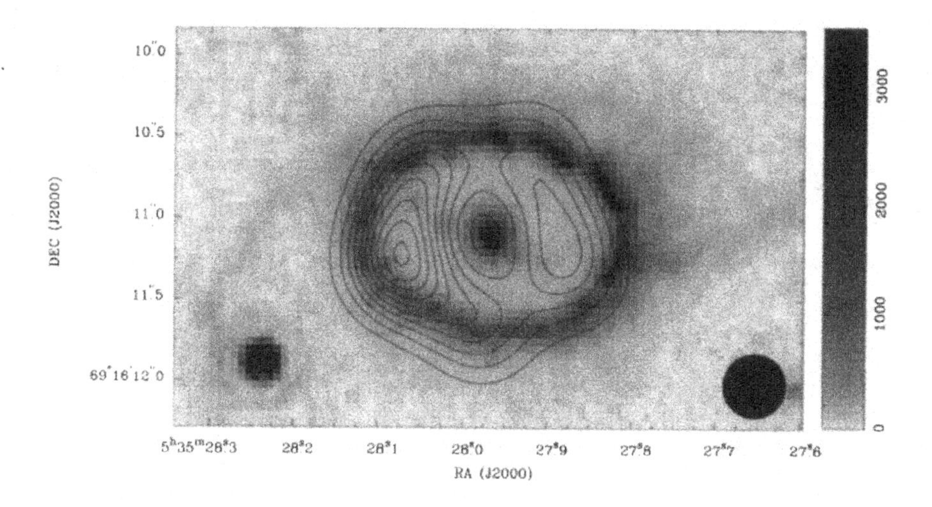

Figure 3. Comparison of radio and optical images of SN1987A in 1995. The dust ring is superposed with the radio shell (Gaensler et al. (1997). Courtesy of the Astrophysical Journal).

3.3. SN 1979C

This supernova was extraordinarily luminous, $M_B^{max} \sim -20$ (Young and Branch 1989). Found as radio supernova by Weiler and Sramek (1980), SN 1979C is, at a distance of 16.1±1.3 Mpc of M100 (Ferrarese et al. 1996), one of the intrinsically strongest radio supernovae. Classified as Type II-L, initial ejecta speeds were estimated at ~9,200 km s^{-1}. From a study of the emission environment, Van Dyk et al. (1999) estimated the mass of the progenitor to be 17-18(±3) solar masses. A modulation in the radio light curves (shown in Figure 4) led Weiler et al. (1992) to suggest that the supernova progenitor was part of a binary system. Schwarz and Pringle (1996) made hydro-dynamical simulations which support the suggestion. As shown in Figure 4, the steady decline of the flux density after maximum turned after ten years into a constant or perhaps increasing trend (Montes et al. 2000) at centimeter wavelengths. This part appears somewhat reminiscent of the radio emission increase seen in SN1987A five years after explosion (see above).

VLBI observations by Bartel et al. (1985) at the wavelengths of 3.6, 6, 13 and 18 cm could not resolve the source but determined a source size and its growth rate for epochs ranging from December 1982 to May 1984. More recent observations by Marcaide et al. (2002) at the wavelength of 18 cm could not resolve the source either but, in combination with those previous

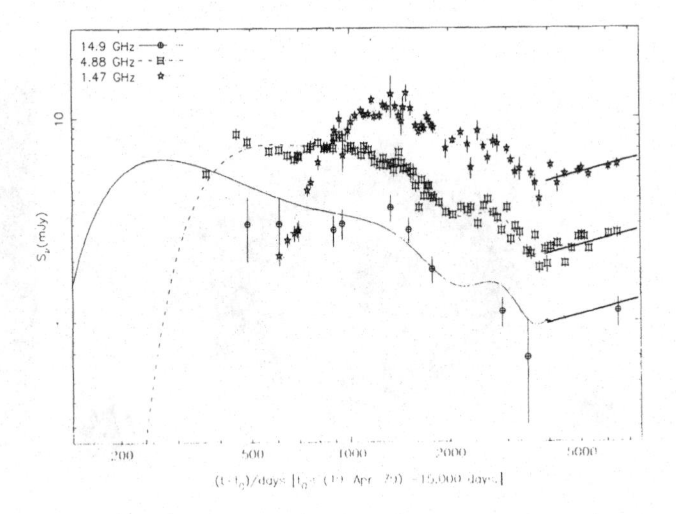

Figure 4. Radio light curves for SN1979C in M100 (Montes et al. 2000. Courtesy of the Astrophysical Journal).

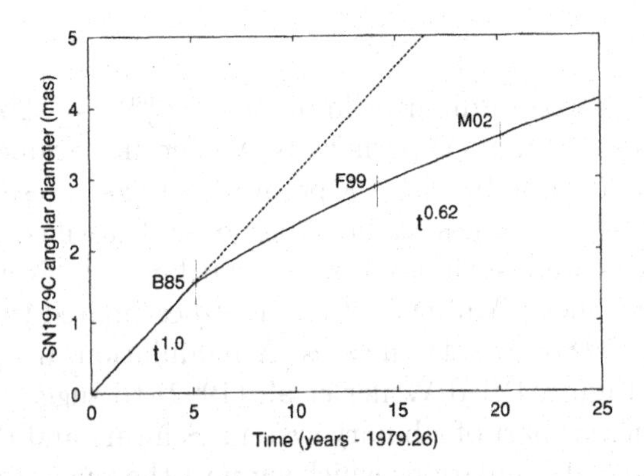

Figure 5. Deceleration of the expansion of SN1979C (Marcaide et al. (2002a). Courtesy of Astronomy and Astrophysics Journal).

VLBI results by Bartel et al (1985) and optical results by Fesen et al. (1999), helped establish evidence that the expansion of SN 1979C suffered a strong deceleration after about 5 years after explosion (see Figure 5). Momentum conservation arguments suggest that the progenitor star was part of a binary system.

3.4. SN 1986J

Discovered in 1986 as a radio source, SN 1986J in NGC891 is one of the most radio luminous supernova ever discovered. At the distance of ~9.6 Mpc (Tully 1998), it had a peak luminosity at 5 GHz about one order of magnitude greater than SN 1979C and SN1993J. SN 1986J was estimated to have exploded around the end of 1982 (Rupen et al. 1987; Chevalier 1987; Weiler, Panagia and Sramek 1990). VLBI observations at 8.4 GHz by Bartel et al. (1991) showed that the radio structure of SN 1986J had the form of a shell, with strong protrusions in the brightness distribution. The difficulty to fit the radio light curves within the SIM model led Weiler et al. (1990) to invoke the existence of a mixed medium of thermal absorbers and non-thermal emitters and/or significant filamentation in the circumstellar medium. In contrast, optical spectra indicated spherical symmetry and velocities much lower than the VLBI results.

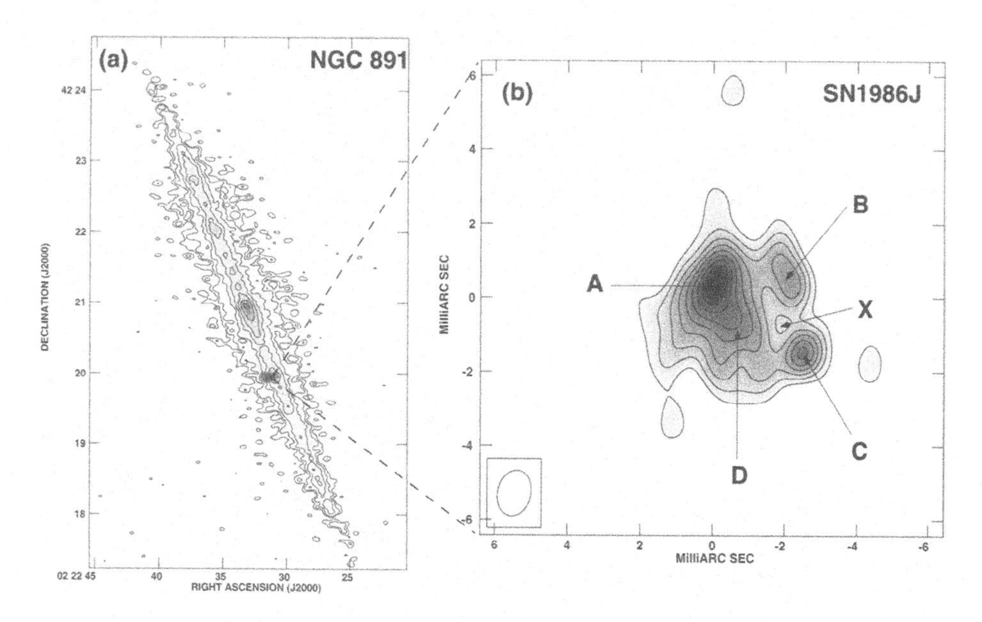

Figure 6. VLBI image of SN1986J at 5GHz (Pérez-Torres et al. (2002). Courtesy of Monthly Notices of the Royal Astronomical Society).

Recently, Pérez-Torres et al. (2002) have obtained with VLBI a new high-resolution image at 5 GHz. The image obtained 16 years after the supernova explosion is shown in Figure 6. The shell appears quite distorted. None of the bright features in the shell or the weak spots outside the shell (which could well be image reconstruction artifacts) seem to correspond to the protrusions reported by Bartel et al. (1991), a result which suggests changing density properties of the presupernova wind. The structure is certainly asymmetric, although the physical reasons for it are not yet clear (see discussion in Pérez-Torres et al. 2002).

Pérez-Torres et al. (2002) obtain an expansion deceleration parameter $m = 0.90 \pm 0.06$, which corresponds to a mild deceleration, in contrast again with the findings in SN1979C and SN1993J, presented elsewhere in this writing. Using momentum conservation arguments, these authors argue in favor of a single massive red supergiant scenario for the supernova progenitor, as suggested earlier by Weiler et al. (1990).

3.5. SN 1993J

Discovered in the galaxy M81 on March 28, 1993, in the optical and two weeks later in the radio, its radio luminosity, sky position in the high northern sky, and distance of 3.6 Mpc combined have made of it the best studied radio supernova ever.

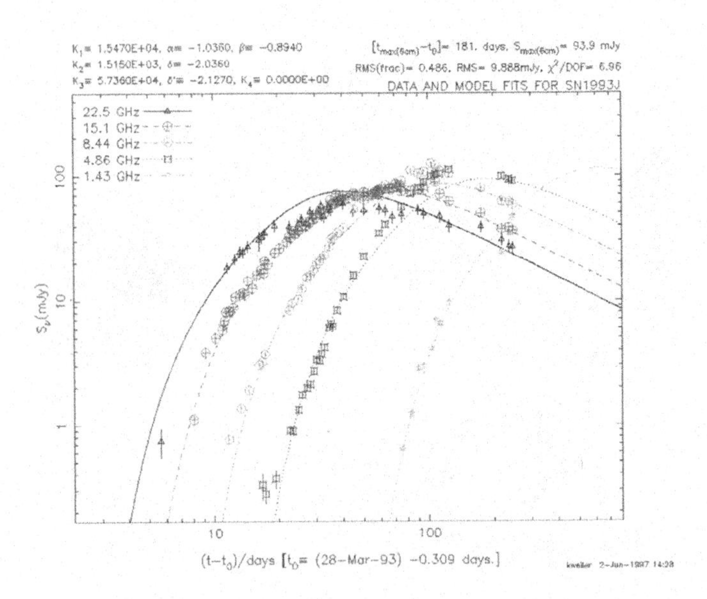

Figure 7. Radio light curves of SN1993J. (Van Dyk, private communication).

Well sampled radio light curves, shown in Figure 7, have been obtained at many wavelengths (Van Dyk et al. 1994). The radio strength of SN1993J has allowed for VLBI studies from the very beginning through the almost 10 years that have elapsed since. Early VLBI results were reported by Marcaide et al (1994) and Bartel et al (1994). A radio shell (Figure 8) was discovered by Marcaide et al. (1995a) with data at 3.6cm corresponding to an epoch 6 months after explosion. Shortly later, Marcaide et al. (1995b) showed the first movie of its expansion (Figure 9) which appeared to be self-similar. Combining the results obtained over 3 years, Marcaide et al. (1997) showed that the expansion was indeed self-similar with a deceleration parameter $m = 0.86 \pm 0.02$. Those results were later confirmed by Bartel et al. (2000) who suggested that the expansion was better characterized by two values of m, an initial one of 0.94 ± 0.02 for the first 300 days and a later one of 0.77 ± 0.01 for days beyond 500. We find a similar result with values of 0.93 and 0.83 with a break time around day 403 (see Figure 10). In Figure 11 it is shown an unpublished recent radio image of SN1993J. Bartel et al. (2000) also found using the nearby radio core of the galaxy M81 that the supernova expansion is very symmetric with respect to a certain point, likely related to the location of the supernova's progenitor star.

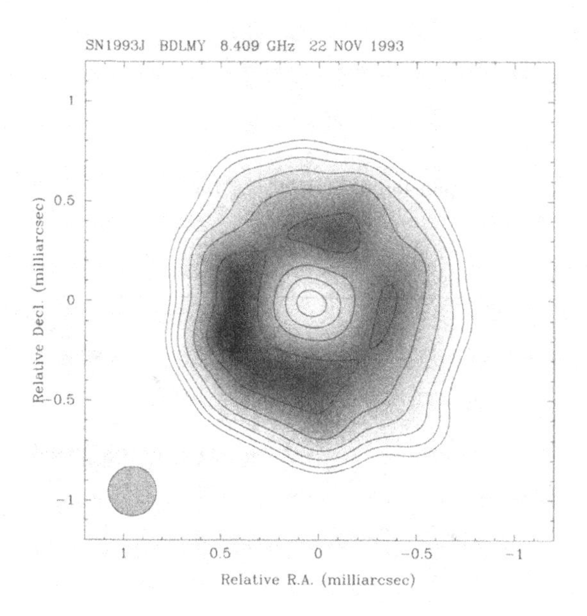

Figure 8. Shell structure of SN1993J discovered. (Marcaide et al. (1995a). Courtesy of Nature.)

Also, computer codes which simulate the radio emission from supernovae have been developed by Pérez-Torres, Alberdi and Marcaide (2001) and Mioduszewski et al. (2001), and applied to the case of SN 1993J. These simulations, together with observational results (radio light curves and VLBI) and theoretical work by Björnsson and Fransson (1998), indicate the presence of very high magnetic fields in the early radio shell and circumstellar medium profiles with index s=1.6. Recent multifrequency VLA measurements spanning the range 0.32 up to 15 GHz (Pérez-Torres, Alberdi

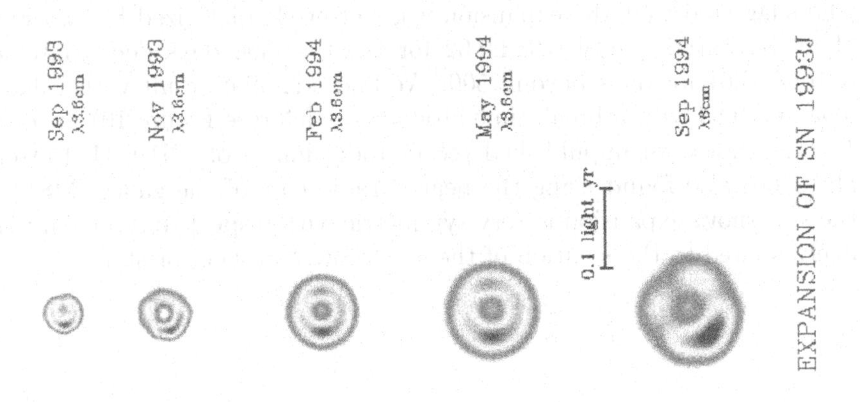

Figure 9. Short movie of the expansion of SN1993J. (Marcaide et al. (1995b). Courtesy of Science.)

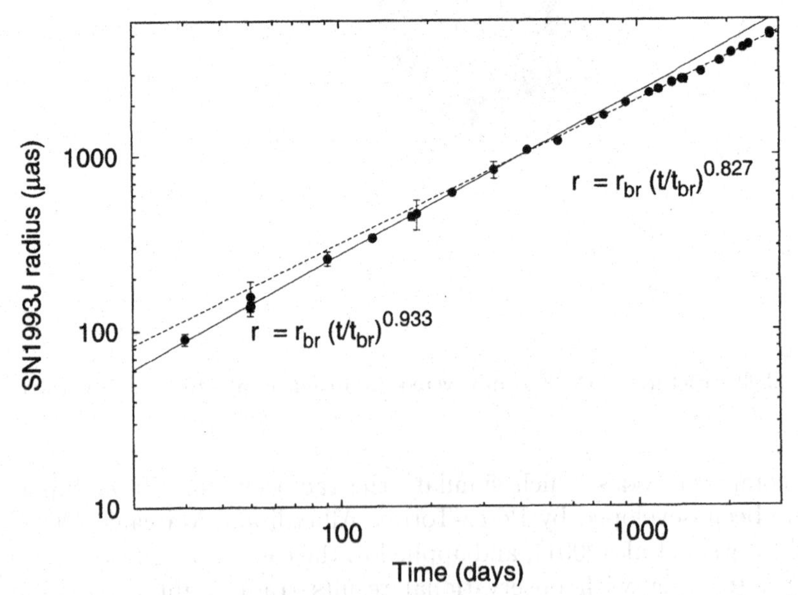

Figure 10. Deceleration of SN1993J (Marcaide et al. (2002b). Note that the scale is logarithmic.

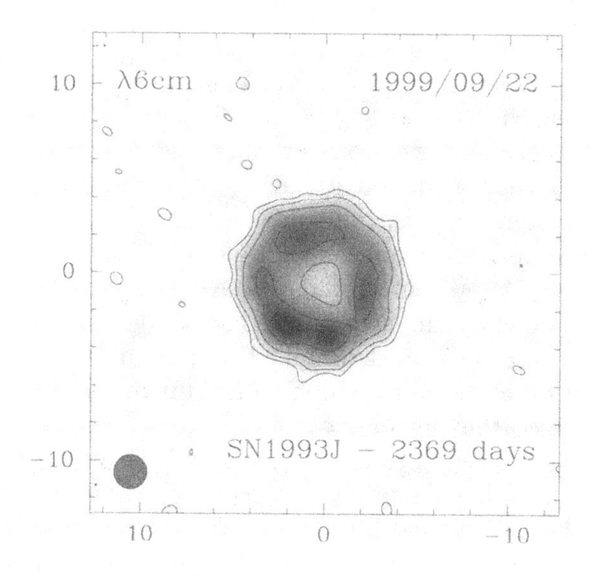

Figure 11. Unpublished image of SN1993J corresponding to day 2369.

and Marcaide 2002) show that the radio spectrum of SN 1993J has been slowly evolving since $t \sim 1000$ days from $\alpha \approx -1$ to $\alpha \approx -0.67$ ($S_\nu \propto \nu^\alpha$) at current epochs.

3.6. OTHER RADIO SUPERNOVAE AND A DISTANCE DETERMINATION

Radio supernovae are not rare as it can be seen in the list of radio supernovae than can be found at http://rsd-www.nrl.navy.mil/7214/weiler, however, as Table 1 shows there are not many that can be studied with the technique of VLBI. Two recent ones that hold a moderate promise for VLBI studies are SN2001gd and SN2001ig. SN2001gd appears to have many things in common with SN1993J, but unfortunately it is at a distance several times larger than SN1993J.

The question of distance itself can be addressed in principle combining VLBI and optical results: If an angular expansion rate can be measured with VLBI and optical results about the radial expansion velocities are available, the distance to the supernova can be determined. In practice, however, the uncertainties involved are too large to be useful. On one hand the angular expansion rates are difficult to measure and on the other hand the velocities involved in the optical do not probably come from the same physical regions as the radio emission, rendering the comparison difficult.

4. Conclusions

The study of radio supernovae by means of VLBI can yield invaluable information about their emission structure. This, in turn, can help understand the characteristics of the progenitor star and of the circumstellar medium. Since supernova expansion speeds are about a thousand times larger than the pre-explosion stellar winds, the study of the circumstellar medium is made in a time-compressed manner as a time machine. Radio supernovae up to the distance of the center of the Virgo Cluster may be accessible to VLBI studies. A few radio supernovae have been studied already and their expansion characteristics determined. Expansion decelerations have been determined for a few cases. Using momentum transfer arguments, it has been determined that in some cases the progenitor star was a single massive star, but in other cases it was a massive star in a binary system. There are indications that the profile of the circumstellar medium may not be in all cases what is expected from a constant pre-explosion stellar wind. The supernova expansion is found not to be spherical in all cases. However, in the best studied case, SN1993J, it is extremely spherical. Models of synchrotron emission with strong radiative losses free-free absorbed by the ionized circumstellar medium account for most of the radio evidence. However, how much the ionized medium is clumpy and how much of the emission is synchrotron self-absorbed remains still an open question.

5. Acknowledgements

I thank very much A. Alberdi and M.A. Pérez-Torres for their substantial contributions to this writing. This work is dedicated to Prof. Domingo González, from the Universidad de Zaragoza, on the occasion of his retirement.

References

Bartel, N.B., Rogers, A.E.E., Shapiro, I.I., et. al., 1985, Hubble's constant determined using very-long baseline interferometry of a supernova *Nature*, **318**, pp. 25–30

Bartel, N.B., Rupen, M.P., Shapiro, I.I., Preston, R.A., Rius, A., 1991, A high-resolution radio image of a young supernova *Nature*, **350**, pp. 212–214

Bartel, N.B., Bietenholz, M.F., Rupen, M.P., et al., 1994, The shape expansion rate and distance of supernova 1993J from VLBI measurements, *Nature*, **368**, pp. 610–

Bartel, N.B., Bietenholz, M.F., Rupen, M.P., et al., 2000, The changing morphology and increasing deceleration of supernova 1993J in M81, *Science*, **287**, pp. 112–116

Chevalier, R.A., 1982a, Self-similar solutions for the interaction of stellar ejecta with an external medium, *Astrophys. J.*,**258**, pp. 790–797

Chevalier, R.A., 1982b, The radio and X-ray emission from type II supernovae, *Astrophys. J.*,, pp. 302-310

Chevalier, R.A., 1987, Circumstellar interaction and a pulsar nebula in the supernova 1986J, *Nature*,**329**, pp. 611–612

Filippenko, A.V., 1997, Optical spectra of supernovae, *Annu. Rev. Astron. Astrophys.*, **35**, pp. 309-355

Ferrarese, L., Freedman, W.L., Hill, R.J., et al., 1996, The Extragalactic Distance Scale Key Project. IV. The Discovery of Cepheids and a New Distance to M100 Using the Hubble Space Telescope, *Astrophys. J.*,**464**, pp. 568–

Fesen, R.A., Gerardy, C.L., Filippenko, A.V., et al., 1999, Late-Time Optical and Ultraviolet Spectra of SN 1979C and SN 1980K, *The Astron. J.*,**117**, pp. 725–735

Fransson, C., Lundqvist, P., Chevalier, R.A. 1996, Circumstellar interaction in SN1993J, *Astrophys. J.*,**461**, pp. 993–1008

Fransson, C., Björnsson,C.-I., 1998, Radio emission and particle acceleration in SN1993J, *Astrophys. J.*,**509**, pp. 861–878

Gaensler, B.M., Manchester, R.N., Staveley-Smith, L. et al., 1997, The asymmetric radio remnant of SN1987A, *Astrophys. J.*,**479**, pp. 845–858

Lundqvist, P., 1994, The circumstellar gas around SN1987A and SN1993J, *in Circumstellar Media in the Late Stages of Stellar Evolution*, Eds. R.E.S. Clegg, I.R.Stevens, & V.P.S. Meikle, Cambridge: Cambridge University Press, pp. 213–220

Marcaide, J.M., Alberdi, A., Elosegui, P., et al., 1994, Radio-size estimates of SN1993J, *Astrophys. J.*,**424**, pp. L25–L27

Marcaide, J.M., Alberdi, A., Ros, E. et al., 1995a, Discovery of shell-like radio-structure in SN1993J,*Nature*, **373**, pp. 44–45

Marcaide, J.M., Alberdi, A., Ros, E., et al., 1995a, Expansion of SN1993J,*Science*, **270**, pp. 1475–1478

Marcaide, J.M., Alberdi, A., Ros, E., et al., 1997, Deceleration in the expansion of SN1993J, *Astrophys. J.*, **486**, pp. L31–L34

Marcaide, J.M., Pérez-Torres, M.A., Ros, E., et al., 2002a, Strongly decelerated expansion of SN1979C, *Astron. & Astrophys.*, **384**, pp. 408–413

Marcaide, J.M., Alberdi, A., Pérez-Torres, M.A., et al., 2002b, How is really decelerating the expansion of SN1993J?, *in the Proceedings of the 6th European VLBI Network Symposium*, Eds. E. Ros, R.W. Porcas, and J.A. Zensus (in press)

McDonald, A.R., Muxlow, T.W.B., Pedlar, A., et al., 2001, Global very long-baseline interferometry observations of compact radio sources in M82, *Mon. Not. R. Astron. Soc.*, **322**, pp. 100–106

Muxlow, T.W.B., Pedlar, A., Wilkinson, P.N. et al., 1994, The structure of young supernova remnants in M82, *Mon. Not. R. Astron. Soc.*, **266**, pp. 455–467

Pedlar, A., Muxlow, T.W.B., Garrett, M.A. et al., 1999, VLBI observations of young supernova remnants in M82, *Mon. Not. R. Astron. Soc.*, **307**, pp. 761–768

Pérez-Torres, M.A., Alberdi, A., Marcaide, J.M., 2001, The role of synchrotron self-absorption in the late radio emission from SN 1993J, *Astron. & Astrophys.*, **374**, pp. 997–1002

Pérez-Torres, M.A., Alberdi, A., Marcaide, J.M., et al., 2002, A distorted radio shell in the young supernova SN1986J, *Mon. Not. R. Astron. Soc.*, in press.

Pérez-Torres, M.A., Alberdi, A., Marcaide, J.M., 2002, The Late Radio Spectrum of SN1993J, *Astron. & Astrophys.*, in press.

Rupen, M.P., Van Gorkom, J.H., Knapp G.R., et al., 1987, Observations of SN 1986J in NGC 891, *The Astron. J.*, **94**, pp. 61–70

Schwarz, D.H., Pringle, T.J., 1996, A self-colliding stellar wind model for SN1979C *Mon. Not. R. Astron. Soc.*, **282**, pp. 1018–1026

Staveley-Smith, L., Manchester, R.N., Kesteven,M.J., et al., 1992, Birth of a radio supernova remnant in supernova 1987A, *Nature*, **355**, pp. 147–149

Staveley-Smith, L., Briggs, D.S., Rowe, A.C.H., et al., 1993, Structure of the radio remnant of supernova 1987A, *Nature*, **366**, pp. 136–138

Tarchi, A., Neininger, N., Greve, A. et al., 2000, Radio supernovae, supernova remnants and HII regions in NGC2146 observed with MERLIN and the VLA, *Astron. & Astrophys.*, **358**, pp. 95–103

Tully, R.B., 1998, Nearby Galaxies Catalogue (Cambridge: Cambridge University Press)

Van Dyk, S.D., Weiler, K.W., Sramek, R.A. et al., 1994, SN1993J: The early radio emission and evidence for a changing presupernova mass-loss rate, *Astrophys. J.*, **432**, pp. L115–L118

Van Dyk, S.D., Peng, C.Y., Bart, A.J., et al., 1999, *Hubble Space Telescope* WFPC2 Imaging of SN1979C and its environment, *Publ. Astr. Soc. Pac.*, **111**, pp. 313–320

Weiler, K.W., Panagia, N., Sramek, R.A., 1990, Radio emission from supernovae. II - SN 1986J: A different kind of type II, *Astrophys. J.*, **364**, pp. 611–625

Weiler, K.W., Van Dyk, S.D., Pringle, J.E., 1992, Evidence for periodic modulation of presupernova mass loss from the progenitor of SN1979C *Astrophys. J.*, **399**, pp. 672–679

Young, T.R., Branch, D., 1989, Absolute light curves of type II supernovae *Astrophys. J.*, **342**, pp. L79–L82

Woosley, S.E., Taam, R.E., Weaver, T.A., 1986, Models for Type I supernova. I - Detonations in white dwarfs, *Astrophys. J.*, **301**, pp. 601–623

ASTROPHYSICAL MASERS

R.S. BOOTH
Onsala Space Observatory, Chalmers University of Technology, Onsala, S-43992, Sweden

AND

P.J. DIAMOND
Jodrell Bank Observatory, University of Manchester, Macclesfield, SK11 9DL, United Kingdom

1. Introduction

Interest in cosmic masers has reached a new peak in recent years because of some remarkable new observations using VLBI. It has been found, for example, that extragalactic water masers provide an accurate dynamical probe of accretion discs around putative black holes. In addition, work on the SiO masers in circumstellar envelopes has reached a new level of sophistication with the production of a movie showing the motion of the masers as the stellar shell expands, and finally, methanol masers have been shown to be very early, perhaps the earliest observational evidence of high mass protostars with dynamical signatures of rotating discs and shocked outflows.

1.1. HISTORICAL OVERVIEW

The first interstellar maser was found at a time when astronomers were still uncertain whether molecules would even survive in the interstellar medium, let alone be detectable. The first molecules (the diatomic free radicals CH, CH+ and CN) had been detected optically in the absorption spectra of hot stars in 1940. However, there was little further interest until the early 1960s when radio astronomers detected the hydroxyl radical (OH) in 2 of its ground state lines, at a wavelength of 18 cm, in absorption against the strong continuum emission from the supernova remnant Cassiopeia A (Weinreb et al. 1963). Subsequent observations towards HII regions (other

F. Mantovani and A. Kus (eds.), The Role of VLBI in Astrophysics, Astrometry and Geodesy, 257–288.

strong continuum emitters) gave surprising results - in several of these objects, the lines were detected but they were seen in emission! Imagine the confusion! The emission spectra consisted of narrow lines peaking at different velocities in the 2 main line frequencies observed (1665 and 1667 MHz) and the line ratios were quite different from the LTE values (which had been confirmed in the CassA absorption spectra). The immediate reaction was that an entirely new species was responsible for the emission line, and it was given the name mysterium (Weaver et al, 1965). But subsequent measurements of the other OH ground state transitions (in addition to 1665 MHz) at 1667, 1720 and 1612 MHz respectively, showed the line frequencies and separations to be consistent with OH, but the line ratios were more like 50:2:1:1 respectively rather than 5:9:1:1 if the molecule is in thermodynamic equilibrium. (Weinreb et al, 1965). This, together with the observation that the lines showed a high degree of polarisation, both linear (Weinreb et al, 1965) and circular (Davies, de Jager & Verschuur, 1966) led to the realisation that the lines were indeed due to OH, but anomalously excited OH. It was subsequently confirmed that the OH emission was the result of maser action, since the sources were unresolved by long baseline interferometers, implying very high brightness temperatures. Radio-linked interferometry was reported by Davies et al (1967), and VLBI by Moran et al (1968). The latter experiments showed that sources were smaller than 0.01 arsec, giving brightness temperatures of 10^{12}K.

Further maser species were discovered in the years following and we now know that there are 4 molecules which produce strong, widespread cosmic masers: OH, water (H_2O), silicon monoxide (SiO) and methanol (CH_3OH), each exhibiting stimulated emission in several transitions. All of these masers are associated with star formation although SiO is not as common in star formation regions as the other three; OH and water and SiO are also associated with evolved stars and OH and water have even been observed in distant galaxies where they are, intrinsically, so bright that they have been termed mega-masers. In addition there are weak masers due to methyladyne (CH), ammonia (NH_3), formaldehyde (H_2CO) and some recombination lines. In this lecture, we will concentrate on the strong masers, their association with star formation, the exciting recent VLBI measurements of the structure and dynamics of methanol masers; the masers in the envelopes of evolved stars and finally will discuss extragalactic megamasers.

2. VLBI and the masers

Since the first observations in the late 1960s VLBI has proved to be an important tool for studying the masers, their structure and dynamics. Through such studies, the masers have proven to be powerful tools in astrophysics. Not only does their intense, beamed radiation signpost regions of star formation but the strong, point-like (barely resolved) maser features can be positioned extremely accurately by the VLBI technique. The relative positions of the individual velocity features may be measured very precisely and information on the dynamics of the regions may be derived on scales of $0.1 \rightarrow 1$ milliarcseconds. This corresponds to $0.1 \rightarrow 1$ astronomical unit (AU) at a distance of 1 kpc. With such precision it has been possible to measure proper motions of the masers, especially the water masers, and hence to determine their distances, using the method of statistical parallax. The maser distance scale has the advantage that it is direct, and independent of the usual chain of intermediate distance indicators and as such, it has been used to correct the general astronomical distance scale through accurate measurements of the distance of masers near the Galactic centre (Reid, 1993).

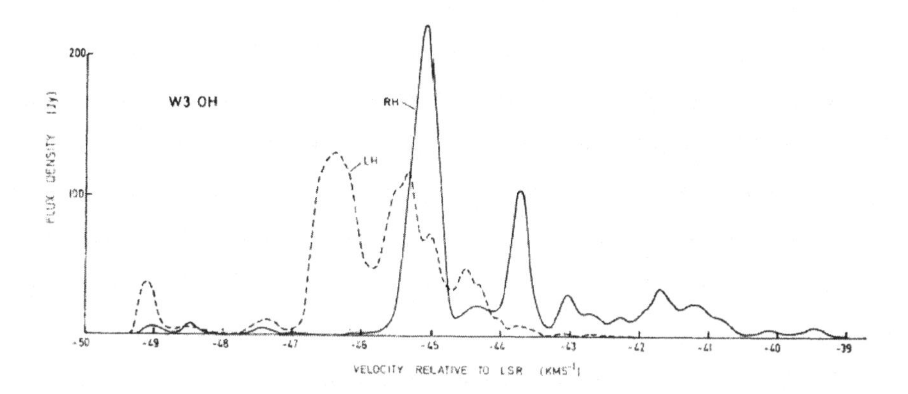

Figure 1. Jodrell Bank spectrum of the 1665 MHz OH maser in W3(OH)in right- and left-hand circular polarisation, epoch 1970. The intensity of the RH feature at -43.8 km/s had decreased by 60% by 1997.

Let us look at the first masers to be studied in detail, the OH masers in W3(OH), a compact HII region associated with the source W3 (third source in the Westerhout catalogue). From the spectrum (Fig.1) we see the typical evidence for the maser process: multiple, overlapping, narrow spectral features with high polarisation and high brightness temperature. The linewidths are usually somewhat narrower than the thermal Doppler widths and in many cases the line amplitudes vary on relatively short time-scales: minutes to years. In the case of W3(OH) the feature at -43.8km/s decreased in flux linearly between 1966 and 1978 by 66%. (Norris and Booth, 1981).

The high degree of circular polarisation seen in some of the spectral features is generally accepted to be the result of Zeeman splitting. The lack of clear Zeeman patterns of RH and LH features is taken to be a result of the maser process: amplification can occur when different pockets of gas along the line of sight amplify the same input signal i.e. a signal with the same shift from the line rest frequency, whether that shift is caused by a local velocity (Doppler) shift or a local magnetic field (Zeeman) shift. Despite the lack of a clear Zeeman signature, the mean frequency shift between the RH and LH features will be the result of an *average* magnetic field in the region. Magnetic fields with strengths of several milli-gauss are inferred from the data.

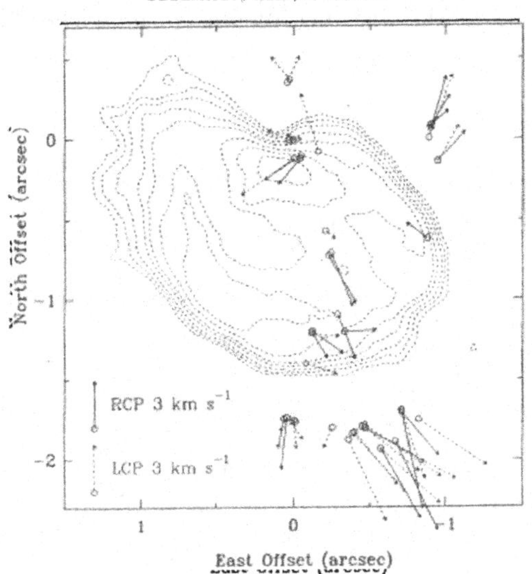

Figure 2. Image showing the OH masers spots of W3(OH) superimposed on the 15 GHz continuum emission. The arrows connected to the circles illustrate the maser proper motions. From Bloemhof et al (1992).

A map of the OH masers in W3(OH) (Bloemhof et al, 1992) is shown in Fig 2. We include it to demonstrate several points, including the measurement of OH maser proper motions (referred to later) but here we point to it as a typical VLBI maser component map. The emission breaks up into individual, largely unresolved sources, each corresponding to a clump of gas with a distinct velocity. The complex spectral features in the single dish spectrum (Fig 1) are blends of several point components with closely similar velocities. Sizes of the features are generally very small, as we have discussed above. The positions of all features are measured relative to a strong reference feature and the measurement accuracy is usually better than 1 mas. The point we wish to emphasize for the moment is that the

2-dimensional spatial distribution of individual velocity components is apparently random. There is no clear dynamical signature in the velocity structure of the region - this is typical for an OH maser map. The same is true for maps of water masers. In both cases, however, when proper motions are measured, the regions are usually found to be expanding, sometimes as bipolar outflows.

Maps of methanol masers often show a clearer dynamical signature — rotation or expansion are indicated by the velocity-position data from the VLBI maps. The VLBI structure of individual methanol maser features also differs from the unresolved, point-like nature of OH and water: methanol masers are frequently observed to be point-like cores surrounded by halos of more diffuse masing gas. This will be discussed further below.

3. Some theory

The acronym MASER stands for microwave amplification by the stimulated emission of radiation. In interstellar space, masers occur naturally when molecules collide, or are radiatively 'pumped resulting in an overpopulation of a higher energy state, or a *population inversion.* Stimulated emission may then occur, with the amplification of an input wave by factors as great as 10^{10}. We can understand this process from basic radiation transfer in a two level system (Fig. 3 and see for example, Cohen 1989). We have also drawn from other good reviews of masers by Reid and Moran (1988) and Downes (1985).

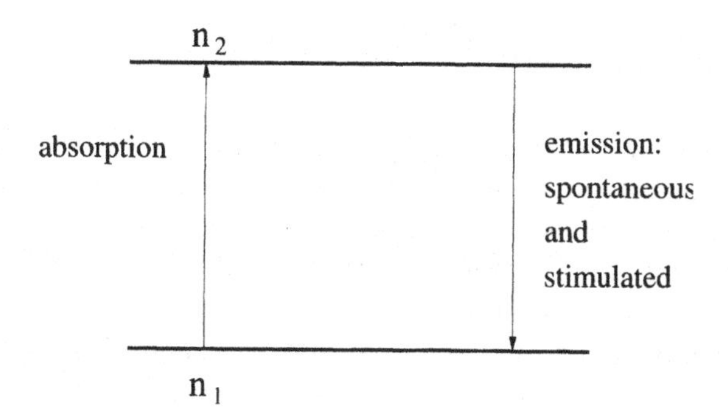

Figure 3. A generic two level system demonstrating absorption and emission.

3.1. INTERSTELLAR MASERS ARE EASY TO EXCITE

The interaction between radio waves and neutral gas in the interstellar medium is governed by the equation of transfer. The change of intensity as the wave traverses a distance, *ds* in the medium is given by:

$$\frac{dI_\nu}{ds} = -\kappa_\nu I_\nu + \epsilon_\nu \tag{1}$$

where: I_ν is the specific intensity of radiation, ϵ_ν is the emissivity, and κ_ν is the absorption coefficient of the interstellar gas

Consider the two level system in Fig 3 where n_1 and n_2 are the number densities in the two states. The emission and absorption can be described in terms of the transition probabilities, as follows. The emissivity is:

$$\epsilon_\nu = n_2 A_{21} \frac{h\nu_{21}}{4\pi} f(\nu) \tag{2}$$

where: A_{21} is the Einstein coefficient giving the probability of spontaneous emission per unit time; ν_{21} is the mean frequency of the emitted photon, and $f(\nu)$ is the line shape function. $\int f(\nu)d\nu = 1; f(\nu) = 1/\Delta\nu; \Delta\nu$ is the line full width at half intensity.

The absorption coefficient is:

$$\kappa_\nu = (n_1 B_{12} - n_2 B_{21}) \frac{h\nu_{21}}{4\pi} f(\nu) \tag{3}$$

where B_{12} is the Einstein coefficient for absorption; B_{21} is the coefficient for stimulated emission. N.B. energy density is sometimes used instead of intensity to define the B coefficients, which leads to definitions differing by $c/4\pi$.

The Einstein coefficients are related by the equations:

$$A_{21} = B_{21} \frac{2h\nu^3}{c^2}; B_{12} = \frac{g_2}{g_1} B_{21} \tag{4}$$

where g_i are the statistical weights of the upper and lower states.

From equations 4 it can be seen that the ratio of stimulated emission to spontaneous emission varies inversely as the frequency cubed; it is of order 10^{18} times greater at radio wavelengths than for visible light. This, in part, explains the paucity of astronomical lasers.

Now, if the molecule is in local thermodynamic equilibrium (LTE), the upper state population, n_2, is smaller than n_1, the ratio of the populations of the two being given by the Boltzmann equation:

$$\frac{n_2}{n_1} = \frac{g_2}{g_1} exp\left(-\frac{hv_{21}}{kT_{ex}}\right) \tag{5}$$

The bracketed term is very small at radio wavelengths (~ 0.1) because the energy difference between the states is very small. So, n_2 and n_1 are actually not very different and the net absorption (eqn. 3) is the difference between two large, almost equal terms. Thus, the radiation transport through the gas is very sensitive to small changes in the level populations from their LTE values. In other words, it does not require too great a perturbation in the state of the gas to cause a population inversion ($n_2/n_1 > g_2/g_1$). In this event, the absorption coefficient becomes negative and the radiation is amplified as it passes through the gas – a MASER is formed! Population inversion, as we have said, arises naturally in interstellar space through radiation or collisions, and masers are relatively common.

3.2. THE RATE EQUATIONS

Let us now return to the transfer equation and substitute for κ and ϵ (at the line centre):

$$\frac{dI_\nu}{ds} = \frac{h\nu}{4\pi}\left[(n_1 - n_2)BI_\nu + n_2 A\right] f(\nu) \quad .(6)$$

We have simplified by writing A for A_{21} and setting $B_{12} = B_{21} = B$ and assuming $g_1 = g_2$.

In the case of a maser, the populations of the upper and lower levels are determined by spontaneous emission, described by the transition rate A, by stimulated emission, and by collisions, described by rate C, (all of which tend to thermalise the population ratio), and by pumping from a third level (or levels) in the molecule, and decay to a reservoir of energy states which is responsible for the population inversion. We define P_1 and P_2 as the net pumping rates from a third level (or levels) to the levels 1 and 2, respectively and $\Gamma_{1,2}$ as the loss rates from the maser energy levels to the reservoir of other unspecified levels.

Note that in detailed balance, the radiative transitions A and BI, and the collisional transitions across the signal levels never produce inversion since they involve *only* exchange between the two maser levels. Inversion can be achieved only when either the pump, or loss rates for the two levels are different. (Note also that, in apparent contradiction, we will invoke collisions as a potential pumping source in cosmic masers. The collisional *pump* involves excitation of other levels of the molecule and radiative decay to the maser levels).

We can now write down the rate equations for the maser:

$$\frac{dn_1}{dt} = -(n_2 - n_1)BI\frac{\Omega_b}{4\pi} - n_2 A + (n_1 - n_2)C + P_2(n - n_{12}) - \Gamma n_2 \quad (7)$$

$$\frac{dn_2}{dt} = -(n_1 - n_2)BI\frac{\Omega_b}{4\pi} - n_2 A + (n_2 - n_1)C + P_2(n - n_{12}) - \Gamma n_1 \quad (8)$$

where is n the total density of the molecule which gives rise to the maser, $n_{12} = n_1 + n_2$ is the total density of the molecule in the upper and lower maser levels; $n - n_{12}$ is thus the density in all other levels of the molecule.

N.B., a characteristic of maser radiation is that the photons are emitted in the direction of those inducing the emission. Therefore, I_ν is reduced by the ratio of the beam solid angle Ω_b to 4π.

Under the typical conditions of a cosmic maser, it may be assumed that A is negligible and we will also, for simplification assume that collisional effects are also negligible compared to B. Thus, in the steady state, these rates are equal to zero and by adding them, we find that:

$$\Gamma n_{12} = (P_1 + P_2)(n - n_{12}) = P_{tot}(n - n_{12}) \tag{9}$$

showing that the number of pumping events per second from the system to levels 1 and 2 is equal to the loss rate from these levels to the system.

Eliminating n, we may solve for the population inversion

$$\Delta n = n_2 - n_1 = (n_1 + n_2)\frac{\Delta P}{P}\frac{\Gamma}{\Gamma + 2BI\Omega/4\pi} \tag{10}$$

where $\Delta P = P_2 - P_1$ and $\Delta P/P$ is called the pump efficiency.

From eqn. 10 it can be seen that stimulated emission reduces the population inversion.

The dependence on I is important here since as the emission increases, the rate of downward transitions increases and then maser pump rate may be insufficient to continue to maintain the population inversion and the maser will saturate, as we will demonstrate below.

3.3. SATURATED AND UNSATURATED MASERS

The general solution of the equation of transfer at the line centre frequency (eqn. 1) is:

$$I(s) = I(0)e^{-\kappa_\nu s} + \frac{\epsilon_\nu}{\kappa_\nu}\left(1 - e^{-\kappa_\nu s}\right) \tag{11}$$

where $I(0)$ is the intensity of the signal entering the maser cloud.

From the definition of the absorption coefficient (eqn. 3), we can write the gain coefficient, or optical depth at the line centre, $\kappa_\nu L$, as

$$\kappa_\nu L = \tau = -\frac{\Delta n \lambda^2 A}{8\pi \Delta\nu_D} \tag{12}$$

where L is the path length through the cloud.

For a maser, Δn is positive and the absorption coefficient becomes negative (stimulated emission is the dominant term in eqn. 3) and the incoming radiation is amplified exponentially. For celestial masers, the optical depth may be of order 25, leading to gain factors around 10^{10}. This enormous amplification occurs in a single pass through the molecular cloud.

While exponential gain occurs, the maser is said to be unsaturated.

This will occur when (eqn. 10)

$$\Gamma \gg 2BI\frac{\Omega}{4\pi} \tag{13}$$

i.e. losses to the reservoir and/or collisions across the signal levels exceed the rate of stimulated transitions.

In terms of brightness temperature, the solution to the equation of transfer is

$$T_b = T_c e^{-\tau} + T_{ex}\left(e^{-\tau} - 1\right) \tag{14}$$

with T_c being the brightness temperature of the input from a background source, and T_{ex} the input due to spontaneous emission in the maser. Note that T_{ex} and τ are negative.

The exponential gain leads to line narrowing, beaming and rapid variability in cosmic masers. Exponential gain amplifies the centre of the line much more than the wings. This leads to an emergent profile which, while still gaussian, is reduced in width by the factor $(1 - \tau_\nu)^{1/2}$ (Goldreich and Kwan, 1974). For a typical strong interstellar maser, the lines are narrowed by a small factor (< 10). Beaming also occurs because of the greater amplification at the line centre because the beam will pick out a smaller and smaller velocity range as it is amplified. Finally, since the gain is exponential, any small physical changes along the maser path will be emphasised.

3.4. SATURATION

Now, when the radiation becomes very intense, the pumping processes are unable to maintain a population inversion against the growing loss of molecules in the higher energy state by stimulated emission. This leads to *saturation* of the maser. The condition for saturation is when the level population is controlled by the induced transitions, or

$$2BI \gg \Gamma \tag{15}$$

and, using this condition in eqn. 10, we can see that the population inversion decreases with the stimulated emission rate:

$$\Delta n = \Delta n_0 \frac{\Gamma}{2BI\Omega/4\pi} \tag{16}$$

This means that (eqn. 12) κ_ν is inversely proportional to intensity and hence, from eqn. 1,

$$\frac{dI_\nu}{ds} = constant. \tag{17}$$

Thus, the saturated maser grows linearly with distance through the cloud. Substituting for κ_ν in equation 11, the solution to the equation of transfer (neglecting spontaneous emission) becomes:

$$I_\nu = \frac{\Delta P}{P}\frac{n_1 \Gamma h\nu}{\Delta\nu\Omega}L \tag{18}$$

In the saturated maser, pumping events are linearly converted to maser photons. The intensity grows linearly with path length through the inverted medium.

The maser brightness temperature is:

$$T_b = \frac{\lambda^2}{2k}I = \frac{\lambda^2}{2k}\frac{\Delta P}{P}\frac{n_1\Gamma}{\Omega}\frac{h\nu}{\Delta\nu}L \tag{19}$$

Despite the discussion of unsaturated masers above, several arguments suggest that the strong masers associated with star formation are, at least partly, saturated. There is, for example, no clear case of the maser intensity reflecting the strength of the background source, as would be expected if the masers were unsaturated. Also the high degree of circular polarisation in OH masers is best explained if the sources are saturated. Finally, the radiation from the strongest star-formation masers is so intense that it is difficult to explain them by the available pumping energy unless they are operating at their maximum efficiency. The most efficient way to convert pumping events to maser photons is by saturation.

The brightness temperature at which the maser saturates is given by:

$$T_s = \frac{h\nu}{2k}\frac{\Gamma}{A}\frac{4\pi}{\Omega} \tag{20}$$

3.5. MASER GEOMETRY

Typical interstellar conditions are characterised by turbulence and supersonic chaotic motions. However, maser action requires paths that maintain close velocity coherence and so will only take place along directions that develop such coherence by chance. We think of masers as long filamentary structures, or tubes through the inverted population in a molecular cloud in which maser amplification may not take place throughout the tube continuously but only in those 'cloudlets which have the right velocity (and/or magnetic field) to maintain coherence. Such maser columns, as well as developing by chance, may arise naturally in shocked regions (Elitzur, Hollenbach & McKee, 1989) or in edge-on discs. The input to the maser may be a spontaneous transition or a background continuum source (more likely in masers associated with HII regions).

The idea of beaming is supported by our rudimentary theory since if the size of the cloud was equal to the measured size of the maser spot, it would be very difficult to achieve the gain required. For example, (Reid & Moran, 1988) for $\Delta n = 0.03$ (e.g. $n_{H_2} \approx 10^5, n_{OH}/n_{H_2} \approx 10^{-4}, \Delta P/P \approx 0.01$ and about 30% of the OH molecules in the ground state) a maser of observed diameter 10^{14}cm would have a gain of only 3. The idea of tubes with $l \gg radius, r$, alleviates this difficulty since the gain length may easily be increased by say, 100 over the spot size (2r). The maser radiation will be beamed into the solid angle $\Omega = \pi(r^2/l^2)$, and an observer would observe only those filaments pointing towards the earth.

3.6. PUMPING

Several mechanisms for inverting the population and causing an overpopulation of the upper state of the maser transition have been proposed. While some of them are species dependent, two basic principles are common to all four common maser species: radiative and collisional pumps. A third mechanism – chemical pumping, has been proposed for water and OH masers. A good discussion of possible pumping mechanisms is given in the book 'Astronomical Masers' (Elitzur, 1992).

In radiative pumps, incoming radiation (usually infrared, given the dusty nature of the medium) excites the molecule to a high energy state and a subsequent cascade down to the upper masing level causes the population inversion. In the case of OH-IR stars a net transfer of population from the upper to the lower hyperfine level of the $^2\Pi_{3/2}$ state is caused by incoming IR radiation at 35 micron, via the J= 5/2 state of the $^2\Pi_{1/2}$ lambda doublet. It is also possible, given the association of OH masers with strong sources of infrared radiation (as shown by IRAS data and other observations) that infrared pumping may occur in star formation regions. ISO observations of some galaxies, eg Arp 220, an ultraluminous galaxy with strong OH maser emission, indicate the presence of strong 35 micron absorption lines and suggest that IR pumping can also excite extragalactic OH masers (see Moorwood, 1999).

However it seems that power considerations rule out radiative pumps of this type for interstellar water masers. In other words, the high luminosity of the strongest water sources causes problems for radiative pumping since the photon rates often far exceed the continuum photon rates in any appropriate continuum band (eg the Orion flare) and so unless line pumping schemes are adopted, radiative pumps are deemed to be unlikely (see Cohen 1989 and references therein).

Most collisional pumping schemes resemble the radiative schemes in as much as collisions excite the molecules to a higher energy state, after which a downward cascade to the upper state of the maser transition occurs. However, again, especially in the case of water, the problem of the maser luminosity implies such high collision rates that the required densities would thermalise the population and quench the maser.

The proposed chemical pumping schemes involve the formation of the molecule in an excited state. Andresen (1986) has shown that photodissociaton of cold water leads to inversion of the 18 cm lambda doublet by radiative cascade but estimates that a strong Galactic OH maser could operate for no longer than an hour before exhausting the probably supply of water, so unless the water can be reconstituted quickly, chemical pumping of this type is unlikely.

4. Masers associated with star formation

The current picture of star formation is that it occurs through the fragmentation of a molecular cloud followed by gravitational collapse. The collapsed fragment forms into a rotating disc and the protostar accretes matter from this disc and sheds angular momentum, which is stabilised by ejection of matter in an expanding outflow – a bipolar jet or wind of gas and dust, along the rotation axis. When the star switches on its radiation will dissociate the molecular gas and eventually it will ionise the surrounding material, forming a compact HII region. Since the masers provide the only radio means of measuring small angular sizes and displacements, with velocity information on seductive scales of 0.1 to 1 AU, it is pertinent to ask if we can learn any more detailed information on their associated star formation regions. What can our VLBI observations tell us? Let us look at the different masers in turn.

4.1. WATER

The water masers at 22 GHz, are the most spectacular masers in the Galaxy. They have the highest luminosities, the most compact sizes, the most rapid variability and the widest velocity ranges. The water maser W49N, the most luminous maser source in the Milky Way, has a total isotropic power of one solar luminosity (i.e. the total power emitted by the Sun over all wavelengths) in its emission at 22 GHz, and it is highly variable on timescales of weeks. The Orion maser is also highly variable and on occasion exhibits a flux density of 10^6 Jy, in a line of order 1 km/s wide. Because of their enormous intensities, water masers are probably the most readily detectable indicators of high mass star formation regions – both high- and low-mass star formation. The Arcetri survey for Galactic water masers at 22 GHz finds lists 259 masers associated with star forming regions (Comoretto et al 1990). Felli et al (1992), using this survey and other data on massive star forming regions, found a correlation between the water luminosity and stellar luminosity, as traced by the far-infrared emission; a similar correlation was found for a sample of low-mass stars with luminosities several orders of magnitude weaker (Wilking et al, 1994).

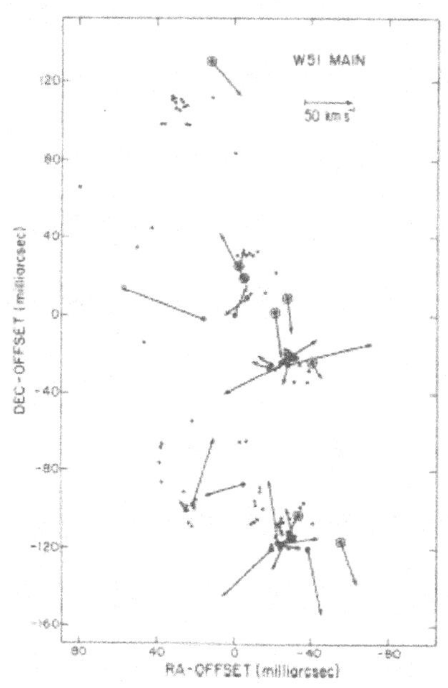

Figure 4. VLBI image of the water masers towards W51M. The arrows show the measured proper motions of the individual components. From Genzel et al. (1981).

As we have seen, high resolution (VLBI) observations of the masers show that each emission feature in the spectrum is a spatially separate component (or group of components). Individual water maser spots are resolved only with transatlantic VLBI and are often less than 1 mas in extent, with physical dimensions of $10^{13} \rightarrow 10^{14}$ cm, typical line-widths of 1 km/s and brightness temperatures of 10^{15}K. The maser spots cluster in groups, or "centres of activty" with physical sizes of $10^{15} \rightarrow 10^{16}$ cm, and the whole maser complex may consist of several such clusters extending over a region an order of magnitude greater (see e.g. Genzel et al, 1981 and Fig. 4).

When multi-epoch VLBI maps of water masers are compared, it is seen that the individual spots of emission have moved. This phenomenon is used to determine proper motions and indeed distances to the masers have been measured using the proper motions and the line of sight velocities (see Genzel et al, 1981). This is a powerful technique in astronomy since it is independent of all other distance indicators. (N.b the masers are used as test particles and although this is not strictly correct, it should be noted that typical maser columns are very small, relative to the full extent of the medium in which they are excited).

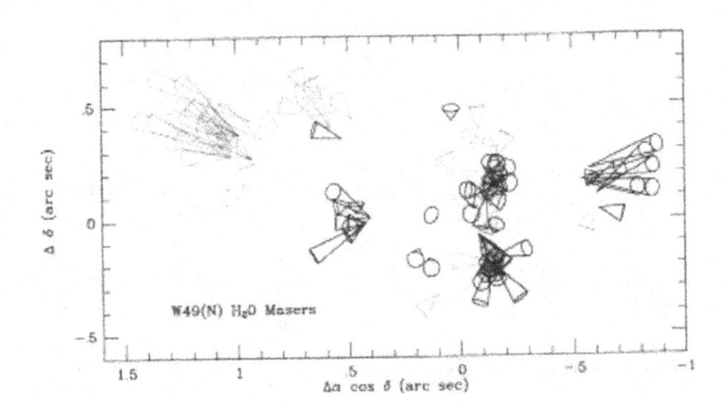

Figure 5. Proper motion diagram for the water masers in W49N. The maser features are located at the apexes of the cones. The cones' lengths and inclinations show how far the masers will travel in 150 years. The motions are clearly dominated by expansion from a common centre. From Gwinn et al, 1992.

A major finding of the proper motion studies was that in many cases, the water maser motions showed clear evidence of collimated bipolar outflows. The first evidence came from VLBI maps of W49N (Walker, Matsakis and Garcia-Barreto, 1982) and since that time outflows have been reported in more than 200 Galactic water maser sources ranging in diameter from 0.01 to a few pc (Liljeström & Gwinn 2002). Fig 5 shows the proper motions of the water masers in W49N based on five epochs of VLBI data (Gwinn, Moran & Reid, 1992, Gwinn 1994). The proper motions across the plane of the sky are combined with the line of sight (Doppler) velocities of the maser features to determine their 3-dimensional space velocity vectors.

From the work described above, a basic model for water masers is emerging. They often appear to stream away from a common centre, probably the site of a newly born star, at velocities approaching 250 km/s. This molecular outflow results from a strong stellar wind from the young massive star (with velocities of several thousand km/s); its interaction with the surrounding dense gas will result in the formation of powerful shocks (see the review by McKee & Hollenbach, 1980). Recent work by Liljeström & Gwinn (2000), combined with theoretical predictions of Hollenbach, McKee & Chernoff (1987), and Elitzur, Hollenbach & McKee, (1989) has started to use the observational parameters of the masers to unravel the details of the shocks and the subsequent excitation of the masers. Their work is based upon long-term single dish monitoring of the maser W49N, together with multi-epoch VLBI observations.

In summary, many water masers seem to be excited in shocks caused when a supersonic wind from a newly formed star interacts with the ambient medium. In the model of Elitzur, Hollenbach & McKee (1989) individual maser features correspond to directions in the shock plane, which attain the required line-of-sight velocity coherence, in an otherwise turbulent medium. Thus, it would seem that water masers are associated with a relatively late phase of star formation, after the production of the supersonic wind. Note, however, the work of Wilking et al (1994) on masers and CO outflows associated with low mass stars. In many cases the young stellar object is still deeply embedded in a dark cloud suggesting an earlier phase of evolution when accretion is still the main energy input. Indeed, some recent observations (Liljeström, Winnberg & Booth, 2002) have detected maser emission from vibrationally excited water at 96 GHz from the water maser 'cocoon in W51M, a 22 GHz maser source which does not show the usual outflow dynamics but which may be an infalling protostellar cocoon (Leppänen, Liljeström & Diamond, 1998), and in the hot FU Orionis-type accretion disc of Z CMa. The excitation temperature of the 96 GHz line is 3065 K and these detections together with the other properties of the related objects do not rule out the presence of water masers at a much earlier phase of the stellar evolution process.

4.2. HYDROXYL (OH)

We have discussed the properties of the hydroxyl masers earlier. The lines of the OH lambda doublet with wavelengths close to 18 cm were the first masers to be detected. They were found to be associated with compact HII regions and set the scene for the association of masers with regions of star formation; they are still among the best signposts of star formation regions although they may not represent the very earliest stages of stellar birth. They were also the first masers to be studied with VLBI and, like water they consist of groups of point sources with angular sizes of a few to a few tens of milli-arcseconds (see e.g. Moran et al, 1968), the most compact corresponding to linear sizes of 10^{14} cm. The measured angular sizes are undoubtedly influenced by interstellar scattering which becomes relatively strong at wavelengths below 8 GHz but it is known e.g. from pulsar studies that the scattering medium (or the interstellar electron density) varies greatly over the sky. The overall dimension of a typical OH maser region is about $10^{16} \rightarrow 10^{17}$cm. The dimensions of typical groups or clumps are an order of magnitude greater. Although the maser spot maps show no clear dynamical signature, the proper motions detected (Bloemhof et al, 1992

(Fig 2) and Migenes et al (1992)) show that they are probably expanding.

The OH masers clearly display circular polarisation and evidence that the lines may be split by the Zeeman effect. Linear polarisation is also observed but is less common; typical values of linear polarisation are 30% (Cohen 1989). OH is a paramagnetic molecule and the Zeeman splitting of the 1665 MHz line is close to 3.2 kHz per milligauss (mG). With typical line widths of 1 km/s or less (about 5.5 kHz), it takes a field of order 1 or 2 mG to give a detectable line splitting. However, the conditions along the maser column that give velocity coherence in one polarisation will generally not apply to the opposite polarisation and so we rarely see clear Zeeman patterns across the spectrum (Fig 1). Nevertheless, VLBI observations show that velocity components with both senses of circular polarisation are often very close to each other, giving the possibility to estimate field strengths, which are of order several mG. The lines of the $^2\Pi_{3/2}$,J=5/2 OH masers at 6 GHz are strongly circularly polarised and frequently show very clear Zeeman patterns across the entire spectrum. VLBI measurements of W3(OH) by Moran et al (1978) identified ten Zeeman pairs and showed that a surprisingly uniform field of order 6 mG exists across the entire source. Similarly, Baudry & Diamond (1998) observed the 13 GHz OH $^2\Pi_{3/2}$,J=7/2 transition in W3(OH) and observed magnetic fields as high as 11 mG. As has been pointed out by Cohen (1989) such fields are of great significance since the magnetic energy density that they imply is comparable with the kinetic and gravitational energy densities in the maser region. The magnetic field will therefore have a significant influence on the dynamics of the region surrounding the young star.

When linearly polarised components are present in the maser spectrum, VLBI measurements in all 4 Stokes parameters enable the full field parameters to be measured – magnitude and direction. An interesting example is provided by the work of Hutawarakorn & Cohen (1999) who measured the field strength in the bipolar outflow source G35.2-0.74N, using the MERLIN interferometer. They found that the masers have an elongated distribution, which they interpreted as a disc surrounding the outflow. The polarisation data shows that the field reverses on opposite sides of the disc.

It is clear from the discussion that OH masers, like their water counterparts, are probably associated with the later stages of protostellar evolution when well developed bi-polar outflows are established. In the case of G35.2 both OH and water masers lie at the centre of the flow, and appear to trace the inner edge of the molecular disc (see Hutawarakorn & Cohen, 1999).

4.3. METHANOL

The methyl alcohol or methanol molecule, CH_3OH, is widespread in the Galaxy; its spectrum is rich and more than 200 interstellar lines have been detected in the radio range between 834 MHz and 350 GHz, many displaying maser emission (Elitzur, 1992). It has strong masing transitions at 12.2 GHz and 6.7 GHz, excited only in star-forming regions, and the 6.7 GHz maser is, after water, the strongest maser in the sky, its intensity reaching thousands of Jy in some cases. Surveys for 6.7 GHz methanol masers, targeted on known star formation indicators like previously known masers or IRAS sources, have revealed 200-300 in the Milky Way.

Methanol masers have been detected only near star forming regions. They have been classified: class 2 or 1, respectively, according to whether they have associated continuum emission, or not, and according to which transitions are excited (see Cragg et al, 1992 for (collisional) pumping schemes which provide a simple explanation for the two classes). Unbiased surveys by using the Hobart telescope (Ellingsen et al, 1998) and the Onsala 25m antenna (Pestalozzi et al, in prep.) are revealing more sources, some of which have no other known star-formation indicator. This, and some of the VLBI results are leading to the suggestion that methanol masers might be among the first, perhaps the very first manifestation of an emerging protostar. Regions around some of the survey sources with no other indicator of star formation have been observed with the SIMBA 1.3 mm wavelength bolometer array on the SEST telescope and find dust continuum emission (Pestalozzi, Humphreys & Booth, 2002). This suggests that the maser is being excited by a dust-embedded protostar. More data is needed to confirm this conclusion.

Because the strong masers at 12.2 GHz and 6.7 GHz were detected only in 1987 and 1991 respectively by Batrla et al (1987) and Menten (1991), their investigation by VLBI is still in its relative infancy. Furthermore, the line frequencies do not lie in the standard radio astronomy bands and, although the 12.2 GHz line was detected in time to be included in the suite of receivers for the VLBA; 6.7 GHz was not. The more flexible EVN is now partially equipped at 6.7 GHz, at the current time only half the telescopes in the array support this frequency, although this is changing. Thus, although some early VLBI was carried out by Menten et al (1988) on W3(OH), it is the Australian group (Norris et al. 1988, 1998) which produced the first really exciting methanol results from a series of observations with the Australia Telescope Compact Array (ATCA) and the Parkes-Tidbinbilla interferometer (TBI). More recent work has been conducted by the Onsala group and two papers have been published by Minier, Conway & Booth (2000, 2001).

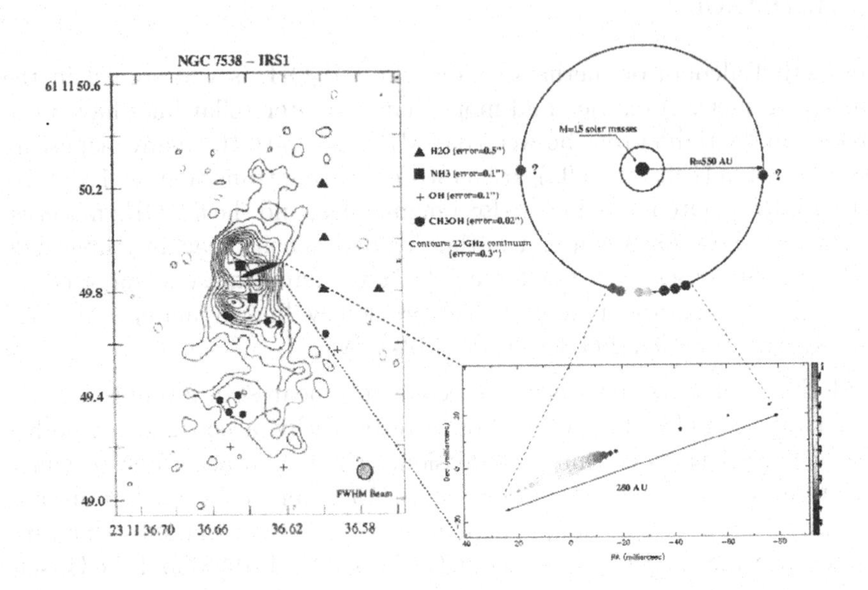

Figure 6. The star-forming region NGC7538 showing the location of various masers. A close-up of the methanol masers is shown in the bottom right with a model of a rotating disc above it.

Let us review the VLBI observations; all have concentrated on Class 2 masers. Observations of W3(OH) by Menten et al (1988, 1992) and Moscadelli et al (1999) show that the 6.7 GHz methanol masers occupy the same region as the OH masers in and around the contours of the compact HII region and the general distribution of the masers resembles closely that of the OH masers, indicating that methanol and OH masers arise from the same gas cloud. Strong 6.7 and 12.2 GHz maser features with the same velocity are coincident within 1-2 mas. This, the authors note, places stringent constraints on the methanol maser excitation model. In the 1999 paper Moscadelli et al showed that the 6.7 GHz maser components were not simple point sources but that there was weak emission connecting the distinct spots found by Menten et al in 1998.

In contrast to this near-conventional picture of the masers, Norris et al (1993, 1998) showed from observations with the ATCA at 6.7 GHz and the PTI at 12.2 GHz, that in 10 out of 16 sources observed, the masers were located along lines or arcs. In 5 of their sources there was a clear velocity gradient along the line. They suggested that this alignment occurred because the methanol masers were located in edge-on discs surrounding the young stars. Assuming Keplarian rotation, Norris et al derived the mass of the central object and the radius of the disc. They found values of

$3 - 100M_{\odot}$ and $500 \rightarrow 2000$AU, respectively – typical of circumstellar discs around high-mass stars (e.g. Lin & Pringle, 1990).

6.7 GHz VLBI observations with the EVN were initiated by the Onsala group in 1997 and results published in a series of papers: Observations were also conducted with the VLBA at 12.2 GHz. 14 star formation regions were observed and in ten of these, the VLBI maps of the centroids of individual spectral features showed aligned structures with linear velocity gradients, consistent with the edge-on discs surrounding protostars. (Minier, Booth & Conway, 2000). Interestingly, fringe visibilities were typically less than unity on the long interferometer baselines and a detailed analysis of three sources has shown that individual spectral features have structures consisting of a compact core of 0.5 – 3mas containing 20%, or less of the total flux, surrounded by a gaussian halo of more diffuse emission with a much larger angular half width – sometimes as many as 50 times the compact structure. (Minier, Conway & Booth, 2001).

If we assume that the full length of the linear compact maser structure represents the diameter of the Keplerian disc, sub-solar central masses are derived, with the exception of W48 when a mass of 6 $M_{\odot}$ is obtained. However, the full extent of the disc is clearly considerably larger, based on the extended structures of the 'halo components, we might not even be able to detect the full disc in the diffuse emission. This implies that the disc-enclosed masses may be much greater, and more consistent with those of high-mass stars e.g. see Fig. 6.

In conclusion, we note that it is common for methanol masers to form in elongated structures with linear velocity gradients. Whether these represent protostellar discs seen edge-on is still open to debate. Linear structures may be produced by shocks or simply by filaments in the interstellar gas. Whatever the outcome of the debate, methanol masers have added an exciting new tool for studying star formation. They may be the first observable sign of a newly emerging star, as seen from the blind surveys and they may be the means to detect protoplanetary discs.

5. Stellar masers

In the previous sections we have described the properties of masers associated with early phases of stellar evolution. However, in 1970, Wilson, Barrett & Moran (1970) discovered 1612 MHz OH maser emission associated with late-type stars. Viewed in terms of the Hertzprung-Russell diagram, which represents stellar luminosity as a function of colour, or effective temperature (the surface temperature a star would have if it radiated like a perfect black body), the OH masers are associated with very red, mostly oxygen-rich stars towards the end of the Asymptotic Giant Branch (red giants and Mira variables). AGB stars are semi-regular variable stars, which have completed their hydrogen-burning phase and are losing mass through stellar pulsations. The ejected material is in the form of heavy elements produced by nuclear fusion in the stellar interiors and it contains profuse amounts of dust (these objects are the great interstellar pollutants) and as it cools and interacts with the interstellar gas, it spawns a large variety of molecules. Among these molecules we find three radiating as masers; i.e. in addition to OH, we find maser emission from water and silicon monoxide (SiO).

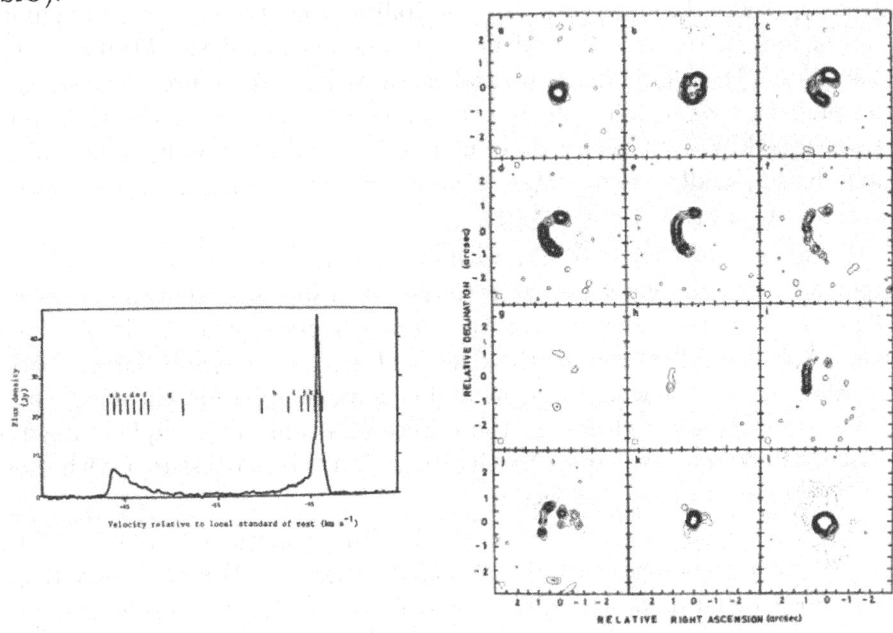

Figure 7. Left: Spectrum of a typical OH/IR star, OH 127.8; Right: Maps of the distribution of the maser emission in the frequency intervals marked in the spectrum. From Norris et al, 1982.

5.1. THE STELLAR OH MASERS

Detailed observations of the OH masers have taught us a lot about the AGB stars: the 1612 MHz OH spectra show two distinct peaks, (see fig 7a) indicative of rotation or expansion. The line emission varies as the star varies but with the phase of the red-shifted peak lagging behind the blue. This suggests that the OH gas is actually in an expanding shell around the star and that the stellar radiation energises the maser. This was confirmed by interferometric observations at Jodrell Bank using the MERLIN interferometer - see Booth et al (1981) and Diamond et al (1985). We see why by inspection of Fig 7b, where maps of the maser emission from the different velocity intervals indicated on the spectrum of Fig 7a are presented. The red and blue OH maser emission peaks, which require the longest constant-velocity-paths are coming from unresolved spots both situated in the centre of the mapped field. This must mean that they are at the near- and far-sides of a shell, in which the material has a velocity relative to the central star equal to the velocity difference between the spectral peaks. In such an expanding shell the other longest maser paths will be along line of sight tangents to the shell and would describe annuli in each velocity interval exactly as we see in the extended arcs delineating the intermediate velocity emission. See Fig 8 for the geometry of the model.

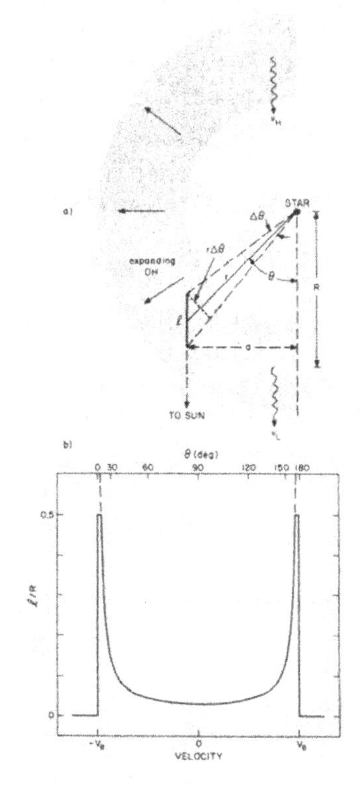

Figure 8. Geometry of the expanding shell model (Diagram from Reid et al, 1977).

Since the maps clearly indicate that the spectral peaks are at the front and back of the shell, relative to the observer, the measured blue-red phase lag, T, gives us a linear measure of the shell diameter – ($Tc = D$) Typical values of D are about 1000 AU. Also, knowing the angular size of the shell (θ) from the maps, the stellar distance may be determined - (D/θ). Finally, with some assumptions about the density of material in the shell, the stellar mass loss may be estimated – typical mass loss rates are $\approx 10^{-5}$ solar masses per year. (see Bowers, 1985)

The location of the blue- and red-shifted maser peaks along the line-of-sight to the central star has been demonstrated by several VLBI observations (Norris et al (1984), Sivagnanam et al (1990)) and then used as a fiducial point to enable measurement of stellar parallax to be carried out (van Langevelde et al, (2000)).

Many evolved stars have been observed and most follow a similar twin-peaked pattern indicative of an expanding shell. Although in a given source, the two peaks may differ in amplitude, there is no statistical difference between the shapes and intensities of the red and blue shifted peaks (Johansson et al, 1977). The two peaks have typical separations of 10 - 50 km/s implying a shell expansion velocity of 5 - 20 km/s.

In addition to the 1612 MHz OH lines, most OH/IR stars also exhibit OH main line maser emission (1665 & 1667 MHz) but usually without the double peak, as well as water and SiO masers. Following their observations of the source VX Sgr and noting that the excitation of the SiO lines (2000K) is much greater than that of water (650K), which again is significantly greater than OH, and considering the maps of the circumstellar gas in all the maser lines, Chapman & Cohen (1986) suggested that the various masers delineate shells as shown in Figure 9. In more recent work on VX Sgr (Szymczak & Cohen (1997), Szymczak, Cohen & Richards (2001)) have measured the strength and structure of the magnetic field in the circumstellar envelope. Other authors have made similar measurements on other sources.

This picture of the circumstellar envelope is now well established. It seems clear that the OH is formed by photo-dissociation of water molecules by UV radiation from the interstellar medium and that OH forms the interface between the expanding shell and the ISM. Si0, on the other hand is found very close to the star – within a few stellar radii.

5.2. STELLAR WATER MASERS

The advent of the VLBA has enabled important observations of the water masers in stars. From multi epoch VLBA observations, Marvel (1966) has measured the structures and kinematics of the intermediate regions of the stellar envelopes between the OH and the SiO. Observations of the water masers in the star S Per have clearly demonstrated that the water shell is expanding e.g. Fig. 10. More recently, Vlemmings, Diamond & van Langevelde (2001) have made high-precision measurements of the polarisation of water masers towards the same star and measured a magnetic field strength of $\approx$ 280mG.

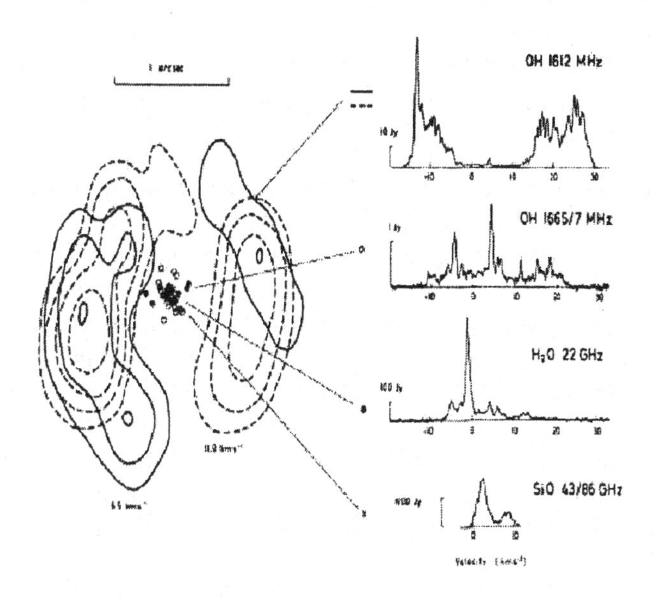

Figure 9. The spectra of the various masing species in VX Sgr and their locations in the circumstellar envelope (after Chapman and Cohen, 1986)

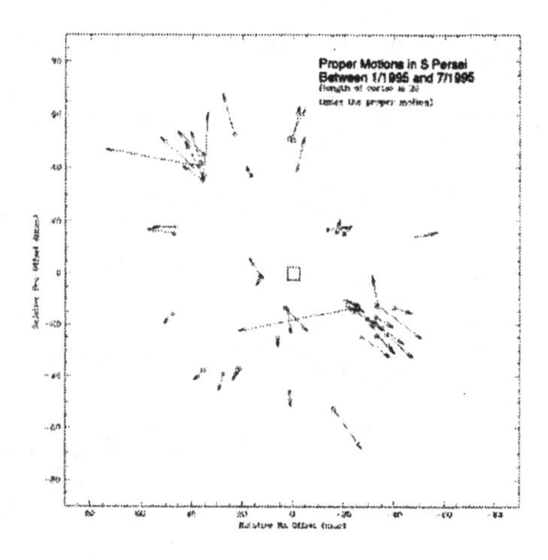

Figure 10. The proper motions of the water masers in the circumstellar envelope of S Per estimated from three epochs of VLBA observations. From Marvel (1996).

5.3. SIO MASERS IN CIRCUMSTELLAR ENVELOPES.

SiO masers lie in a very interesting region of the stellar atmosphere. They are rotational transitions of vibrationally excited molecules with excitation temperatures > 2000 K and hence are populated only close to the star, in a region extending from the stellar photosphere to the dust formation point, where one expects intense activity and violent mass motions in convective cells or other complex structures. The SiO masers also show periodic variations linked to the stellar light curves but delayed in phase.

Colomer at al (1992) showed that the SiO masers have compact structures that may be studied by VLBI and observations by Diamond et al (1994) with the VLBA produced the first images of the maser distribution showing that in TX Cam and U Her, the maser features lie in ordered rings around the star and were not distributed randomly. The masers were shown to be beamed tangentially to the stellar envelope (see also Miyoshi et al, 1994, Greenhill et al, 1995a). Since that time, Diamond et al (in prep.) have monitored the v=1, J=1-0 transition at 43 GHz over more than one stellar cycle and show very interesting but complex motions within the maser region. They also showed that the masers in TX Cam are linearly polarised with the E-vectors predominantly aligned tangentially to the maser ring (Kemball & Diamond, 1997), indicating a radial magnetic field. Such observations are of great importance for our understanding of this complex and normally observationally inaccessible region of the stellar atmosphere.

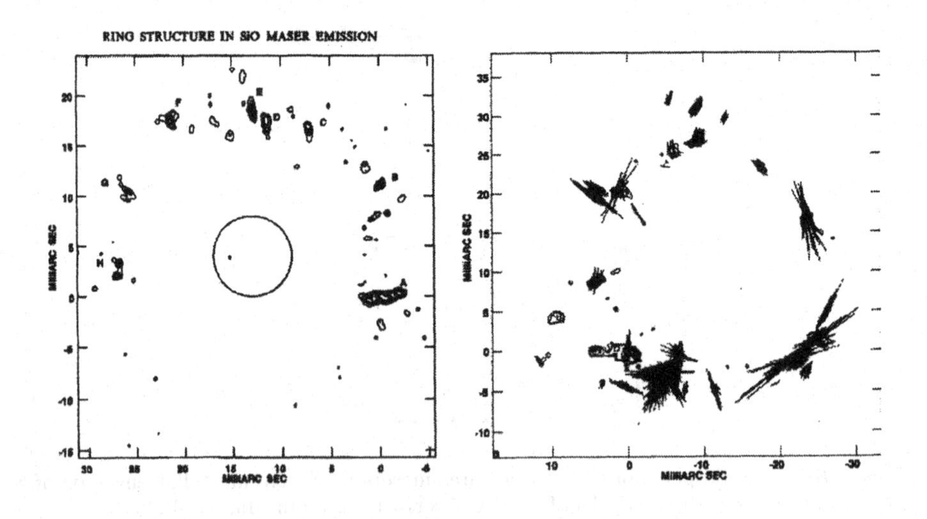

Figure 11. Left: The first image of SiO masers around the Mira variable TX Cam (Diamond et al, 1994). Right: the polarisation structure at a later epoch (Kemball & Diamond, 1997).

6. Extragalactic Masers

Soon after the discovery of strong masers in the Milky Way astronomers turned their telescopes to other galaxies to search for similar objects. However, in order for the subject to advance the receivers and other equipment needed to be improved. After much work Churchwell et al (1977) discovered water masers in M33; these were later demonstrated to be related to regions of star-formation. The first so-called 'nuclear maser' was discovered by Dos Santos and Lepine (1979) in the galaxy NGC4945. This was soon followed by the discovery of several other objects (Claussen, Heiligman and Lo (1984)) including the now famous maser in NGC4258. Meanwhile, searches for OH masers had yielded nothing until, serendipitously and while searching for OH absorption with the Arecibo telescope, Baan et al (1982) discovered the strong OH maser towards the starburst galaxy Arp220. It was clear from the start that both species of maser were different in nature from those observed in our own galaxy; the principal difference was the luminosities of the emission lines, they were several orders of magnitude more intense than galactic masers and hence were termed "megamasers".

Major single-dish surveys have been conducted searching for both species of maser. At the time of writing a total of 31 galaxies exhibit confirmed water maser emission; 22 of them are associated with active galactic nuclei (AGN), the remainder with regions of star-formation. Approximately 90 galaxies are known to contain OH megamasers (OHM) and one, Arp220, contains H_2CO maser emission (Baan & Haschick, 1995). The vast majority of the OHM have been discovered with Arecibo (e.g. Darling & Giovanelli, 2001). A small number of OH masers are associated with regions of star-formation in nearby galaxies (e.g. M82) and are presumably analogues of similar regions in the Milky Way. The rest of this section will discuss only the water masers associated with AGN and the OHM.

6.1. H_2O MEGAMASERS AND VLBI

Table 1 lists the water masers known to be associated with AGN at the time of writing. This list was kindly supplied by Lincoln Greenhill (priv. comm.)

VLBI studies have been attempted of most of these sources. Unfortunately, it appears that at the current time it is only possible to make useful images of sources with fluxes greater than ≈ 0.3Jy. However, existing studies have demonstrated that there appear to be two principal classes of object: discs and jet/outflow sources.

The archetypal disk maser is the well-known source NGC4258. Nakai, Inoue & Miyoshi (1993) serendipitously discovered high-velocity maser lines bracketing the previously discovered systemic velocity masers. This discov-

TABLE 1. H_2O megamasers in AGN

Galaxy	Distance (Mpc)	Flux (Jy)	Galaxy	Distance (Mpc)	Flux (Jy)
NGC4945	3.7	4	IC2560	38	0.4
Circinus	4	4	NGC2639	44	0.2
NGC4258	7.3	4	NGC5793	50	0.4
M51	9.6	0.2	ESO 103-G35	53	0.7
NGC1386	12	0.9	Mrk 1210	54	0.2
NGC3079	16	6	ESO 103-G12	57	0.2
NGC1068	16	0.6	Mrk 348	63	0.04
NGC1052	20	0.3	Mrk 1	65	0.1
NGC5506	24	0.6	IC1481	83	0.4
NGC5347	32	0.1	NGC6240	98	0.03
NGC3735	36	0.2	IRAS F22265-1826	100	0.3

ery was rapidly followed up by VLBA observations of the maser system (Miyoshi et al, 1995) which demonstrated quite convincingly that the maser lay in a rapidly rotating disc around a super-massive black-hole with a mass of $3.9 \times 10^7 M_\odot$. Fig. 12 shows a montage of the velocity and spatial structure of the maser and a model of the warped, Keplerian rotation disc that provides the best fit to the observed structure.

It is interesting to briefly review the simple mathematics that describe Keplerian rotation. For a particle of mass m, orbiting a larger mass M at a radius r and with a velocity v, the gravitational attraction is matched by the centripetal acceleration and the simple equation of motion applies:

$$\frac{GmM}{r^2} = \frac{mv^2}{r} \tag{21}$$

in which G is the gravitational constant. Simplifying this equation results in the following:

$$M = \frac{rv^2}{G} \tag{22}$$

$$v = \sqrt{\frac{GM}{r}} \tag{23}$$

Important consequences of this for the water maser VLBI results are that once the angular size of the disc has been measured, the linear orbital radius can be estimated from the distance to the galaxy, and since the disc orbital velocity is known from the maser spectrum the central mass can

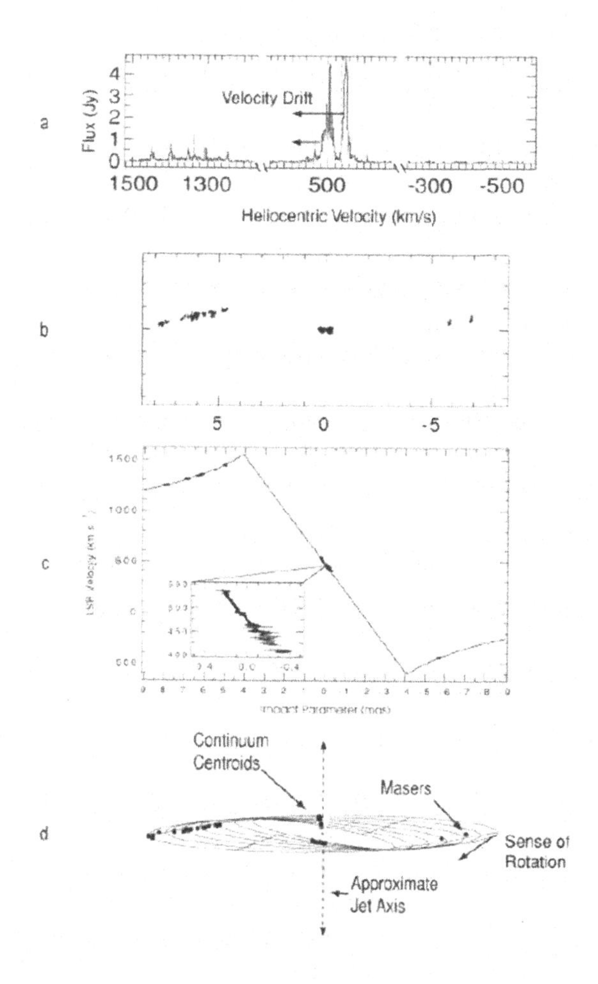

Figure 12. The principal observations of the NGC4258 water maser. a) is the single dish spectrum; b) is the VLBA map of January 1995; c) is the maser velocity vs. impact parameter, the solid line represents a fit of a Keplerian rotation curve; d) a schematic of the warped disc with the maser spots superimposed. From Bragg et al, 2000.

be calculated. Alternatively, r may be determined from single-dish monitoring of the maser spectrum. Since the masers lie in a rotating disc their centripetal acceleration will be $a = v^2/r$.

Later observations by Herrnstein et al (1997) revealed the presence of a continuum jet visible along the rotation axis of the disc, a beautiful confirmation of the jet-blackhole hypothesis. Nakai et al (1995) and Greenhill et al (1995b) detected acceleration of maser features of 9.5km/s/yr, in accordance with expectations from the rotating disc model. Then, Herrnstein et al (1999) used proper motion measurements to determine the distance to NGC4258 to be 7.2 ± 0.4Mpc, the most accurate extragalactic distance ever determined.

Other objects, e.g. one of the NGC1068 masers, Circinus, IC2560 and NGC5793, also exhibit structure or spectra that suggest the presence of discs. However, none is as obvious a case as NGC4258.

The other well-defined class of maser is exemplified by NGC1052 (Claussen et al, 1998). In this case, the galaxy exhibits a strong, well-defined continuum jet. The maser emission has a broad, featureless spectrum, very unlike that seen towards NGC4258. VLBI of the masers demonstrate that they lie in two groups offset from, but close to the central engine and appear to be closely associated with the jets. Claussen et al suggested that they might be embedded in gas being dragged along by the jet. Mrk348 (Peck et al 2001) exhibits a similar structure.

Clearly, VLBI observations of extragalactic water masers can improve our understanding of the environments close to the super-massive blackholes in the centre of AGN and also may be able to address the physics of the jets. What we lack is a far greater number of objects to study. Surveys are currently underway that will attempt to rectify this.

6.2. OH MEGAMASERS AND VLBI

The recent surveys by Darling and Giovannelli (2000, 2001) have tripled the number of known OHM, ≈ 90 are now known. All known OHM are associated with ULIRGs (Ultra-luminous IR galaxies) and all ULIRGs are believed to be associated with mergers. As described earlier, IR radiation often plays a role in the pumping of masers, especially OH masers; therefore it is probably the case that the presence of such vast levels of IR play some role in the occurrence of the OHM.

Although there are a significant number of OHM, the first is still the best. Any models of OHM physics will always have to be able to explain the increasingly detailed structures observed in Arp220. Early single-dish and low-resolution interferometric observations of Arp220 led to the suggestion by Baan (1985) that the OH was distributed on scales of hundreds of parsecs within the galaxy, was pumped by IR and was amplifying the background continuum emission arising within the central areas of the merging system. Early VLBI by Diamond et al (1988) showed that much of the OHM emission was resolved except for a few compact spots, somewhat

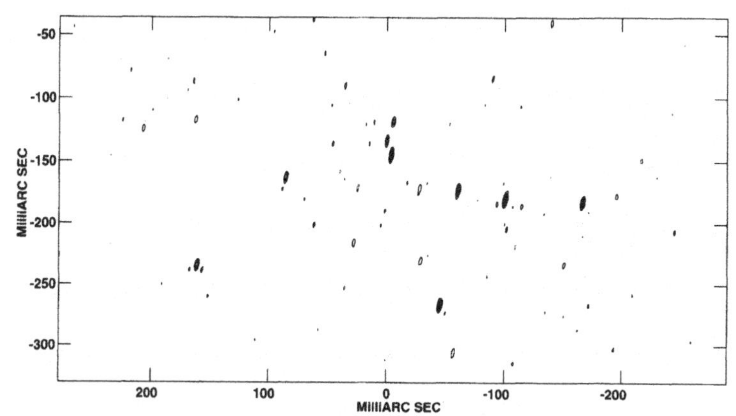

Figure 13. Continuum image of the radio supernovae in Arp220, Smith et al 1998.

in contradiction to the model of Baan. However, these data did not have the spectral coverage or the dynamic range to be able to offer a definitive picture of the source.

In 1994, Lonsdale et al (1994) demonstrated serendipitously that there existed very compact maser emission in Arp220. This result was followed up by full-scale global VLBI imaging which produced two surprising results. First (Lonsdale et al 1998), the OHM emission in Arp220 was revealed to have a two-component structure: diffuse emission on size-scales of 10s - 100s of parsecs formed from all of the 1665 MHz masers and 1/3 of the 1667 MHz emission; compact components in several regions comprising the remaining 2/3 of the 1667 MHz emission. This latter component could not be easily explained as a result of IR pumping and often exhibited large velocity gradients. As yet there is no satisfactory explanation of these structures. The second surprise (Smith et al, 1998) was that the continuum emission was seen to break up into $\approx$ 12 point sources of 1 mJy or less (Fig 13). These were interpreted as luminous radio supernovae and convincingly demonstrated the starburst nature of Arp220.

Other galaxies have been studied in some detail with VLBI. Number two in the rankings is clearly IIIZw35. Various groups have studied this source (Trotter et al, 1997; Diamond et al, 1999; Pihlström et al, 2001). The latter result is potentially the most interesting, showing as it does a ring of OH maser emission with a radius of 22pc and hints that there might be weak continuum point sources similar to those seen in Arp220 (Fig 14).

7. Summary

We hope that this review which, due to limited space, cannot do full justice to the subject of astrophysical masers has at least given you a flavour of

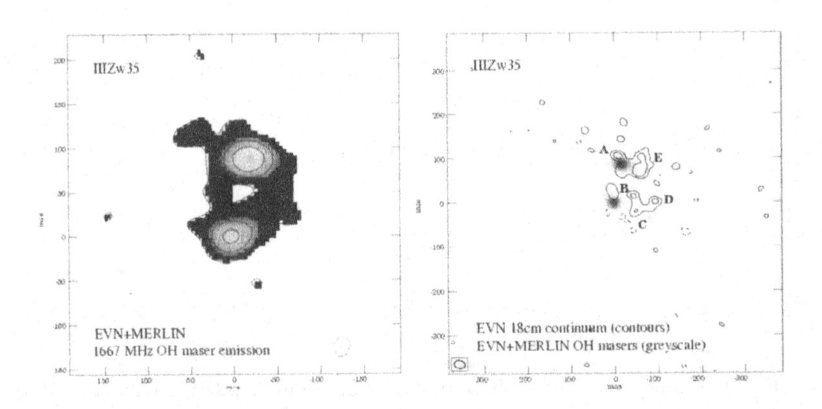

Figure 14. Left: OH maser emission from IIIZw35; right: continuum image. Both from Pihlström et al, 2001.

the breadth of the subject area and the exciting science that can be done with VLBI. We apologize to those of our colleagues whose work we have not referred to and suggest to those interested that a good starting point to learn more about masers is the book "Cosmic Masers: From Protostars to Blackholes", which contains the proceedings of IAU Symposium 206.

8. References

References

Andresen, P. 1986, Astron. Astrophys., 154, 42
Baan, W.A., Wood, P.A.D. & Haschick, A.D., 1982, Astrophys. J., 260, 49
Baan, W.A., 1985, Nature, 315, 26
Baan, W.A. & Haschick, A.D., 1995, Astrophys. J., 454, 745
Batrla, W., Matthews, H.E., Menten, K.W. & Walmsley, C.M. 1987, Nature, 326, 49
Baudry, A. & Diamond, P.J., 1998, Astron. Astrophys., 331, 697.
Bloemhof, E.E., Reid, M.J. & Moran, J.M. 1992, Astrophys. J., 397, 500
Booth, R.S., Norris, R.P., Porter, N.D. & Kus, A.J., 1981, Nature, 290, 382
Bowers, P.F., in "Mass-loss from red-giants", p189, Reidel, 1985
Bragg, A.E., Greenhill, L.J., Moran, J.M. & Henkel, C., 2000, Astrophys. J., 535, 73
Chapman, J.C. & Cohen, R.J., 1986, MNRAS, 220, 513
Churchwell, E., Witzel, A., Pauliny-Toth, I. et al, 1977, Astron. Astrophys. 54, 969
Claussen, M.J., Heiligman, G.M. & Lo, K.-Y., 1984, Nature, 310, 298
Claussen, M.J., Diamond, P.J., Braatz, J.A., Wilson, A.S. & Henkel, C., 1998, Astrophys. J., 500, L129
Cohen, R.J. 1989, Rep. Prog. Phys., 52, 881.
Colomer, F., Graham, D.A., Krichbaum, T.P. et al, 1992, Astron. Astrophys., 254, L17
Comoretto, G., Palagi, F., Cesaroni, R., et al, 1990, Astron. Astrophys. Suppl., 84, 179.
Darling, J. & Giovanelli, R., 2000, Astron. J., 119, 3003
Darling, J. & Giovanelli, R., 2001, Astron. J., 121, 1278
Davies, R.D., de Jager, G. & Verschuur, G.L., 1966, Nature, 209, 974

Davies, R.D., Rowson, B., Booth, R.S. et al, 1967, Nature, 213, 1109
Diamond, P.J., Norris, R.P., Rowland, P.R., Booth, R.S. & Nyman, L-Å, 1985, MNRAS, 212, 1
Diamond, P.J., Norris, R.P., Baan, W.A. & Booth, R.S., 1988, Astrophys. J., 340, L49
Diamond, P.J., Kemball, A.J., Junor, W., Zensus, J.A., Benson, J. & Dhawan, V., 1994, Astrophys. J., 430, L61
Diamond, P.J., Lonsdale, C.J., Lonsdale, C.J. & Smith, H.E., 1999, Astrophys. J., 511, 178
Downes, D. 1985, in Birth and Infancy of Stars (Les Houches Session XLI) ed R.Lucas, A. Omont & R.Stora (Elsevier)
Dos Santos, P.M. & Lépine, J.R.D., 1979, Nature, 278, 34
Elitzur, M. 1992, Astronomical Masers, (Kluwer Academic Publishers).
Elitzur, M., Hollenbach, D.J. & McKee, C.F. 1989, Astrophys. J., 346, 1983
Ellingsen, S.P., Norris, R.P. & McCulloch, P.M. 1996, MNRAS, 279, 101
Felli, M., Palagi, F. & Tofani, G. 1992, Astron. Astrophys, 255, 293
Genzel, R., Reid, M.J., Moran, J.M. & Downes, D. 1981, Astrophys. J., 244, 884
Goldreich P. & Kwan, J., 1974, Astrophys. J., 1190, 27
Greenhill, L.J., Colomer, F., Moran, J.M. et al, 1995a, Astrophys. J., 449, 365
Greenhill, L.J., Henkel, C., Becker, R., Wilson, T.L. & Wouterloot, J.G.A., 1995b, Astron. Astrophys., 304, 21
Gwinn, C.R., Moran, J.M. & Reid M.J. 1992, Astrophys. J., 393, 149
Gwinn, C.R. 1994, Astrophys. J. 429, 253
Herrnstein, J.R., Moran, J.M., Greenhill, L.J, Diamond, P.J. et al, 1999, Nature, 400, 539
Hollenbach, D.J., McKee, C.F. & Chernoff, D. 1987, in Star Forming Regions, ed M. Peimbert & J. Jugaku (Dordrecht: Reidel) 334
Hutawarakorn, B. & Cohen, R.J. 1999, MNRAS, 303, 845
Johansson, L.E.B., Andersson, C., Goss, W.M. & Winnberg, A., 1977, Astron. Astrophys, 54, 323
Kemball, A.J. & Diamond, P.J., 1997, Astrophys. J., 481, L111
van Langevelde, H.J., Vlemmings, W., Diamond, P.J., Baudry, A. & Beasley, A.J., 2000, Astron. Astrophys., 357, 945
Leppänen, K., Liljeström, T. & Diamond, P. 1998, Astrophys. J. 507, 909
Liljeström, T. & Gwinn, C.R. 2000, Astrophys. J., 534, 781
Liljeström, T., Winnberg, A. & Booth, R.S. 2002, in From Protostars to Black Holes. Proc IAU Symp 206, in press Lin, D.N.C.
Lin, D.N.C. & Pringle, J.E. 1990 Astrophys. J., 358, 515
Lonsdale, C.J., Diamond, P.J., Lonsdale, C.J. & Smith, H.E., 1994, Nature, 370, 117
Lonsdale, C.J., Diamond, P.J., Smith, H.E. & Lonsdale, C.J., 1998, Astrophys. J., 493, L13
Marvel, K.B., 1996, Ph.D Thesis, New Mexico State University
Menten, K.M. 1991, Astrophys. J., 380, L75
Menten, K.M., Reid, M.J., Moran, J.M., et al, 1988, Astrophys. J. 333, L83
Menten, K.M., Reid, M.J., Pratap, P. & Moran J.M., 1992, Astrophys. J. 401, L39
McKee, C.F. & Hollenbach, D.J. 1980, Ann. Rev., Astron. Astrophys., 18, 219
Migenes, V., Cohen, R.J. & Brebner, G.C. 1992, MNRAS, 254, 501
Minier, V., Booth, R.S. & Conway, J.E. 2000 Astron. Astrophys., 362, 1093
Minier, V., Conway, J.E. & Booth, R.S. 2001 Astron. Astrophys., 369, 278
Miyoshi, M., Matsumoto, K., Kameno, S., Takaba, H. & Iwata, T., 1994, Nature, 371, 395
Miyoshi, M., Moran, J.M., Herrnstein, J.R., Greenhill, L.J., Nakai, N., Diamond, P.J. & Inoue, M., 1995, Nature, 373, 127
Moorwood, A.F.M., in "The Universe as seen by ISO" p. 825. ESA SP-427, 1999, Eds. P. Cox and M.F. Kessler.
Moran, J.M., Burke, B.F., Barrett, A.H., Rogers, A.E.E., Ball, J.A., Carter, J.C. &

Cudaback, D.D., 1968, Astrophys. J., 152, L97
Moran, J.M., Reid, M.J., Lada, C.J. et al, 1978, Astrophys. J., 224, L67
Moscadelli, L., Menten, K.M., Walmsley, C.M. & Reid, M.J. 1999, Astrophys., J. 519, 244
Nakai, N., Inoue, M. & Miyoshi, M., 1993, Nature, 361, 45
Nakai, N., Inoue, M., Miyazawa, K., Miyoshi, M. & Hall, P., 1995, PASJ, 40, 139
Norris, R.P. & Booth, R.S, 1981, MNRAS, 195, 213
Norris, R.P., Diamond, P.J. & Booth, R.S., 1982, Nature, 299, 131
Norris, R.P., Booth, R.S., Diamond, P.J., et al, 1984, MNRAS, 208, 435
Norris, R.P., McCutcheon, W.H., Caswell, J.L., et al. 1988, Nature, 335, 149
Norris, R.P., Whiteoak, J.B., Caswell, J.L., et al, 1993, Astrophys. J., 412, 222
Norris, R.P., Byleveld, S.E., Diamond, P.J., et al, 1998, Astrophys. J., 508, 275
Peck, A.P., Falcke, H., Henkel, C. et al, 2001, in Proc. IAU Symp. 206., p434
Pestalozzi, M., Humphreys, E. & Booth, R. S. 2002, Astron. Astrophys., 384, L15
Pihlström, Y., Conway, J.E., Booth, R.S., Diamond, P.J. & Polatidis, A.G., 2001, Astron. Astrophys., 377, 413
Reid, M.J., Muhkeman, D.O., Moran, J.M., Johnston, K.J. & Schwartz, P.R., 1977, Astrophys. J., 214, 60
Reid, M.J. 1993, Ann Rev. Astro. & Astrophys., 31, 345
Reid, M.J. & Moran, J.M. 1988 in Galactic and Extragalactic Radio Astronomy (2nd edition) ed G.L Verschuur & K.I Kellermann (Berlin: Springer)
Sivagnanam, P., Diamond, P.J., Le Squeren, A-M., Biraud, F., 1990, Astron. Astrophys., 229, 171
Smith, H.E., Lonsdale, C.J., Lonsdale, C.J. & Diamond, P.J., 1998, Astrophys. J., 493, L17
Szymczak, M. & Cohen, R.J., 1997, MNRAS, 288, 954
Szymczak, M., Cohen, R.J. & Richards, A.M.S., 2001, MNRAS, 371, 1012
Trotter, A.S., Moran, J.M., Greenhill, L.J., Zheng, X.-W. & Gwinn, C.R., 1997, Astrophys. J., 485, L79
Vlemmings, W., Diamond, P.J. & van Langevelde, H.J., 2001, Astron. Astrophys., 375, L1
Walker, R.C., Matsakis, D.N. & Garcia-Barreto, J.A. 1982, Astrophys. J., 255, 128
Weaver, H., Williams, D.R.W., Dieter, N.H. & Lum, W.T. 1965, Nature, 208, 29.
Weinreb. S., Barrett, A.H., Meeks, M.L. & Henry, J.C. 1963, Nature, 200, 829
Weinreb, S., Meeks, M.L., Carter, J.C., Barrett, A.H. & Rogers, A.E.E., 1965, Nature, 208, 440.
Wilking, B.A., Claussen, M.J., Benson, P.J. et al 1994, Astrophys. J., 431, L119
Wilson, W.J., Barrett, A.H. & Moran, J.M., 1970, Astrophys. J., 160, 545

RADIO AND OPTICAL INTERFEROMETRY

W. D. COTTON
National Radio Astronomy Observatory
520 Edgemont Road
Charlottesville, VA 22903
USA

1. Introduction

This lecture will cover the basics of astronomical interferometers as imaging devices. The basic response of such an instrument to celestial signals is covered as well as the relationships between what can be measured and what is desired, i.e. the brightness distribution on the sky. This topic is covered in much detail in [11], [10].

Both radio and optical/IR techniques will be covered, although the emphasis will be on the radio since that is the main topic of this school. The physical realizations of radio and optical/IR interferometers are quite different as are many of the detailed technical problems, but the underlying physical principles are the same. This lecture primarily covers these underlying physical principles so a distinction is usually not needed. There is a section at the end of this lecture that discusses some of the special issues for optical/IR interferometers. For a recent review of developments in optical interferometry see [6].

2. The Quantum Limit

When the received photon rate is high, i.e. many photons in a time equal to the reciprocal of the bandwidth of the instrument, then an interferometer measures the statistics of the photons and the properties of an individual photon are relatively unimportant. However, when the photon rate becomes low, the properties of individual photons, i.e. quantum effects do become important.

F. Mantovani and A. Kus (eds.), The Role of VLBI in Astrophysics, Astrometry and Geodesy, 289–305.

The Heisenberg Uncertainty Principle states that the energy and arrival time of a photon cannot be both known to arbitrary precision [11]:

$$\Delta E \times \Delta t = \frac{h}{2\pi}$$

An alternate way of stating this is that knowledge of the phase (arrival time) of a photon costs the knowledge of its energy. Written in terms of the uncertainty of phase ($\Delta\phi$) and the uncertainty in the number of photons (Δn) received this becomes:

$$\Delta\phi \times \Delta n = 1$$

This is a fundamental limit for phase sensitive astronomical detection systems. In the language of radio engineers, the effect on a phase sensitive detection system can be expressed as adding an additional "noise" to the system which, expressed in terms of the equivalent physical temperature [11], is

$$T_{sys}^{QM} = \frac{h\nu}{k\Delta\phi}$$

This effect is relatively minor at low frequencies (radio) where the photons have low energy and are numerous, but it is very serious at high frequencies (optical/IR) where the photons are energetic but sparse. The quantum noise at representative frequencies is shown in Table 1.

TABLE 1. Quantum Noise added by Phase Sensitive Detection

Frequency	Temperature Equivalent Quantum Noise
1.4 GHz	0.07 K
230 GHz	11 K
900 GHz	43 K
500 THz (optical)	30,000 K

As a result of this effect, phase sensitive systems are practical in the radio through mid IR, but are impractical in the near IR and higher frequencies where the noise of the quantum effects overwhelm all other noise sources. Thus, most radio/sub mm interferometers use phase sensitive (heterodyne) detection systems while optical/IR interferometers use incoherent (direct) detectors. These will be discussed further in the following sections.

3. Basic Interferometer Types

There are numerous ways to form an interferometer, that is the way the incoming waves are combined to form the interference fringes. First, consider the mathematical operation of the combination; the waves can be added as in the simple Young double–slit interferometer, or cross-correlated (multiplied) as in modern radio interferometers. A further way of distinguishing interferometers is whether the electromagnetic field is sampled before interference is formed (heterodyne), or after the fringes are formed (direct). These possibilities are considered in the following sections.

3.1. ADDING INTERFEROMETER

The simplest way of interfering signals is to add their electric fields. A schematic representation of such a system with a very small bandpass is shown in Figure 1. In this representation, the electric field received by each element of the interferometer is composed of an independent part (a and b) and a common part due to the unresolved source being observed (S_t). This common part of the signal has a phase that rotates at the frequency of the radiation (ν) but with a phase lag at the more remote telescope of $-2\pi\nu\tau$ where τ is the additional time it takes a wavefront passing the first telescope to reach the second. The detection in such a system is mathematically described as the square modulus of the sum of the electric fields averaged in time. The response of this interferometer is

$$\text{output} = < a^2 > + < b^2 > + < 2S^2(1 + cos[2\pi\nu\tau]) >$$

There are three components in this response, the first two being the independent powers received by the two telescopes and the third is the power received from the source multiplied by a trigonometric term in τ, usually referred to as delay. This final term varies between 0 and S^2 as a cosine function and constitutes the "fringe". In this approximation, except for the periodic variations, the fringe does not decrease in amplitude with τ. One difficulty with this type of interferometer at long wavelengths where the background is high, is that the $< a^2 >$ and $< b^2 >$ may dominate the response and can be variable as well, making the detection of the fringe more difficult.

3.2. CORRELATION INTERFEROMETER

Some of the problems of adding interferometers can be eliminated by means of a correlation interferometer. In this type of interferometer, the time lag cross correlation function of the electric fields is formed rather than the simple sum. A quasi-monochromatic interferometer of this type is shown in Figure 2. In this case, the addition becomes a multiplication and the response becomes:

$$\text{output}_{\text{real}} = < 2S^2 cos[2\pi\nu\tau] >$$

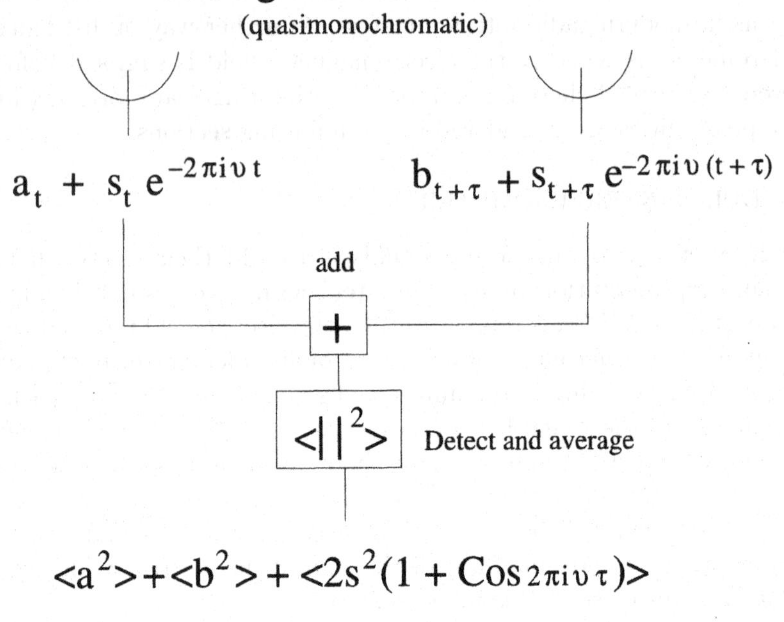

Figure 1. A simple adding interferometer

The independent terms average out, leaving only a response proportional to the power due to the source. This interferometer can be converted into a full complex system by the addition of a second correlation where the retarded signal has an extra 1/4 of a wavelength, equivalent to a phase shift of $\pi/2$, added:

$$\text{output}_{\text{imaginary}} = < 2S^2 cos[2\pi\nu\tau + \frac{\pi}{2}] > = < 2S^2 sin[2\pi\nu\tau] >$$

An adding interferometer could be made to produce a complex output using a similar scheme. However, since most adding interferometers also use the direct interference method, such a system is not used, since it would require another division of the number of photons from each telescope.

3.3. DIRECT INTERFEROMETER (ADDING)

At high frequencies, the quantum noise added by a phase sensitive detection of electromagnetic waves mandates that the waves be interfered before the resultant fringes are detected. Thus, each photon is interfered with itself before it is detected. This type of interferometer is called a "direct" interferometer. All optical interferometers and most IR interferometers are

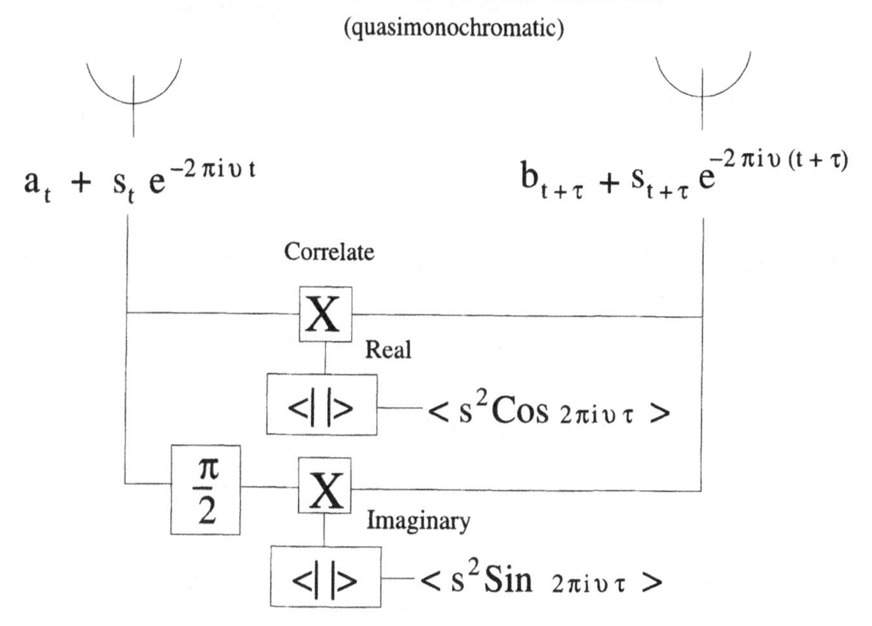

Figure 2. A basic complex correlation interferometer. Two multiplies are done, one for the real part of the correlation, the second with a $\pi/2$ phase shift for the imaginary part.

of this type. The two incident waves are first interfered, either in free space, air, or a fiber optics waveguide and then projected onto an incoherent detector such as a CCD chip. Since the interference is formed directly from the photons, these interferometers are of the adding type.

The inability to duplicate photons has serious consequences for direct interferometry. Multiple simultaneous baselines require that the photons from each telescope be divided to form the different interferometers. Multiple baselines cost sensitivity. Likewise, only a single delay or part of the complex correlation function can be measured without a similar loss of sensitivity. On the positive side, direct interferometers can have tremendous bandwidth.

3.4. HETERODYNE INTERFEROMETER (CORRELATION)

At longer wavelengths, the quantum noise added by phase sensitive detection is not severe and it is practical to sample the electric field and then form the interference from the sampled voltages. This has the advantage that once the field has been sampled, the resultant signal can be replicated as needed with almost no loss in sensitivity. It is then possible to form a

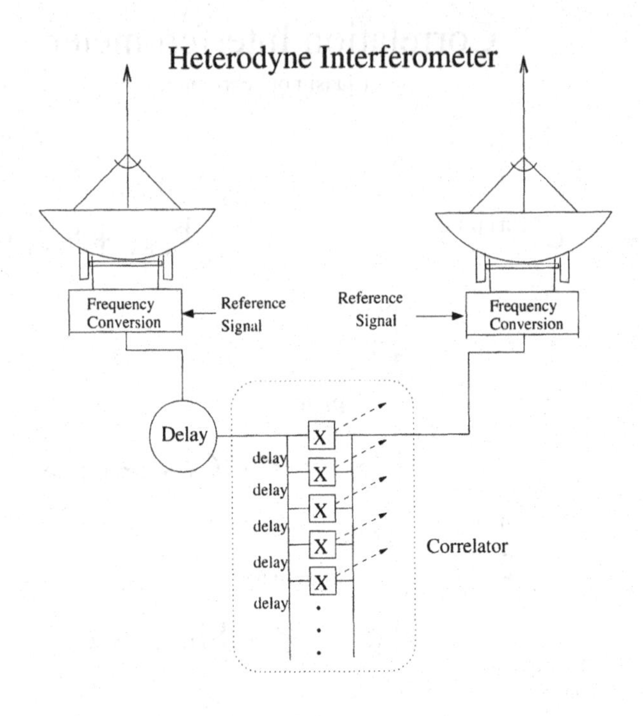

Figure 3. A simplified multi-lag heterodyne correlation interferometer. Each correlation may be complex.

correlation interferometer and correlators are used on modern heterodyne interferometers.

As the name heterodyne implies, this type of interferometer usually (but not always) involves a frequency conversion using the heterodyne method. In this case, a reference tone is "mixed" with the celestial signal resulting in "images" of the celestial signal at the sum and differences of the celestial and reference signal frequencies. This converts the signal to a frequency that is easier to manipulate and also allows very large amplification with reduced risk of the amplified signal feeding back into the receiver. One (or both) of these images are then selected and used to form the interference fringes. A simplified heterodyne interferometer is shown in Figure 3.

In a practical system, there are usually multiple heterodyne stages. A heterodyne interferometer can be either of the adding or correlation type, and so modern systems are almost exclusively correlation. One disadvantage of the heterodyne/correlation interferometer versus the direct interferometer at high frequencies is that the bandwidth which can be observed, hence the sensitivity, is limited by the speed of electronics to something on the order of a few 10s of GHz. This should improve as Moore's Law, as

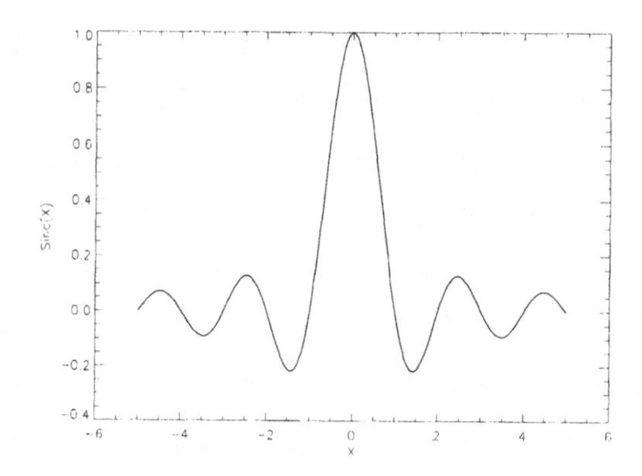

Figure 4. The Sinc function near zero argument.

the technology is the same as for computers. The effects of bandwidth are explored further in the following.

4. Finite Bandwidth and Delay Compensation

In the preceding, it was assumed that the total bandwidth of the system was negligible. In practice, this is not very useful as the sensitivity of a continuum measurement is proportional to the square root of the bandwidth. For spectroscopy, a finite bandwidth is needed as well; spectral lines have an intrinsic width, and the Doppler effect can shift them over a considerable range of frequency.

If the interferometer responses given in previous sections are integrated over a finite (assumed rectangular) bandpass $\Delta\nu$, the results become

$$\text{output} = < a^2 > + < b^2 > + < 2S^2(1 + cos[2\pi\nu\tau])sinc(\Delta\nu\tau) >$$

for an adding interferometer and

$$\text{output}_{\text{real}} = < 2S^2(cos[2\pi\nu\tau])sinc(\Delta\nu\tau) >$$

$$\text{output}_{\text{imaginary}} = < 2S^2(sin[2\pi\nu\tau])sinc(\Delta\nu\tau) >$$

for a correlation interferometer where $sinc(x) = \frac{sin(\pi x)}{\pi x}$.

The portion of the Sinc function near zero argument is shown in Figure 4. This function is unity near zero argument and oscillates away from zero argument but asymptotically approaches a value of zero. As shown above, the interferometer response is multiplied by the Sinc of the bandwidth $\Delta\nu$ times the delay difference τ. Coherence will be lost if the value of τ becomes much larger than the inverse of $\Delta\nu$.

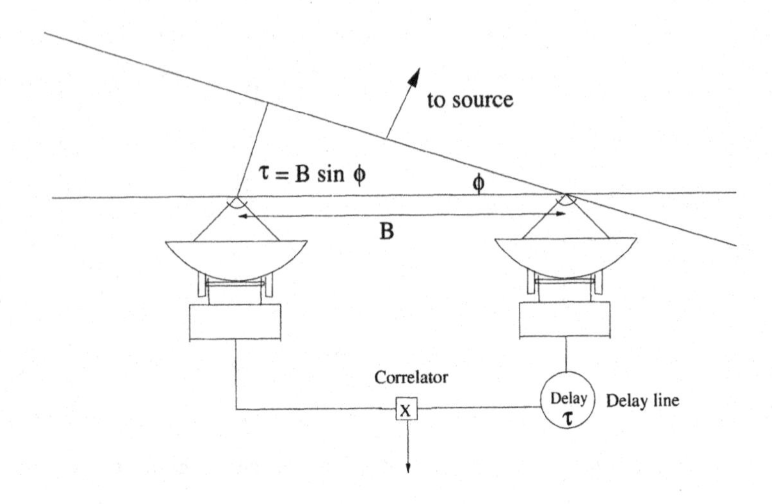

Figure 5. A correlation interferometer with delay compensation

The difference in delay experienced by a given wavefront at the point where the interference is formed has a number of components:

$$\tau_{total} = \tau_{geom} + \tau_{atmos} + \tau_{instr} + \dots$$

where τ_{geom} is due to the source–array geometry, τ_{atmos} is the atmospheric delay difference and τ_{instr} is the difference in delay through the interferometer prior to interference. It is the τ_{total} that needs to be kept near zero to keep the interferometer coherent.

Thus, one of the critical features of a practical interferometer is "Delay Compensation" to keep τ_{total} close to zero. With a correlation interferometer, as illustrated in Figure 5, it is possible to measure the correlation at multiple values of τ_{total} to determine the relative delay which maximizes the correlated output.

5. Time Lag - Sky Frequency Relationship

One of the important relationships in interferometry is that between the measured time–lag cross correlation function ($\rho(\tau)$)and the sky frequency correlation function ($\rho(\nu)$). This relation is a simple Fourier transform:

$$\rho(\nu) = c \int \rho(\tau) e^{-2\pi i \nu \tau} \, d\tau$$

or

$$\rho(\tau) = \frac{1}{c} \int \rho(\nu) e^{2\pi i \nu \tau} \, d\nu$$

Several of the uses of this relationship are discussed below.

5.1. SPECTROSCOPY

As seen above, a sky spectrum can be obtained by transforming a measured lag function. The resolution of the spectrum is determined by the range of lags measured. For a heterodyne system such as those used in radio interferometers, the range of the lags (delays) for which the signals are correlated is arbitrary, implying that an arbitrary frequency resolution is theoretically possible. Most spectrometers on radio interferometers, and a large number of single dishes, use this principle.

5.2. DELAY TRACKING

In both VLBI and optical/IR interferometers, the *a priori* delay model used by the interferometer is not sufficiently accurate to avoid at least some coherence loss. As explained above, the precision needed for the delay compensation depends on the bandwidth used. A wide total bandwidth is desirable for sensitivity reasons. If the output of the interferometer has some spectral dispersion, the the accuracy requirement on the *a priori* delay model is set by the bandwidth of the individual frequency channels rather than the total bandwidth. It is the usual practice in VLBI to produce data with sufficient spectral resolution that the coherence loss due to delay errors is negligible. Then the relationship above can be used to determine the actual delay; this process is called "fringe fitting." For a direct interferometer, a similar approach is to use a prism or grism during the observations to disperse the light before the detector.

6. Source Structure

In the preceeding it has been assumed that the source is a point, more exactly, that the size is small enough not to effect the response significantly. If the source is not spatially coherent, that is, the signal from one part of the source is completely independent from all other parts, then we can consider a resolved source to be the sum of a number of independent sources. In practice, almost all astronomical sources are spatially incoherent, even when the emission is from a coherent mechanism as in masers.

The response to a resolved source is the integral of the response over the brightness distribution on the sky $I(l, m)$ [9]:

$$V(u, v, w) = \int\int A(l, m) I(l, m) e^{-2\pi i [ul + vm + w(\sqrt{1-l^2-m^2}-1)]} \frac{dl\, dm}{\sqrt{1 - l^2 - m^2}}$$

where A is the primary telescope power pattern, (l, m) are coordinates on the sky, (u, v, w) is the baseline vector in the coordinate frame of the source.

This relationship is difficult to invert to recover $I(l, m)$. However, if the array is intrinsically two dimensional (e.g. east-west array) or the source is sufficiently small, this can be approximated by:

$$V(u,v) = \int\int A(l,m)I(l,m)e^{-2\pi i[ul+vm]} \frac{dl\ dm}{\sqrt{1-l^2-m^2}}$$

which is a 2-D Fourier transform which can be readily inverted:

$$\frac{A(l,m)I(l,m)}{\sqrt{1-l^2-m^2}} = \int\int V(u,v)e^{2\pi i[ul+vm]} du\ dv$$

Thus, it is possible to recover the sky brightness from measurable values.

7. Aperture Sampling and Imaging

The difficulty with performing the final integral in the previous section is that it requires measurements of all values of $V(u, v)$. In practice, it is only possible to measure a subset of these values. The value of (u, v) for a given measurement is given by the apparent separation and orientation of the two ends of the interferometer as seen by the source. For earth rotation synthesis interferometers, the position in u–v space of a given interferometer traces an ellipse as the earth rotates. If the function being observed is real (always the case for total intensity), then its Fourier transform is Hermitian, that is conjugate points in the u–v plane are symmetric. An example is shown in Figure 6. The sampling in the u–v plane is determined by the relative geometry of the telescopes in the array and the source at the times at which the source was observed. This only incompletely covers the u–v plane. Thus, the integral needed to recover the brightness distribution cannot be performed.

However, we know where the data has been sampled and can introduce a "Sampling function" $S(u, v)$ which is 1 where data was sampled and 0 elsewhere. We can create a "sampled visibility function" $V'(u, v)$ defined by

$$V'(u,v) = V(u,v)\ S(u,v)$$

which is defined everywhere. It is then possible to derive the "Dirty" image, $D(l, m)$:

$$D(l,m) = \int\int V'(u,v)e^{2\pi i[ul+vm]} du\ dv$$

which is related to the true sky brightness.

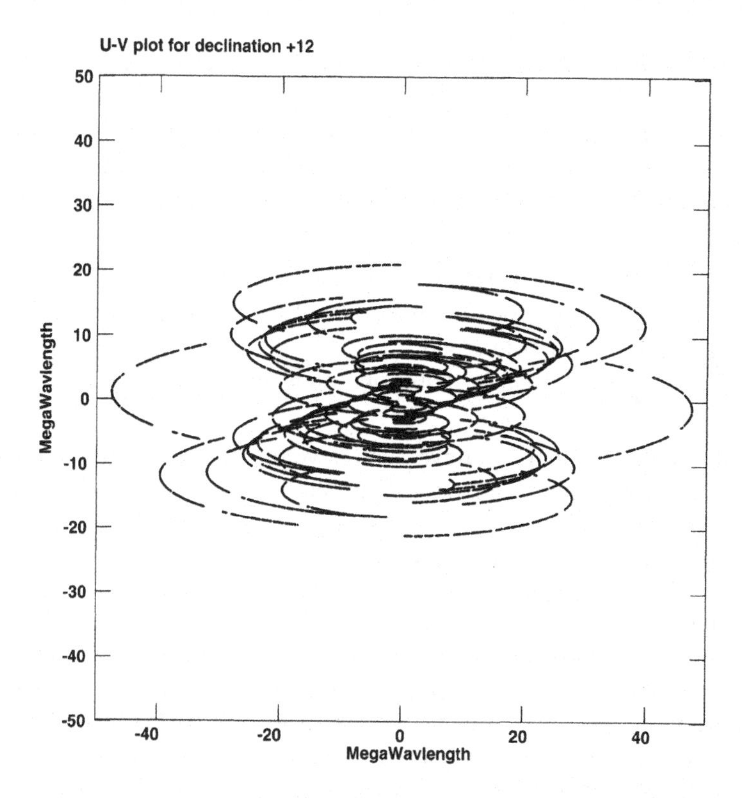

Figure 6. Typical UV coverage for VLBA+VLA, 11 antennas, 55 baselines at $+12°$ declination.

7.1. DECONVOLUTION

The dirty image is not the true image of the sky since the sampled visibility is not the true visibility function. However, the Fourier convolution theorem states that the Fourier transform of the product of two functions is the convolution of the Fourier transforms of the two functions. Therefore, the dirty image is related to the true image by

$$D(l,m) = A(l,m)I(l,m) \star b(l,m)$$

where $\star$ denotes convolution and $b(l,m)$ is the Fourier transform of $S(u,v)$. The following will assume the primary antenna power pattern $A(l,m)$ is 1.

In order to recover $I(l,m)$, a deconvolution is needed. This deconvolution needs to be nonlinear as the Fourier transform of the dirty image is full of zeroes. The two most popular deconvolution methods in radio astronomy are CLEAN and maximum entropy [2]. The CLEAN technique is discussed in the following.

7.2. CLEAN

The CLEAN algorithm [2] makes use of the fact that the sky is mostly empty and decomposes its representation into a set of delta functions on the vertices of the image grid. The method is very robust and always converges to an answer which is consistent with the data, although not always to an answer that agrees with our prejudice of how the sky ought to look. The original Högbom algorithm [5] is very simple:

1. 1) Find the peak brightness pixel in image.
2. 2) Subtract dirty beam times a fraction of the peak in the image.
3. 3) Repeat until converged.
4. 4) Convolve "CLEAN" components with a nice beam and add to residuals.

Variations on this method have been developed for computational efficiency and to reduce artifacts introduced in the imaging process [1] [8].

8. Atmospheric (and other) Delays

In practice, the *a priori* interferometer delay model is never perfect and there are residual delay, hence phase, errors. The principle source of these uncertainties is the earth's atmosphere although errors in the knowledge of the geometry and instrumental instabilities will contribute as well.

Which part of the atmosphere is most the responsible depends on the observing wavelength. At meter wavelengths, the ionosphere dominates while at cm. and mm. wavelengths, cells of water vapor in the lower troposphere dominate. At optical and IR wavelengths, the principle source of delay fluctuations is thought to be from pressure fluctuations in the shear layers between moving components of the atmosphere, generally at altitudes on the order of 10 km.

8.1. PHASE CALIBRATION

Under many circumstances, the phase errors in interferometric measurements can be considered to depend only on the individual telescope. This is clear in the case of water vapor cells which are only over one telescope or clock errors in a VLBI interferometer.

There are two basic schemes that are frequently used for phase calibration: referencing phases to an external source and "self–calibration". The phase referencing method uses measurements of a calibrator of accurately known position in the same isoplanatic patch as the target (same phase error) to determine the phase errors on this calibrator and then to remove them from the target observations. In the radio, the coherence time of the atmosphere is usually minutes or longer allowing switching between the

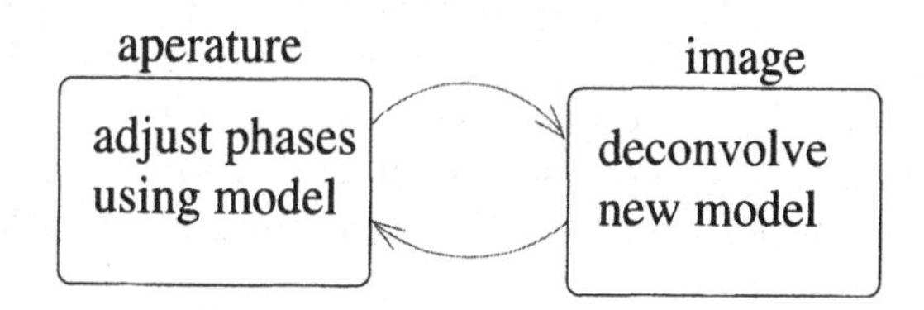

Figure 7. The self–calibration iterative procedure.

calibrator and target. In the optical/IR, the coherence time is milliseconds and the observations must be simultaneous.

8.2. SELF–CALIBRATION

If the target source is sufficiently strong that it is detected in the coherence time, then the process of self–calibration (sometimes called "hybrid mapping") can be applied. This jointly solves for the telescope based phase errors and the source structure [7] [4] [3]. This process makes use of the vast over-determinacy of the telescope phases when many telescopes are used simultaneously. There are $\frac{n(n-1)}{2}$ baselines and only n-1 phases to be determined per coherence time. Self–calibration is an iterative process whereby constraints are first applied in the aperture domain to determine better estimates of telescope phases based on a model of the source. Then in the image domain, constraints (positivity, finite support, etc.) are applied using CLEAN to improve the model of the source. This process is illustrated in Figure 7 For optical and IR observations, the source is usually not detectable in the coherence time so the related technique of bispectrum analysis is used.

9. Optical Interferometry

Optical/IR interferometry differs from radio interferometry in a number of important respects. These are due to the different technology needed, the very serious atmospheric problems, and historical/cultural differences between the optical and radio communities. See [6] for a recent review.

As described earlier, the quantum mechanical limitations of the low photon rate in the optical and IR dictate the use of direct interferometers and this also implies adding interferometers. A practical technical consideration is also that the current speed of computer hardware makes it infeasible to build a correlator with enough bandwidth to be interesting for most optical/IR applications.

Direct interference of the photons imposes a sensitivity penalty for simultaneously observing on many baselines, so current optical/IR interferometers tend to have few baselines, many only one or three. Due to the small number of baselines and the difficulty in phase calibration, many current optical interferometers only measure visibility amplitudes and as such aren't imaging instruments. Other optical interferometers measure "closure phase" and use bispectrum analysis to obtain images. A future optical/IR development is to use phase referencing techniques.

Since the interference is formed using photons, the delay compensation must also be done to the photons. This is usually done with delay lines consisting of mirrors mounted on piezoelectric stacks, and sometimes acoustic coils, which are mounted on movable carts. Since the delay must be maintained to a small fraction of the light travel time of a wavelength, the mechanical requirements on the delay line, as well as the entire system, are severe.

Atmospheric phase corruptions are extremely severe in the optical; phase variations across the primary elements have limited the resolution of optical systems almost since the beginning of the era of telescopes. Even for small apertures, rapid "tip-tilt" corrections are needed to to correct for the apparent motion of the source due to the atmosphere and to stabilize the image of the target on the interferometer. Use of large apertures requires adaptive optics systems to remove the phase distortions across the aperture. The technology needed to do serious optical interferometry is only now becoming available; this is a field which has great potential for development in the coming years.

9.1. JARGON

Historically, radio and optical astronomers have not communicated well with each other. This has lead to the development of different jargon. Table 9.1 gives the corresponding terms for the two fields. It should be noted that these are functional equivalents in some sense and not always different words for the same thing.

9.2. BEAM COMBINATION

Optical/IR interferometers are usually direct interferometers and require a "beam combiner" to interfere the photons. These are generally either a set of optics to superimpose the beams from the two telescopes allowing the interference in vacuum (or air), or forming the interference in fiber optics waveguides which allow the electric fields to interfere. The basic designs are shown in Figure 8. In all cases, there are multiple interferometric products, usually two with a 180° phase difference. Due to the rapid atmospheric fluc-

TABLE 2. Translation Guide for Interferometric Jargon

Radio	Optical/IR
Delay, lag	Optical path difference (OPD)
Delay residual	Differential piston
Correlator	Beam Combiner
Antenna gain	Strehl ratio
System temperature	Background level
Phase referencing	Fringe tracking
Antenna	Telescope
Feed	Detector
Dirty (or CLEAN) beam	Point spread function (PSF)
Log (flux density)	Magnitudes
Confusing band designations[1]	Obscure band designations

[1] Radio band designations were inverted during World War II by American and British radar engineers to confuse the Germans.

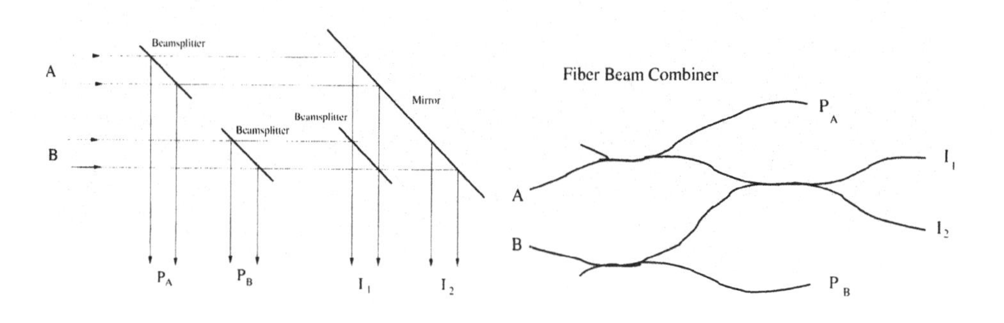

Figure 8. Left) beam combination in vacuum or air. Right) Fiber optics beam combiner.

tuations, beam combiners frequently further divide the photons to measure photometric fluctuations.

9.3. ATMOSPHERIC PHASE

Atmospheric phase fluctuations in the optical and IR are fast in time and occur on small physical scales. This decorrelates the wavefront across the primary apertures which breaks the image up into "speckles". An example of such a speckle pattern is shown in Figure 9. The spatial size of the

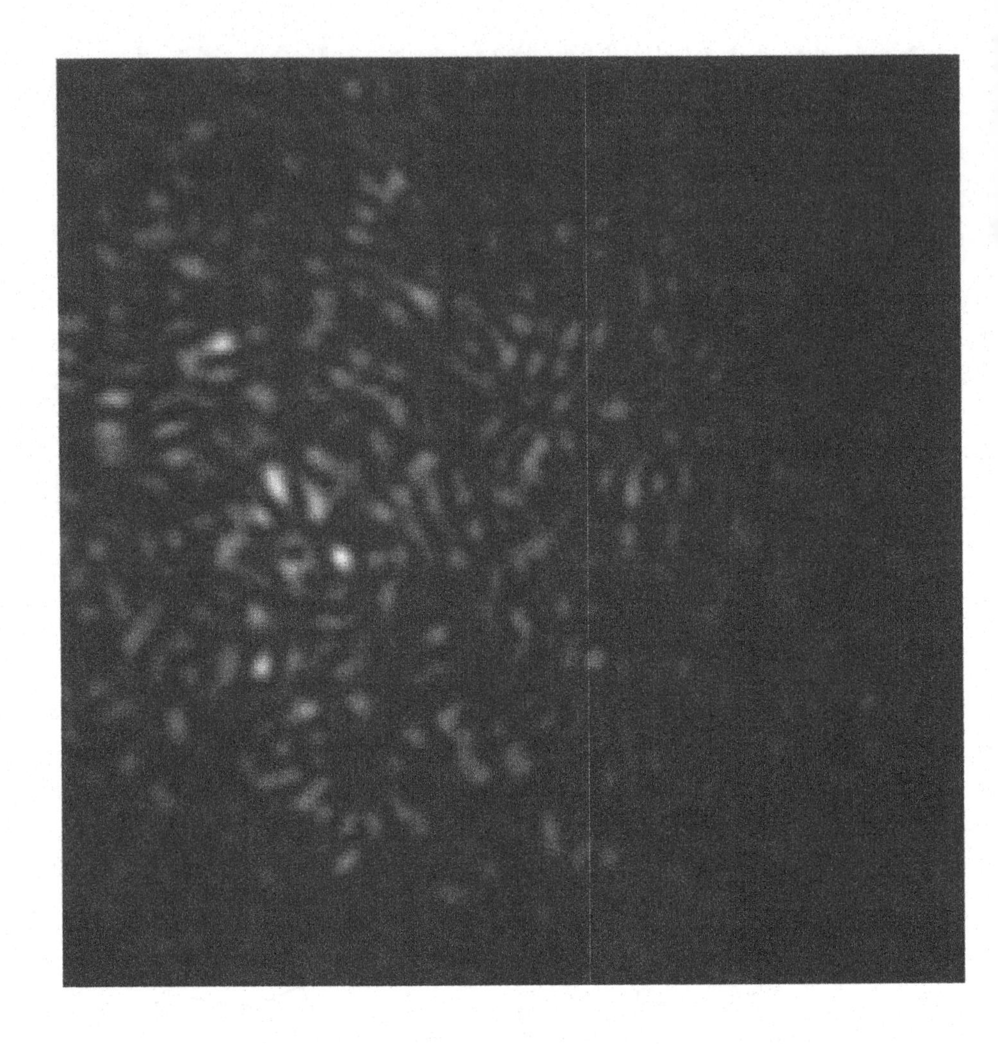

Figure 9. Simulated speckle pattern r_0=0.2m, telescope = 8m

coherence across the aperture is called the "Fried" parameter and is usually denoted r_0. At least a tip-tilt system is needed to center the speckle pattern on the interferometer, although an adaptive optics system is even better for putting most of the power into a single speckle. The fraction of the power in the brightest speckle is called the "Strehl" ratio. Correction of the effects of atmospheric phase fluctuations is an ongoing development.

References

1. Clark, B.G. (1980), An Efficient Implementation of the Algorithm "CLEAN", *Astron. Astrophys.*, **89**, pp. 377–378.
2. Cornwell, T.J., Braun, R. and Briggs, D. S. (1999) "Deconvolution" in Taylor, G., Carilli, C. and Perley, R. eds. **Synthesis Imaging in Radio Astronomy II**, Astronomical Society of the Pacific, pp. 11–36
3. Cornwell, T.J. and Fomalont, E.B. (1999), "Self–Calibration" in Taylor, G., Carilli, C. and Perley, R. eds. **Synthesis Imaging in Radio Astronomy II**, Astronomical Society of the Pacific, pp. 11–36
4. Cotton, W.D. (1979) A Method of Mapping Compact Structure in Radio Sources using VLBI Observations, *Astron. J.*, **84**, pp. 1122.
5. Högbom, J. (1974), Aperture synthesis with a non–regular distribution of interferometer baselines, **Astron. Astrophys. Suppl.**, **15**, pp. 417–426.
6. Quirrenbach, A. (2001) Optical Interferometry, *Ann. Rev. Astron. Astrophys.*, **39**, pp. 353–401
7. Readhead, A.C.S. and Wilkinson, P.N. (1978), The mapping of Compact Radio Sources from VLBI Data, *Astrophys. J.*, **223**, pp. 25–36.
8. Schwab, F.R. (1984), Relaxing the isoplanatism assumption in self calibration; applications to low frequency radio interferometry, *Astron. J.*, **89**, pp. 1076–1081
9. Thompson, A. R. (1999) Fundamentals of Radio Interferometry, in Taylor, G., Carilli, C. and Perley, R. eds. **Synthesis Imaging in Radio Astronomy II**, Astronomical Society of the Pacific, pp. 11–36
10. Taylor, G., Carilli, C. and Perley, R. eds. (1999) **Synthesis Imaging in Radio Astronomy II**, Astronomical Society of the Pacific
11. Thompson, A.R., Moran, J.M. and Swenson, G.W. Jr. (2001) **Interferometry and Synthesis in Radio Astronomy**, John Wiley & sons, Inc., New York

OPTICAL INTERFEROMETRY: ASTRONOMICAL APPLICATION

W. D. COTTON
National Radio Astronomy Observatory
520 Edgemont Road
Charlottesville, VA 22903
USA

1. Introduction

At the present time, observations made with optical interferometers are limited to large bright stars. This limitation is largely due to the small apertures, short baselines and short coherence times allowed by atmospheric phase fluctuations. These effects reduce the sensitivity of optical and IR interferometers, restricting observations to large bright stars and multiple star systems. In the next few years, this situation should change dramatically with a new generation of interferometers designed to deal with atmospheric effects. Among these advances are use of adaptive optics to allow larger apertures, up to 10 meters, longer baselines for higher resolution and phase referencing techniques to increase the atmospheric coherence time. These enhancements will allow observations of fainter, smaller stars as well as nearby active galactic nuclei (AGN).

This lecture will cover past and current interferometers including those still in the development stage. Following this is a discussion of the types of science accessible to current interferometers or those which are anticipated in the immediate future. A more detailed discussion of the techniques can be found in [27]. A review of recent scientific results with optical/IR interferometers is [26].

2. Past and Current Interferometers

The first efforts on astronomical interferometry were made by Michelson in the early 1920s using the Mt. Wilson 100 inch telescope [21] [22] [23]. While it was possible to measure the diameters of a handful of stars, the

F. Mantovani and A. Kus (eds.), The Role of VLBI in Astrophysics, Astrometry and Geodesy, 307–317.

technology of the day was much too primitive for interferometry to become a practical science. At radio wavelengths, interferometric techniques became widespread in the 1960s while optical and IR developments lagged far behind. Twiss and Hanbury–Brown experimented with optical intensity interferometry in Australia in the 1960s and early 1970s [11][12] [13] but the first practical implementations of coherent optical interferometry was begun by Laberie in France in the 1970s [17] [18]. The following is a discussion of the major efforts at optical and IR interferometry up to the present.

2.1. MICHELSON

The first attempts at astronomical interferometry were made by Michelson using the 100 inch telescope on Mt. Wilson, CA, USA [21] [22] [23]. These efforts culminated in the 1920s using the mount of the 100 inch telescope to support outrigger mirrors with a baseline of 20 feet (6.1 meter). This interferometer was able to measure the diameters of Betelgeuse, Arcturus and a few other nearby, large stars in the early 1920s.

2.2. INTENSITY INTERFEROMETER

After an initial demonstration of optical intensity interferometry [11], Hanbury-Brown and collaborators built an operational intensity interferometer at Narrabri, NSW, Australia in the 1960s and 1970s [12] [13]. This was a novel instrument that correlated fluctuations of the arrival time statistics of the photons which are insensitive to atmospheric fluctuations. This type of interferometer is also relatively insensitive to stars, and despite the 6.5 meter mirrors used, the Narrabri interferometer could only measure the diameters of a few dozen stars. This array had baselines up to 200 meters and operated from 1963 to 1975.

2.3. GRAND INTERFEROMETRE A 2 TELESCOPES (GI2T)

This project was started as the Interferometre a 2 Telescopes [17] [18] [24] in the 1980s and has been developed since then. It specializes in high resolution stellar spectroscopic interferometry. This interferometer which is at Calern, France has two telescopes of 1.5 m aperture and baselines to 35 meters.

2.4. SYDNEY UNIVERSITY STELLAR INTERFEROMETER (SUSI)

The SUSI interferometer at Narrabri, NSW, Australia is under development by Sydney University [8]. It consists of thirteen 0.14 m apertures and has baselines to 640 m.

2.5. U.S. NAVY MKIII

The MkIII is one of a series of optical interferometers developed by the U.S. Navy for astronomical and astrometric purposes [28]. It was sited on Mt. Wilson, CA, USA and had three 5 cm apertures and baselines to 31.5 m. This project was started in 1982 and was in operation from 1986 until the early 1990's when it was replaced by the NPOI interferometer.

2.6. NAVY PROTOTYPE OPTICAL INTERFEROMETER (NPOI)

NPOI is the current U. S. Navy optical interferometry project for astrometric and astronomical measurements [1]. It is located near Flagstaff, AZ, USA and consists of six 0.5 m apertures arraigned in a "Y" shape with baselines up to 435 m. This array can do closure–phase imaging. Work on this project began in 1989 and operations started in 1994.

2.7. INFRARED SPATIAL INTERFEROMETER (ISI)

The ISI works in the wavelength range of 9–12 microns and is the only heterodyne infrared interferometer in operation [16] [3] [10]. This interferometer is operated by the University of California at Berkeley and is located on Mt. Wilson, CA, USA. It consists of two 1.65 m telescopes with baselines up to 75 m and is devoted to spectroscopic observations.

2.8. CAMBRIDGE OPTICAL APERTURE SYNTHESIS TELESCOPE (COAST)

The COAST array is run by the University of Cambridge, in Cambridge, UK [14] [2]. It consists of five 0.4 m apertures in a "Y" configuration with baselines up to 48 m. This instrument can make optical and near–IR closure phase images.

2.9. INFRARED AND OPTICAL TELESCOPE ARRAY (IOTA)

This instrument is a collaboration involving the Smithsonian Astrophysical Observatory and the Observatoire de Paris, Meudon section and is located on Mt. Hopkins, AZ, USA [4] [7]. It has three 0.45 m apertures with baselines up to 38 m and works in the optical and near infrared.

2.10. PALOMAR TESTBED INTERFEROMETER (PTI)

The PTI is a JPL test bed for the Keck and Space interferometers [5]. It is located on Mt. Palomar, CA, USA and consists of two 0.4 m apertures with a baseline of 110 m. It is used for astronomy and astrometry in the optical and near infrared.

2.11. MITIAKA OPTICAL AND INFRARED ARRAY (MIRA)

The MIRA project is the optical and infrared interferometry development project of the National Astronomical Observatory of Japan [25]. The MIRA-I array consists of two 0.3 m apertures with a 4 m baseline.

2.12. KECK INTERFEROMETER

The Keck Interferometer is run by the University of California on Mauna Kea, HI, USA [6]. This interferometer is one of a new generation that uses large apertures with adaptive optics for improved sensitivity in the optical and near IR. The interferometer currently consists of two 10 m telescopes with a separation of 85 m and has plans to expand with a number (four?) outrigger 1.8 m telescopes. The Keck Interferometer currently operates in the near infrared.

2.13. VERY LARGE TELESCOPE INTERFEROMETER (VLTI)

The VLTI is another of the new generation of large interferometer projects being developed by European Southern Observatory on Cerro Paranal in Chile [9]. The array consists of four 8 m telescopes and a system of three or more 1.8 m outrigger telescopes is under construction with baselines up to 200 m.

2.14. CENTER FOR HIGH ANGULAR RESOLUTION ASTRONOMY (CHARA)

The CHARA array is being developed by the Georgia (USA) State University on Mt. Wilson, CA USA [19]. The array will consist of six 1 m apertures.

2.15. LARGE BINOCULAR TELESCOPE (LBT)

The LBT is being built by a Consortium of Italian, German and American institutions [15]. This instrument will be sited on Mt. Graham, AZ, USA and will consist of two 8.4 m mirrors on a common mount giving effective baselines up to 22.8 m. Operation of the LBT is anticipated in 2003-4.

3. What is Observable by Optical Interferometry?

Consider for the moment the types of emission that are observable by optical interferometry. At milliarcsecond resolution in the radio, only very bright, nonthermal emission is observable. In the Rayleigh-Jeans portion of spectrum, a given baseline has the same relative resolution of a 1 Jy source independent of wavelength. This can be seen from the expression for the brightness temperature (the equivalent physical black body temperature) of a source:

$$T_B = \frac{c^2 S}{2k_B \nu^2 \theta^2} = \frac{S D^2}{2k_B}$$

where T_B is the brightness temperature, c the speed of light, S the source flux density, k_B Boltzman's constant, ν the observing frequency, θ the apparent diameter of the source and D the projected length of the interferometer baseline. Since the size (θ) of a source with a given fractional resolution scales as ν^{-2}, the frequency dependence of the apparent brightness temperature cancels. This implies that for a given sensitivity, the minimum observable brightness temperature depends only on the baseline length. At optical frequencies baselines of only a few hundred meters are required to get to milliarcsecond resolution. This simple relationship breaks down at frequencies approaching or above the peak in the black body spectrum but indicates that at baselines of a few hundred meters, optical interferometers can observe thermal objects.

The current state of the art with optical/IR interferometers is that crude closure phase images are possible with a few interferometers. Most interferometers cannot measure phase and are restricted to measuring visibility amplitudes. These measurements can be used to fit models, e.g. limb darkened disks of stars, to the observations. The following sections describe in more detail some of the areas of research to which optical interferometry can be applied.

4. Exoplanets

The detection of extrasolar planets is currently a major industry in astronomy. At the moment, all detections of planets around normal stars (i.e. not pulsars) use accurate radial velocity measurements to detect the reflex motion of the star due to the planet. The radial velocity technique gives no information about the inclination of the plane of the planetary orbit to the plane of the sky, so all that can be measured is V sin(i) where V is the orbital velocity and i is the inclination of the plane of the orbit to the sky. Complementary astrometric measurements can allow the solution for all the components of the orbit, giving planetary masses rather than limits.

The necessary astrometric precision requires the use of interferometry and such observations are being planned.

Astrometric planet detection is complementary to radial velocity detections in another way. For a given planetary mass, the reflex radial velocity of the host star is larger for planets whose orbits have a small semi-major axis while the total apparent angular motion of the host star is larger for planets with large orbits. Thus, the radial velocity technique is biased towards planets close to the star, while astrometric measurements are biased towards planets with larger orbits.

Interferometric techniques have the further advantage that the planets can be imaged as well as detected. Astrometric techniques have sufficient sensitivity to detect Jupiter–like planets from ground based observations but the detection of an Earth–like planet requires a space–based interferometer working at mid–IR wavelengths to reduce the star–planet contrast.

A further exciting possibility is that spectroscopic imaging of earth–like planets, while being a very difficult measurement, can detect the presence of earth–like life. This is through the detection of the ozone bands in the planetary atmosphere. Ozone requires the presence of free oxygen in the planetary atmosphere and free oxygen is chemically unstable; unless the oxygen is constantly replenished it will disappear from the atmosphere. On the earth, the oxygen is replenished by plant activity, especially cynobacteria. Both NASA and ESA are planning extensive efforts in the search for life on other planets.

5. Effective Temperatures

Effective (or bolometric) temperatures are needed to verify theoretical stellar models. The presence of emission and absorption lines in the envelopes of stars complicates spectroscopic temperature measurements. Especially for cool stars, line blanketing by molecular absorption bands causes serious redistribution of the spectrum so that it is no longer describable by a single black body. In this case, the angular diameter and bolometric flux of the star are needed to determine an effective temperature. Except for the very largest stars, interferometric resolution is needed to measure the diameter. In this case, model fits to visibility amplitudes are sufficient.

6. Limb darkening

For many stars a uniform disk model is insufficient to describe the photospheric disk. This is especially true for cool stars which have extensive envelopes. For these stars, there may be significant center to limb brightness changes (limb darkening or limb brightening). These effects may be modeled using sufficient visibility amplitude data.

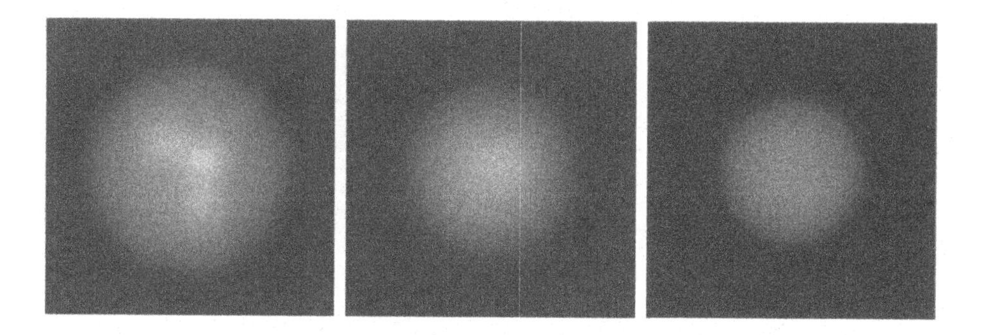

Figure 1. Betelgeuse at 700, 905, 1290 nm as observed by COAST(700, 905 nm) and WHT (1290 nm) (Young et al., 2000a).

7. Multiple Star Systems

Close multiple star systems need interferometric resolution to separate the different components. For spectroscopic binaries this is necessary to obtain all the orbital elements needed to determine the masses of the individual stars. For sufficiently close stars, imaging can show the tidal distortions of the stellar envelopes. An example of the image of a close binary system can be seen in [2].

8. Star Spots

Many giant stars are sufficiently large that is possible to image small scale structures on their surfaces. Many cooler stars can have large star spots which can affect the energy transport in the star.

Multi-wavelength imaging of the cool super giant Betelgeuse is shown in Figure 1. These features are interpreted by [29] as being due to TiO absorption bands in the red whereas in the IR, the opacity is less and the emission from inner, hotter regions is visible. Their model is shown in Figure 2.

9. Pulsating AGB stars

The pulsation mechanisms of stars are not well understood and the diameter and effective temperatures as a function of pulsation phase are needed to validate models. An example of the variations in angular diameter of χ Cygni is shown in Figure 3 [30].

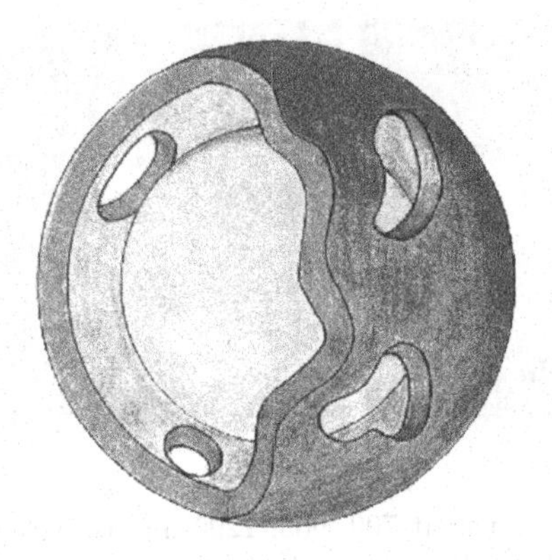

Figure 2. Model of Betelgeuse spots as being due to TiO absorption in the red in the cooler, outer parts of the envelope (Young et al., 2000a).

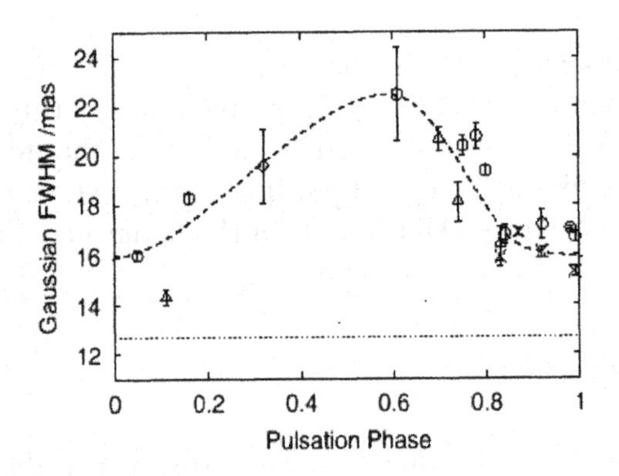

Figure 3. Variation of diameter of the Mira χ Cygni as observed with COAST (Young et al., 2000b).

10. Ae / Be stars

These are relatively hot stars which have extended equatorial disks with strong emission lines. An image of such a system in H_α is shown in Figure 4.

11. Young Stellar Objects (YSOs)

Protostars show structure on a wide variety of size scales, some can only be studied interferometrically [20]. The earliest phases of stellar formation take place in the cores of dense molecular clouds which are heavily obscured

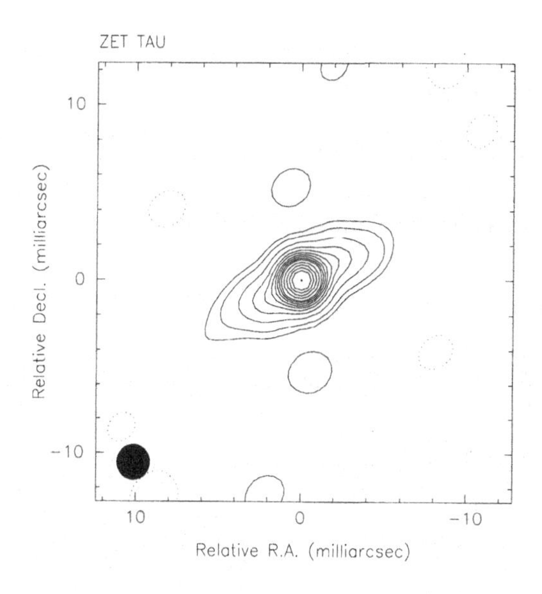

Figure 4. ζ Tauri as seen in the H_α line by COAST. Image courtesy of Amanda George.

by dust. This limits observations to the longer wavelengths for which the extinction is less important. Very good resolution or excellent calibration are needed to study the pre–main sequence phase when the stars are quite small. However, protoplanetary dust disks will be more extended.

12. Dust disks / shells

Much of the dust in the interstellar medium comes from the winds of evolved stars in a red giant phase. The winds in some of these stars are quite dense and dust shells form when the escaping molecular gas becomes cool enough to condense into grains. Dust formation in pulsating stars may be episodic. The dust around stars can be well studied in the IR using interferometric techniques.

13. AGN

An exciting area of research that is about to open up with the new generation of interferometers is the study of Active Galactic Nuclei (AGNs). Optical interferometers should be able to probe the nonthermal core and inner jets in regions that are optically thick in the radio. Infrared measurements should also allow the study of the dusty obscuring torus thought responsible for many of the orientation–dependent observational effects in

AGNs.

14. Acknowledgements

The author would like to thank John Young and Amanda George for examples shown in this lecture. The author would also like to thank Prisse Könönen for comments and suggestions.

References

1. Armstrong, J.T., Mozurkewich, D., Rickard, L.J., Hutter, D.J., Benson, J.A., Bowers, P.F., Elias II, N.M., Hummel, C.A., Johnston, K.J., Buscher, D.F., Clark III, J.H., Ha, L., Ling, L.-C., White, N.M. and Simon, R.S. (1998) The Navy Prototype Optical Interferometer, *Astrophys. J.*, **496**, pp. 550–571
2. Baldwin, J.E., Beckritt, M.G., Boysen, R.C., Burns, D., Buscher, D.F., Cox, G.C., Haniff, C.A., Mackay, C.D., Nightingale, N.S., Rogers, J., Scheuer, P.A.G., Scott, T.R., Tuthill, P.G., Warner, P.J., Wilson, D.M.A. and Wilson, R.W. (1996), The First Images from an Optical Aperture Synthesis Array: Mapping of Capella with COAST at Two Epochs, *Astron. Astrophys.*, **306**, pp. L13–L16
3. Bester, M., Danchi, W.C. and Townes, C.H. (1990), Long baseline interferometer for the mid–infrared, *Proc SPIE*, **1237**, pp. 40–48
4. Carleton, N.P., Traub, W.A., Lacasse, M.G.,Nisenson, P., Pearlman, M.R., Reasenberg, R.D., Xu, X., Coldwell, C., Panasyuk, A., Benson, J.A., Papaliolios, C., Predmore, R., Schloerb, F.P., Dyck, H.M. and Gibson, D. (1994), Current Status of the IOTA Interferometer, *Proc SPIE,*, **2200**, pp. 152–165
5. M.M. Colavita, Wallace, J.K., Hines, B.E., Gursel,Y., Malbet, F., Palmer, D.L., Pan, X.P., Shao,M., Yu, J.W., Boden, A.F., Dumont, P.J., Gubler, J., Koresko, C.D., Kulkarni, S.R., Lane, B.F., Mobley, D.W. and van Belle, G.T. (1999) The Palomar Testbed Interferometer, *Astrophys. J.*, **510**, pp. 505–521
6. Colvita, M.M. and Wizinowich, P.L. (2000) Keck Interferometer: progress report, *Proc SPIE*, **4006**, pp. 310–320
7. Coude du Foresto, V.,Perrin, G.,Ruilier, C., Mennesson, B.P., Traub, W.A. and Lacasse, M.G. (1998), FLUOR fibered instrument at the IOTA interferometer, *Proc SPIE*, **3350**, pp. 856–863
8. Davis, J., Tango, W.J., Booth, A.J., ten Brummelaar, T.A., Minard, R.A. and Owens, S.M. (1999), The Sydney University Stellar Interferometer - I. The Instrument, *Mon. Not. R. Astron. Soc.*, **303**, pp. 773–782
9. Glindemann, A., Delplancke, F., Derie, F., Ferrari, M., Gennai, A., Gitton, P., Kervella, P., Koehler, B., Leveque, S. A., de Marchi, G., Menardi, S., Michel, A., Paresce, F., Richichi, A., Schoeller, M., Tarenghi, M. and Wallander, A. (2000), The VLT Interferometer: a unique instrument for high–resolution astronometry, *Proc SPIE*, **4006**, pp. 2–12
10. Hale, D.D.S., Bester, M., Danchi, W.C., Fitelson, W., Hoss, S., Lipman, E.A., Monnier, J.D., Tuthill, P.G. and Townes, C.H. (2000), The Berkeley Infrared Spatial Interferometer: A Heterodyne Stellar Interferometer for the Mid–Infrared, *Astrophys. J.*, **537**, pp. 998–1012
11. Hanbury Brown, R. and Twiss, R.Q. (1956), A Test of a New Type of Stellar Interferometer on Sirius, *Nature*, **178**, pp. 1046–1048
12. Hanbury Brown, R., Davis, J. and Allen, L.R. (1967), The Stellar Interferometer at Narrabri Observatory - I, *Mon. Not. R. Astron. Soc.*, **137**, pp. 375–392
13. Hanbury Brown, R., Davis, J. and Allen, L.R. (1974), *Mon. Not. R. Astron. Soc.*, **167**, pp. 121–136

14. Haniff, C.A., Mackay, C.D., Titterington, D.J., Sivia, D., Baldwin, J.E. and Warner, P.J. (1987), The First Images from Optical Aperture Synthesis, *Nature*, **328**, pp. 694–696
15. Hill, J.M. and Salinari, P. (2000) The Large Binocular Telescope Project, *Proc SPIE*, **4004**, pp. 36–46
16. Johnson, M.A., Betz, A.L., and Townes, C.H. (1974), 10μm Heterodyne Stellar Interferometer, *Phys. Rev. Lett.*, **33**, pp. 1617–1620
17. Labeyrie, A. (1970) Interference Fringes Obtained on Vega with Two Optical Telescopes, *Astrophys. J.*, **196**, pp. L71–L75
18. Labeyrie, A., Schumacher, G., Dugu, M., Thom, C., Burlon, P., Foy, F., Bonneau, D. and Foy, R. (1986) Fringes obtained with the large boules interferometer at CERGA, *Astron. Astrophys.*, **162**, pp. 359–364
19. McAlister, H.A., Bagnuolo, W.G., ten Brummelaar, T.A., Cadman, R., Hopper, C.H., Ridgeway, S.T., Simison, E.J., Shure, M.A., Sturmann, L. and Turner, N.H. (2000) The Chara Array on Mt. Wilson, California, *Proc SPIE*, **4006**, pp. 465–471
20. Men'shchikov, A.B., Henning, T. and Fischer, O. (1999) Self–consistent model of the Dusty Torus Around HL Tauri, *Astrophys. J.*, **519**, pp. 257–278
21. Michelson, A. A., (1890) On the application of Interference Methods to Astronomical Measurements, *Philos. Mag.*, Ser. 5, **30**, pp. 1–21
22. Michelson, A. A., (1920) On the application of Interference Methods to Astronomical Measurements, *Astrophys. J.*, **51**, pp. 257–262
23. Michelson, A. A. and F. G. Pease, (1921) Measurement of the Diameter of α Orionis with the interferometer, *Astrophys. J.*, **53**, pp. 249–259
24. Mourard, D., Tallon–Bosc, I., Blazit, A., Bonneau, D., Merlin, G., Morand, Vakili, F., and Laberie, A. (1994), The GI2T Interferometer on Plateau de Calern, *Astron. Astrophys.*, **283**, pp. 705–713
25. Nishikawa, J., Sata, K., Yoshizawa, M., Fukushima, T., Machida, Y., Honma, Y., Matsuda, K., Kudo, K., Iwashita, H., Suzuki, S., Kubota, Y., Shimazaki, K. and Nemoto, Y. (2000) Mitaka Optical and InfraRed Array first stage (MIRA-I.1) instruments, *Proc SPIE*, **4006**, pp. 681–687
26. Quirrenbach, A. (2001) Optical Interferometry, *Ann. Rev. Astron. Astrophys.*, **39**, pp. 353–401
27. Shao, M. and Colvita, M.M. (1992) Long–Baseline Optical and Stellar Interferometry, *Ann. Rev. Astron. Astrophys.*, **30**, pp. 457–498
28. Shao, M., Colavita, R.S., Hines, B.E., Staelin, D.H., Hutter, H.J., Johnston, K. J., Mozurkewich, D., Simon, R.S., Hershey, J.L., Hughes, J.A. and Kaplan, G.H. (1988), The Mark III Stellar Interferometer, *Astron. Astrophys.*, **193**, pp. 357–371.
29. Young, J.S., Baldwin, J.E., Boysen, R.C., Haniff, C.A., Lawson, P.R., Mackay, C.D., Pearson, D., Rogers, J., St.-Jacques, D., Warner, P.J., Wilson, D.M.A. and Wilson, R.W. (2000a), New views of Betelgeuse: multi–wavelength surface imaging and implications for models of hotspot generation, *Mon. Not. R. Astron. Soc.*, **315**, pp. 635–645
30. Young, J.S., Baldwin, J.E., Boysen, R.C., Haniff, C.A., Pearson, D., Rogers, J., St.-Jacques, D., Warner, P.J. and Wilson, D.M.A. (2000b), Cyclic variations in the angular diameter of chi Cygni, *Mon. Not. R. Astron. Soc.*, **381**, pp. 381–386

DATA ANALYSIS

W. D. COTTON
National Radio Astronomy Observatory
520 Edgemont Road
Charlottesville, VA 22903
USA

1. Introduction

This lecture will cover some of the practical details of the calibration, editing and imaging of radio interferometer data. Both continuum as well as spectral line and polarization sensitive measurements are included. Analysis of the derived images depends strongly on the scientific objectives of the observations and will not be covered here. Many of the topics covered here are also discussed in [15]. The following discussion is explicitly for Very Long Baseline Interferometer (VLBI) data but the processing for connected element interferometers is very similar.

Following a general discussion of calibration issues, the details of continuum calibration are explored. This is calibration which must generally be done for both continuum and spectral line observations. Then issues of relevance mainly to spectral line observations are covered. This discussion is a high level guide to the calibration procedures and the reader is referred to more detailed descriptions of individual procedures. In all discussions, polarization measurements are implicitly or explicitly included.

2. Calibration Theory

Interferometers can't directly measure what we really want, images of the sky. In practice, an interferometer samples the Fourier transform of the sky brightness distribution. A further complication is that atmospheric and instrumental effects corrupt the measured visibilities. These corruptions must be corrected through the process of calibration before a sky brightness distribution can be derived.

F. Mantovani and A. Kus (eds.), The Role of VLBI in Astrophysics, Astrometry and Geodesy, 319 336.

2.1. THE SIMPLE PICTURE

Under certain simplifying assumptions, the relation between the desired sky brightness distribution for the Stokes parameters I,Q,U and V and the Stokes's visibilities (R_I, R_Q, R_U, R_V), which can be sampled, is a simple Fourier transform [8]:

$$
\begin{aligned}
R_I(u,v) &= \int\int I(l,m)e^{-2\pi j(ul+vm)}dl\ dm \\
R_Q(u,v) &= \int\int Q(l,m)e^{-2\pi j(ul+vm)}dl\ dm \\
R_U(u,v) &= \int\int U(l,m)e^{-2\pi j(ul+vm)}dl\ dm \\
R_V(u,v) &= \int\int V(l,m)e^{-2\pi j(ul+vm)}dl\ dm
\end{aligned}
$$

Stokes' I is the total intensity, Q and U give the linear polarization and V the circular polarization. If sufficient samples of the visibility function are measured, then the equations above can be inverted to derive the sky brightness distribution. This process is covered in another lecture [7].

2.2. THE UGLY TRUTH

In practice, the situation is far more complicated. For VLBI observations, the geometry of the of the array and the source is not well determined, the wavefronts at each telescope are time–tagged with independent clocks and atmospheric effects are uncorrelated. Radio telescopes do not respond directly to the Stokes' parameters, but rather to a specific polarization state. The usual practice is to measure the response in two nominally orthogonal polarizations. In VLBI usage, dual circular polarization systems are used for this purpose. All of the following assumes that the telescopes respond to circular polarizations. Crossed linear, or other polarization combinations, could be used as well at some increase in the complexity.

Correlations of the various combinations of right (R) and left (L) circular polarizations fully measure the polarization state of the visibilities. A first order expansion in time and frequency of the actual response of an interferometer is given by:

$$
\begin{aligned}
R_{RR}^{ikt\nu} &= g_{iR}g_{kR}^{*}b_{iR\nu}b_{kR\nu}^{*}e^{-2\pi j[(\tau_{iR}^{0}-\tau_{kR}^{0})(\nu-\nu_0)+(\dot{\tau}_{iR}-\dot{\tau}_{kR})(t-t_0)]} \\
&\quad \times[(R_I+R_V)e^{-j(\chi_i-\chi_k)} + (R_Q+jR_U)D_{Rk}^{*}e^{-j(\chi_i+\chi_k)} + \\
&\quad (R_Q-jR_U)D_{Ri}e^{j(\chi_i+\chi_k)} + (R_I-R_V)D_{Ri}D_{Rk}^{*}e^{j(\chi_i-\chi_k)}]
\end{aligned}
$$

$$R_{RL}^{ikt\nu} = g_{iR}g_{kL}^{*}b_{iR\nu}b_{kL\nu}^{*}e^{-2\pi j[(\tau_{iL}^{0}-\tau_{kL}^{0})(\nu-\nu_{0})+(\dot{\tau}_{iR}-\dot{\tau}_{kL})(t-t_{0})]}$$
$$\times[-(R_I+R_V)D_{kL}^{*}e^{-j(\chi_i-\chi_k)} + (R_Q+jR_U)e^{-j(\chi_i+\chi_k)} -$$
$$(R_Q-jR_U)D_{iR}D_{kL}^{*}e^{j(\chi_i+\chi_k)} + (R_I-R_V)D_{Ri}e^{j(\chi_i-\chi_k)}]$$
$$R_{LR}^{ikt\nu} = g_{iL}g_{kR}^{*}b_{iL\nu}b_{kR\nu}^{*}e^{-2\pi j[(\tau_{iL}^{0}-\tau_{kR}^{0})(\nu-\nu_{0})+(\dot{\tau}_{iL}-\dot{\tau}_{kR})(t-t_{0})]}$$
$$\times[-(R_I+R_V)D_{iL}e^{-j(\chi_i-\chi_k)} - (R_Q+jR_U)D_{iL}e^{-j(\chi_i+\chi_k)} -$$
$$(R_Q-jR_U)D_{kL}^{*}e^{j(\chi_i+\chi_k)} + (R_I-R_V)e^{j(\chi_i-\chi_k)}]$$
$$R_{LL}^{ikt\nu} = g_{iL}g_{kL}^{*}b_{iL\nu}b_{kL\nu}^{*}e^{-2\pi j[(\tau_{iL}^{0}-\tau_{kL}^{0})(\nu-\nu_{0})+(\dot{\tau}_{iL}-\dot{\tau}_{kL})(t-t_{0})]}$$
$$\times[(R_I+R_V)e^{-j(\chi_i-\chi_k)} + (R_Q+jR_U)D_{Rk}^{*}e^{-j(\chi_i+\chi_k)} +$$
$$(R_Q-jR_U)D_{Ri}e^{j(\chi_i+\chi_k)} + (R_I-R_V)D_{Ri}D_{Rk}^{*}e^{j(\chi_i-\chi_k)}]$$

where g_{iX} is the complex gain of antenna i, polarization X, $b_{iX\nu}$ is the complex bandpass function of antenna i, polarization X, τ_{iX} is the group delay error of antenna i, polarization X, $\dot{\tau}_{iX}$ is the fringe rate error of antenna i, polarization X, D_{iX} is the polarization "leakage" term for antenna i, polarization X and χ_i is the parallactic angle of antenna i.

The g, D, b and τ terms need to be estimated so that their effects can be removed from the data. This process is know as calibration.

A further complication is that not all data is good. Equipment malfunction, radio frequency interference (RFI) and poor weather can corrupt some data to the point that they cannot be recovered. Such data must be removed or "flagged"; a process knowing generically as editing.

3. Continuum Calibration

Most of the parameters needed to calibrate the visibility data are derived from measurements of continuum sources. The following sections go through the various stages of calibration derived from continuum data. These operations are generally needed for spectroscopic data as well. Steps which are only needed for polarization sensitive observations are marked "[Polarization]" in the section heading. The details of performing the operations described here in a real software package evolve with time; such details are left as an exercise for the reader. See also [9].

3.1. CORRECT PHASES FOR PARALLACTIC ANGLE

For alt–az mounted telescopes, the apparent orientation of the telescope feeds as seen by the source (parallactic angle) rotates as the telescope tracks the source. This rotation of the feed will have a corresponding rotation of the phase of the signal received. For polarization or phase referenced data, it is important to remove this effect before deriving any phase–like values directly from the data. This can be expressed as a modification of the antenna gains:

$$g_{iR} = g'_{iR}\, e^{-j\chi_i}$$

$$g_{iL} = g'_{iL}\, e^{j\chi_i}$$

This causes the source response to be fixed with time and the instrumental polarization to rotate with parallactic angle.

3.2. APPLY SYSTEM TEMPERATURE AND ANTENNA GAIN

A correlator measures the fraction of the power received by the telescopes which is correlated, and is thus sensitive to receiver and sky noise contributions. That is, a correlator measures a correlation coefficient (ρ) but we want the visibility (R):

$$|R| = |\rho||g_i||g_k|$$

For a point source of flux density $S(Jy)$ and a perfect correlator:

$$\rho = S(Jy)\sqrt{\frac{T_{Si}}{T_{Ai}}}\sqrt{\frac{T_{Sk}}{T_{Ak}}}.$$

Here T_{Si} or "System temperature" is the temperature equivalent of the total power received by telescope i and T_{Ai} or "Antenna temperature" is the temperature equivalent power received by telescope i from the source. This is related to the antenna gain (G_A) and source flux density by

$$T_A = G_A(K/Jy) * S(Jy)$$

and

$$|g_i| = \sqrt{\frac{S(Jy)T_{Si}}{G_{Ai}(K/Jy) * S(Jy)}} = \sqrt{\frac{T_{Si}}{G_{Ai}(K/Jy)}}$$

An estimate of the amplitude of the gain (g) can thus be made using measurements of the 'system temperature and the antenna gain. System temperature measurements should be made at the same time as the interferometric measurements. At higher frequencies, the gain of the antenna will be a function of elevation and possibly other things. The gain of the antennas as a function of elevation is usually measured independently.

At sufficiently high frequencies, the atmospheric opacity may be an issue. Weather measurements may be useful in estimating the opacity in order to correct the data. Opacity corrections can be applied as corrections to the antenna gain.

3.3. APPLY PHASE CALS [OPTIONAL]

When the telescope signals are divided up into multiple polarizations or sections of bandpass, the different electronics will have different effects on the data. The most troublesome of these are differences in phase and group delay. These may be measured and removed using "phase cals", tones injected into the signal path that can be used to measure the phase and delay through the rest of the system. A phase correction can be added by

$$g_{ti} = g'_{ti} \, e^{-j\phi^{PC}_{ti}}$$

and the delay correction by

$$\tau_{ti} = \tau'_{ti} - \tau^{PC}_{ti}.$$

While a phase cal system can correct for drifts in the instrumental phase and delay, it should not be used during spectroscopic observations.

3.4. "MANUAL" PHASE CAL OR CORRECTION

If an electronic phase cal is not used, or to make corrections to a phase cal measurement, observations of a strong continuum source can be used. These corrections can be determined by a "Fringe fit" to a single segment of data on a strong calibrator when all telescopes were observing. Fringe rates from this solution should be set to zero before applying them to the data. This will also refer all phase–like quantities to a "reference" antenna. Residual delay errors at the time of the calibration data will be corrected.

3.5. FRINGE FITTING

The variations of phase with time and frequency ($\dot{\tau}$,τ) for each telescope and data stream is determined by a process called "Fringe Fitting". There are several variations on this process [14], but the one described here is referred to as "Global Fringe Fitting" as all data obtained at a single time are used jointly. Global fringe fitting is basically a multi–dimensional self calibration. This procedure estimates the phase error (ϕ), group delay (τ) and fringe rate ($\dot{\tau}$) in time segments of data using the model for baseline ik:

$$\phi^{ik}_{\nu t} = (\phi^{i}_{\nu t} - \phi^{k}_{\nu t}) + 2\pi(\tau^{i}_{\nu t} - \tau^{k}_{\nu t})(\nu - \nu_0) + 2\pi(\dot{\tau}^{i}_{\nu t} - \dot{\tau}^{k}_{\nu t})(t - t_0).$$

where ν is frequency and t is time. The time segments are picked such that this linear approximation is valid. The Fourier transform of R(t,ν) gives $R(\tau, \dot{\tau})$ in which the response to a source is localized. The position of the peak amplitude in $R(\tau, \dot{\tau})$ gives the τ and $\dot{\tau}$ needed to correct the data.

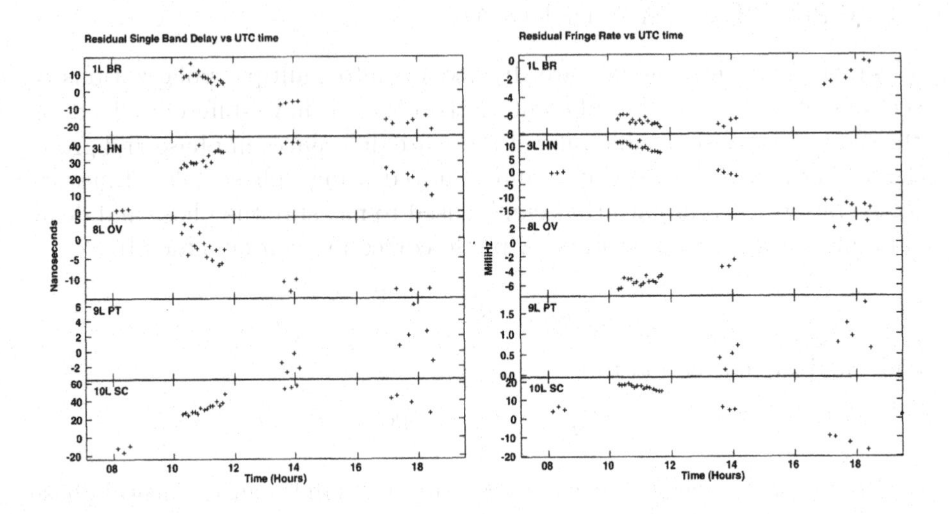

Figure 1. Left) Residual delays determined from a fringe fit. Right: Residual rates determined from a fringe fit.

If a model of the source is available, it can be used to remove structure phase before fringe fitting. In the case of phase referencing, no fringe fit is needed for the target source but the phase reference calibrator should be fully self–calibrated before the final fringe fit.

Since only differential phases are measured, these values are respect to a "reference" antenna. The Right– and Left–handed polarizations systems are frequently evaluated independently; the resultant fits should be examined to be sure the right–left phase stability is not disturbed. Factorizing the solutions into antenna components is important for maintaining the closure relations used by self calibration.

Example results from fringe fitting are shown in Figures 1 and 2. Figure 1 shows the residual delays and rates (τ, $\dot{\tau}$) for a set of VLBA data. Figure 2 shows the time dependence of the difference in phases fitted to the Right– and Left–handed polarizations for several antennas of an 18 cm VLBI experiment. All show the same drift in time which indicates a small amount of ionospheric Faraday rotation at the reference antenna.

3.6. RIGHT–LEFT DELAY CALIBRATION [POLARIZATION]

In the calibration scheme described in this lecture, the Right– and Left–handed polarization systems are calibrated independently. This allows for offsets in the phase–like quantities, especially phase and delay, between the two systems. These must be determined and corrected in order to use the

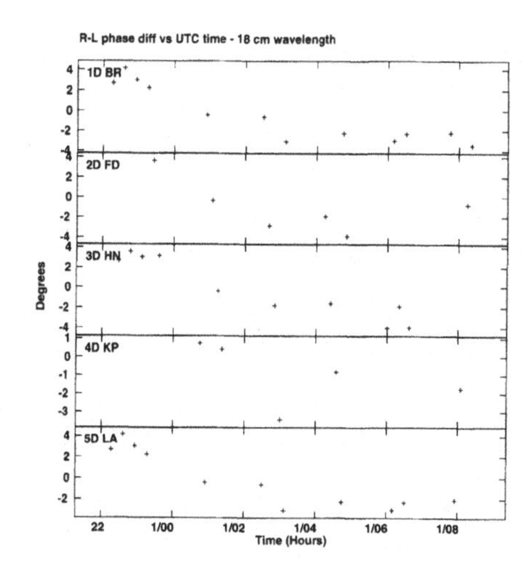

Figure 2. Residual right-left phase differences at 18 cm showing slight ionospheric Faraday rotation.

cross–polarized (RL, LR) correlations to measure linear polarization. These offsets are referred to as the R-L phase and delay offsets and are measured by separate operations. The R-L delay (τ_{R-L}) must be determined from cross polarized fringes on a strong polarized calibrator by a fringe fitting procedure. Once determined the correction can be applied to the delay estimates:

$$\tau_{iL} = \tau'_{iL} - \tau_{R-L}.$$

If the interferometer is stable, and in the absence of ionospheric Faraday rotation, τ_{R-L} should be constant, hence its time derivative, $\dot{\tau}_{R-L}$ is zero.

3.7. EDITING

Not all data are created equal. Bad weather, interference and equipment, etc. failures will render some data uncorrectable. This data needs to be identified and discarded, or "flagged". This is actually a continuous process as the calibration steps need to be monitored to be sure they work correctly. Bad data can corrupt the calibration solutions, or cause there to be no solutions at all. There are a number of techniques for determining bad data, either looking for the effects in calibration procedures or direct examination or comparison with a model.

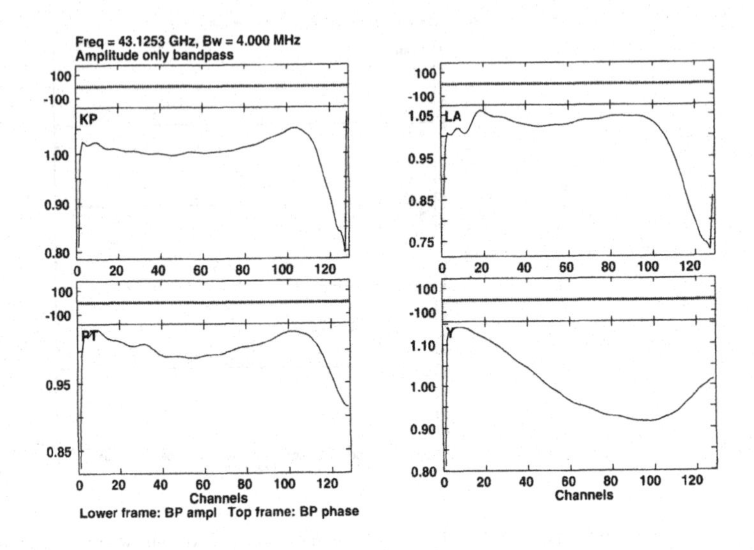

Figure 3. Example amplitude only bandpass functions.

3.8. BANDPASS CALIBRATION

Receiver systems rarely, if ever, have frequency responses that are flat in gain and constant in phase. Deviations from a perfect bandpass should be determined and removed from the data. This is clearly needed for spectroscopic observations where frequency dependent amplitude and phase errors will degrade the resultant images. Even for continuum measurements, bandpass mismatches will introduce non-closing, or baseline dependent, artifacts which cannot be removed by self calibration. If uncorrected, these will limit the dynamic range of the resultant images.

To remove these bandpass effects, the bandpass function $b_{i\nu}$, needs to be determined for each telescope/polarization/frequency band using observations of a strong continuum calibrator. The observations need enough SNR to get a good determination in each channel. The amplitude part of the bandpass function $b_{i\nu}$ can be derived from autocorrelation measurements although these are sensitive to interference. Examples of bandpass amplitude functions derived from autocorrelation data are shown in Figure 3. The advantage of this is that autocorrelations usually have much higher SNR than cross-correlations. However, to get the phase of the bandpass function requires cross-correlation data on a strong continuum source.

3.9. AVERAGE IN TIME AND FREQUENCY

After the basic calibration is done, it is possible to average continuum data in time and frequency. This step is not really necessary but can dramatically reduce the volume of data needed for further processing. Any averaging should consider the desired field of view and subsequent self calibration [1]. Spectral line data can be averaged in time subject to the same field of view and self calibration constraints.

3.10. SELF–CALIBRATION

After the basic calibration, the final corrections for the time variations of the phase, and possibly amplitude, of the gains are made through the process of "Self Calibration" [3]. In this procedure the source image and calibration corrections are jointly estimated from the data. This requires that the source be detected to each antenna in each coherence time.

If the phase referencing technique is being used, then the target source need not be self–calibrated. However, the phase reference source should be completely self–calibrated before the final fringe fit, unless it is completely unresolved.

3.11. INSTRUMENTAL POLARIZATION [POLARIZATION]

Telescope feeds never respond exactly as intended. In particular, a feed designed to respond to one polarization state will have a non–negligible response to the orthogonal polarization as well. This effect can be modeled in terms of "leakage" terms denoted D above. The effect of these leakage terms is to cause an apparently polarized response even to an unpolarized source, thus, this is referred to as "instrumental" polarization. If the data is to be used for deriving polarization images, these terms must be estimated and corrected. Even for total intensity images, the instrumental polarization terms can limit the dynamic range, especially for heterogeneous arrays. For more details see [6] and [10].

The instrumental polarization terms must be derived from measurements of astronomical sources. Since the compact extra-galactic radio sources used for calibration are synchrotron emitting sources, they may also be polarized. In the general case, both the instrumental and calibrator polarizations must be determined simultaneously. Ignoring second and higher order terms, the cross–polarized correlations are:

$$R^{ik}_{RL} = gain \times [R_I(D_{iR}e^{j(\chi_i-\chi_k)} + D^*_{kL}e^{-j(\chi_i-\chi_k)}) - (R_Q + jR_U)e^{-j(\chi_i+\chi_k)}]$$
$$R^{ik}_{LR} = gain \times [R_I(D_{iL}e^{-j(\chi_i-\chi_k)} + D^*_{kR}e^{j(\chi_i-\chi_k)}) - (R_Q - jR_U)e^{j(\chi_i+\chi_k)}]$$

Instrumental Source

As can be seen, the first order response consists of two terms, the first, the spurious polarized response and the second, the true response to the source. Note however, that the source response varies with the sum of the parallactic angles while the instrumental term varies with the difference.

Observations over a range of parallactic angles will help separate the source and instrumental terms. This procedure works best for unresolved or unpolarized calibrators.

3.12. IMAGE CALIBRATORS FOR R-L PHASE [POLARIZATION]

The proceeding procedure will have allowed an arbitrary phase offset between the right– and left–handed polarization systems. This is equivalent to an arbitrary offset in the polarization angle of derived images, but will not de-polarize the images. This offset must be determined and corrected for the derived linear polarization images to have the correct angle on the sky. This correction must be determined from observations of a polarized calibrator with known polarization.

Since polarized compact calibrators tend to be blazars, they are likely to be highly time variable. Frequently, the polarization angle calibration depends on measurements made with an instrument on which it is unresolved so only its integrated polarization is known. If all of the emission from the source is represented in the VLBI image, then the apparent polarization angle can be derived from the integrated Q and U images, in practice the sum of the CLEAN components, from the deconvolution:

$$\phi^{apparent}_{R-L} = Tan^{-1}(\frac{\Sigma U}{\Sigma Q})$$

If the true $R-L$ phase offset is known (twice the true integrated polarization angle) then the gains can be corrected:

$$g_{iLt} = g'_{iLt}\, e^{-j(\phi^{true}_{R-L} - \phi^{apparent}_{R-L})}.$$

3.13. IMAGE/DECONVOLVE

Once all the calibration parameters are known, the corrected stokes visibilities can be derived from the measured correlations. These visibilities can then be used to derive images in the various Stokes parameters using the usual techniques [7], [2] [4].

4. Things that go Bump in the Night

There are numerous conditions that complicate the calibration process. Several of these are discussed in the following.

4.1. IONOSPHERIC FARADAY ROTATION

At wavelengths longer than about 10 cm, the ionosphere can cause significant, time variable Faraday rotation. If uncorrected this will will corrupt Stokes' Q and U images. The detailed Faraday rotation depends on the path the signal has taken through the ionosphere and varies with time and observing geometry. It is possible to use GPS measurements to correct for this, or "self–calibrate", by estimating the variations in Faraday rotation from the target source data. This self–calibration will lose the absolute polarization angle.

4.2. RESOLVED, POLARIZED INSTRUMENTAL POLARIZATION CALIBRATORS

If the calibrator used to determine the instrumental calibration is both resolved and polarized, then solving for the instrumental and source polarization is more complex. One can invoke the "Similarity" approximation, that Q and U are scaled version of the Stokes' I and then determining the source polarization consists of determining the scaling factors. Unfortunately, this approximation breaks down fairly quickly as the source becomes resolved. An example is shown in Figure 4 where the polarized structure is quite different from that of total intensity.

One solution to this problem is to iteratively refine the model of the source polarization using improved solutions of the instrumental polarization. A more elegant solution is given by [11] in which the total intensity source is broken up into components, and the similarity approximation applied separately to each.

5. Spectroscopy

The preceding discussion is relevant to both continuum and spectroscopic measurements. For spectroscopy, there are additional considerations and calibrations to be applied. The following discussion first considers issues relevant to spectral line observations and then describes the additional processing needed.

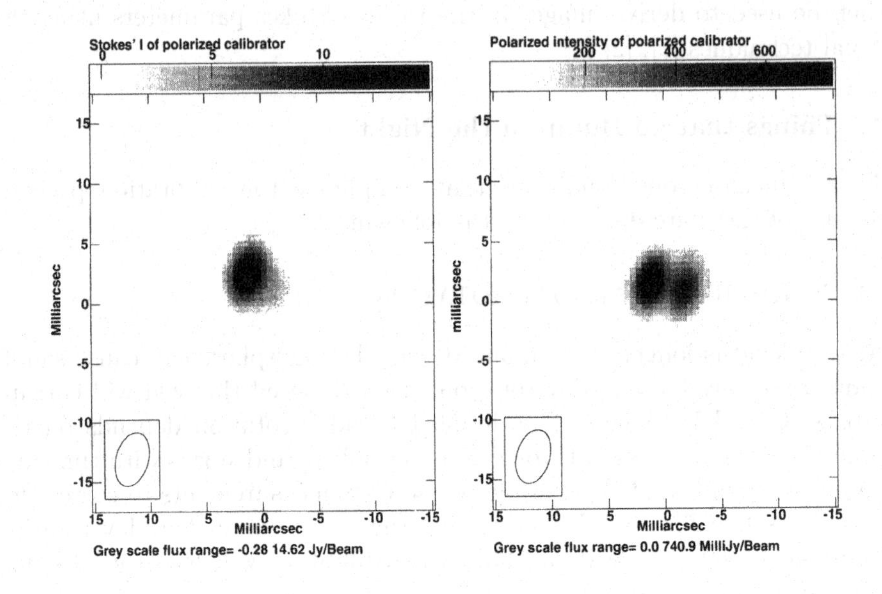

Figure 4. Total and polarized intensity of resolved, polarized calibrator.

Since VLBI observations are limited to extremely bright objects, only non-thermal emission can be observed. This limits VLBI spectroscopic measurements to interstellar masers and absorption lines in front of non-thermal continuum sources. The calibration of absorption line data is largely the same as continuum data except that the channels in the absorption line are not used. Most of the discussion in the following concerns maser sources for which there is usually no detectable continuum emission. See also [12].

5.1. LAG – SKY FREQUENCY RELATIONSHIP

An important relation is the Fourier transform relation between a measured lag function and the (sky) frequency spectrum [12]. Using this relationship it is possible to derive spectra from a lag correlator by Fourier transforming:

$$R_\nu = c \int R_\tau \, e^{-2\pi\nu\tau} \, d\tau$$

5.2. GIBBS PHENOMENON

Features such as spectral lines which are narrow in the frequency domain are broad in the time, or lag, domain. For practical reasons, lag functions

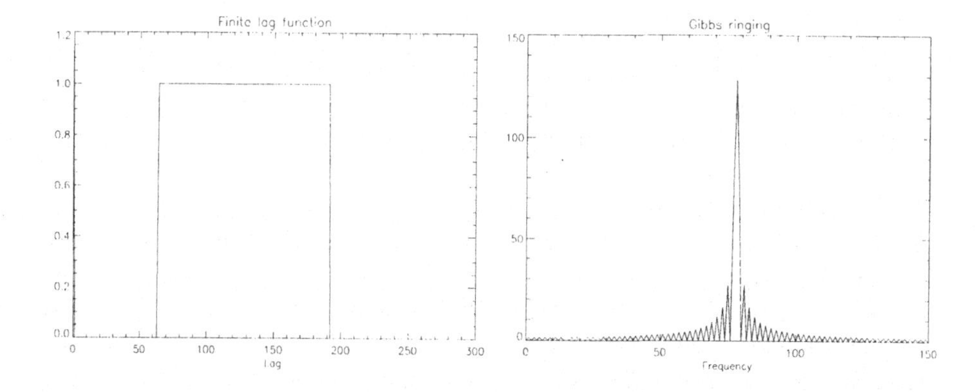

Figure 5. Left) Flat, finite lag function, Right) the Fourier transform showing the Gibbs ringing.

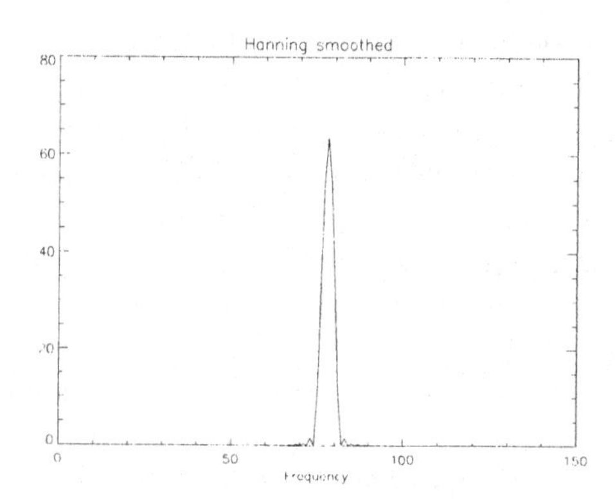

Figure 6. Spectrum after Hanning smoothing

are measured over a finite range and are thus truncated. This truncation in the lag domain will lead to a ringing in the frequency domain, an effect know as the Gibbs Phenomenon. This is illustrated in Figure 5; on the left is illustrated a flat, truncated function. On the right is the Fourier transform of this function which shows a strong ringing. This effect can be greatly reduced by a filter known as "Hanning" smoothing [17] [13] at the cost of a factor of two in spectral resolution. The function shown in Figure 5 right, after Hanning smoothing, is shown in Figure 6. The ringing is dramatically reduced.

5.3. DIFFERENCES WITH CONTINUUM:

There are a number of considerations which are important in spectroscopic observations which are either irrelevant or different for continuum only observations. These are considered in the following.

5.3.1. *Doppler tracking*

The Doppler velocities of sources due to motions of the Earth and Sun are not small compared with the spectroscopic resolution, and vary continuously during the day and year. A further complication is that observations are usually done with fixed frequencies to simplify the observations and the hardware requirements. In order to properly image this data, the spectra must be shifted such that a given LSR velocity corresponds to a given channel. This should be done after bandpass calibration as the bandpass function is fixed in the reference frame of the telescope.

5.3.2. *Amplitude Calibration with Autocorrelations*

Observations of strong maser sources have the advantage that the source can easily be seen in the autocorrelation spectra. This is especially important at high frequencies where antenna pointing errors and variable weather make amplitude calibration difficult. The response of the source in the autocorrelation spectra includes the effects of pointing and weather, although not LO noise and other sources of decorrelation.

The relative effective gain of the various telescopes can be determined from the autocorrelation spectra through a process known as "Template fitting" [12]. In this procedure, a template source spectrum is derived from a good set of data at a single telescope, and then a scaling is determined to make the spectra at other times and telescopes most closely match the template spectrum. These scaling parameters give the relative effective gains of the telescopes.

5.3.3. *Phase Reference on Spectral Feature*

Maser regions consist of a number of independent spots. Different spots may appear at different frequencies. If a self–calibration is determined for a single channel, then this calibration can be applied to all channels causing them to be focused and registered on the same coordinate frame [12]. Note, this procedure only corrects the relative positions, the absolute positions are lost using self–calibration.

5.3.4. *Delay Calibration*

For phase referencing to work, other frequency dependent phase effects, such as delay errors have to be corrected. Numerous effects such as the independent clocks used as frequency standards, atmospheric and geomet-

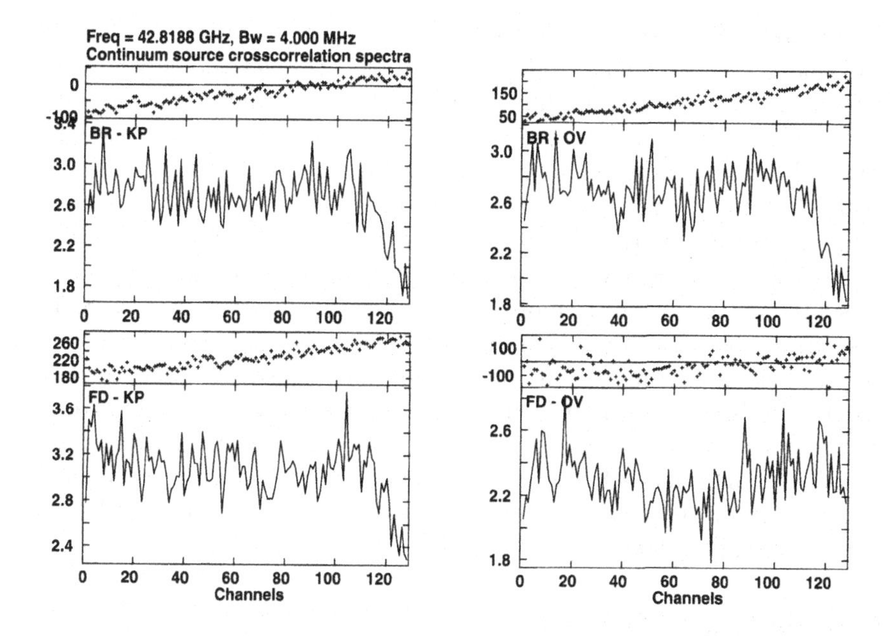

Figure 7. Cross–correlation spectra of a continuum source, the linear drift of phase with frequency indicates delay residuals. In each panel, amplitudes are shown on the bottom and phases on the top.

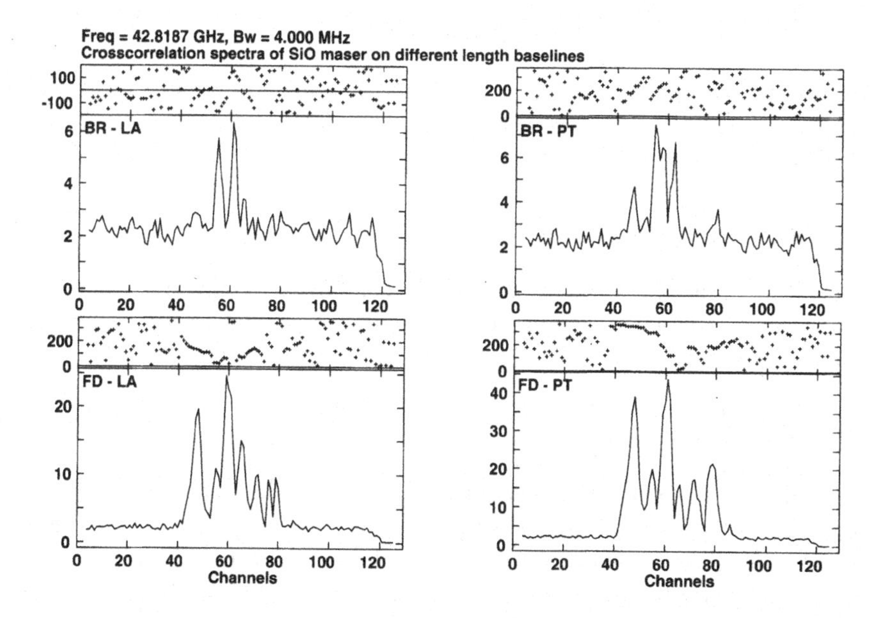

Figure 8. Example cross-correlation spectra of SiO maser source.

ric errors cause delay errors. The signature of a group delay is a linear slope of phase with frequency. While such slopes are relatively straightforward to measure with continuum observations, this is not the case with maser emission. This is illustrated in Figures 7 and 8. Figure 7 shows the uncorrected cross–correlation spectra of a continuum source on a number of baselines. Since there is emission in all channels, which has the same distribution on the sky, there is observed power in all channels with a linear drift of phase with frequency due to uncorrected delay errors. For a maser source the situation is more complex, as is shown in Figure 8. In these spectra, there is only power in some channels, and since the spots are not co-located on the sky, the phases are determined by the frequency dependent sky brightness distribution as well as any residual delay errors. If delay errors are uncorrected, they will introduce phases errors away from the phase reference channel which will defocus the images. Observations of continuum calibrators close in time and location on the sky are needed to measure the residual delays.

6. Spectral Line Calibration

The following sections describe the additional calibration and processing steps needed for spectral line data after the basic calibration has been determined from continuum calibrators.

6.1. APPLY CONTINUUM CALIBRATION

The calibration parameters derived from the continuum calibrators for amplitude, phase, delay, fringe rate, bandpass and polarization should be applied to the spectral line data.

6.2. DOPPLER CORRECTION

After the bandpass corrections have been applied, the data need to be shifted in frequency to remove the diurnal and annual Doppler shift of the telescopes relative the the LSR coordinate system.

6.3. TEMPLATE FITTING

The relative gains of the telescopes with time is then determined from the template fitting procedure using the spectral features visible in the autocorrelation data. This does not work if there is significant radio frequency interference (RFI). If dual polarization measurements are being made, care must be taken to preserve the relative amplitude scaling of the two polarizations systems.

6.4. SELF–CALIBRATION ON LINE CHANNEL

The final time dependent phase calibration is determined from the self–calibration of a single (or several) channels. Ideally a channel containing a single strong source is used, or at least a channel with relatively simple structure. The phases needed to calibrate this channel are then applied to all channels. If the source is strongly circularly polarized, the case of OH masers in star forming regions, it is necessary to determine the phases in one polarization, and then transfer them to the other.

It is desirable to use a channel with a strong signal in order to obtain a good calibration. Channels with a simple structure may have only a weak signal. However, calibration derived from a channel with simple, but weak structure can be used to bootstrap the calibration of a channel with more complex, but brighter, structure.

6.5. IMAGE/DECONVOLVE

Once the calibration is completed, each spectral channel is imaged and deconvolved as in the continuum case.

References

1. Bridle A.H. and Schwab, F.R. (1999) Bandwidth and Time–Average Smearing, in Taylor, G., Carilli, C. and Perley, R. eds. **Synthesis Imaging in Radio Astronomy II**, Astronomical Society of the Pacific, pp. 187–199
2. Briggs, D.S. Schwab, F.R. and Sramek, R.A. (1999) Imaging, in Taylor, G., Carilli, C. and Perley, R. eds. **Synthesis Imaging in Radio Astronomy II**, Astronomical Society of the Pacific, pp. 127–148
3. Cornwell, T.J. and Fomalont, E.B. (1999) Self–Calibration, in Taylor, G., Carilli, C. and Perley, R. eds. **Synthesis Imaging in Radio Astronomy II**, Astronomical Society of the Pacific, pp. 187–199
4. Cornwell, T.J., Braun, R. and Briggs, D.S. (1999) Deconvolution, in Taylor, G., Carilli, C. and Perley, R. eds. **Synthesis Imaging in Radio Astronomy II**, Astronomical Society of the Pacific, pp. 151–170
5. Cotton, W.D. (1993) Calibration and Imaging Techniques for Polarization Sensitive VLBI Observations, *Astron. J.*, **106**, pp. 1241–1248
6. Cotton, W.D. (1999) Polarization in Interferometry, in Taylor, G., Carilli, C. and Perley, R. eds. **Synthesis Imaging in Radio Astronomy II**, Astronomical Society of the Pacific, pp. 111–124
7. Cotton, W.D. (2002), Radio and Optical Interferometry, this volume
8. Cotton, W.D. (2002), Radio Polarimetry, this volume
9. Fomalont, E.B. and Perley, R.A. (1999) Calibration and Editing, in Taylor, G., Carilli, C. and Perley, R. eds. **Synthesis Imaging in Radio Astronomy II**, Astronomical Society of the Pacific, pp. 79–109
10. Kemball, A. (1999) VLBI Polarimetry, in Taylor, G., Carilli, C. and Perley, R. eds. **Synthesis Imaging in Radio Astronomy II**, Astronomical Society of the Pacific, pp. 499–511
11. Leppänen, K.J. (1995), *22 GHz Polarimetric Imaging with the Very Long Baseline Array*, thesis Helsinki University of Technology, Finland

12. Reid, M.J. (1999) Spectral Line VLBI, in Taylor, G., Carilli, C. and Perley, R. eds. **Synthesis Imaging in Radio Astronomy II**, Astronomical Society of the Pacific, pp. 481–497
13. Rupen, M.P. (1999) Spectral Line Observing II: Calibration ad Analysis, in Taylor, G., Carilli, C. and Perley, R. eds. **Synthesis Imaging in Radio Astronomy II**, Astronomical Society of the Pacific, pp. 229–274
14. Schwab, F.R. and Cotton, W.D. (1983) Global fringe search techniques for VLBI, *Astron. J.*, **88**, pp. 688-694
15. Taylor, G., Carilli, C. and Perley, R. eds. (1999) **Synthesis Imaging in Radio Astronomy II**, Astronomical Society of the Pacific
16. Thompson, A.R., Moran, J.M. and Swenson, G.W. Jr. (2001) **Interferometry and Synthesis in Radio Astronomy**, John Wiley & sons, Inc., New York.
17. Westpfahl, D.J. (1999) Spectral–Line Observing I: Introduction , in Taylor, G., Carilli, C. and Perley, R. eds. **Synthesis Imaging in Radio Astronomy II**, Astronomical Society of the Pacific, pp. 201–227

DEEP FIELDS: THE FAINT SUB-MJY AND MICROJY RADIO SKY

- A VLBI Perspective

M.A. GARRETT
Joint Institute for VLBI in Europe
Postbus 2, 7990 AA Dwingeloo, The Netherlands

1. Introduction

Until recently, VLBI targets have been drawn almost exclusively from the brightest and most compact radio sources in the sky, with typical flux densities well in excess of a few tens of mJy. These sources are predominantly identified with Active Galactic Nuclei (AGN), located at cosmological distances ($z \sim 1$). Exotic but also rather rare, these luminous AGN systems have been studied in great detail by VLBI over the last 3 decades, producing many front-line discoveries along the way (see this volume and references therein).

However, in recent years, connected arrays (such as the VLA, WSRT, ATCA and MERLIN) have also began to focus a significant fraction of their time towards understanding the nature of the faint radio sky — sometimes observing the same field for many days or even weeks at a time. At these microJy noise levels, the radio sky literally "lights up", and a new population of vigorous star forming galaxies begin to dominate the radio source counts (Fomalont et al. 1997, Richards et al. 1998, Muxlow et al. 1999, Richards 2000, Garrett et al. 2000a, Norris et al. 2000). For many astronomers (usually radio astronomers!) it comes as some surprise that a well calibrated VLBI array, composed of the largest telescopes in the world, can also contribute to our understanding of this sub-mJy and microJy radio source population. Nevertheless, these are the facts, as recently demonstrated by the simultaneous detection of 3 sub-mJy radio sources in the Hubble Deep Field North (HDF-N) by the European VLBI Network (Garrett et al. 2001).

In this lecture I will attempt to summarise what is currently known about the general properties of the faint sub-mJy and microJy radio source population, as determined from deep multi-wavelength studies of the HDF-N. In particular, I will try to provide a VLBI perspective, describing the

F. Mantovani and A. Kus (eds.), The Role of VLBI in Astrophysics, Astrometry and Geodesy, 337–347.

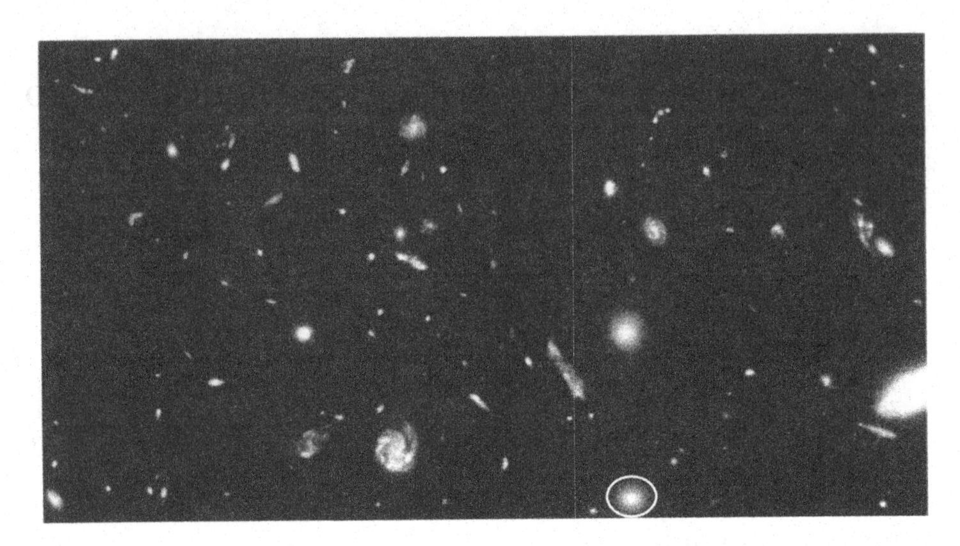

Figure 1. Close-up of a small region of the HDF-N (Williams et al. 1996). More distant galaxies also tend to be more morphologically disturbed (NASA/STScI). One of the radio sources (VLA J123644+621133) detected by the European VLBI Network is circled (see section 4).

first deep, wide-field, VLBI pilot observations of the HDF, together with a summary of the main results. The role VLBI can play in future high resolution studies of faint radio sources will also be addressed.

2. The Hubble Deep Field and Galaxy Formation

The Hubble Deep Field North (HDF-N) is one of the most important, publicly available resources ever generated in the history of astronomy (see Ferguson, Dickinson & Williams 2000 and references therein). What initially appeared to be a small, undistinguished $2.5' \times 2.5'$ patch of the celestial sphere was transformed during a 10-day long Hubble Space Telescope (HST) deep field integration (Williams et al. 1996).

The key advance of the HDF images was not only the depth which they reached ($I < 29^m$) but perhaps more crucially, the superb angular resolution that accompanied them. On inspection it immediately became clear (see Fig. 1), that the most distant galaxies in the field were also the most morphologically disturbed. For example, familiar "grand design" spirals observed locally, all but disappear beyond $z \sim 0.3$. In terms of morphology, these distant, disturbed systems are most akin to nearby Ultra Luminous Infrared Galaxies (ULIG) and interacting starburst systems (Abraham & van den Bergh 2001).

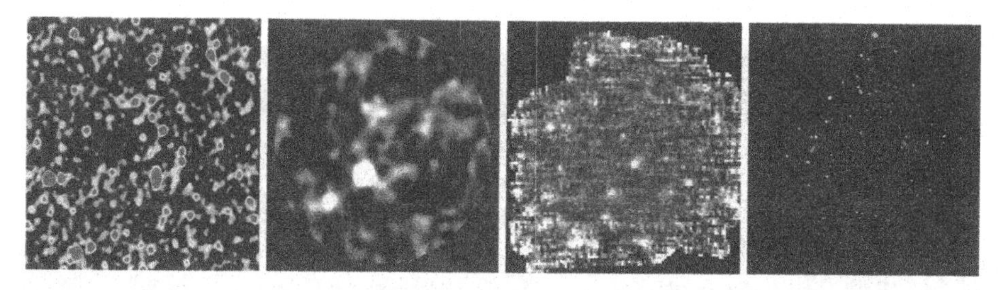

Figure 2. Multi-wavelength images of the HDF-N in the radio (Garrett et al. 2000a – includes the HFF), sub-mm (Hughes et al. 1998), Mid-IR (Rowan-Robinson et al. 1997 – includes part of the HFF) and x-rays (Brandt et al. 2001 -includes the HFF).

2.1. MULTI-WAVELENGTH FOLLOW-UP OBSERVATIONS OF THE HDF

The HDF represented a huge investment in HST observing time, and this was matched by an equally ambitious multi-wavelength follow-up campaign. The latter included every part of the e-m spectrum accessible via both ground and space based instrumentation. In many cases, extremely long integrations produced the very deepest view of the Universe in a given waveband. In addition, extensive efforts were made to obtain spectroscopic and photometric redshifts of sources in the field (including the adjacent Hubble Flanking Fields – HFF). A pictorial summary of a limited subset of these multi-wavelength follow-up observations is shown in Fig 2.

It comes as no surprise that the source counts in the HST images (both in the optical and Near-IR) are considerably greater than that observed at other wavelengths. What is surprising, however, is that despite this fact, a considerable fraction (10-20%) of the faint radio, Mid-IR and x-ray detections, appear to be heavily obscured in the optical i.e. they have no obvious optical counterparts (R> 25^m). An even larger fraction of the faint SCUBA source population fall into this category, suggesting that the sub-mm observations reveal a completely different (unobscured) view of the high redshift, dusty, star-forming Universe (Hughes et al. 1998). We will return to the nature of these faint sub-mm sources in section 3.3.

3. Deep Radio Imaging of the HDF

In order to detect even a handful of radio sources in the HDF-N, noise levels of a few microJy must be achieved. These in turn require integration times ranging from a few days – in the case of the WSRT and VLA, to many days in the case of MERLIN.

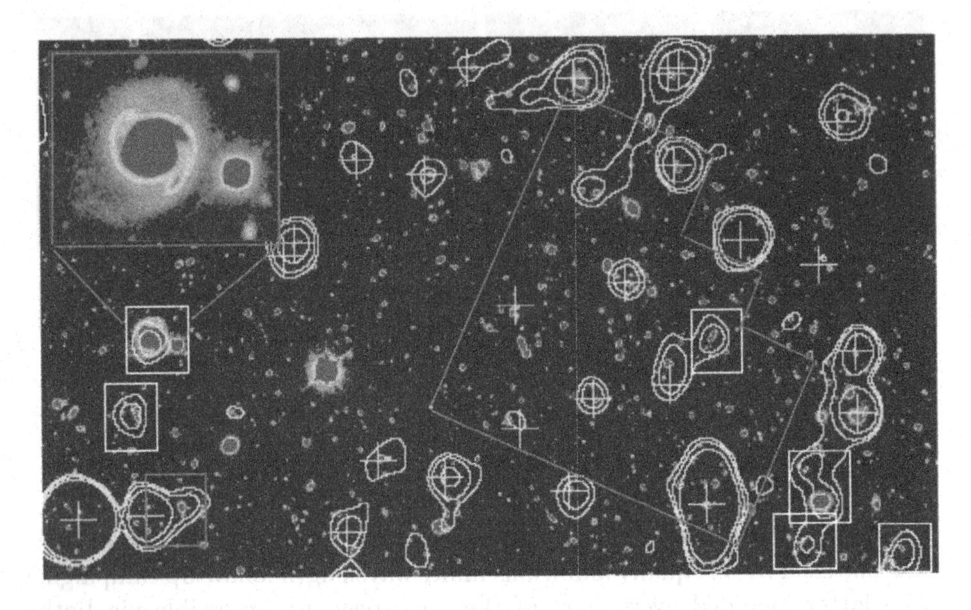

Figure 3. The WSRT 1.4 GHz contour map of the HDF (see incomplete rotated square) and part of the HFF superimposed upon a deep CFHT optical image of the HDF region made by the Canada-France-Hawaii telescope (courtesy Amy Barger). Crosses indicate previously known VLA detections, boxes indicate new WSRT detections. The detection of a nearby, extended star forming galaxy is highlighted (upper left).

3.1. VLA 8.3 GHZ, VLA-MERLIN & WSRT 1.4 GHZ OBSERVATIONS

Deep VLA observations of the HDF (including the HFF) have been conducted at both 8.3 and 1.4 GHz (Fomalont et al. 1997, Richards et al. 1998, Richards 2000). The 8.3 GHz observations reach noise levels of a few microJy per beam (several times better than the 1.4 GHz observations) but more sources are actually detected at 1.4 GHz where the source counts are steeper and the VLA field of view wider. Perhaps the most "complete" radio view of the HDF (see Fig. 3) is provided by the WSRT 1.4 GHz observations (Garrett et al. 2000a). These are sensitive to very extended radio structures, although for the WSRT (and indeed the VLA), the vast majority of the microJy radio source population remain barely or completely unresolved at arcsecond resolution. Combined VLA-MERLIN 1.4 GHz observations with a resolution of 0.2″ (Muxlow et al. 1999) begin to resolve most of these sources but the detailed morphology of the microJy radio source population still remains unknown. The main results of the VLA, MERLIN and WSRT data can be summarised as follows:

- Most of the radio sources are identified with relatively bright ($R < 25^m$), moderate redshift, optical counterparts – often identified as interacting, irregular or peculiar morphological type (Richards et al. 1998)

- The 1.4 GHz VLA source sample is steep spectrum in nature (typically $\alpha \sim -0.85$ – Richards 2000). Sources selected by the VLA at 8.3 GHz are significantly flatter specutrum ($\alpha \sim -0.35$)
- There is a strong correspondence between the Mid-IR ISO detections and the radio detections (see Fig. 4 right). Indeed the majority of radio sources (after applying appropriate k-correction factors) appear to closely follow the FIR-radio correlation (Garrett 2002).
- Of the 91 sources detected with the combined MERLIN-VLA 1.4 GHz array (see Fig. 4 left), the majority show radio structure on *sub-galactic* scales (Muxlow et al. 1999). About 50% of the MERLIN-VLA 1.4 GHz detections show extended structure aligned with the optical isophotes of the galaxy.
- The WSRT detects a small but significant population of both star forming galaxies and AGN, some of which are resolved by the higher resolution VLA and MERLIN observations (Garrett et al. 2000a; Muxlow priv. comm).
- A comparison between the VLA and WSRT 1.4 GHz images (separated by several years) shows evidence for significant variability (factors of 2 or more) for a few percent of the sub-mJy radio source population (presumably low-luminosity AGN).
- Around 10-20% of the microJy source population (see Fig. 4 -left and Richards et al. 1999) are optically faint or completely unidentified ($R > 25^m$).

3.2. SUMMARY OF RESULTS

These results, in particular the general correlation between radio and FIR luminosity, suggest that the radio emission arising in the faint sub-mJy and microJy radio source population is largely associated with massive star formation in distant star forming galaxies. However, a significant fraction of all the sub-mJy sources ($\sim 30\%$) are also identified with low-luminosity AGN. The remaining 10-20% of the faint radio source population are assoiciated with either extremely faint ($R > 25$) optical identifications, or remain unidentified altogether - even in the HDF-N itself ($I > 28$). Note that these conclusions are dominated by the faintest (and therefore more numerous) microJy radio sources – for example, the AGN fraction increases rapidly at higher (sub-mJy) flux density limits. In addition, simply labeling sources as pure "starbursts" or pure "AGN" is a little misleading; it is quite possible (even likely) that both phenomena co-exist in some of these faint systems. When we label sources in this way, we are only identifying

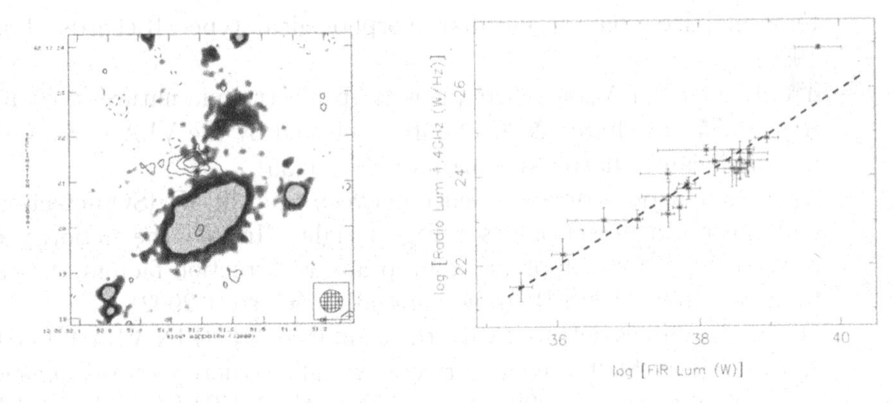

Figure 4. Left: MERLIN-VLA contour map (super-imposed on the HST gray-scale image) of J123651+621221 – an optically faint dust-obscured starburst system in the HDF with a total flux density of 49μJy (Muxlow et al. 1999). (b) Right: The strong correlation between FIR and radio luminosity for high-z galaxies detected by both ISO and the WSRT in the region of the HDF-N (Garrett 2002)

the *dominant* phenomenon that gives rise to the bulk of the observed radio emission.

Studies of nearby (nuclear) starburst galaxies such as M82 and Arp 220 give us some idea of how the radio emission arises in these star forming galaxies. According to conventional theories, the chief ingredients are supernova (SN) events associated with massive star formation (see Marcaide this lecture series), and in particular, the global acceleration of cosmic ray electrons via shocks associated with these events. The total radio luminosity of a "normal" galaxy is therefore a direct measure of the SN event rate, and in turn, the star formation rate (SFR) of massive stars (e.g. Condon 1992). In this scenario, the tight correlation between the FIR and radio luminosity of star forming galaxies is explained by the FIR emission arising from the absorption and re-radiation (via dust) of the intense uv emission also associated with massive stars. By assuming an Initial Mass Function (e.g. a Salpeter IMF) for the stellar population, and some scaling factors based on local observations of the Milky Way, radio observations provide unbiased estimates of the SFR that are largely unaffected by extinction due to dust. The SFR inferred in this way for M82 and Arp 220 are $\sim$ 10 and 100 $M_{\odot}$/yr respectively. The levels of radio and FIR emission observed for the more most distant radio sources in the HDF implies much higher star formation rates: $\sim$ 1000 $M_{\odot}$/yr.

3.3. THE OPTICALLY FAINT MICROJY RADIO SOURCE POPULATION

There is now good evidence to suggest (e.g. Barger et al. 2000; Chapman et al. 2001) a close correspondence between the optically faint microJy radio sources and the faint (SCUBA) sub-mm source population. The bulk of the observational evidence suggests that these sources are located at cosmological distances, and are enveloped in thick, opaque dust. Since it is estimated that a substantial fraction of the total radiation in the Universe is emitted from these obscured systems, the detailed nature of these sources is a key topic in astronomy today. However, the study of these dusty systems is severely hampered by the fact that they are so difficult to detect in almost all wave-bands, except the sub-mm and radio domains. It is thought that the sub-mm emission is associated with FIR emission (again absorption and re-radiation of uv emission by dust) that is redshifted into the sub-mm wavelength range. However, the source of the original uv emission is unclear – it could be generated purely by massive stars in dense star forming regions, purely by embedded AGN, or some mixture of both phenomena. Similarly in the radio, it is not known whether the synchrotron emission is generated by massive star formation processes or by accretion associated with a central AGN. As we have seen, if massive star formation is responsible, radio flux densities imply SFRs $\sim 1000\ M_{\odot}/yr$, and an inferred global SFR that is completely "at odds" with previous optical and ultraviolet studies (Haarsma et al. 2001, and references therein).

Distinguishing between the AGN and starburst phenomena in these systems is clearly of fundamental importance. In the radio, only VLBI provides sufficient angular resolution to distinguish between the two cases. In particular, radio emission generated by star formation processes should be resolved by current (sensitivity limited) VLBI observations. AGN, on the other hand, should remain considerably more compact and readily detectable.

4. Deep Field VLBI observations of the HDF

On 12-14 November 1999 the EVN conducted the first "pilot" VLBI "blank field" observations of the radio sky. The field chosen was the HDF-N – an area that is just about as "blank" and undistinguished as the radio sky gets. The brightest source in the ~ 2 arcminute radial field of view was an FR-I radio galaxy with a total WSRT 1.4 GHz flux density of ~ 1.6 mJy.

The data were recorded at a rate of 256 Mbits/sec for 32 hours – a sustained capability that is unique to the EVN (and has recently been extended to 512 Mbits/sec). Observing in phase-reference mode, a total of ~ 14 hours of "on-source" data were collected. With a resolving beam area 1 million times smaller than the WSRT HDF-N observations (see sec-

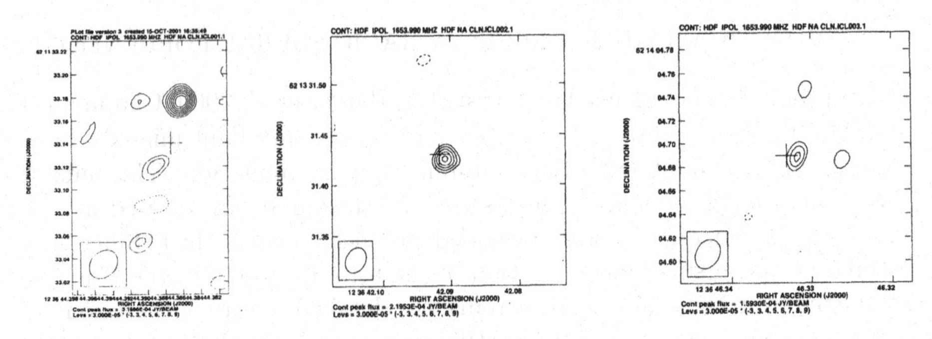

Figure 5. EVN detections in the HDF: the distant z=1.01 FRI (left), the z=4.4 dusty obscured starburst hosting a hidden AGN (middle) and the z=0.96 AGN spiral (right).

tion 3), the EVN imaged an area of about 12 arcmin^2. Six HDF-N radio sources were thus targeted simultaneously (using wide-field imaging techniques – see Garrett et al. 1999). The final naturally weighted images have an r.m.s. noise level of $\sim$ 33 μJy/beam – much larger than that expected from thermal noise considerations ($\sim$ 11 μJy/beam). The images are probably limited by the inclusion of poorly calibrated or completely corrupt data - difficult to identify in this case, simply by inspection.

Nevertheless, the EVN simultaneously detected three radio sources above the 165 μJy ($\sim 5\sigma$) limit, in the inner part of HDF-N region (see Fig 5). These include: VLA J123644+621133 (a $z = 1.013$, low-luminosity and extremely distant FR-I radio source which is resolved by the EVN into a core plus hot-spots associated with the larger scale jet), VLA J123642+621331 (a dust enshrouded, optically faint, $z = 4.424$ starburst system – Waddington et al. 1999) and the faintest detection, VLA J123646+621404 (a face-on spiral galaxy at $z = 0.96$ with a total EVN flux density of 180 μJy/beam). The diversity of optical type is interesting but the real surprise is the detection of a radio-loud AGN in the dust obscured, $z = 4.4$, starburst system. This argues that at least some fraction of the optically faint radio source population harbour hidden AGN (this may be similar to the same obscured population detected by Chandra). These AGN powered systems might be quite difficult to detect with SCUBA, assuming the dust temperatures are higher than that associated with pure star forming systems. In any case, the detection of this system highlights the use of VLBI as a powerful diagnostic – able to distinguish in principle (via brightness temperature arguments) between radio emission generated by nuclear starbursts and AGN.

5. Future Prospects for deep, wide-field VLBI studies

The EVN observations of the HDF suggest that deep, wide-field VLBI studies are not only possible, but in principle they can deliver important

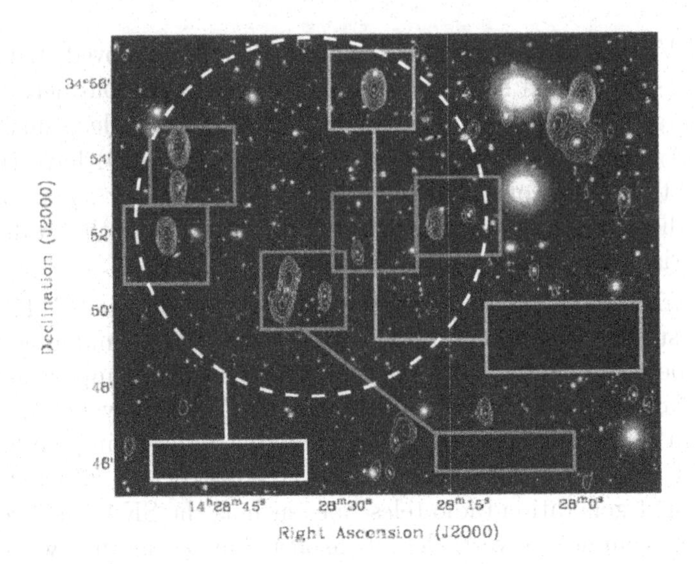

Figure 6. A small portion of the NOAO-DWFS with a shallow WSRT 1.4 GHz contour map superimposed (image courtesy of Raffaella Morganti, de Vries et al. 2001, first radio contour at 30μJy/beam). The dashed circle shows the FWHM of a 70-m telescope primary beam. Assuming modest r.m.s. noise levels of $\sim$ 10 μJy/beam, 5 (6σ) target fields (of extent $2' \times 2'$) can be identified. These fields are all located within the primary beam, and can be correlated simultaneously, and mapped out in their entirety. A (faint) VLBI calibrator to the North of the field permits accurate and continuous phase calibration to be applied to the target fields.

astronomical results. So far we have only scratched the surface. In a sense, we are just beginning to appreciate the fact, that VLBI has reached a sensitivity level where we can expect to detect many discrete radio sources in a single field of view (*irrespective* of where you point the telescopes!). This is quite a departure from the traditional role of isolated VLBI observations of very compact, and often very bright radio sources.

In the short term the use of *in-beam* phase-calibration techniques (Fomalont et al. 1999) should permit us to reach the expected thermal noise level of only a few microJy/beam (assuming a global VLBI array operating at 256 Mbits/sec and an on-source integration time of 24 hours). The real advance, however, will be in making full use of the raw data i.e to map out the primary beam response of individual VLBI elements in their entirety. Simultaneous multiple-field correlation, coupled with incredibly fast data output rates, is now being pursued at the EVN MkIV Data Processor at JIVE. When complete, this development will provide astronomers with the ability to image dozens of faint sub-mJy radio sources – all observed simultaneously with microJy sensitivity, full uv-coverage *and* milliarcsecond resolution (Garrett 2000b). Fig. 6 summarises the concept of deep, in-beam,

wide-field VLBI but with current sensitivity limits employed. Large areas of the sky (such as that shown in Fig. 6) are now being routinely surveyed in great detail by optical and near-IR instruments. These deep surveys (e.g. the NOAO Deep Wide-Field Survey, Januzi & Dey 1999) have the great advantage that they cover enormous areas of sky (many square degrees) and thus there is always some region of the survey area that will include an appropriate "in-beam" VLBI calibrator.

What fraction of these target sources can be detected ? The HDF-N results suggest that about 1/3 of the targets in a randomly selected field will be AGN, and that most of these will have compact structure. The remaining distant starburst systems will most likely be resolved by VLBI (even with microJy sensitivity) – detecting and imaging these systems with VLBI scale resolution, must await the construction of a much more sensitive, next generation radio telescope, such as the SKA (see Lecture by A. Kus, this volume). Nevertheless, I suspect that by the time we return to Castel St. Pietro Therme for the next NATO-ASI school, the use of VLBI as a deep, wide area survey instrument will already be well established.

References

Abraham, R.G. & van den Bergh, S. 2001, *Science*, **293**, p. 1273-1278.
Barger, A.J., Cowie, L.L. & Richards, E.A. 2000, *AJ*, **119**, 2092.
Brandt, W.N., Alexander, D.M., Hornsheimeier, A.E., et al. 2001, *AJ*, **122**, 2810.
Condon, J.J. 1992, *ARA&A*, **30**, 575
Chapman, S.C., Richards, E.A., Lewis, G.F., Wilson, G., & Barger, A.J. 2001, *ApJ*, **548**, L147.
de Vries, W.H., Morganti, R., Rottgering, H.J.A. et al. 2002, *AJ* submitted (astro-ph//0111543)
Ferguson, H.C., Dickinson, M.W, Williams, R. 2000, *ARA&A*, **38**, pp. 667-715.
Fomalont, E.B., Kellermann,K.I., Richards, E.A., Windhorst, R.A. & Patridge, R. B. 1997, *ApJ Letters*, **475**, L5.
Fomalont, E.B., Goss, W.M., Beasley, A.J. & Chatterjee, S. *AJ*, **117**, 3025.
Garrett, M.A., Porcas, R.W., Pedlar, A. et al. 1999, *NewAR*, **43**, 519 (astro-ph/9906108).
Garrett, M.A., de Bruyn, A.G., Giroletti, M., Baan, W.A., Schilizzi, R.T. 2000a, *A&A Letters*, **361**, L41 (astro-ph/0008509).
Garrett, M.A. 2000b, *Procs. of the VSOP Symposium*, Eds.: H. Hirabayashi, P.G. Edwards, and D.W. Murphy, Published by the ISAS, pp 269-272 (astro-ph/0003073).
Garrett, M.A., Muxlow, T.W.B., Garrington, S.T. et al. 2001, *A&A Letters*, **366**, L5 (astro-ph/0008509).
Garrett, M.A., 2002, *A&A Letters*, in press (astro-ph/0202116).
Haarsma, D.B., Partridge, R.B., Windhorst, R.A., Richards, E.A. 2000, *ApJ*, **544**, 641.
Hughes, D.H., Serjeant, S., Dunlop, J., Rowan-Robinson, M., et al. 1998, *Nature*, **394**, 241.
Januzi, B.T. and Dey, A. *BAAS*, **31**, 1392.
Muxlow, T.W.B., Wilkinson, P.N., Richards, A.M.S., et al. 1999, *NewAR*, **43**, 623.
Norris, R.P, Hopkins, A., Sault, R.J., et al. 2000, *Procs. of the ESO workshop – Deep Fields*, eds. Cristiani, Renzini & Williams, Springer pp 135-138.
Richards, E.A., Kellermann, K.I., Fomalont, E.B., et al. 1998, *AJ*, **116**, 1039.
Richards, E.A., Fomalont, E.B., Kellermann, K.I. et al. 1999, *ApJ*, **526**, L73.

Richards, E.A. 2000, *ApJ*, **533**, 611.
Rowan-Robinson, M., Mann, R.G., Oliver, S.J. et al. 1997, *MNRAS*, **289**, 490.
Waddington, I., Windhorst, R.A, Cohen, S.H., Partridge, R.B. Spinrad, H. & Stern, D. 1999, *ApJ Letters*, 526, L77.
Williams, R.E., Blacker, B, Dickinson, M. et al. 1996, *AJ*, **112**, 1335.

SPACE VLBI

SÁNDOR FREY
Institute of Geodesy, Cartography and Remote Sensing (FÖMI)
Satellite Geodetic Observatory
P.O. Box 585, H-1592 Budapest, Hungary

Abstract. Space VLBI (SVLBI) is an ambitious new technique, which, in the past couple of years, has proven its usefulness in high angular resolution imaging of bright compact radio sources. An overview of the principles of SVLBI, the extension of the VLBI technique with an orbiting interferometer element, is given. The first dedicated SVLBI satellite HALCA was launched as a part of the Japanese-led VLBI Space Observatory Programme (VSOP) in February 1997. The VSOP operations and the scientific program are reviewed. The mission's highlights are presented. Finally, the current plans for future SVLBI missions are briefly outlined.

1. Space VLBI: motivation, principles and brief history

For ground-based VLBI observations, the practical limit of the baseline length (about 10000 km) is set by the Earth diameter. Consequently this limits the angular resolution achievable at a given observing frequency. At 5 GHz, this angular resolution limit is about 1 milli-arcsecond (mas). However, many radio sources are known to be unresolved or only slightly resolved by ground-based VLBI observations.

A straightforward solution to increase the baseline lenghts by putting an interferometer element into Earth orbit has been proposed in the early days of VLBI. The feasibility of SVLBI was successfully demonstrated first using a communications satellite at 2.3 and 15 GHz in 1986–1988 ([25] and references therein). The practical realization of SVLBI, however, poses several technical difficulties. First of all, there is a striking difference between SVLBI and other space astronomy missions. The orbiting antenna is just an element of a VLBI network, thus an SVLBI mission requires an extensive support of globally distributed and inhomogeneous ground-based facilities (radio telescopes, tracking stations, orbit determination systems and VLBI

F. Mantovani and A. Kus (eds.), The Role of VLBI in Astrophysics, Astrometry and Geodesy, 349–358.

data processing centers). An important restriction is that the data taken by the orbiting element have to be down-linked to the Earth in real time, using a wide-band communications link to one of the (preferably many) satellite tracking ! stations. Also, the on-board electronics has to be synchronized from the ground to provide the frequency and timing accuracy necessary for later correlation of data. Unlike ground-based VLBI antennas, the position of the orbiting antenna is not known to high accuracy, and the position changes rapidly. This necessitates a special orbit determination system and poses serious additional requirements on the SVLBI correlators. The present-day space technology makes possible a launch into Earth orbit a 10-m class parabolic antenna with sufficient accuracy of its reflecting surface. However, such an aperture is much smaller than that of a typical ground-based radio telescope (25–100 m in diameter). This fact, together with the long baselines, limits the sensitivity of the space–ground interferometer. The image fidelity is somewhat degraded by the uneven spatial sampling, i.e. the lack of intermediate baseline lengths between ground-only and space–ground baselines. In many cases, the "holes" in the (u, v) coverage limit the practically achievable image dynamic range to $\sim 100:1$ [26].

The difficulties above explain why it took 30 years after the first successful ground-based VLBI experiments to build the first dedicated SVLBI mission. Recent and more detailed reviews of the realization of space–ground radio interferometry can be found in e.g. [12, 14, 40].

2. The VLBI Space Observatory Programme

The first satellite dedicated to SVLBI was launched in Japan by the Institute of Space and Astronautical Science (ISAS) on 12 February 1997 [15]. After the successful launch, the MUSES-B satellite developed as an orbiting element of the VLBI Space Observatory Programme (VSOP) was renamed to Highly Advanced Laboratory for Communications and Astronomy (HALCA). The HALCA satellite carries an on-board 8-m diameter antenna. Its eccentric orbit has an apogee height of ~ 21400 km and perigee height of ~ 560 km above the Earth's surface. In the standard observing mode, SVLBI data from two adjacent 16-MHz-wide base-band channels are digitized with 2-bit sampling and formatted. The digital data stream is then transmitted to a ground tracking station at the aggregate bit rate of 128 Mbit/s, in real time. The carrier frequency of this data down-link is 14.2 GHz. The observing frequency bands of VSOP are 1.60–1.73 GHz, 4.7–5.0 GHz and 21.9–22.3 GHz. The performance of the latter receiving system did not meet the specifications, due to too high attenuation between the main antenna and the low-noise amplifier [24]. Although the high system temperature made routine observing impossible at 22 GHz, the bursting Orion-KL water maser with flux density of $\sim 10^5$ Jy allowed the mission to detect fringes on baselines to HALCA in February 1998 [23].

More than 40 radio telescopes from all over the world take occasionally part in the VSOP observations, forming a truly global interferometric network. Five dedicated tracking stations (at Goldstone and Green Bank in the USA, Robledo in Spain, Tidbinbilla in Australia, and Usudsa in Japan) are used to establish the two-way communications link to HALCA to transmit a stable reference signal to the satellite and to receive and record the digitized astronomical data. Three VLBI data correlating facilities (at Mitaka in Japan, Penticton in Canada, and Socorro in the USA) participate in the program. More information on the tracking and orbit determination of HALCA, the co-observing ground radio telescopes and the correlators can be found in e.g. [18] and references therein. The nominal lifetime of HALCA was 3 years. However, at the time of writing this review, almost five years after the launch, the mission is continuing to conduct successful observations.

3. VSOP science program and selected results

Significant fraction of the in-orbit time is devoted to observations based on peer-reviewed proposals. Observing possibilities open for the entire scientific community are announced several times a year. Some of the mission's scientific highlights published to date in the proceedings of the COSPAR symposium on VSOP [19], and the VSOP Symposium [16], held in 1998 and 2000, respectively, and in other journals are briefly reviewed here.

3.1. KEY SOURCES

One of the key science programs determined by the VSOP mission prior to launch was to image relatively nearby, bright active galactic nuclei (AGNs). At low redshift, high angular resolution corresponds to high linear resolution. At 5 GHz, the close vicinity of the central engine in the elliptical galaxy M87 (the host for the bright radio source Virgo A) can be studied with a resolution corresponding to ~ 300 Schwarzschild radii of the central super-massive black hole. VSOP monitoring observations at 5 GHz revealed no proper motion in the inner jet over 1.5 years, as opposed to the apparent superluminal motion seen in the outer jet on VLA and HST scales [20].

The effect of improved angular resolution is spectacularly demonstrated in the case of the quasar 3C 273. The transverse structure of the compact jet at 5 GHz is resolved by VSOP observations. Emission profiles taken accross the jet show regular oscillating patterns, which can be understood in terms of Kelvin-Helmholtz instabilities developing and propagating in the relativistic plasma along the jet [28, 30].

3.2. POLARIZATION OBSERVATIONS

Polarization SVLBI observations offer uniquely high resolution imaging of compact polarization structures in AGNs. With HALCA, there are some difficulties such as the limited sensitivity compared to ground-based antennas, and the limited polarization calibration strategies. HALCA receives only left-circularly polarized radiation. However, with dual-polarization observations in the ground network, VSOP can provide polarization images for sources with sufficiently high correlated polarized flux density [22]. The parsec-scale linear polarization structures in a number of BL Lac objects have been studied. The source 1803+784 shows a smoothly bent jet structure, with the magnetic field transverse all along the jet [6]. The 5-GHz VSOP observation of OJ 287 was complemented with the 22-GHz VLBA data. The magnetic field in the core is transverse and becomes longitudinal in the inner jet. Further out from the core, the magnetic field orientation becomes transverse, presumably due to a shock. The nearly simultaneous observations at two different frequencies but with comparable angular resolution enabled the detection of $\sim 90°$ rotation in polarization position angle [7]. The comparison of the 5-GHz VSOP and 15-GHz VLBA observations of 0735+178 with similar angular resolution revealed a good agreement in polarization structures, but a different jet path at the two frequencies. This can be interpreted as an evidence for free-free absorption in the jet [8].

3.3. PHASE-REFERENCE OBSERVATIONS

The phase-reference technique is based on rapidly alternating (or simultaneous) observations of two sources close in direction. The visibility phase fluctuations of atmospheric origin are nearly the same for both sources. If one of the sources is bright and compact enough, this method can be used to image the other source which may otherwise be too weak for self-calibration. On the other hand, the visibility phase difference can be used for relative astrometry at high precision (see J.M. Marcaide's lecture on phase-referencing in this volume). Although HALCA, due to technical restrictions, cannot switch rapidly between different sources, phase-reference observations are possible with two objects close enough to lie within the primary beam of the 8-m antenna on board [33]. The $4\farcm8$ angular separation of the quasar pair 1342+662 and 1342+663 is extremely favorable for VSOP phase-reference test observations, since the sources are within the primary beam of HALCA and the VLBA antennas. The analysis of phase-reference images of the sources suggest that the HALCA orbit uncertainty was not larger than ~ 3 m at the time of the observations [9]. The improved orbit determination accuracy of the next-generation SVLBI satellites will facilitate phase-reference observations, and thus provide a tool for astrometry as well as enhance the imaging sensitivity.

3.4. SPECTRAL INDEX MAPPING

VSOP provides a unique tool for high resolution radio spectral index mapping of AGNs. The angular resolution at 1.6 and 5 GHz can be matched with ground-only VLBI observations at ~ 3 times higher observing frequencies. The spectral index map of the γ-ray blazar 3C 279, one of the key VSOP target sources, shows highly inverted spectrum of the core between 1.6 and 5 GHz [32]. In the case of the X-ray and TeV γ-ray source Mkn 501, the spectral index map also reveals the flat-spectrum core and the steep-spectrum inner jet. There is also significant spectral variation across the complex jet. Surrounding the inner jet, there are lower surface brightness regions with unexpectedly flatter spectra, possibly due to processes in a slower outer shear layer [2]. The central 10-pc region of NGC 6251, a nearby radio galaxy was studied with VSOP at 5 GHz and with the VLBA at 15 GHz, at two close epochs. A sub-parsec-scale counterjet was discovered based on i! ts optically thin spectrum derived from the spectral index image [37].

3.5. EXTREMELY HIGH REDSHIFT QUASARS

Among the most distant ($z > 3$) radio-loud quasars, about 20 are bright enough to be imaged with VSOP. Due to the extremely high redhifts, mas-scale radio structures at emitted frequencies up to ~ 30 GHz can be studied. For steep-spectrum jet components, the high resolution of SVLBI is unique and cannot be supplemented with ground-based VLBI observations at higher frequencies [10]. The quasar 2215+020 ($z = 3.57$) shows an unexpectedly prominent jet structute at 1.6 GHz, out to ~ 80 mas angular distance from the core. One of the jet components seen in the ground-only VLBI image is resolved out in the VSOP image, leading to an estimate of the transverse dimension of the jet, and the mass of the central black hole (4×10^9 solar mass) [29]. Based on the VSOP image taken at 5 GHz observing frequency, the jet in the quasar 1351$-$018 ($z = 3.71$) bends sharply ($> 120°$) between the sub-mas and 10 mas scale [3]. Such jet misalignments are f! ound in other high-redshift sources as well.

3.6. HIGH BRIGHTNESS TEMPERATURE SOURCES

Equipartition and inverse Compton theories limit the brightness temperature ($T_{\rm B}$) of incoherent synchrotron sources to $\sim 10^{11} - 10^{12}$ K [34, 21]. Curiously, the ability of ground-based VLBI for diretcly measuring radio source brightness temperatures is also constrained to this limit. The improved resolution of SVLBI makes this technique a unique tool to directly measure $T_{\rm B}$ values apparently violating the theoretical limits. Brightness temperatures in excess of 10^{12} K have often been measured with VSOP for individual AGNs, e.g. NRAO 530 [1] and 1921−293 [35]. This indicates that Doppler boosting of the radiation, which enhances the apparent $T_{\rm B}$, is indeed a common phenomenon in the cores of radio-loud AGNs. The highest rest-frame brightness temperature measured to date with VSOP is $T_{\rm B} > 6 \times 10^{13}$ K for the compact, violently variable BL Lac object 0235+164 [4].

3.7. THE PEARSON-READHEAD SURVEY FROM SPACE

A 27-element sub-sample of the well-studied Pearson-Readhead survey [31] sources was imaged with high angular resolution and high dynamic range using VSOP [26]. New information on the parsec-scale radio stucture of this statistically complete core-selected sample of AGNs, together with extensive data taken at various wavelengths allowed studies of the relativistic beaming effects in compact radio sources. Many previously known correlations between different source properties have been confirmed. Several new trends that support the beaming model have been discovered [27]. A significant fraction of the sources studied have brightness temperature in excess of 10^{12} K. A relationship between the brightness tempearuture and the intra-day variability type has also been found [39].

3.8. THE VSOP SURVEY PROGRAM

About a quarter of the HALCA observing time is used by the mission-led VSOP Survey Program [17]. The major goal of the Survey is to collect data on sub-mas scale structural properties of a complete flux density-limited sample of 402 extragalactic radio sources at 5 GHz. The program was designed to observe using typically 3–4 ground-based radio telescopes during ~ 1 orbital period of HALCA, and thus to make effective use of the in-orbit time when less ground resources are available. The Survey observations and the data reduction are still in progress. One of the most important preliminary results is that the brightness temperature distribution of the sources shows a clear tail well above 10^{12} K [17]. When complete, the public Survey data base will most likely facilitate studies that lead to some of the major scientific achievements of VSOP.

3.9. HYDROXIL MASER OBSERVATIONS

Angular resolution can only be improved by means of SVLBI observations in the case of maser sources that radiate at well defined frequencies. The VSOP observations of OH34.26+0.15 maser in two main-line hydroxil transitions at 1665 and 1667 MHz led to images of individual maser spots at ~ 1 mas resolution. The maser spots were only partially resolved even on the long ground-to-space baselines [36].

3.10. VSOP OBSERVATIONS OF STELLAR OBJECTS

Due mainly to the limited sensitivity, only a few stellar studies have been done with VSOP. A radio outburst of the X-ray binary star LSI+61°303 was observed at 5 GHz at multiple epochs. The images reveal a stationary-looking, symmetric emission structure out to 2 mas on both sides of the central source. An expansion velocity of ~ 700 km s^{-1} was measured for the central source [38]. The speckle pattern caused by the interstellar scattering of the Vela pulsar was observed with VSOP, which could eventually allow imaging the emission region with hundreds of km linear resolution [11].

3.11. GEODETIC APPLICATION OF SVLBI

A possible geodetic application of SVLBI has also been investigated. While the technique in principle would be useful for e.g. directly connecting celestial and terrestrial reference frames, the limited bandwidth and satellite re-pointing capability, and a number of other restrictions prevent to obtain accurate results [5].

4. Future Space VLBI missions

The success of VSOP, in both technical and scientific senses, provides a firm foundation for planning next generation SVLBI missions. One of the key issues is certainly the increase of sensitivity, which would enable to study many more sources and astrophysical phenomena. In principle, this can be achieved in number of ways, by e.g. increasing the antenna collecting area, the down-link data rate, or reducing the receiver system temperature. The planned increase in sensitivity with respect to VSOP is a factor of ~ 10 for the VSOP-2 mission being considered at ISAS (Japan) [13]. Furthermore, one technically challenging but scientifically promising possibility is to make VSOP-2 able to conduct phase-reference observations routinely. A general trend is to increase the observing frequencies in the planned SVLBI missions, up to 43 GHz and 86 GHz in the case of VSOP-2 and ARISE [41], respectively. The latter mission is being proposed to NASA in the USA, to carry a 25-m inflatable antenna. This increase in frequency implies improvement in the angular resolution (down to $\sim 8\mu$as for ARISE at the highest observing frequency of 86 GHz).

There is a real chance to meet the technical challenges (to build a large, accurate, deployable antenna and advanced receiver cryogenics, to develope high bit-rate communications and precise orbit determination methods), and to launch at least one next-generation SVLBI satellite within ten years from now. For a recent review of the future SVLBI missions see [14] and references therein.

Acknowledgements

The VSOP Project is led by the Institute of Space and Astronautical Science (Japan) in cooperation with many organizations and radio telescopes around the world. This work was partially supported by the Netherlands Organization for Scientific Research (NWO), the Hungarian Scientific Research Fund (OTKA, grant. no. N31721 and T031723) and the Hungarian Space Office. The author wishes to thank Leonid Gurvits, Zsolt Paragi and István Fejes for valuable comments on the manuscript.

References

1. Bower, G.C. and Backer D.C. (1998) Space VLBI Observations Show $T_B > 10^{12}$ K in the Quasar NRAO 530, *Astrophys. J.*, **507**, L117–L120
2. Edwards, P.G., Giovannini, G., Cotton, W.D., Feretti, L., Fujisawa, K., Hirabayashi, H., Lara, L. and Venturi, T. (2000) A Spectral Index Map from VSOP Observations of Markarian 501, *Publ. Astron. Soc. Japan*, **52**, 1015–1019
3. Frey, S. (2000) VLBI Studies of Extremely Distant Quasars, PhD Thesis, Eötvös University, Budapest
4. Frey, S., Gurvits, L.I., Altschuler, D.R., Davis, M.M., Perillat, P., Salter, C., Aller, H.D., Aller, M.F. and Hirabayashi H. (2000) Dual-Frequency VSOP Observations of AO 0235+164, *Publ. Astron. Soc. Japan*, **52**, 975–982
5. Frey, S., Meyer U., Fejes I., Paragi Z., Charlot P. and Biancale R. (2002) Geodetic space VLBI: the first test observations, *Adv. Space Res.*, in press
6. Gabuzda, D.C. (1999) VSOP observations of the compact BL Lacertae object 1803+784, *New Astron. Rev.*, **43**, 691-694
7. Gabuzda, D.C. and Gómez, J.L. (2001) VSOP polarization observations of the BL Lacertae object OJ 287, *Mon. Not. R. Astron. Soc.*, **320**, L49–L54
8. Gabuzda, D.C., Gómez, J.L. and Agudo, I. (2001) Evidence for parsec-scale absorption from VSOP observations of the BL Lacertae object 0735+178, *Mon. Not. R. Astron. Soc.*, **328**, 719–725
9. Guirado, J.C., Ros, E., Jones, D.L., Lestrade, J.-F., Marcaide, J.M., Pérez-Torres, M.A. and Preston, R.A. (2001) Space-VLBI phase-reference mapping and astrometry, *Astron. Astrophys.*, **371**, 766–770
10. Gurvits, L.I (2000) Why Space VLBI is of Special Value for Studies of High-Redshift Radio Sources, *APRSV* [16], 151–154

11. Gwinn, C.R., Reynolds, J.E., Jauncey, D.L., Tzioumis, A.K., Carlson, B., Dougherty, S.M., Del Rizzo, D., Hirabayashi, H., et al. (2000) *APRSV* [16], 117–120
12. Hirabayashi, H. (1999) Space-to-Ground Interferometry for Radio Astronomy, *Modern Radio Science 1999*, ed. Stuchly, M.A., Oxford University Press, 51–62
13. Hirabayashi, H. (2000) The Japanese Space VLBI Mission After VSOP, *Adv. Space Res.*, **26**, 751–756
14. Hirabayashi, H. (2001) Space VLBI, *Galaxies and Their Constituents at the Highest Angular Resolutions, IAU Symp. 205*, eds. Schilizzi, R.T., Vogel, S., Paresce, F. and Elvis, M., Kluwer, Dordrecht, 422–427
15. Hirabayashi, H., Hirosawa, H., Kobayashi, H., Murata, Y., Edwards, P.G., Fomalont, E.B., Fujisawa, K., Ichikawa T., et al. (1998) Overview and Initial Results of the Very Long Baseline Interferometry Space Observatory Programme, *Science*, **281**, 1825–1829
16. Hirabayashi, H., Edwards, P.G. and Murphy, D.W. (eds.) (2000) *Astrophysical Phenomena Revealed by Space VLBI*, Institute of Space and Astronautical Science, Sagamihara, Japan [*APRSV*]
17. Hirabayashi, H., Fomalont, E.B., Horiuchi, S., Lovell, J.E.J., Moellenbrock, G.A., Inoue, M., Burke, B.F., Dewdney, P.E., et al. (2000) The VSOP 5 GHz AGN Survey I. Compilation and Observations, *Publ. Astron. Soc. Japan*, **52**, 997–1014
18. Hirabayashi, H., Hirosawa, H., Kobayashi, H., Murata, Y., Asaki, Y., Avruch, I.M., Edwards, P.G., Fomalont, E.B., et al. (2000) The VLBI Space Observatory Programme and Radio-Astronomical Satellite HALCA, *Publ. Astron. Soc. Japan*, **52**, 955–965
19. Hirabayashi, H., Preston, R.A. and Gurvits, L.I. (eds.) (2000) *VSOP Results and the Future of Space VLBI, Adv. Space Res.*, **26**, No. 4
20. Junor, W., Biretta, J.A., Owen, F.N. and Begelmann, M.C. (2000) Multi-Epoch Global+VSOP/HALCA Observations of Virgo A at λ6 cm, *APRSV* [16], 13–16
21. Kellermann, K.I. and Pauliny-Toth, I.I.K. (1969) The Spectra of Opaque Radio Sources *Astrophys. J.*, **155**, L71–78
22. Kemball, A., Flatters, C., Gabuzda, D., Moellenbrock, G., Edwards, P., Fomalont, E., Hirabayashi, H., Horiuchi, S., et al. (2000) VSOP Polarization Observing at 1.6 GHz and 5 GHz, *Publ. Astron. Soc. Japan*, **52**, 1055–1066
23. Kobayashi, H., Shimoikura, T., Omodaka, T. and Diamond, P.J. (2000) Monitoring of the Orion-KL Water Maser Outburst, *APRSV* [16], 109–112
24. Kobayashi, H., Wajima, K., Hirabayashi, H., Murata, Y., Kawaguchi, N., Kameno, S., Shibata, K.M., Fujisawa, K., Inoue, M. and Hirosawa, H. (2000) HALCA's Onboard VLBI Observing System, *Publ. Astron. Soc. Japan*, **52**, 967–973
25. Linfield, R.P., Levy, G.S., Edwards, C.D., Ulvestad, J.S., Ottenhoff, C.H., Hirabayashi, H., Morimoto, M., Inoue, M., et al. (1990) 15 GHz Space VLBI Observations Using an Antenna on a TDRSS Satellite, *Astrophys. J.*, **358**, 350–358
26. Lister, M.L., Tingay, S.J., Murphy, D.W., Piner, B.G., Jones, D.L. and Preston, A.R. (2001) The Pearson-Readhead Survey of Compact Extragalactic Radio Sources from Space. I. The Images, *Astrophys. J.*, **554**, 948–963
27. Lister, M.L., Tingay, S.J. and Preston, A.R. (2001) The Pearson-Readhead Survey of Compact Extragalactic Radio Sources from Space. II. Analysis of Source Properties, *Astrophys. J.*, **554**, 964–980
28. Lobanov, A.P., Zensus, J.A., Krichbaum, T.P. and Witzel A. (2000) Space VLBI of Parsec-Scale Jets: the Impact of VSOP, *APRSV* [16], 239–244
29. Lobanov, A.P., Gurvits, L.I., Frey, S., Schilizzi, R.T., Kawaguchi, N. and Pauliny-Toth, I.I.K. (2001) VLBI Space Observatory Programme Observation of the Quasar PKS 2215+020: A New Laboratory for Core-Jet Physics at z=3.572, *Astrophys. J.*, **547**, 714–721
30. Lobanov, A.P. and Zensus, J.A. (2001) A Cosmic Double Helix in the Archetypical Quasar 3C273, *Science*, **294**, 128–131
31. Pearson, T.J. and Readhead, A.C.S. (1988) The milliarcsecond structure of a com-

plete sample of radio sources. II - First-epoch maps at 5 GHz, *Astrophys. J.*, **328**, 114–142
32. Piner, B.G., Edwards, P.G., Wehrle, A.E., Hirabayashi, H., Lovell, J.E.J. and Unwin, S. C. (2000) Space VLBI Observations of 3C 279 at 1.6 and 5 GHz, *Astrophys. J.*, **537**, 91–100
33. Porcas, R.W., Rioja, M.J., Machalski, J. and Hirabayashi, H. (2000) Phase-Reference Observations with VSOP, *APRSV* [16], 245–252
34. Readhead, A.C.S. (1994) Equipartition brightness temperature and the inverse Compton catastrophe, *Astrophys. J.*, **426**, 51–59
35. Shen, Z.-Q., Edwards, P.G., Lovell, J.E.J., Fujisawa, K., Kameno, S. and Inoue, M. (1999) High-Resolution VSOP Imaging of the Southern Blazar PKS 1921-293 at 1.6 GHz, *Publ. Astron. Soc. Japan*, **51**, 513–518
36. Slysh, V.I., Voronkov, M.A., Migenes, V., Shibata, K.M., Umemoto, T., Altunin, V.I., Val'tts, I.E., Kanevsky, B.Z., et al. (2001) Space-VLBI observations of the OH maser OH34.26+0.15: low interstellar scattering, *Mon. Not. R. Astron. Soc.*, **320**, 217–223
37. Sudou, H., Taniguchi, Y., Ohyama, Y., Kameno, S., Sawada-Satoh, S., Inoue, M., Kaburaki, O. and Sasao, T. (2000) Sub-Parsec-Scale Acceleration of the Radio Jet in the Powerful Radio Galaxy NGC 6251, *Publ. Astron. Soc. Japan*, **52**, 989–995
38. Taylor, A.R., Dougherty, S.M., Scott, W.K., Peracaula, M. and Paredes, J.M. (2000) VSOP Imaging of the Unusual X-Ray Binary Star LSI+61°303, *APRSV* [16], 223–226
39. Tingay, S.J., Preston, R.A., Lister, M.L., Piner, B.G., Murphy, D.W., Jones, D.L., Meier, D.L., Pearson, T.J., et al. (2001) Measuring the Brightness Temperature Distribution of Extragalactic Radio Sources with Space VLBI, *Astrophys. J.*, **549**, L55–L58
40. Ulvestad, J.S. (1999) Space Very Long Baseline Interferometry, *Synthesis Imaging in Radio Astronomy II*, eds. Taylor, G.B., Carilli, C.L. and Perley R.A., ASP Conference Series **180**, 513–536
41. Ulvestad, J.S. (2000) The ARISE Space VLBI Mission, *Adv. Space Res.*, **26**, 735–738

VLBI FOR GEODESY AND GEODYNAMICS

JAMES CAMPBELL
Geodetic Institute of the University of Bonn
Nussallee 17, D-53115 Bonn, Germany

Abstract.
In geodetic and geodynamic applications, Very Long Baseline Interferometry is first and foremost appreciated for its capability to provide direct geometrical ties to the quasi-inertial system of extragalactic radio sources. Geodetic VLBI can measure Earth rotation and orientation as well as precise station positions and their velocities without involving the gravity field of the Earth. In the broader context of Earth observation and the monitoring of geodynamic processes, these and many more unique features have allowed the VLBI-technique to achieve pioneering feats such as the determination of present-day plate tectonic motions, post glacial rebound, subdaily Earth rotation variations and parameters of general relativity. The principal elements of the geodetic approach to VLBI as well as its relation to other space techniques will be at the center of this chapter.

1. Introduction

To astronomers, a radio interferometer is an instrument to observe the sky, just as much as an optical telescope, and the Very Long Baseline Interferometer is just a very large telescope for observing radio signals. For geodesists (and other geoscientists) the interferometer is an instrument for measuring distances to a very high level of accuracy. The elements of a Very Long Baseline Interferometer form a three- dimensional geometric structure composed of baseline vectors. If we imagine the wavefronts from distant radio sources passing through this structure it is quite straightforward to imagine that the observed time delays between the contacts of the wavefronts at the vertices of the structure can be used to determine the geometry of the interferometer (Fig. 1).

F. Mantovani and A. Kus (eds.), The Role of VLBI in Astrophysics, Astrometry and Geodesy, 359–381.

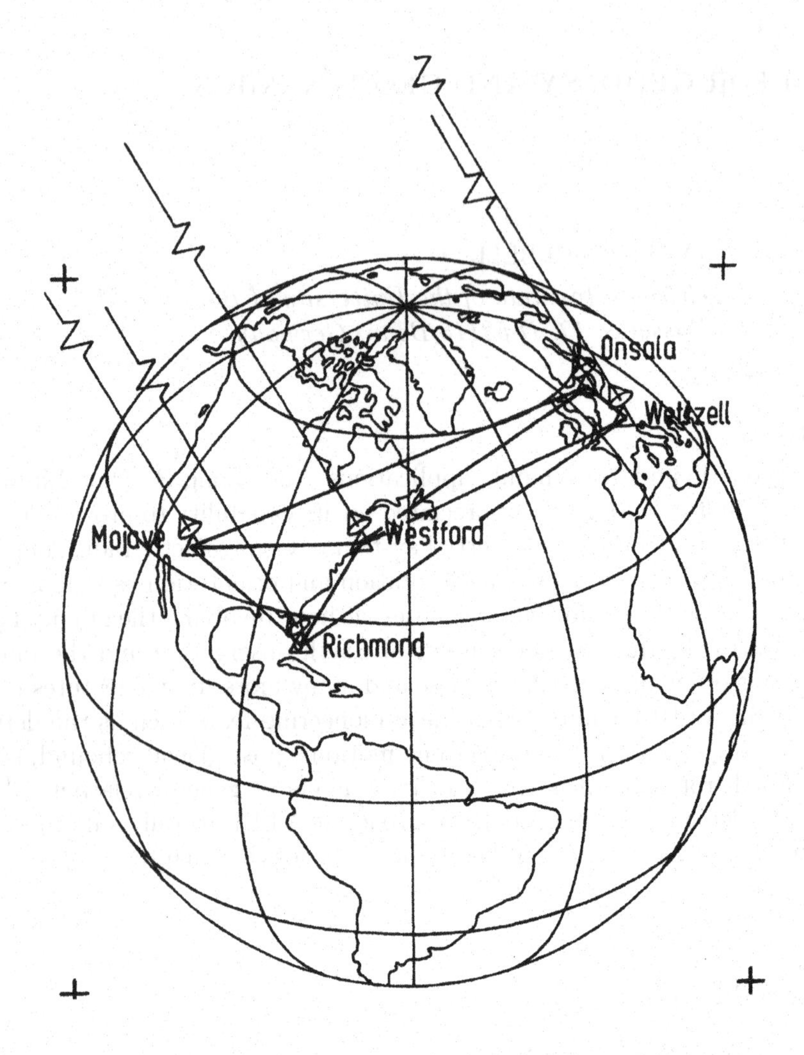

Figure 1. A Very Long Baseline Interferometer attached to the revolving Earth

The enormous potential of the VLBI-technique for geodetic baseline measurements has been recognised already in the early years of its development (Shapiro, Knight, 1970). With the invention of the bandwidth synthesis technique (Rogers, 1970), the accuracy of group delay measurements could be increased to a level of 30 picoseconds, i.e. 1 centimeter in the late seventies. The continued development of the MkIII data acquisiton an correlation system and its Canadian and Japanese counterparts, the S2 and K3 systems, has led to a new generation of extremely powerful tools for high precision geodetic VLBI (Clark et al., 1985; Cannon et al., 1997; Koyama et al., 1998). In this situation, the development of the MkIV system as a natural follow-up of the MkIII and MkIII-A systems is a consistent and consequent step to maintain as well as improve the operational capabilities of VLBI (Whitney, 2000a).

A look at one of the recent fringe plots obtained at the Bonn copy of the new MkIV correlator illustrates the state of development (Fig.2). A newly designed coloured version now replaces the 25-year old alphanumeric fringe plot (Whitney et al., 1976), that has served generations to find fringes and sort out problems. The 8-channel-700MHz-multiband delay function shown in blue in the upper plot has a main peak halfwidth of 1.4 ns, which at an SNR of 108.2 (4C39.25 is a strong source) leads to a group delay resolution of 5.8 ps (first of the bottom lines) or 1.5 millimeter.

Further developments beyond the MkIVsystem include new concepts for affordable, yet high data rate VLBI recording and playback systems, such as the MkV system which uses a set of hard disks (Whitney, 2000b) and the realization of a VLBI Standard Interface (VSI) that will allow the data from different systems, such as the Japanese K4 and the Canadian S2 to be processed together at the same correlator(Whitney, 2000a). There are of course hopes that in the more distant future direct transmission of RF-data via fibre links to the correlator will become affordable.

2. Geodetic VLBI observables

The geodetic and geophysical interest in VLBI is based on the use of an inertial reference frame formed by a given set of extremely compact extragalactic radio sources. VLBI measures very accurately the angles between the Earth-fixed baseline vectors and these space-fixed radio sources. Thus, even the most subtle changes in the baseline vectors and in the angles between the reference systems can be detected. The main geodynamical phenomena such as polar motion, UT1-variations, nutation and precession, Earth tides and tectonic plate motions can be monitored with unprecedented accuracy.

The first VLBI experiments that were explicitly aimed at achieving geodetic accuracy on long baselines were conducted by the Haystack/MIT group on the 845km baseline between the Haystack Observatory in the north of Massachusetts and the National Radio Astronomy Observatory of Greenbank, West Virginia (Hinteregger et al., 1972). The key to the high group delay resolution of about 1ns attained in these experiments was the invention of the so-called bandwidth synthesis technique (Rogers, 1970), which helped to overcome the limitations of tape recording equipment in terms of bandwidth.

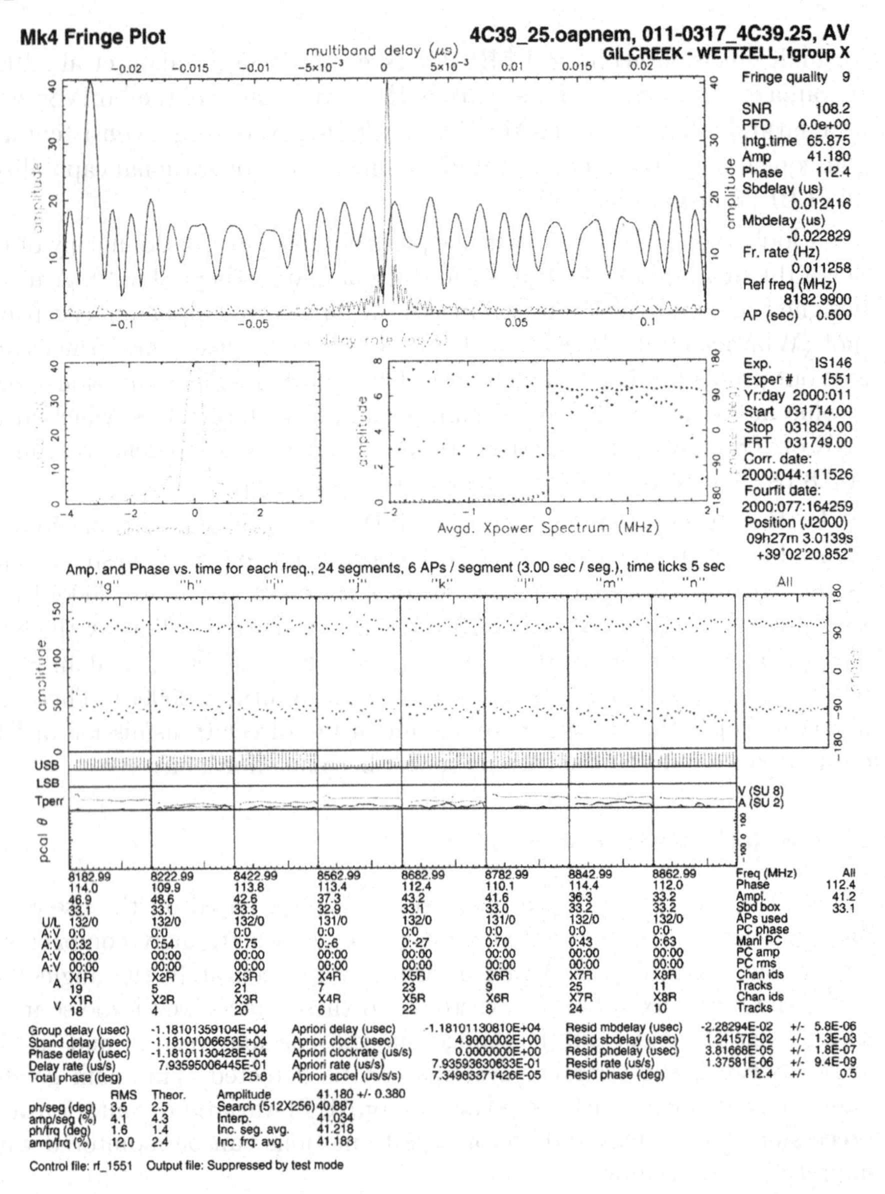

Figure 2. Fringe plot generated by the fringe processing software at he correlator of the new MkIV VLBI system

Considering the successful use of the phase observable in GPS, one could wonder why it is so important to strive for high bandwidth. After all, the fringe phase is among the VLBI observables that are derived from the correlator output. One 2-MHz-channel would be enough to detect and measure the phase of a source (observing strong sources with large dishes and cooled receivers). Two channels in two different bands would take care of the ionospheric phase delay, so one would just have to follow the GPS-recipe and solve the ambiguities by forming double differences.

Obviously this is not the way taken in geodetic VLBI. One important reason is that sources in different parts of the sky cannot be observed simultaneously. Radiotelescopes have to be steered and pointed at the sources one after the other to collect enough signal strength to permit their detection during correlation. During telescope slewing, several instrumental and environmental effects have ample time to destroy the phase coherence between scans.

This situation evidences an intrinsic division between VLBI and GPS: no double differencing means no elimination of timing errors between receiving stations. Therefore VLBI has to rely on 'error-free' time keepers at the stations to bind the sequential observations of the radio sources together. The tempting idea to use omnidirectional antennas at the VLBI stations and forget about H-maser oscillators has to remain a dream because quasars are at least 6 orders of magnitude weaker than the signals emitted by the GPS-satellites (MacDoran et al., 1982).

Thus several factors are combining to obstruct the use of the phase observable in geodetic VLBI (Herring, 1992, Petrov, 1999). In any case, the natural ultra wide band continuum radiation is offered free of charge by the majority of the compact radio sources and provides the means to use the essentially unambiguous wide band group delay as the prime geodetic VLBI observable.

The group delay resolution is proportional to the inverse of the SNR and the spanned bandwidth (Rogers, 1970). So, if we increase the spanned bandwidth at a given SNR by a factor of ten, the group delay uncertainty will be reduced by the same factor, a relation with tremendous potential. There are virtually no limitations to further improve the geodetic group delay performance, except (evolving) technological constraints and costs. Phase delay, on the other hand, is still an issue of research and does provide very high accuracy on very short baselines (Herring, 1992; Hase, Petrov, 1999).

3. Models for geodetic VLBI data analysis

The most commonly used way to extract significant and meaningful information from nature's ongoing processes is to try to model these processes by applying the laws of physiscs to the best of our knowledge. The less well known parts of our model will be equipped with a set of well seasoned unknown parameters. The least squares adjustment will then reveal the degree of correspondence achieved between theory and observation.

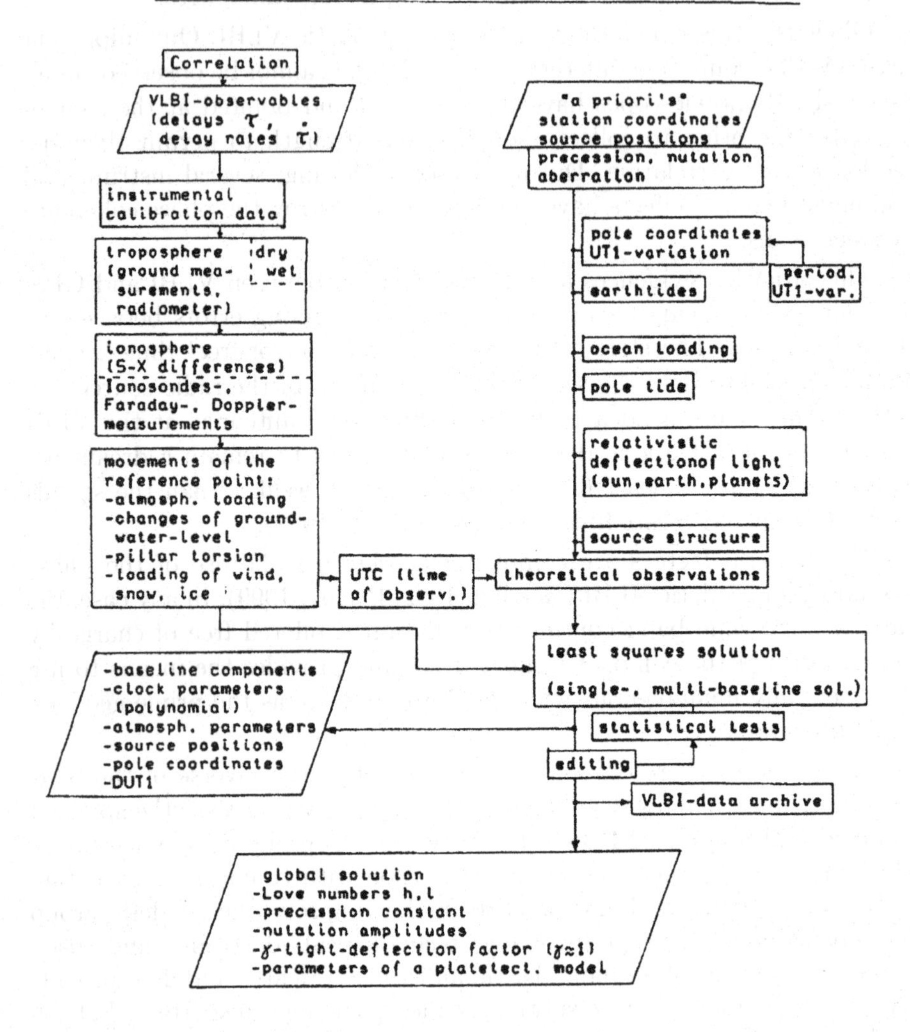

Figure 3. Flow diagram of geodetic VLBI data analysis

In VLBI, just like most other measuring techniques, the data analysis system has two main branches (Fig. 3), one that takes in the raw observations and provides a number of instrumental and environmental corrections and the other one that produces the 'model observations' or 'theoreticals'.

The fundamental geometric model of the time delay forms the heart of the system. This model has evolved from its simple initial form (Fig. 4) in a geocentric system to the highly complex relativistic formulation in the solar system barycenter (Preuss and Campbell, 1992; Sovers et al., 1998). The complete formulation includes both the effects of special relativity (part of what is known as aberration in spherical astronomy) and of general relativity, which describes the curvature of space-time. The effect of gravity on the propagation of electromagnetic waves is considerable: even at an angle of 180° away from the sun the differential delay effect (for a 6000 km baseline) is still 0.4 ns, a non-negligible quantity. VLBI observations have been used to verify Einstein's theory (PPN-formulation) to an accuracy of 0.1% (Robertson and Carter, 1984; Preuss and Campbell, 1992).

In principle, the description of the Earth's orientation with respect to the celestial system (precession, nutation) and the motion of the Earth's axis with respect to the crust (polar motion)(Fig. 5) has to reach the same level of accuracy as all the other model components, which means roughly better than one milliarcsecond. The same holds for the rotational speed of the Earth about its axis: to compute the phase angle of the Earth's rotation at any epoch to better than 1 mas, the UT1-variations have to be known to better than 0.1 ms of time. Of course, our understanding of the origin of all these variations is still far behind these levels of accuracy and this is why at present we still have to regard all angles involved in the transformation between the terrestrial and the celestial systems as unknown parameters in our solution (Herring et al., 1986b; Herring, 2000). It is due to the intrinsic strength of the geodetic VLBI observing strategy that we are still able to solve the system of observation equations, without ending up in singularity.

Already in the early seventies the periodic deformations of the Earth's crust could be seen in the VLBI observations: solid Earth tides cause diurnal and semidiurnal oscillations with vertial amplitudes of about 40 cm and horizontal displacements of up to 10% of the vertical effect. Although good models are available, the relevant parameters can be estimated from larger sets of data (Herring et al., 1983). More difficult to model are the tidal loadig effects of the oceans, which amount to as much as a decimeter on some coastal or island sites (Scherneck, 1991).

The fact that the VLBI stations are tied to the solid crust reveals itself helas as a rather deceptive assumption. Apart from the periodic convulsions of the Earth there are all sorts of aperiodic motions, the most prominent of which are the horizontal and vertical motions associated with plate tectonics (Minster, Jordan, 1978). The obvious problem that arises for the definition of a terrestrial reference frame is akin to the problem of proper motions in the optical celestial reference frames: how do we fix the origin? Here we have to resort to the concept of a priori constraints, e.g. the 'no net(to) rotation'(NNR), 'no net translation' (NNT) constraints (Argus and Gordon 1996), but rigorously speaking there is no solution to this problem if

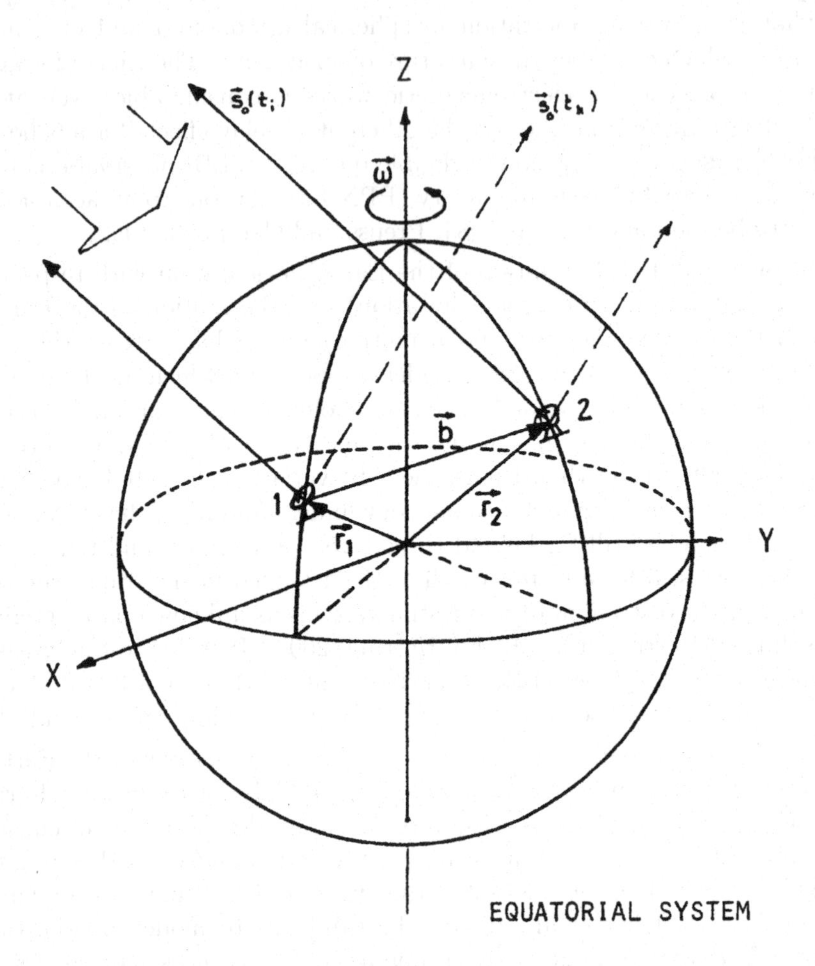

Figure 4. Single baseline interferometer in its basic form

we have different sets of defining stations in the global networks (Altamimi 2000, Altamimi et al., 2001)).

In global solutions with large data sets precise source positions can be determined simultaneously with the other parameters (Ma et al., 1993). The accuracy of the celestial reference frame may now be estimated to be around 0.3 arcsec on short as well as on longer time scales, although individual sources show greater variations (Feissel and Gontier, 2000). The

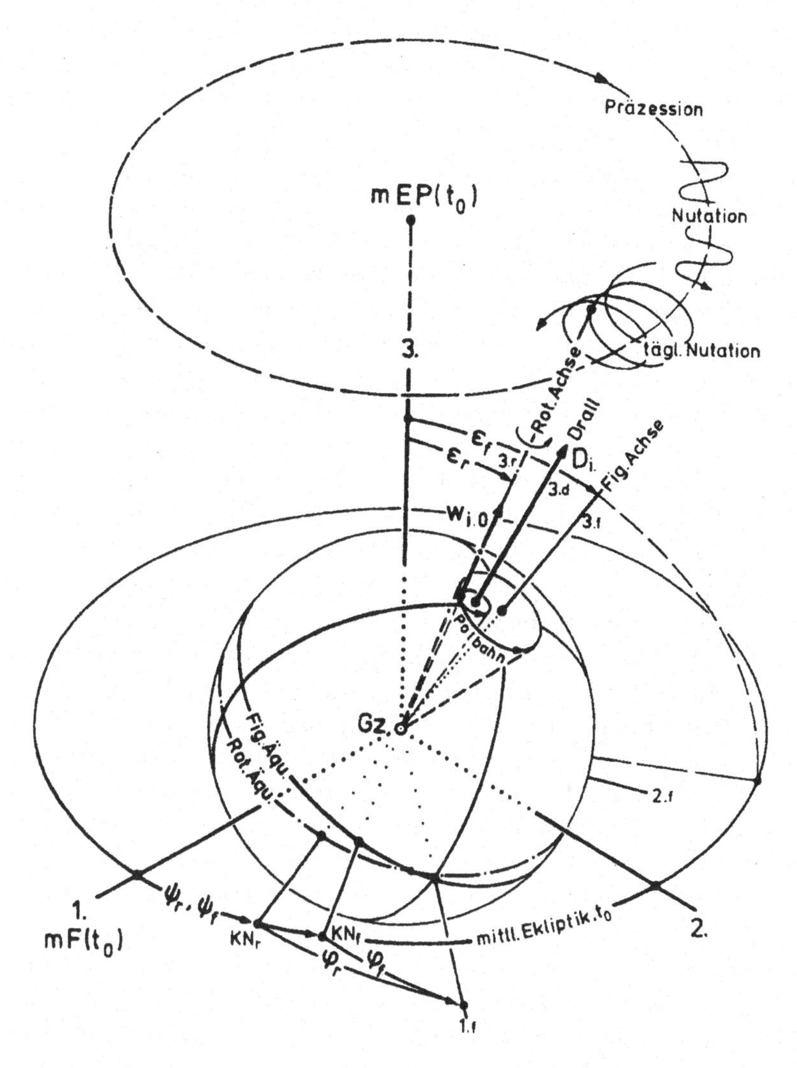

Figure 5. The Earths rotation in the Earth-fixed and space-fixed reference systems (Heitz, 1976)

physical nature of quasars is still under debate, although models have been developed that are able to explain several of the observed features, such as the core-jet structure. For the geodesist the fact remains that most of the observed compact sources are indeed showing structure at the level of several mas. This effect, in particular any changes in the structure, poses a limit on the accuracy of the radio-reference system. However, permanent monitoring of the structure, which is accomplished in part by the same VLBI data, can be done in parallel to the geodetic analysis, thus providing a means to correct for the structure effects (Fey et al., 2000). To maintain the extragalactic reference frame as the ultimate representative of the Celestial Reference System (CTS) remains one of the major challenges of VLBI in the future (Walter and Sovers, 2000).

The systematic instrumental effects include clock instabilities, electronic delays in cables and circuitry and deformations of the telescope structure. Usually as a clock model a second order polynomial is introduced and occasionally a break has to be allowed for. Clock modelling is still very much an interactive procedure and belongs to the editing session. The instrumental delay changes are, or at least should be, monitored by the phase and delay calibration system, which is part of the MkIII-system (Petrov 2000). In the telescope the distance between the feed horn and the axis intersection should be constant; in this case it becomes part of the clock offset parameter (Sovers et al., 1998). Large telescopes such as the Effelsberg 100m antenna exhibit direction dependent changes that have to be measured by local geodetic surveying techniques (Nothnagel, 1999).

The effect of the atmosphere on VLBI-observations is considered to be the most serious problem, because at widely separated stations the elevations of the telescopes during a scan differ greatly as well as the meteorological conditions themselves (Mathur et al., 1970). But while the ionosphere can be readily eliminated to first order by using two different observing frequencies, the neutral atmosphere, essentially the troposphere, presents the same problems in VLBI as in GPS observations. Its influence on radio signals adds up to an extra zenital path of 1.8 to 2.5 meters. The contribution of the dry part is rather stable, although care has to be taken to choose a proper mapping function for the lower elevations (Davis et al. 1985, Niell, 2000).

In more recent analyses with a large number of observing sessions, it could be shown that there is also an azimuthal asymmetry in the refractivity of the tropospheric layers around each station. This is in a way a rediscovery of the inclination of refractive layers experienced by optical observers in positional astronomy (latitude and longitude observation with astrolabes and theodolites). The effect can be absorbed by expanding the tropospheric model adding corresponding parameters (MacMillan and Ma, 1997).

The wet component, although the smaller part of the total tropospheric effect, changes rapidly and has to be monitored either by more sophisticated stochastic modeling, such as Kalman Filter techniques or by some external means. Still today one promising - albeit costly - method appears to be the radiometer technique, which consists of measuring the microwave thermal emission from water vapour near 22GHz in the line-of-sight (Elgered et al. 1982, Resh et al., 2000). Other approaches include numerical weather prediction models and estimation from permanent GPS data (Behrend et al., 2000).

In VLBI data processing there are two levels of least squares solutions, one in which only the "local" unknowns are estimated (such as clocks and atmospheric parameters) thus creating a first data base version of a particular experiment, and another one which collects all available experiments for a comprehensive solution including the "global" unknowns (such as station and source positions, Earth rotation parameters, etc.). Among the various VLBI software systems the MkIII-Data-Analysis-System should be mentioned, which is built around the CALC/SOLVE software system developed jointly by the US-East-Coast VLBI groups and has become a standard against which the other systems can be compared (Ma et al., 1989).

4. Geodetic observing programs and early results

A detailed description of the great potential of VLBI for geophysical applications has been presented as early as 1969 at a conference held in London, Canada on "Earthquake displacement fields and the rotation of the Earth" (Shapiro, Knight, 1970). In subsequent years virtually all of the goals mentioned there were to be achieved. The first results of baseline measurements with cm accuracy of the average of measurements over several years were published in the early eighties and the first estimate of plate tectonic motion measured by VLBI appeared in 1986 (Fig. 6).

In the meantime, uncertainties at the cm level and even less have been demonstrated by thousands of VLBI experiments in networks connecting almost all major continents of the globe. In order to realise the initial goals and many more, the efforts in different countries around the world have been combined in setting up several programs that include international cooperation, e.g. the NASA Crustal Dynamics Project (CDP, 1979-1993)(Coates et al., 1985), mainly intended for the measurement of global plate tectonic motion, and Project IRIS (International Radio Interferometric Surveying 1983-1990), a progam of regular weekly observations of the rotation of the Earth (precession, nutation, polar motion and UT1-variarions)(Carter et al., 1979; Carter et al., 1985; Robertson et al., 1985).

In this context, it is necessary to mention that the growing importance of regular monitoring of geodynamic processes required a substantion increase of observing time at the existing radio telescopes, leading to conflicts with the ongoing astronomical observing programs. In this situation, the need for dedicated geodetic radio telescopes became apparent. A dedicated telescope for geodetic work would be a compromise between sensitivity and fast slewing between sources, i.e. a 20m-dish built on stable concrete foun-

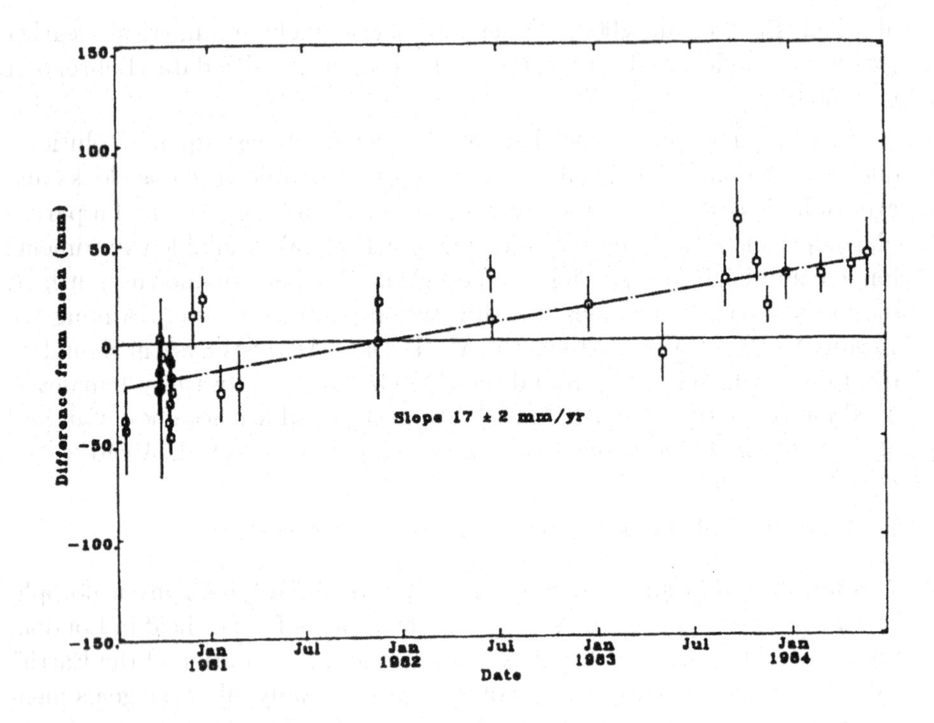

Figure 6. First estimate of intercontinental baseline rate (length) on the transatlantic baseline Haystack (Mass.,USA) - Onsala (Sweden)(Herring et al. 1986)

dations with a strong azimuth and elevation drive. The telescope reference point would be well defined and easy to access for local geodetic control measurements.

In Wettzell, the construction of a dedicated geodetic radio telescope was part of the broader concept of a geodynamic "Fundamental Station" that would assemble the different space techniques such as Satellite Laser Ranging (SLR), Lunar Laser Ranging (LLR), Optical, Doppler and GPS systems, as well as supplementary geophysical monitoring intrumentation at one site to be able to compare and combine the data gathered as well as the results obtained from the different networks (Schneider et al., 1982). In recent years, the idea has been extended to build a transportable duplicate of the Fundamental Station, the TIGO (Transportable Integrated Geodetic Observatory) and operate the system at a site in the southern hemisphere, in order to improve the global coverage (Hase, Petrov, 1999).

Unfortunately, there are even today only very few stations in the world that have reached the status of a true Fundamental Station. The station of Matera in southern Italy is one more example of a perfect geodetic site assembling the complete range of space geodetic measuring systems (Colucci et al., 2001). The northernmost VLBI station in the world, Ny Ålesund on the Archipelago of Spitsbergen (Norway) is on the way of becoming an important multi-technique geodetic observatory (Digre and Plag, 2001). Still, the fact that a major part of the astronomical VLBI observatories have been very positive to provide part of their resources to the global and regional geodetic observation programs, has allowed for a significant contribution to the geodetic and geodynamic data sets.

5. Recent results and achievements of VLBI in Geodesy and Geodynamics

An example from the set of long time series of geodetic VLBI measurements is the baseline Wettzell-Westford, which has the densest coverage of observing sessions reached so far. It is interesting to compare the Wettzell-Westford baseline length evolution from one of the most recent global solutions (Fig. 7) with the earliest determinations of the transatlantic baseline rates (Fig. 6). The extraordinary smoothness and linearity of the length evolution postulates an absolutely uniform crustal motion, and by the same token a perfectly uniform spreading process at the oceanic ridges. In their extensive analysis of the global baseline changes and site motions observed by VLBI, Argus and Gordon (1996) find no really significant departures from continuous motion over 10 to 20 years, except of course examples of coseismic displacements in very locally defined areas (Clark et al., 1990).

Another example of the potential of VLBI on longer time scales is the observed time series of the nutation angles, shown as corrections to the IAU 1980 reference model (Fig. 8 and 9). In these plots, we already clearly see the departure in the main 18.6 yr nutation term. After some twenty years of intensive theoretical work by many groups in different parts of the world, the resulting new and thoroughly improved model will soon be published (Herring, 2000; Mathews et al. 2000). It is quite certain that the continued time series of VLBI observations at well defined stations will contribute by their integrity to the solution of many more of the very intriguing and challenging questions in geodynamic research.

At the other extreme, VLBI observations have also lend themselves to investigate very short period, even down to sub-diurnal phenomena, such as the ocean tide induced variations in Earth rotation (Brosche et al. 1991, Clark et al., 1998). In this respect, the ultimate goal is of course a truly continuous observation program which will optimise the resolution power at the short periods down to a few hours. To achieve this goal, the CORE project has been devised at NASA/GSFC to combine the strengths of different station configurations on the globe and cover the seven days of a week without placing too much observational burden on any single

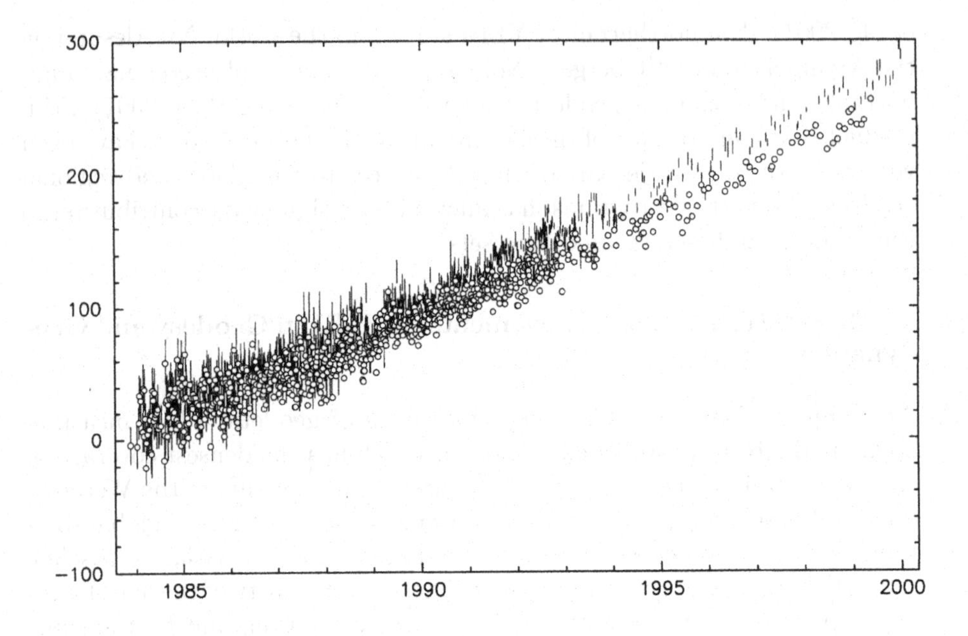

Figure 7. Wettzell-Westford baseline length rate from recent global solution (L. Petrov, NASA Goddard Space Flight Center, Greenbelt, MA)

one of the VLBI stations (Clark et al., 1998; MacMillan et al., 1999). This project has begun in an initial phase with only two days per week and will be stepped up with the full deployment of the capabilities of the new MkIV VLBI system.

Most of the presently active observing stations, analysis groups and technical development centers are listed and described in the first Annual Report (Vandenberg, 1999) of the newly created

- IVS (International VLBI Service for Geodesy and Astrometry).

A strong incentive to introduce a more formal organisation that would provide the means for a better coordination of the activities in the global geodetic VLBI community towards a common goal and a more production-oriented approach, was given by the example of the GPS-community in their International GPS Service for Geodynamics (IGS). The IVS was set up during 1998 and began its operations on the 1st of March 1999 (Campbell, Vandenberg, 1999; Vandenberg, Baver, 2001). The 1st General Meeting at Kötzting gives proof of the successful start of a new, more internationally oriented phase of VLBI with even closer cooperation between its members (Schlüter, 2000).

At present geodetic VLBI may be seen to have matured and regular observing campaigns such as the programs under the recently established IVS service are providing data that are showing many interesting details

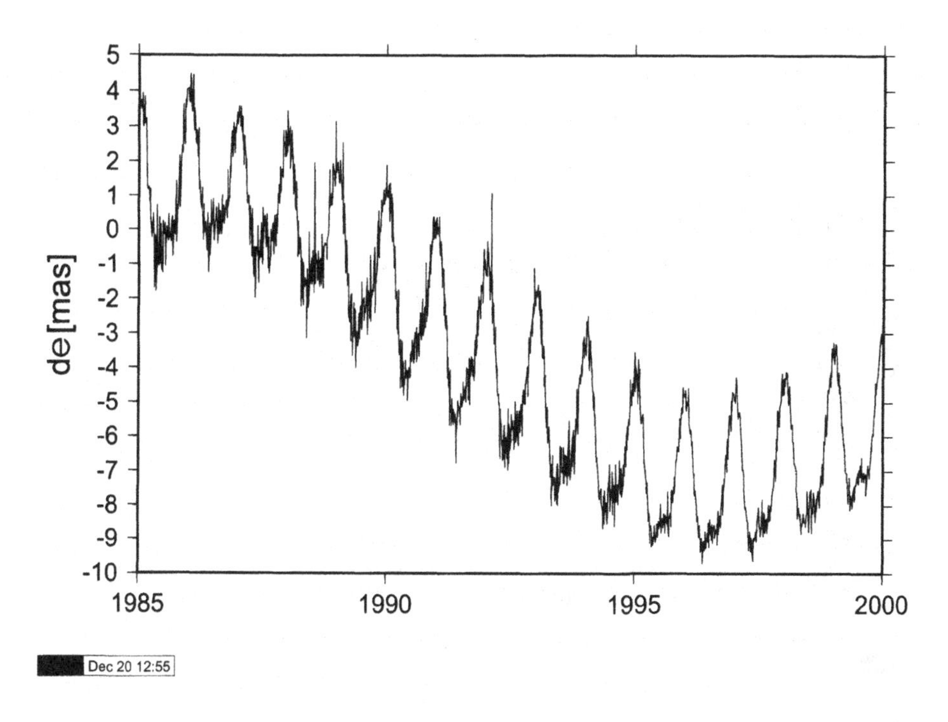

Figure 8. Differences between observed and computed nutation angles: longitude (reference: IAU 1980 model)

of the wide spectrum of geodynamical effects. In particular, VLBI has produced the first significant evidence of present-day plate motion across the Atlantic. The measured motions of about 80% of the Earths solid surface are a surprisingly good agreement with plate models derived from geological and geophysical data. The Earth rotation data on the other hand are clearly showing that the main factor causing the irregularities in the rotational speed, i.e. the changes in the length-of-day, can be identified with atmospherical processes, primarily the zonal wind pattern. Finally, to-day VLBI continues to be the prime technique to determine corrections to the standard nutation model (Fig. 8 and 9).

The future of VLBI for geodesy and geodynamics relies primarily on a continued interest in a stable reference frame for the monitoring of Earth orientational dynamics and for the alignment of the global geodetic terrestrial systems. In this respect, the 20yr-time VLBI series of nutation determinations presents an invaluable basis for the further improvement of the nutation theory.

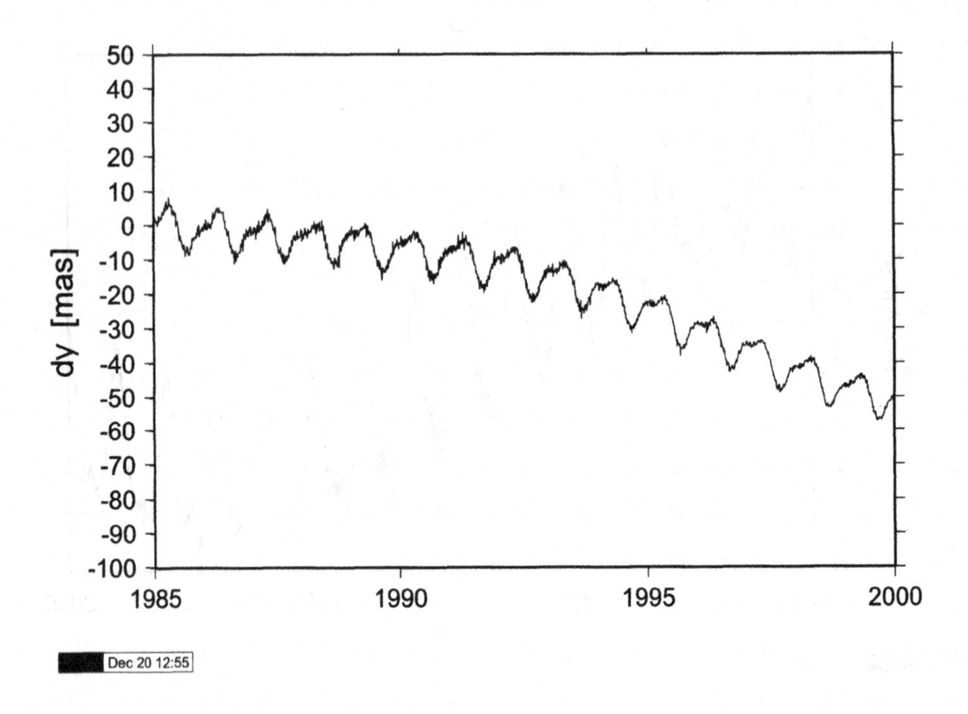

Figure 9. Differences between observed and computed nutation angles: obliquity (reference: IAU 1980 model)

6. VLBI vs. GPS as a tool for monitoring global change

Since the early nineties, when the GPS satellite constellation reached its full deployment with 24 satellites (21 + 3 spare) in three different orbits, the geodetic community worked hard to build a global network of permanently observing GPS stations (Seeber 1993). In 1994, the IGS (International GPS Service for Geodesy and Geodynamics) was officially established to provide service and support for geodetic and geophysical research activities through GPS data products:

- GPS satellite ephemerides
- Earth rotation parameters
- IGS tracking station coordinates and velocities
- GPS satellite and IGS tracking station clock information

Extensive information is available at the IGS website:

http://igscb.jpl.nasa.gov

In principle, every sytem of orbital bodies may serve as a spacial reference frame, provided the modelling of the orbital motion is perfect in the sense that it includes all forces acting on these bodies. The modelling of the

Earth's gravity field has made enormous progress and its present accuracy would be high enough to describe the orbits of the GPS satellites to an accuracy beteer than 1 cm at the average altitude of 20,000 km above the Earth's surface. However, the non-gravitational forces, such as solar winds and residual atmospheric particles are extremely hard to model and cause deviations in the adjusted orbits of several decimeters. In spite of the fact that a great deal of averaging occurs in the analysis of the continuously collected GPS data from many satellites and stations, small residual motions of the orbital system remain and can only be detected by comparison with independent techniques, such as VLBI.

If we consider successive realisations of the terrestrial reference frame by GPS, we may see discontinuities that are related to the introduction of new refined orbital models, and the increase of the number of stations used in forming the data sets (see e.g. Rothacher, 1998, p. 48, Fig.8). It is also obvious that the quality of the frame realization was actually substantially augmented in 1995/1996, when the number of global reference stations rose above 50 (today there are almost 300 IGS stations around the world). Therefore, we may expect a convergence between the system realisations of the individual space techniques towards a level of better than 1 cm in three dimensional position and to better than 1 mas in orientation.

Presently, the IERS (International Earth Rotation Service) is performing the combination of the results from different techniques providing

- time series of daily values of UT1-UTC, x and y coordinates of the pole
- daily variations of the celestial pole relative to the IAU model.

The IERS Bulletin A is currently published at the website of the US Naval Observatory in Washington D.C.:

http://maia.usno.navy.mil/index.html

The IERS website is presently in the process of reorganisation:

http://www.iers.org

Detailed expositions on the theoretical background and the mathematical formulation of the relations between different reference frames are found in Sovers et al. (1998), Bock et al. (1998), and Seeber (1993).

If we consider the question whether any one of the space techniques will be sufficient on its own to carry the task of realising and maintaining the fundamental reference frames, it is useful to compare the characteristic profile of each of the techniques:

Both tables provide many instances of complementary potential which makes a strong case for the combined use of all three of the most familiar of the space techniques. Of course, there are other space systems, such as DORIS (the French microwave beacon system), GLONASS (the Russian

TABLE 1. Comparison of techniques VLBI, SLR (Satellite Laser Ranging) and GPS

Potential for	VLBI	SLR	GPS
Site coordinates	X	X	X
site velocities	X	X	X
baseline evolution	X	X	X
Inertial system	X	-	-
Geocenter	-	X	X
Scale	X	X (GM)	X (GM)
Polar motion	X	X	X
UT1 variations (long per.)	X	-	-
(short period)	X	X	X
Precession	X	-	-
Nutation (long period)	X	-	-
(short period)	X	X	X
Tropospheric monitoring (dry)	X	X	X
(wet)	X	-	X
Ionospheric monitoring	X	-	X
Precise time transfer	X	X	X

GM: In satellite geodesy the scale is also affected by the zero order term of the gravity field.

TABLE 2. Strengths and weaknesses of the techniques

Strengths & weaknesses	VLBI	SLR	GPS
All weather capability	X	-	X
Low cost equipment	-	-	X
Mobility	(X)	(X)	X
Multipath-free	X	X	-
less affected by wet trop.	-	X	-
tracking of other satellites	(X)	X	X

GPS equivalent), and GALILEO (the planned European Satellite Navigation Sytem), which may be added to the combination. In the ongoing effort to obtain most accurate and reliable results, there can be no doubt that VLBI, as a fundamental support technique, will continue to play an essential role in the monitoring of geodynamical processes.

7. References

Altamimi, Z. (2000): ITRF Status and Plans for ITRF20000. N.R. Vandenberg and K.D. Baver (eds.): IVS 2000 General Meeting Proceedings, NASA/CP-2000-209893, p. 57-61.

Altamimi, Z., D. Angermann, D. Argus, C. Boucher, B. Chao, H. Drewes, R. Eanes, M. Feissel, R. Ferland, T. Herring, B. Holt, J. Johannson, C. Larson, C. Ma, J. Manning, C. Meertens, A. Nothnagel, E. Pavlis, G. Petit, J. Ray, J. Ries, H.-G. Scherneck, P. Sillard, M. Watkins (2001): The Terrestrial Reference Frame and the Dynamic Earth, EOS, Transactions, Vol. 82, No. 25, p. 275 and 278.

Argus, D.F., R.D. Gordon (1996): Tests of the rigid-plate hypothesis and bounds on ntraplate deformation using geodetic data from very long baseline interferometry, J. Geophys. Res. 101, 13,555-13,572.

Behrend, D., L. Cucurull, J. Vila, R. Haas (2000): An Inter-comparison Study to Estimate Zenith Wet Delays Using VLBI, GPS and NWP Models. Earth Planets Space, Journal of the Japanese Societies for Earth and Planetary Sciences, 52, 759-764.

Beutler, G., I.I. Mueller, R.E. Neilan (1994): The International GPS Service for Geodynamics (IGS), Bull. Godsique 68, 39-70.

Bock, Y., C.H. and I.M. Green (1998): Reference Systems, Chapter 1 in "GPS for Geodesy", A. Kleusberg, P.J.G. Teunissen (eds.) p. 1-41, Springer.

Brosche, P., J. Wünsch, J. Campbell, H. Schuh (1991): Ocean Tide Effects in Universal Time detected by VLBI. Astronomy & Astrophysics 245, 676-682.

Campbell, J., N. Vandenberg (1999): Very Long Baseline Interferometry (VLBI) - Subcommission Report - , IAG, CSTG-Bulletin No. 15, Progress Report 1998, Eds. G. Beutler, H. Drewes, H. Hornik, DGFI, Munich, 16-27.

Cannon, W.H., Baer, D., Feil, G., Feir, B., Newby, P., Novikov, A., Dewdney, P., Carlson, B., Petrachenko, W.T., Popelar, J., Mathieu, P., Wietfeldt, R.D. (1997): The S2 VLBI System, Vistas in Astronomy 41, No 2, 297-302.

Carter, W.E., D.S. Robertson, M.D. Abell (1979): An Improved Polar Motion and Earth Rotation Monitoring Service Using Radio Interferometry, AGU-Symposium No. 82, Time and the Earth's Rotation, San Fernando, Spain, 8-12 May 1978, Reidel Publ. Co., Dordrecht, 191-198.

Carter, W.E., Robertson, D.S., MacKay, J.R. (1985): Geodetic Radiointerferometric Surveying: Applications and Results. J. Geophys. Res. 90, 4577-4587.

Carter, W.E., D.S. Robertson, A. Nothnagel, G.D. Nicolson, H. Schuh, J. Campbell (1988): IRIS-S: Extending Geodetic Very Long Baseline Interferometry Observations to the Southern Hemisphere. J. Geophys. Res. 93, 14,947-14,953.

Clark, T.A., B.E. Corey, J.L. Davis, G. Elgered, T.A. Herring, H.F. Hineregger, C.A. Knight, J.L. Levine, G. Lundqvist, C. Ma, E.F. Nesman, R.B. Phillips, A.E.E. Rogers, B.O. Rönnäng, J.W. Ryan, B.R. Schupler, D.B. Shaffer, I.I. Shapiro, N.R. Vandenberg, J.C. Webber, A.R. Whitney (1985): Precision Geodesy using the Mk-III Very-Long-Baseline Interferometer System. IEEE Transactions on Geoscience and Remote Sensing GE-23, No.4, 438-449.

Clark, T.A., C. Ma, J.M. Sauber, J.W. Ryan, D. Gordon, D.B. Shaffer, D.S. Caprette, N.R. Vandenberg (1990): Geodetic measurement of deformation in the Loma Prieta, California, Earthquake with Very Long Baseline Interferometry. Geophys. Res. Letters 18, 1215-1218.

Clark, T.A., C. Ma, J.W. Ryan, B.F. Chao, J.M. Gipson, D.S. MacMillan, N.R.

Vandenberg, T.M. Eubanks, A.E.Niell (1998): Earth Rotation Measurement Yields Valuable Information About the Dynamics of the Earth System. EOS 79, No. 17, 205

Cohen, M.H., D.B. Shaffer (1971): Positions of Radio Sources from Long Baseline Interferometry. Astronomical Journal 76, 91-101.

Coates, R.J., H. Frey, G.D. Mead, J. Bosworth (1985): Space-Age Geodesy: The NASA Crustal Dynamics Project. IEEE Transactions on Geoscience and Remote Sensing GE-23, No.4, 360-368.

Colucci,G., D. Del Rosso, F. Vespe (2001): Matera CGS VLBI Station. IVS 2000 Annual Report, N.R. Vandenberg and K.D. Baver (eds.), NASA Publication No. NASA/TP-2001-209979, p. 98-101.

Davis, J.L. et al. (1985): Geodesy by Radio Interferometry: Effects of Atmospheric Modeling Errors on Estimates of Baseline Length. Radio Science 20, 1593-1607.

Digre, H.G., H.-P. Plag (2001): NYAL Ny Ålesund 20 Metre Antenna. IVS 2000 Annual Report, N.R. Vandenberg and K.D. Baver (eds.), NASA Publication No. NASA/TP-2001-209979, p. 106-109.

Elgered, G., Rönnäng, B., Askne, J. (1982): Measurements of Atmospheric Water Vapour with Microwave Radiometry. Radio Science 17, 1258-1264.

ESA (1979): Satellite Linked VLBI, Phase-A Study. European Space Agency, Document SD/20545/AFB/MZ, Paris, December 1979

Feissel, M., A.-M. Gontier (2000): Stability of ICRF, a Time Series Approach. N.R. Vandenberg and K.D. Baver (eds.): IVS 2000 General Meeting Proceedings, NASA/CP-2000-209893, p. 280-284.

Fey, A.L., D.A. Boboltz, R.A. Gaume, K.J. Johnston (2000): Improving the ICRF using the Radio Reference Frame Image Database. N.R. Vandenberg and K.D. Baver (eds.): IVS 2000 General Meeting Proceedings, NASA/CP-2000-209893, p. 285-287.

Hase, H., L. Petrov (1999): The first Campaign of Observations with the VLBI-Module of TIGO. Proceedings of the 13th Working Meeting on European VLBI for Geodesy and Astrometry, Viechtach/Wettzell Feb. 12-13, 1999, Eds. W. Schlüter and H. Hase, 19-24.

Heitz, S. (1976): Mathematische Modelle der geodätischen Astronomie. Deutsche Geodätische Kommission, Reihe A, Nr. 85, Frankfurt am Main 1976.

Herring, T.A., B.E. Corey, C.C. Counselman III, I.I. Shapiro, A.E.E. Rogers, A.R. Whitney, T.A. Clark, C.A. Knight, C. Ma, J.W. Ryan, B.R. Schupler, N.R. Vandenberg, G. Elgered, G. Lundquist, B.O. Rönnäng, J. Campbell, P. Richards (1983): Determination of Tidal Parameters from VLBI Observations. Proc. Ninth Int. Sympos. on Earth Tides, New York, Aug. 17-22, 1981, ed. J. Kuo, p. 205-214, E. Schweizerbart'sche Verlagsbuchhandlung, Stuttgart 1983.

Herring, T.A., I.I. Shapiro, T.A. Clark, C. Ma, J.W. Ryan, B.R. Schupler, C.A. Knight, G. Lundquist, D.B. Shaffer, N.R. Vandenberg, B.E. Corey, H.F. Hinteregger, A.E.E. Rogers, J.C. Webber, A.R. Whitney, G. Elgered, B.O. Rönnäng, J.L. Davis (1986a): Geodesy by Radio Interferometry: Evidence for Contemporary Plate Motion. J. Geophys. Res. 91, 8341-8347

Herring, T.A., Gwinn, C.R., Shapiro, I.I. (1986b): Geodesy by Radio Interferometry: Studies of the Forced Nutations of the Earth, I. Data Analysis, J. Geophys. Res. 91, 4745-4754.

Herring, T.A. (1992): Submillimeter Horizontal Position Determination Using Very Long Baseline Interferometry. J. Geophys. Res. 97, p. 1981-1990.

Herring, T.A. (2000): Geophysical Applications of Earth Rotation Measur4ments.

N.R. Vandenberg and K.D. Baver (eds.): IVS 2000 General Meeting Proceedings, NASA/CP-2000-209893, p. 62-68.
Hinteregger, H.F., I.I. Shapiro, D.S. Robertson, C.A. Knight, R.A. Ergas, A.R. Whitney, A.E.E. Rogers, J.M. Moran, T.A. Clark, B.F. Burke (1972): Precision Geodesy via Radio Interferometry. Science 178, 396-398.
Koyama, Y., N. Kurihara, T. Kondo, M. Sekido, Y. Takahashi, H. Kiuchi, K. Heki (1998): Automated geodetic Very long Baseline Interferometry observation and data analysis system, Earth Planets and Space, 50, 709-722.
Kunimori, H., F. Takahashi, M. Imae, Y. Sugimoto, T. Yoshino, T. Kondo, K. Heki, S. Hama, Y. Takahashi, H. Takaba, H. Kiuchi, J. Amagai, N. Kurihara, H. Kuroiwa, A. Kaneko, Y. Koyama, K. Yoshimura (1993): Contributions and Activities of Communications Research Laboratory under the Cooperation with Crustal Dynamics Project. Contributions of Space Geodesy to Geodynamics: Technology, AGU Geodynamics Series, Vol. 25, p. 65-80.
Ma, C., Ryan, J.W., Caprette, D. (1989): Crustal Dynamics Project Data Analysis - 1988, VLBI Geodetic Results 1979-87. NASA Technical Memorandum 100723, Greenbelt, Md., February 1989.
Ma, C., J.W. Ryan, D. Gordon, D.S. Caprette, W.E. Himwich (1993): Reference Frames from CDP VLBI Data. Contributions of Space Geodesy to Geodynamics: Earth Dynamics, AGU Geodynamics Series, Vol. 24, p. 121-145.
MacDoran, P.F., D.J. Spitzmesser, L.A. Buennagel (1982): SERIES: Satellite Emission Range Inferred Earth Surveying. Proc. 3rd Int. Sympos. on Satellite Doppler Positioning, Defense Mapping Agency and National Ocean Survey, Vol. 2, 1143-1164.
MacMillan, D.S., W.E. Himwich, N.R. Vandenberg, C.C. Thomas, J.M. Bosworth, B. Chao, T.A. Clark, C. Ma (1999): CORE: Continuous, High-Accuracy Earth Orientation Measurements. Proc. 13th Working Meeting on European VLBI for Geodesy and Astrometry, Viechtach/Wettzell Feb. 12-13, 1999, Eds. W. Schlüter and H. Hase, p. 166-171.
MacMillan, D.S., C. Ma (1997): Atmospheric gradients and the VLBI terrestrial and celestial reference frames, Geophys. Res. Letters 24, 4530.
Mathews, P.M., T.A. Herring, B.A. Buffet (2000): Modeling of nutation-precession: New series for non-rigid Earth, and insights into the Earths interior, sbmitted to J. Geophys. Res. March 2000.
Mathur, N.C., M.D. Grossi, M.R. Pearlman (1970): Atmospheric Effects in Very Long Baseline Interferometry. Radio Science 5, No. 10, 1253-1261.
Minster, J.B., Jordan, T.H. (1978): Present-Day Plate Motions. J. Geophys. Res. 83, 5331-5354.
Nothnagel, A. (1999): Local Survey at the Effelsberg Radio Telescope 1997 - Preliminary Results. Proc. 13th Working Meeting on European VLBI for Geodesy and Astrometry, Viechtach/Wettzell Feb. 12-13, 1999, Eds. W. Schlüter and H. Hase, p. 25-31.
Niell, A. (2000): Improved Mapping Functions for GPS and VLBI. N.R. Vandenberg and K.D. Baver (eds.): IVS 2000 General Meeting Proceedings, NASA/CP-2000-209893, p. 263-266.
Petrov, L. (1999): Steps towards phase delay VLBI. Proc. 13th Working Meeting on European VLBI for Geodesy and Astrometry, Viechtach/Wettzell Feb. 12-13, 1999, Eds. W. Schlüter and H. Hase, p. 144-151.
Petrov, L. (2000): Instrumental Errors of Geodetic VLBI. IVS 2000 General Meeting, Proceedings ed. by N.R. Vandenberg and K.D. Baver, NASA Publication

No. NASA/CP-2000-209893, p. 230-235.

Preuss, E., J. Campbell (1992): Very-Long-Baseline Interferometry in Astro-, Geo- and Gravitational Physics. In J. Ehlers and G. Schaefer (Eds.) Relativistic Gravity Research, Lecture Notes in Physics, No. 410, p. 100-130.

Ray, J., A. Whitney (1999): Real-Time VLBI Forum, IAG, CSTG-Bulletin No. 15, Progress Report 1998, Eds. G. Beutler, H. Drewes, H. Hornik, DGFI, p. 28-31, Munich 1999.

Resh, G., C. Jacobs, S. Keihm, G. Lanyi, C. Naudet, A. Riley, H. Rosenberger, A. Tanner: Calibration of atmospherically induced delay fluctuations due to water vapor. N.R. Vandenberg and K.D. Baver (eds.): IVS 2000 General Meeting Proceedings, NASA/CP-2000-209893, p. 274-279.

Robertson, D.S. (1975): Geodetic and astrometric measurements with very long baseline interferometry. Ph.D. Thesis, MIT, Cambridge, Mass., Doc. No. X-922-77-228, NASA/GSFC, Greenbelt, Maryland, 1975

Robertson, D.S., W.E. Carter (1984): Relativistic Deflection of radio signals in the solar gravitational field measured with very-long-baseline interferometry, Nature, 310, 572-574.

Robertson, D.S., Carter, W.E., Campbell, J., Schuh, H. (1985): Daily Earth Rotation Determinations from IRIS Very Long Baseline Interferometry. Nature 316, 424-427.

Rogers, A.E.E. (1970): Very Long Baseline Interferometry with Large Effective Bandwidth for Phase Delay Measurements. Radio Science 5, 1239-1247.

Rothacher, M. (2000): Towards an Integrated Global Geodetic Observing System. IAG Symposia, Vol. 120, IGGOS-Symposium, held at Munich, Oct. 5- 9, 1998, R. Rummel, H. Drewes, W. Bosch, H. Hornik (Eds.), p. 41-52, Springer.

Ryan, J.W., C. Ma, D. Caprette (1993): NASA Space Geodesy Program - GSFC Data Analysis - 1992, Final Report of the Crustal Dynamics Project, VLBI Geodetic Results 1979-91. NASA Technical Memorandum 104572, Greenbelt, Md. 1993.

Ryan, J.W., T.A. Clark, R.J. Coates, C. Ma, W.T. Wildes, C.R. Gwinn, T.A. Herring, I.I. Shapiro, B.E. Corey, C.C. Counselman, H.F. Hinteregger, A.A.E. Rogers, A.R. Whitney, C.A. Knight, N.R. Vandenberg, J.C. Pigg, B.R. Schupler, B.O. Rönnäng (1986): Geodesy by Radio Interferometry: Determinations of Baseline Vector, Earth Rotation, and Solid Earth Tide Parameters with the Mark I Very Long Baseline Radio Interferometry System. Journ. Geophys. Res. 91, p. 1935-1945.

Scherneck, H.-G. (1991): A parameterized Solid Earth Tide Model and Ocean Tide Loading Effects for Global Geodetic Baseline Measurements, Geophys. J. Int. 106, 677-694.

Schlüter, W. (2000): Report of the IVS Chair. IVS 2000 General Meeting, Proceedings ed. by N.R. Vandenberg and K.D. Baver, NASA Publication No. NASA/CP-2000-209893, p. 11-15.

Schneider, M., R. Kilger, K. Nottarp, E. Reinhart, J. Campbell, H. Seeger (1982): Concept and Realization of a 20m Radiotelescope for the Satellite Observation Station Wettzell. Proceedings of the IAG Symposium No. 5: Geodetic Applications of Radio Interferometry, Tokyo, Japan, May 7-8, 1982, NOAA Techn. Rep. NOS 95 NGS 24, p. 266-284.

Seeber, G. (1993): Satellite Geodesy, Foundations, Methods, and Applications, De Gruyter.

Shapiro, I.I., C.A. Knight (1970): Geophysical applications of long baseline radio interferometry. In: Earthquake Displacement Fields and the Rotation of the Earth, ed. L. Mansinha, D.E. Smylie and A.E. Beck, pp. 284, Springer, New York 1970.

Sovers, O.J., J.L. Fanselow, C.S. Jacobs (1998): Astrometry and geodesy with radio interferometry: experiments, models, results. Reviews of Modern Physics 70, No. 4, 1393-1454.

Vandenberg, N.R. (Ed.)(1999): IVS 1999 Annual Report, NASA Publication TP-1999-209243, Greenbelt, MD

Vandenberg, N.R. and K.D. Baver (Eds.)(2001): IVS 2000 Annual Report, NASA Publication TP-2001-209979, Greenbelt, MD

Walter, H.-G., O.J. Sovers (2000): Astrometry of Fundamental Catalogues. Springer Verlag, Berlin, Heidelberg

Whitney, A.R., A.E.E. Rogers, H.F. Hinteregger, C.A. Knight, J.I. Levine, S. Lippincott, T.A. Clark, I.I. Shapiro, D.S. Robertson (1976): A Very-Long-Baseline Interferometer for Geodetic Applications. Radio Science 11, 421-432.

Whitney, A.R.(2000a): Technology Coordinator Report. IVS 2000 General Meeting, Proceedings ed. by N.R. Vandenberg and K.D. Baver, NASA Publication No. NASA/CP-2000- 209893, p. 93-97.

Whitney, A.R.(2000b): Concept for an affordable high data rate VLBI recording and playback system. IVS 2000 General Meeting, Proceedings ed. by N.R. Vandenberg and K.D. Baver, NASA Publication No. NASA/CP-2000- 209893, p. 103-110.

TROPOSPHERIC AND IONOSPHERIC PHASE CALIBRATION

J.-F. LESTRADE
Observatoire de Paris/DEMIRM-CNRS
77 avenue Denfert Rochereau
F75014 -Paris - France
lestrade@obspm.fr

1. Introduction

Radio interferometry is traditionally sketched by a plane wavefront, from an infinitely distant radio source, that "hits" successively two antennae located on the Earth surface. The interferometer measures the time-delay τ for the passage of this wavefront from one antenna to the other. In reality, the assumption of a plane wavefront is not correct because the atmosphere imprints corrugations upon the wavefront (Figure 1). Consequently, propagation in the atmosphere induces fluctuations of the time-delay τ on a wide range of time scales. The upper part of the atmosphere is the ionosphere at a mean altitude of 350 km. This ionised shell around the Earth causes delay fluctuations that are dispersive ($\propto \frac{1}{\nu^2}$) and become severe at observation frequencies below $\sim$ 2 GHz. The lower part of the atmophere is the troposphere made of moist air above the ground. Inhomogeneities in the troposphere cause delay fluctuations that are non-dispersive and might be as large as $\sim$ 1 millimeter leading to loss of coherence for observations at frequencies > 43 GHz. Only in the intermediate range, between 2 and 43 GHz, both of these effects are relatively mild and can be measured or modeled relatively easily to correct the observed interferometer phase.

The time-delay τ of the interferometer has the general form :

$$\tau = \tau_{geo} + \tau_{ion} + \tau_{trop} + \tau_{GR} + \tau_{struc} + \tau_{clk} \qquad (1)$$

where the geometric delay $\tau_{geo} = \frac{\vec{B}.\vec{\sigma}}{c}$, $\vec{B}$ and $\vec{\sigma}$ are the baseline vector and the direction of the source, and the subscripts $_{ion}$ and $_{trop}$ stand for ionosphere and troposphere, $_{GR}$ for General Relativity, $_{struc}$ for source structure

F. Mantovani and A. Kus (eds.), The Role of VLBI in Astrophysics, Astrometry and Geodesy, 383–401.

(morphology) and $_{clk}$ for the clock and electronics offsets and drifts that are *a priori* unknown in VLBI.

The tropospheric and ionospheric delays can be split into two parts : a non or very slowly varying part (termed static) and a rapidly fluctuating part on time scales of less than half-hour :

$$\tau_{trop} = \tau_{trop,static} + \delta\tau_{trop,fluctuations}$$

$$\tau_{ion} = \tau_{ion,static} + \delta\tau_{ion,fluctuations}$$

We shall examine these effects and shall stress that very short baseline connected-element interferometry and VLBI are not affected similarly. These atmospheric delays are discussed extensively in Thompson, Moran & Swenson (1986) or latter edition. The main sections below describe the static tropospheric delay, the tropospheric delay fluctuations and the ionospheric delay.

2. Static tropospheric delay

The atmosphere is characterized by the index of refraction n(z) that depends mainly on level z above the ground. In the lower atmosphere of total height h, the phase velocity of a wavefront $v_\phi = \frac{c}{n}$ is less than c by about 0.03 % and the excess time to traverse the medium is :

$$\tau_{trop,static} = \frac{1}{c}\int_0^h \Big(n(z) - 1\Big)dz \qquad (2)$$

and is $\sim$ 7 nanoseconds.

The Smith-Weintraub equation (Smith & Weintraub 1953) is an empirical formula for the refractivity $N(z) = 10^6\Big(n(z) - 1\Big)$ which depends on the partial pressures p_D of "Dry air" and p_V of "Water vapor" (millibar) and on tempereture T (K):

$$N(z) = 77.6\frac{p_D(z)}{T(z)} + 64.8\frac{p_V(z)}{T(z)} + 3.776\ 10^5\frac{p_V(z)}{T(z)^2}$$

This equation is accurate to about 1 % below 100 GHz. This equation and Eq. (2) yield the excess tropospheric delay in form of its dry and wet parts. A complete discussion and bibliographical references are in Thompson, Moran & Swenson (1986) or latter editions.

2.1. ESTIMATION OF DRY TROPOSPHERIC DELAY AT ZENITH

The dry portion of the troposphere is $\sim$ 10 km thick above the Earth surface and is electrically neutral. It is very nearly in hydrostatic equilibrium and

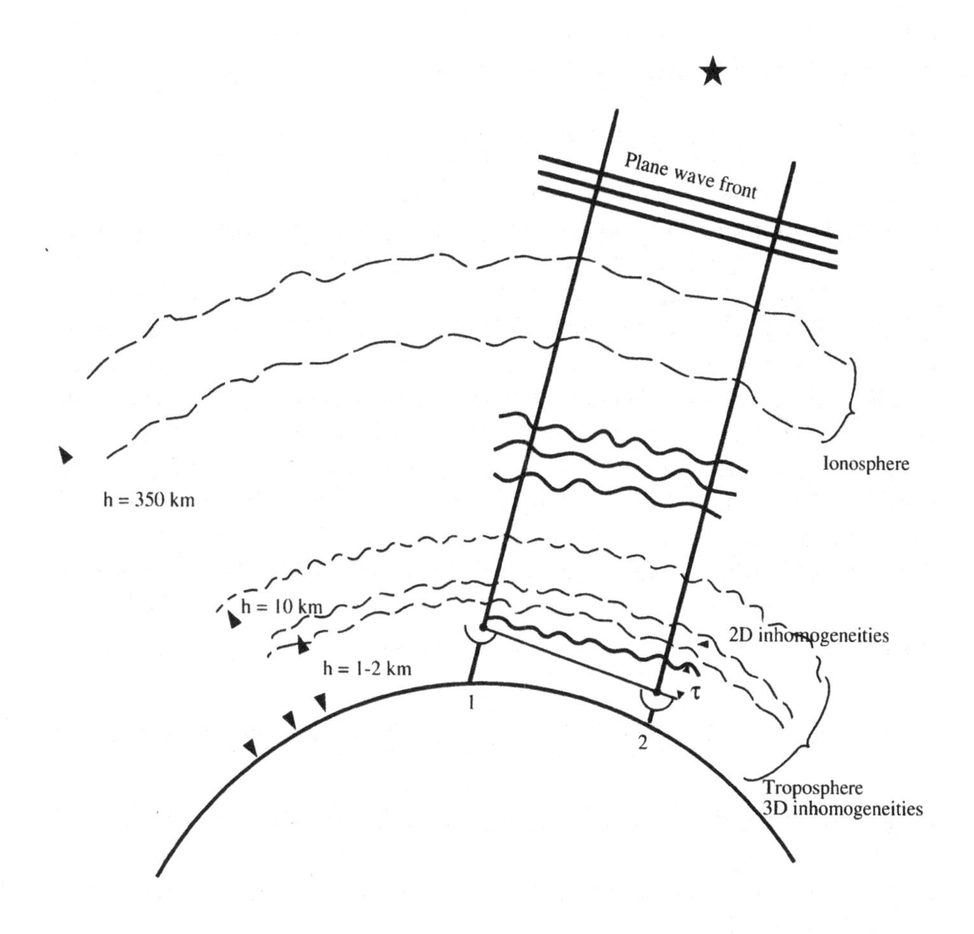

Figure 1. The ionosphere and troposphere imprint corrugations upon the wavefront from a distant radio source, initially plane in outer space.

as a result the dry delay at zenith does not depend on the details of p_D and T along the signal path but only on the ground surface pressure.

The dry path length at zenith Z_D (equivalently the dry delay at zenith τ_D) can be accurately estimated with the formula of Saastamoinen (1972) by simply measuring the barometric pressure p at the radiotelescope:

$$Z_D = c\tau_D = \frac{2.2768\ 10^{-3}\ p}{(1 - 0.00266 \cos 2\phi - 0.00028\ h)}$$

ϕ and h are the geodetic latitude and altitude of the station, p is in millibar and Z_D in meters. The excess path length Z_D is comprised between ~ 1.70 m at high altitude site (*e.g.* VLA) and $\sim$ 2.30 m at sea level, *i.e.* 5.7 and 7.7 nanosecond of delay τ_D.

2.2. ESTIMATION OF WET TROPOSPHERIC DELAY AT ZENITH

On average, the water vapor density has an exponential distribution with a scale height of 2 km. The wet path length along any line of sight can be determined by measuring the brightness temperature from the H_2O rotational line at 22 GHz with Water Vapor Radiometers (Marvel & Woody, 1998) and also at 183 GHz (Lay *et al* 1998). Water Vapor Radiometer measurements are carried out at mm-λ observation sites but are rarely available at radiotelescopes conducting cm-λ observations. So, for these observations, one usually resorts to the approximate formula of Saastamoinen (1972) which is based on surface measurements of temperature and humidity readily available. This might be quite approximative because the water vapor is not well mixed in the troposphere and, consequently, is not well correlated with ground-based meteorological data. Nonetheless, this approximate formula for the zenith wet path length Z_W (m) provides an useful indication and is :

$$Z_W = c\tau_W \sim 0.002277 \times \frac{1255}{T} \times \frac{RH}{100} \times 1013.246 \times 10^{(17.443 - \frac{2796}{T} - 3.868 \mathrm{Log}_{10} T)}$$

where temperature T is in K and relative humidity RH is in %. The path length Z_W is comprised between $\sim$ 0.05 m at very dry sites (West USA) and 0.40 m at very damp sites (St Croix, Virgin Islands), *i.e.* 0.17 and 1.4 nanosecond of delay τ_W.

2.3. MAPPING ZENITH DELAY TO ANOTHER LINE OF SIGHT

The tropospheric delay τtrop, static in the direction of a source at elevation E integrates a longer column density through the atmosphere than at zenith. $\tau_{\mathrm{trop,static}}(E)$ at elevation E can be scaled from the zenith path length above the radiotelescope by appropriate mapping functions for the dry and wet components :

$$\tau_{\mathrm{trop,static}}(E) = \mathbf{M_D}(E) \times \frac{Z_D}{c} + \mathbf{M_W}(E) \times \frac{Z_W}{c}$$

The simplest mapping function assumes an azimuthally symmetric slab covering a flat Earth portion and is :

$$\mathbf{M_D}(E) = \mathbf{M_W}(E) = \frac{1}{\sin E}$$

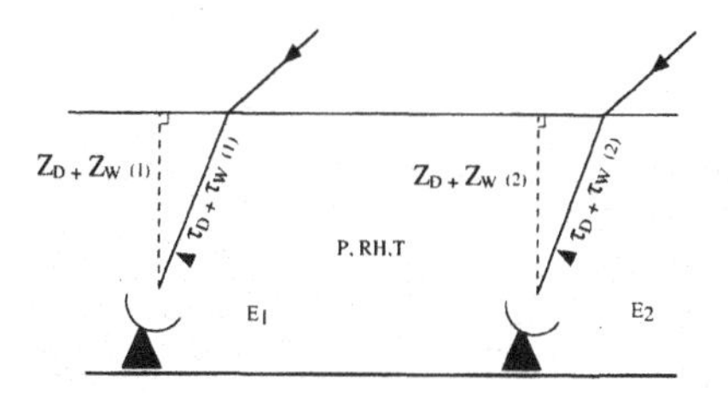

Figure 2. Flat and uniform portion of the atmosphere above a short baseline interferometer.

The literature is rich in more sophisticated mapping functions including the form devised by Marini(1972) to account for the curvature of the Earth:

$$\mathbf{M}(E) = \frac{1}{\mathrm{sinE} + \frac{\mathrm{a}}{\mathrm{sinE} + \frac{\mathrm{b}}{\mathrm{sinE+c}}}}$$

and other forms proposed by Lany (1984), Davis (1985) and Niell (1996).

2.4. CONNECTED INTERFEROMETRY (SHORT BASELINES)

Over very short baselines, the relevant portion of the troposphere is confined to a **flat and uniform** slab (Figure 2). The flatness makes the elevations of the source identical at the two radiotelescopes. Uniformity makes pressure p, relative humidity RH and temperature T identical over the whole slab yielding identical zenith path lengths ($Z_D(1) = Z_D(2)$ and $Z_W(1) = Z_W(2)$). In these conditions, the two dry delays τ_D at stations 1 and 2 cancel out in the interferometric delay τ between the two stations. Similarly, the two wet delays at both ends of the baseline cancel in τ.

This is true only for very short baseline intererometer on leveled ground as the first interferometers were built. Note that a barometric variation of ~ 10 millibars is expected between two antennae with an altitude difference of only 100 m at typical altitude and slow variations of the interferometer phase occur. For instance, the VLA, in its extended configuration, usually exhibits phase variation of tens of degrees over a few hours at 8.4 GHz. This phase drift is calibrated by observations of a phase reference source about every 20 minutes at the VLA.

2.5. VERY LONG BASELINE INTERFEROMETRY (VLBI)

In VLBI, the remoteness of the stations makes the relevant portion of the troposphere confined to a **curved and non-uniform** slab over a large portion of the Earth (Figure 3). Consequently, pressures, temperatures and

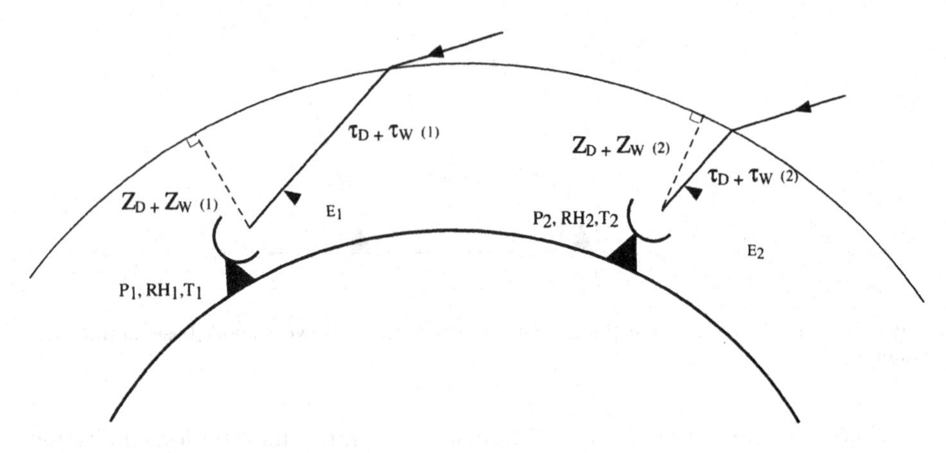

Figure 3. Curved and non-uniform portion of the Earth above a long baseline array (VLBI)

relative humidities at the 2 stations are quite different owing to weather and altitudes changes yielding different zenith path lengths at the stations ($Z_D(1) \neq Z_D(2)$ and $Z_W(1) \neq Z_W(2)$). In addition, the elevations of the observed source are different at the 2 stations ($E_1 \neq E_2$) and the mapping functions yield different scaling factors at both ends of the baseline. In these conditions, the two tropospheric delays τ_D and τ_W at stations 1 and 2 are, respectively, quite different and do contribute to the interferometric delay $\tau = \tau_{geo} + \tau_{D(2)} - \tau_{D(1)} + \tau_{W(2)} - \tau_{W(1)}$.

2.6. ACCURACY OF SURFACE METEOROLOGICAL DATA

Finally, it is useful to have an idea of the errors of the tropospheric delays modeled from surface meteorological data.

The dry delay depends on pressure p through $\tau_D \propto \frac{p}{\sin E}$. Realistically, barometers in the field are rarely calibrated to better than $\sigma_p = 1$ millibar. The corresponding errors for the dry delays at zenith and at the low elevation E = 6° are in Table 1. The wet delay depends on relative humidity RH and temperature T through $\tau_W \propto \frac{RH}{T \times \sin E}$. The accuracy on RH is $\sigma_{RH} \sim 10$ % and the accuracy on T is $\sigma_T \sim 1\,K$. The corresponding errors for the wet delays are also in Table 1 but might actually be larger since the Saastamoinen (1972) model itself can be unaccurate by as much as 100 % depending on the mixing of water vapor and air over head. We remind the reader that these errors apply to the static delays and are constant or only slowly varying in the course of the observations.

TABLE 1. Errors in the estimation of the static tropospheric delays

Component	σ surface meas.	σ zenith length (delay)	σ length (delay) at E=6°
dry	pressure $\sigma_P = 1$ millibar	$\sigma_{Z_D} \sim 0.2$ cm (7 picosec)	$\sigma_{\tau_D} \sim 2$ cm (70 picosec)
wet	Rel. Hum. $\sigma_{RH} = 10$ %	$\sigma_{Z_W} \sim 4$ cm (140 picosec)	$\sigma_{\tau_W} \sim 40$ cm (1.4 nanosec)
wet	Temp. $\sigma_T = 1$ K	$\sigma_{Z_W} \sim 0.5$ cm (18 picosec)	$\sigma_{\tau_W} \sim 5$ cm (180 nanosec)

2.7. STRATEGIES TO ACCOUNT FOR THE STATIC TROPOSPHERIC DELAY

In VLBI, the difficulty in modelling the wet component of the atmosphere has been circumvented by various strategies.

In *VLBI global astrometry and geodesy*, the parameter estimation softwares (JPL-MODEST and GSFC-CALC-SOLV) solve for zenith delays above each radiotelescope for consecutive time segments of about 1 or 2 hours each (Sovers & Fanselow, 1998). The observed sources must be well distributed over the sky, including low elevation observations, to strongly constrain the fit.

In *VLBI mapping*, phases of individual baselines are not used directly but through closure phases over triangles of baselines. Advantageously, closure phase is immune to tropospheric (and ionospheric) fluctuations (Jennison *et al*, 1958; Rogers *et al*, 1974) but looses the position information of the source on the celestial sphere.

Finally, *phase-referencing* used in VLBI is a technique to get rid in large part of the atmosphere while retaining the positional and structural information of the source (Shapiro *et al*, 1979). The static troposphere delay (equivalently phase) cancels in large part in the differenced phase between two angularly close sources. Precisely, an unmodeled error $\delta\phi_{trop,static}$ along one line of sight is scaled down by the source separation $\Delta\theta$ in radian and reduces to $\delta\Delta\phi \propto \delta\phi_{trop,static} \times \Delta\theta$ in the differenced phase $\Delta\phi$. For instance, a 1° source separation yields the significant reduction factor $\frac{1}{57}$ that must be applied the path length errors of Table 1. After this reduction, the unmodeled path length error is usually much less than the wavelength for cm-λ observations. Observations of the two sources are usually not simultaneous but interpolation of the phases between adjacent observations a few minutes apart is easy most of the time. Delay rate ($\dot{\tau}$ in s/s) can be of help for this phase connection. Phase-referenced VLBI observations provides both

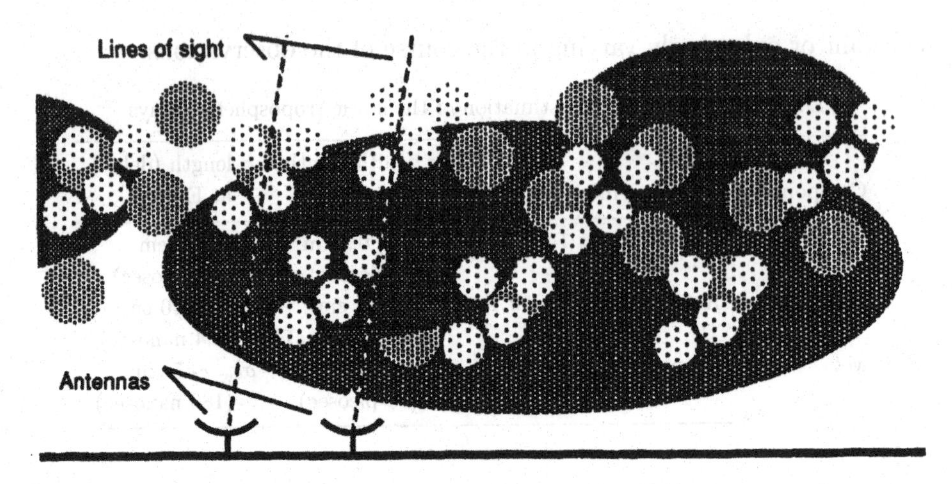

Figure 4. Inhomogeneities (blobs) of water vapor blown by the wind above the ground. These inhomogeneities are responsible for refractive index changes and delay fluctuations on time scales from a few seconds to a few minutes. Picture from Masson (1994).

the map and the precise astrometric position of the target source relative to the reference source (Rioja & Porcas, 2000). Phase referencing works for strong sources even at separation as large as 15° (Pèrez-Torres *et al*, 2000) and for very weak radio sources, at the level of 1 milliJansky or less (Lestrade *et al*, 1990) and (Garrett *et al*, 2001).

3. Fluctuations of tropospheric delay

In the troposphere, inhomogeneities of the water vapor content are blown above the ground by the wind. The resulting changes in the refractive index cause delay fluctuations with typical time scales of a few seconds to a few minutes. The rms of these fluctuations is highly dependent on the altitude of the site. These fluctuations are relatively unimportant for observations at $\lambda > 7$mm but are a major issue for millimeter$-\lambda$ observations.

3.1. PHASE STRUCTURE FUNCTION :

Any random fluctuations f(x) on a one-dimensional screen can be characterized by :

i) its power spectrum $P_f(q)$ where q is the spatial frequency 1/a, corresponding to inhomogeneity of scale a.

ii) its autocorrelation function $B_f(b)$:

$$B_f(b) = < [f(x+b) - \overline{f(x)}] \times [f(x) - \overline{f(x)}] >$$

where b is a distance-lag in the screen and <> is the ensemble average over all pairs ($f(x_i)$, $f(x_i+b)$)
iii) its structure function $D_f(b)$:

$$D_f(b) = < [f(x+b) - f(x)]^2 >$$

which is the variance of f(x) over a range of distance-lag b.

The latter function was introduced by A. Kolmogorov to study turbulence. It describes how the variance of some medium property depends on the separation b between 2 points of the screen. The structure function has the merit to be defined for non-stationary series and is also better behaved than the autocorrelation function for data spans which are shorter than or of the order of a few fluctuation spatial scales. In these circonstances, estimates of the autocorrelation function are prone to errors that arise from a poorly estimated mean. The relationship between the structure function and the autocorrelation function is :

$$D_f(b) = 2[B_f(0) - B_f(b)]$$

And, in general, the autocorrelation function itself can be computed from the power spectrum with the relation : $B_f(b) = \frac{2}{b} \int_0^\infty q P_f(q) \sin(qb) dq$. These simple 1-D formulae can be generalized for a 3-D medium (Tatarski, 1962).

The atmosphere is turbulent in nature. It is an inhomogeneous medium that can be thought of as made of blobs with a range of sizes "a" (Figure 4), *i.e.* a range of spatial frequencies q=1/a. It is often considered isotropic and its fluctuation power spectrum is in form of the power-law $P_{3N} = C_n^2 q^{-\beta}$. The structure coefficient C_n^2 is the strength of the turbulence and the Kolmogorov index β is 11/3 for a 3D medium (noted by the subscript 3N). This power spectrum has a high frequency cutoff q_i that corresponds to the inner scale a_i (the smallest blob). The structure function saturates - the variance stops growing with b - at some distance-lag that is called the outer scale a_o (the largest blobs). The outer scale a_o is of the order a few km. This is fundamental to make VLBI work across intercontinental baselines.

The most complete phase structure fonction above a radioastronomy site has been measured for the VLA at 22 GHz (Figure 5) by Carilli and Holdaway (1997) for a limited period of 90 minutes on January 27 1997 in the A configuration with baselines ranging from 200 m to 37 km in the direction of a calibrator (0748+240). Three regions of the Kolmogorov turbulence theory are apparent in this plot; for baseline length $b < 1.2$ km turbulence expands in 3D space; while for 1.2 km $< b <$ 6 km, turbulence

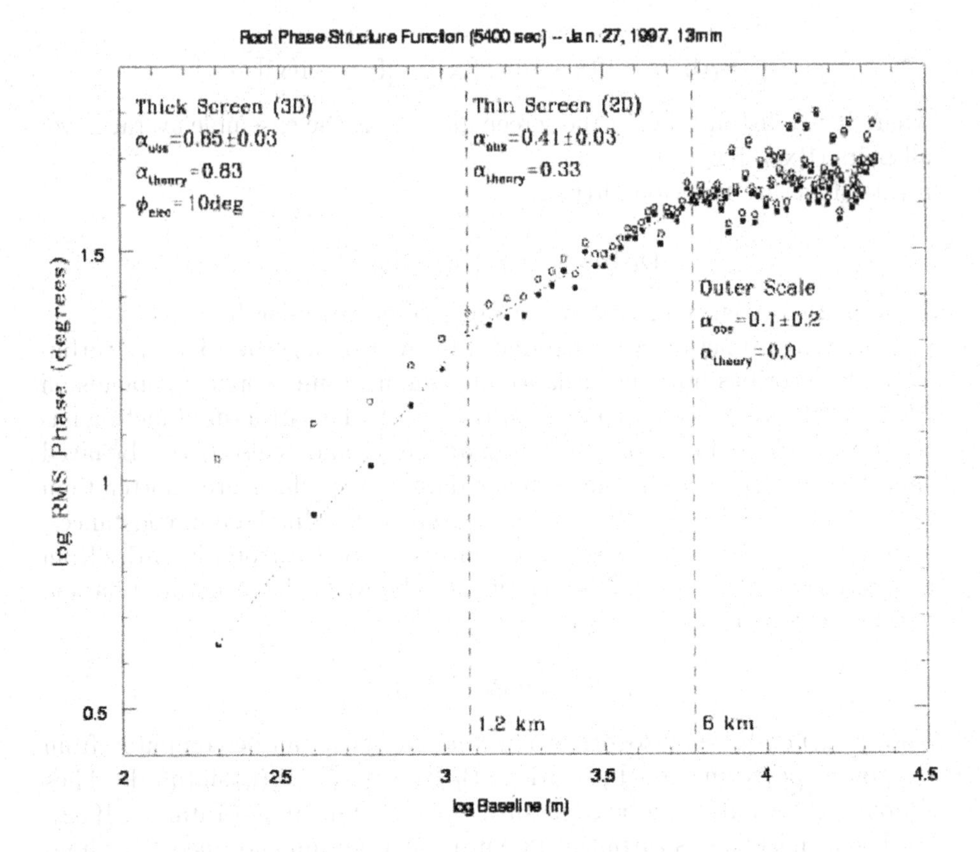

Figure 5. Phase structure function measured at 22 GHz over the VLA (configuration A) in the direction of the calibrator 0748+240 for 90 minutes on January 27 1997 by Carilli & Holdaway (1997)

is confined to a 2D layer at some height above the ground; for b > 6 km, the fluctuations tend to saturate, *i.e.* the outer scale for these observations is $\sim$ 6 km. It is not clear whether or not this function is typical since the observations were limited to a very short period. But it certainly conforms to the model expected for Kolmogorov turbulence in the atmosphere.

3.2. TROPOSPHERIC FLUCTUATIONS AT SEVERAL MILLIMETER$-\lambda$ SITES

Studies of propagation fluctuations have been conducted at several sites around the world. As just mentioned, a phase structure function was measured at 22 GHz over the VLA (Carilli & Holdaway, 1997). A phase structure function was measured at 86 GHz over the IRAM baselines 24m$-$288m during 49 days in Feb$-$August 1990 (Olmi & Downes, 1992). A path length rms versus time was measured using phases at 11.198 GHz over two 300m baselines at Chajmantor and Pampa la Bola in Chili (ALMA) (Delgado & Nyman, 2001). A path length rms versus time was measured over a 100m BIMA baseline for 14 days in April–May 1998 (Akeson, 1998 & http://bima2.astro.uiuc.edu/phasemon/phasemon.html). A phase structure function was measured at 98 GHz at OVRO over the baselines 50–500m (courtesy of John Carpenter). A path length rms versus local hours and months was measured at SMA (Hawai) over a 100 m baseline (Masson, 1994).

In Table 2, we summarize those measurements and convert, when needed, from phase to path length for easier comparison of fluctuations between these millimiter observations sites used for interferometry.

TABLE 2. Tropospheric path length fluctuations of millimeter$-\lambda$ observing sites used for interferometry

Name of site	altitude (m)	baselines b (m)	atmospheric path length fluctuations range (microns)	references
BIMA (CA, USA)	1050	100	100 - 1200	Akeson (1998)
OVRO (CA, USA)	1220	50-500	85 - 1200	Carpenter (2001)
VLA (NM, USA)	2124	200m - 37km	110 - 1150	Carilli & Holdaway (1997)
IRAM (Alps, France)	2500	24-288	150 - 1000	Olmi & Downes (1992)
SMA (Hawai)	4200	100	50 - 600	Masson (1994)
ALMA (Chili)	5000	300	10 - 40	Delgado & Nyman (2001)

3.3. INTERFEROMETRY AT MILLIMETER WAVELENGTHS

Delay fluctuations due to spatial inhomogeneities in the troposphere is a limiting factor for interferometry at millimeter-wavelengths, i.e. $\lambda < 7$mm. There are two types of solutions : continuous water vapor radiometry or phase-referenced observations with fast switching of the antennae. We shall discuss only phase referencing and refer the reader to Hills & Richer (2000) and Lay *et al* (1998) for an outline of radiometry at 183 GHz for ALMA and Marvel & Woody (1998) at 22 GHz for OVRO.

Phase referencing is the use of a calibrator source whose position is accurately known to establish a reference phase that tracks variations due to atmosphere and instrumentation. The reference and target sources are alternately observed to form the differenced phase between the two sources.

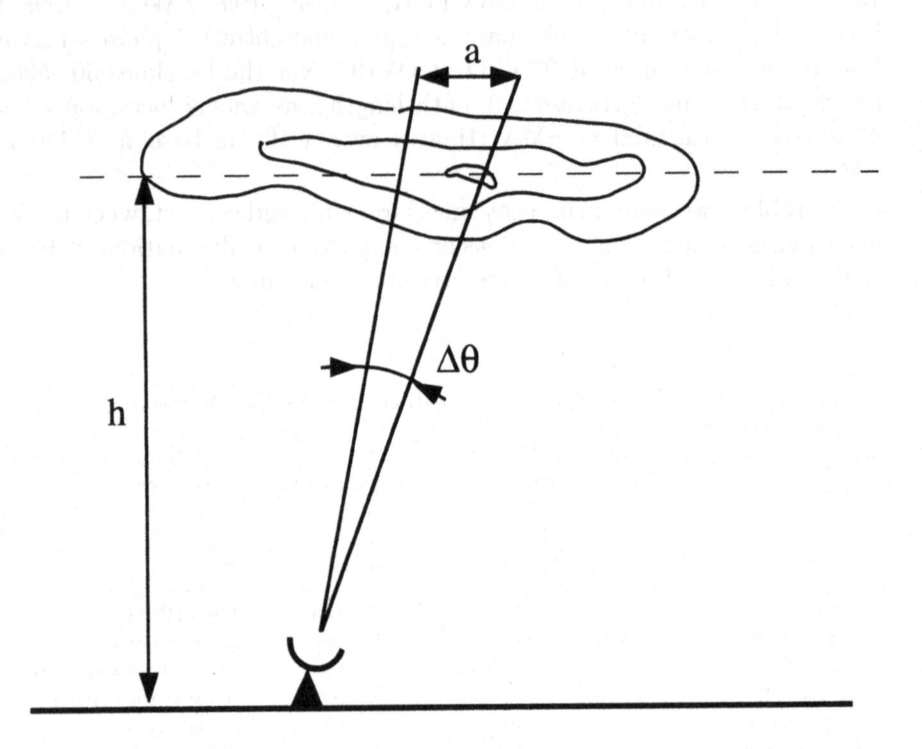

Figure 6. Scale "a" of the tropospheric inhomogeneities sampled by two lines of sight separated by $\Delta\theta$

Troposheric fluctuations are largely removed from this differenced phase if the angular separation ($\Delta\theta$) between sources is small enough and if the switching cycle time t_{switch} between consecutive observations is short enough as emphasized by Carilli & Holdaway (1997).

a) Condition on the angular separation $\Delta\theta$:

The troposphere is usually described as frozen-in, *i.e.* the spatial inhomogeneities are stationnary and blown by the wind at a constant speed V (1-10 m/s). The scale of the tropospheric inhomogeneity sampled over $\Delta\theta$ (rd), as shown in Figure 6, is :

$$a \sim \frac{h}{\sin E} \times \Delta\theta = b_{eff}$$

This scale "a" is an effective baseline b_{eff} that can be used to estimate

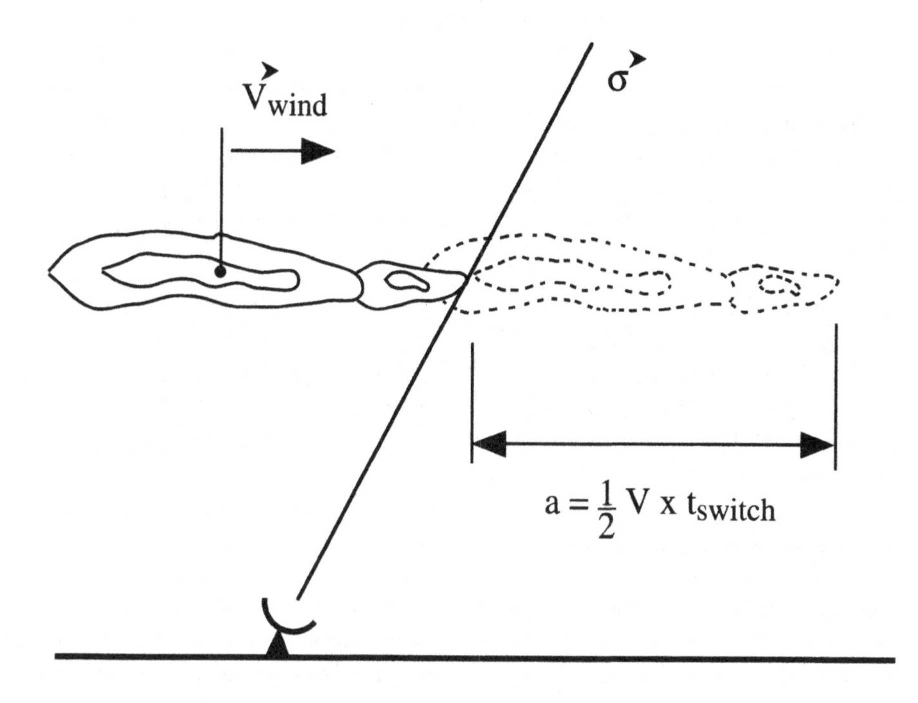

Figure 7. Scale "a" of the tropospheric inhomogeneities sampled by two consecutive observations separated by the switching cycle time t_{switch}

the rms phase fluctuation between the two sources from the phase structure function measured at the site. For instances, if troposphere height h = 2km, elevation $E = 20°$ and $\Delta\theta = 5°$, then scale "a" of the inhomogeneity sampled is 600 m. This yields a path length rms $\sigma_{c\tau}(600\text{m}) \sim 350\mu\text{m}$ with the phase structure function of the VLA (Figure 5). For the future array ALMA, in assuming a phase structure function consistent with the Kolmogorov turbulence spectrum normalised by the 300m baseline variance of Delgado & Nyman (2001), the path length rms would be $\sigma_{c\tau}(600\text{m}) \sim 80\mu\text{m}$ for the worst conditions and $\sigma_{c\tau}(600\text{m}) \sim 30\mu\text{m}$ for the best conditions.

b) Condition on the switching cycle time:

Observations of the target and reference sources are usually not simultaneous and the antennae of the array must be switched back and forth between the two sources over the cycle time t_{switch}.

The scale of the inhomogeneity sampled during t_{switch}, as shown in Figure 7, is :

$$a = \frac{1}{2} \times V \times t_{switch} = b_{eff}$$

This scale "a" is an effective baseline b_{eff} that can be used to estimate the rms phase fluctuation between consecutive observations from the phase structure function measured at the site. For instance, if the wind speed is $V = 10m/s$ and $t_{switch} = 2min$, then the scale or effective baseline is 600m and the resulting delay rms $\sigma_\tau(b_{eff})$ can be estimated from the phase structure function of the site as above.

Lim *et al* (1998) have used the fast-switching observing mode at the VLA (extended A configuration) to detect and resolve for the first time the photosphere of the supergiant star Betelgeuse at λ=7 mm. Their reference source was 4.2° away and t_{switch} = 2.5 minutes was sufficient to remove in large part the atmospheric fluctuations in the differenced phase at this wavelength.

4. Ionospheric delay

The ionised shell around the Earth is caused by the UV radiation of the Sun and is inflated in its direction. The electronic column density peaks at an height of $\sim$ 350 km above the ground but spreads by several tens of kilometers around this altitude. An example of the dependence of the electronic column density on longitude, or on local hour, is shown in Figure 8. At night, the level of ionisation is low with a column density at zenith of $\sim$ 2 10^{16} $\bar{e}/m^2$. At daytime, it typically peaks at $\sim$ 50 10^{16} $\bar{e}/m^2$ but it can be 10 times larger in extreme conditions. Variations have time scale of several hours and induce a static ionospheric delay $\tau_{ion,static}$ comprised between $\sim$ 0.1 and 2 nanosecond at zenith at 8.4 GHz. In addition, short term variations (minutes) induce the fluctuation $\tau_{ion,fluctuation} \sim 0.5$ nanosecond at 8.4 GHz. The ionosphere is dispersive and so τ_{ion} is 13 times larger at 2.3 GHz. The ionosphere column density can be mapped to the line of sight by a simple geometric slant factor (Sovers & Fanselow, 1998).

Several strategies have been developped to calibrate the ionosphere in VLBI data.

4.1. REMOVAL OF IONOSPHERE IN DELAY BY DUAL FREQUENCY OBSERVATIONS

Dispersive property of the ionosphere makes its contribution to the group delay to be $\tau_{gd,ion} = \frac{Q}{\nu 2}$, with $Q = cr_eI_e/2\pi$, the electron radius r_e and the electronic column density I_e, *i.e.* the total number of electrons per unit area along the integrated line of sight. In geodesy and global astrometry,

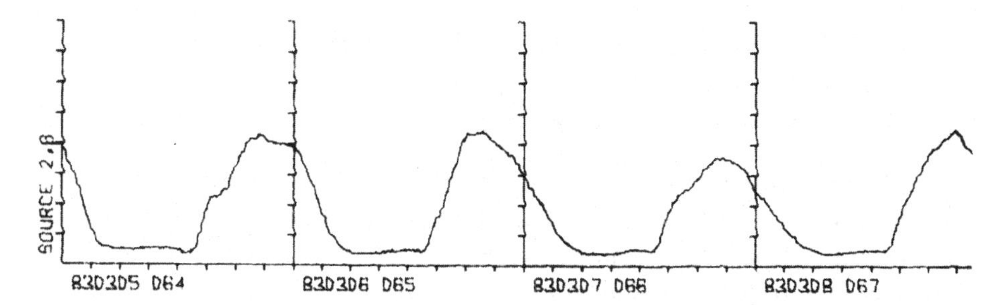

Figure 8. Example of the daily variation of the ionosphere Total Electronic Content above Goldstone (CA, USA) during a period of 4 days in March 1983 (5th, 6th, 7th, 8th). The peak is $\sim 50\ 10^{16}\ \bar{e}/m^2$ at $\sim$ 14 hour local time. Note the low plateau at night.

the ionosphere is calibrated out by simultaneously measuring two group delays, τ_x and τ_s, at X band ($\nu_x = 8.4$ GHz) and S band (ν_s =2.3 GHz) :

$$\tau_x = \tau + \frac{Q}{\nu_x^2}$$

$$\tau_s = \tau + \frac{Q}{\nu_s^2}$$

A linear combination of these group delays yields straightforwardly the ionosphere-free delay τ (Sovers & Fanselow, 1998) :

$$\tau = \frac{\nu_x^2}{\nu_x^2 - \nu_s^2}\tau_x - \frac{\nu_s^2}{\nu_x^2 - \nu_s^2}\tau_s$$

A caveat is that if the source has different morphologies at 2.3 and 8.4 GHz, the removal of the ionosphere is biased. Nevertheless, in practice, this technique has been very successful for VLBI global astrometry and geodesy where sources of interest are chosen to be as close as possible to a point source.

4.2. COMBINATION OF PHASE AND GROUP DELAYS

The definitions of the group delay ($\tau_{gd} \propto \frac{\partial\phi}{\partial\nu}$) and of the phase delay ($\tau_{pd} \propto \frac{\phi}{\nu}$) make the contribution of the ionosphere to be of opposite signs:

$$\tau_{pd} = \tau - \frac{\mathrm{Q}}{\nu^2}$$

$$\tau_{gd} = \tau + \frac{\mathrm{Q}}{\nu^2}$$

So, summation of these delays yields the ionosphere-free delay τ :

$$\tau = \frac{\tau_{pd} + \tau_{gd}}{2}$$

There are two difficulties in applying this technique to VLBI data. First the phase delay must be made unambiguous by the proper strategy of observation. Second the group delay is measured with a precision which is at least one order of magnitude lower than the phase delay so that the sum above is ill-determined. However, this technique has been studied in great details and used successfully at 5 GHz by Guirado *et al* (1995) for sub-milliarcsecond astrometry of strong radio sources.

4.3. PHASE REFERENCING

Phase-referenced observations at a single frequency allow removal of ionosphere by differencing phases between target and reference sources. This kind of observations yields differential astrometry or a map of the source referenced to a predefined phase center. This is similar to the cancellation of the tropospheric contributions discussed above (Figue 6 with h=350 km).

For ionosphere removal by phase-referencing, lower the frequency of observation is, more severe perturbations are and angularly closer the sources must be. When it happens that the target and reference sources are close enough to be within the antenna beam, no antenna switching is required. This becomes more likely at lower frequency where antenna beams are larger. Fomalont *et al* (1999) have shown that weak in-beam calibrators are better than strong but off-beam calibrators for 1.5 GHz VLBI observations of two pulsar. They have determined their motions with a sub-milliarcsecond precision using this strategy.

4.4. GPS IONOSPHERIC MAP

Erickson *et al* (2001) have conducted an experiment to evaluate the usefulness of ionospheric data produced using the Global Positioning System (GPS) for making Faraday rotation and interferometer phase corrections at the NRAO Very Large Array (VLA). Four GPS receivers were installed at the VLA site - one at the array center and one at the end of each arm. A simple ionospheric model consisting of a vertical Total Electronic Content (TEC), a horizontal gradient, and the azimuth of that gradient was developed and fitted to the GPS TEC data from each receiver. The model was then used to predict the TEC in the observing direction. Ionospheric Faraday rotation and phase gradients were then estimated and compared with VLA measurements taken at frequencies of 322 and 333 MHz. They find that Faraday rotation corrections can be predicted to an accuracy of $\leq 2°$, in general. The interferometer phase shifts caused by large-scale ($>$ 1000 km) ionospheric structures can be predicted by their model. However the phase shifts caused by smaller ($<$ 100 km) structures can be estimated only when the direction of observation lies within a few degrees of one of the GPS satellites.

The Jet Propulsion Laboratory provides daily global map of the ionosphere (http://iono.jpl.nasa.gov).

4.5. REMOVAL OF IONOSPHERE AT LOW FREQUENCY BY MULTI-BAND OBSERVATION

Brisken *et al* (2000) have developed a technique to remove the ionosphere distorting effects from low-frequency VLBI observations ($<$ 2 GHz). They used the phases from several sub-bands spread over $\sim$ 10 % of the observation frequency. The dispersive and nondispersive components in the phase are :

$$\phi(\nu) = \left[\frac{1}{c}(Ul + Vm) + \delta A\right]\nu + \left[\delta B\frac{1}{\nu}\right]$$

where U and V are the projected baseline lengths, l and m are the pulsar vector distance from the phase center, and $\delta A\nu$ is the phase that is due to the unmodeled wet troposphere and is small ($< 45°$ of phase). The term $\delta B/\nu$ is the phase that is due to uncorrected ionosphere and is dispersive. The first three terms are nondispersive and separate from the last term $\delta B/\nu$ when phases at several frequencies are measured. Brisken *et al* (2000) have used eight different 8 MHz bands around frequencies ranging from 1410 to 1730 MHz and effectively removed ionosphere from the data without the use of *a priori* calibration information. This technique was used to perform accurate astrometry of the pulsar B0950+08 at 1.4 GHz, resulting in a sub-milliarcsecond precision measurement of its proper motion (Brisken *et al*, 2000). In essence, this technique is the same as removal of ionosphere by S and X band observations. However, the phase delay being more precise than the group delay, the spread of sub-bands required for the Brisken *et al* 's approach need to be less and fits nicely across current receiver bandpass of 100-400 MHz.

Acknowledgements : I am in debt to Robert A. Preston and Ojars J. Sovers at Jet Propulsion Laboratory and to Alan E.E. Rogers and Robert B. Phillips at Haystack Observatory for many enlightening discussions on VLBI in the course of more than a decade during the Hipparcos/VLBI link project.

References

Akeson, R.L., 1998, Atmospheric phase conditions for the Hat Creek 1997-98 observing season, *BIMA Memoranda Series No 68, Berkeley University*

Carilli, C.L. Holdaway, M. A., 1997, Application of Fast Switching Phase Calibration at mm Wavelengths on 33 km Baselines, *MMA/ALMA Memorandum Series No 173, National Radio Astronomy Organisation NRAO*

Brisken, W. F., Benson, J. M.; Beasley, A. J., Fomalont, E. B.; Goss, W. M.; Thorsett, S. E., 2000, Measurement of the Parallax of PSR B0950+08 Using the VLBA, *Astroph. J.* , **Vol. 541**, pp. 959

Davis, J.L., Herring, T.A., Shapiro, I.I., Rogers, A.E.E., Elgered, G., 1985, Geodesy by radiointerferometry: Effects of Atmospheric Modelling Errors on Estimated Baseline Length, *Radio Science*, **Vol. 20**, pp. 1593-1607

Delgado, G. , Nyman, L-A, 2001, Velocity of the Effective Turbulence Layer at Chajnantor Estimated From 183 GHz Measurements, *ALMA Memorandum Series No 363, National Radio Astronomy Organisation NRAO*

Erickson, W. C., Perley, R. A., Flatters, C., Kassim, N. E., 2001, Ionospheric corrections for VLA observations using Local GPS data, *Astron. & Astrophys.*, **Vol. 366**, pp. 1071

Fomalont, E. B., Goss, W. M., Beasley, A. J., Chatterjee, S., 1999, Sub-Milliarcsecond Precision of Pulsar Motions: Using In-Beam Calibrators with the VLBA, *Astron. J.*, **Vol. 117**, pp. 3025

Garrett, M. A., Muxlow, T. W. B., Garrington, S. T., Alef, W., Alberdi, A., van Langevelde, H. J., Venturi, T., Polatidis, A. G., Kellermann, K. I., Baan, W. A., Kus, A., Wilkinson, P. N., Richards, A. M. S., 2001, AGN and starbursts at high redshift: High resolution EVN radio observations of the Hubble Deep Field, *Astron. & Astroph.*, **Vol 366**, pp. L5-L8.

Guirado, J. C., Marcaide, J. M., Elosegui, P., Ratner, M. I., Shapiro, I. I., Eckart, A., Quirrenbach, A., Schalinski, C. J., Witzel, A., 1995, VLBI differential astrometry of the radio sources 1928+738 and 2007+777 at 5 GHz, *Astron. & Astrophys.*, **Vol. 293**, pp. 613

Jennison, R.C., 1958, A phase sensitive interferometer technique for the measurement of the Fourier Transforms of spatial brightness distributions of small angular extent, *Mon. Not. Roy. Astron. Soc.*, **Vol. 118**, pp. 276

Hills, R., Richer, J., 2000, Water Vapor radiometers for ALMA, *NRAO/ESO ALMA memo series*, **No 303**

Lestrade, J.-F., Rogers, A. E. E., Whitney, A. R., Niell, A. E., Phillips, R. B., Preston, R. A., 1990, Phase-Referenced VLBI Observations of Weak Radio Sources - Milliarcsecond position of Algol, *Astron. J.*, **Vol. 99**, pp. 1663

Lanyi, G.E., 1984, Tropospheric Delay Effects in Radio Interferometry, *Telecommunication and Data Acquisition Prog. Rept Jet Propulsion Laboratory, Pasadena, California,* , **42-78**, pp. 152-159.

Lay, O. P., Wiedner, M. C., Carlstrom, J. E., Hills, R. E., 1998, CSO-JCMT Interferometer and 183-GHz radiometric phase correction, *SPIE–The International Society for Optical Engineering*, **Vol. 3357**, pp. 254

Lim, J., Carilli, C.L., White, S.M., Beasley, A.J., Marson, R.G., 1998, Large convection cells as source of Betelgeuse's extended atmosphere, *Nature*, **Vol. 392**, pp. 575-577

Marini, J.W., 1972, Correction of Satellite Tracking Data for an Arbitrary Tropospheric Profile, *Radio Science*, **Vol. 7, No. 2**, pp. 223-231.

Marvel, K. B., and Woody, D. P. 1998, Phase correction at millimeter wavelengths using observations of water vapor at 22 GHz, *SPIE–The International Society for Optical Engineering*, **Vol. 3357**, pp. 442

Masson, 1994, Atmospheric Effects and Calibrations, *Astronomy with Millimeter and Submillimeter Wave, Interferometry, IAU Colloquium 140, ASP Conference Series*, **Vol. 59** , edided M. Ishiguro & J. Welch, pp. 87

Niell, A.E., 1996, Global Mapping Functions for the Atmosphere Delay at Radio Wavelengths, *J. Geophys. Res.*, **Vol. 101**, pp. 3227-3246

Olmi, L., Downes, D, 1992, Interferometric measurement of tropospheric phase fluctuations at 86 GHz on antenna spacings of 24 M to 288 M, *Astron. Astroph.*, **Vol. 262**, 634

Pèrez-Torres, M. A., Marcaide, J. M., Guirado, J. C., Ros, E., Shapiro, I. I., Ratner, M. I., Sardòn, E., 2000, Towards global phase-delay VLBI astrometry: observations of QSO 1150+812 and BL 1803+784, *Astron. & Astrophys.*, **Vol. 360**, pp. 161

Rioja. M.J., Porcas, R.W., 2000, A phase-reference study of the quasar pair 1038+528A,B, *Astron. Astrophys.*, **Vol. 355**, pp. 552-563.

Rogers, A.E.E., Hinteregger, H.F., Whitney, A.R., Counselman, C.C., Shapiro, I.I., Wittels, J.J., Klemperer, W.K., Warnock, W.W., Clark, T.A., Hutton, L.K., Marandino, G.E., Ronnang, B.O., Rydbeck, O.E.H., Niell, A.E., 1974, The structure of radio sources 3C273 and 3C84 deduced from the "closure" phases and visibility amplitudes observed with three-element interferometers *Astroph. J.*, **Vol. 193**, pp. 293-301

Saastamoinen, J, 1972, Atmospheric Correction for the Troposphere and Statosphere in Radio Ranging Satellite, *Geographical Monograph Series* , **Vol. 15**, ed. by S.W. Henriksen et al, AGU, Washington, D.C., pp. 247-251.

Shapiro, I. I., Wittels, J. J., Counselman, C. C., III, Robertson, D. S., Whitney, A. R., Hinteregger, H. F., Knight, C. A., Clark, T. A., Hutton, L. K., Niell, A. E., 1979, Submilliarcsecond astrometry via VLBI. I - Relative position of the radio sources 3C 345 and NRAO 512, *Astron. J.*, **Vol. 84**, pp. 1459

Smith, E.K., Jr., Weintraub, S., 1953, The Constants in the Equation for Atmospheric Refractive Index at radio frequency, *Proc. IRE*, **Vol. 41**, pp. 1035-1037

Sovers, O. J., Fanselow, J. L., 1998, Astrometry and geodesy with radio interferometry : experiments, models, results, *Rev. Modern Phys.*, **Vol. 70**, pp. 1393-1454

Tatarski, V.I., 1962, Wave Propagation in a Turbulent Medium, *Mc Graw-Hill Book Co, Inc, Publisher, New-York*

Thompson, A.R., Moran, J.M., Swenson, G.W., (1986), Interferometry and Synthesis in Radio Astronomy, *John Wiley & Sons Publishers* .

GROUND BASED VLBI FACILITIES – THE EUROPEAN AND GLOBAL VLBI NETWORK

M.A. GARRETT
Joint Institute for VLBI in Europe
Postbus 2, 7990 AA Dwingeloo, The Netherlands

1. Introduction

There are at least four well known Very Long Baseline Interferometer (VLBI) networks that provide open access to astronomers around the world. These include:

- European VLBI Network, www.evlbi.org
- Very Long Baseline Array, www.aoc.nrao.edu/vlba/html/VLBA.html
- Coordinated Millimeter VLBI Array, web.haystack.mit.edu/cmva
- Australian Long Baseline Array, www.atnf.csiro.au/vlbi & the Asia-Pacific Telescope, www.atnf.csiro.au/apt.

For a comprehensive guide to the various array characteristics, corresponding correlator capabilities, range of diverse observing modes, different proposal submission procedures, user support *etc.*, I refer the reader to the on-line web pages highlighted above. In this paper I have chosen to focus on a sub-set of these facilities, in particular the European VLBI Network (EVN) both in stand-alone mode but also as a major component of a Global VLBI Network, involving MERLIN (see www.merlin.ac.uk) and the Very Long Baseline Array (VLBA – see Zensus, Diamond & Napier 1995, for a comprehensive review).

2. The European VLBI Network

The EVN is a "part-time" VLBI network that observes in 3 "block sessions" per year. Each of these block sessions is 3-4 weeks long, and usually 3 or more different observing frequencies are available within any given session. Observing sessions are scheduled in February-March, May-June and October-November of each year and often involve both Global and EVN-only observations. The EVN and Global VLBI "Call for Proposals" is issued

F. Mantovani and A. Kus (eds.), The Role of VLBI in Astrophysics, Astrometry and Geodesy, 403–413.

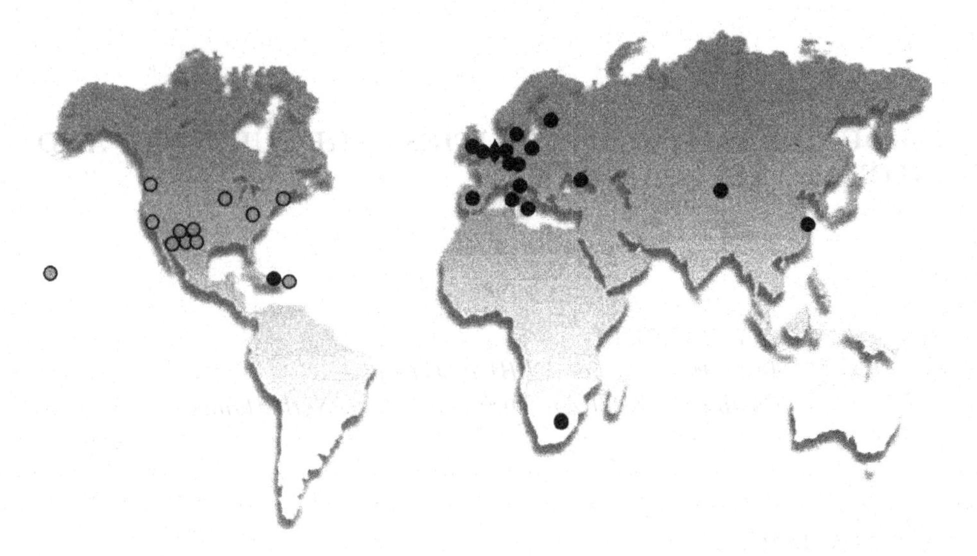

Figure 1. The Global VLBI Network of radio telescopes, including the European VLBI Network (dark dots, including the new 64-m telescope In Sardinia) and the VLBA (lighter dots, including the GBT and VLA). The EVN and VLBA often co-observe forming a very sensitive, high resolution global VLBI network.

three times per year with deadlines of February 1, June 1, and October 1. Please refer to the web-based *EVN User Guide* at `www.evlbi.org` for more details on how to apply for EVN observing time.

The locations of the telescopes, and in addition the EVN MkIV Correlator at JIVE (Joint Institute for VLBI in Europe), are shown in Figure 1. Members of the EVN with radio telescopes are listed in table 1. There are several categories of EVN membership – these recognise the different levels of commitment by the various participating institutes – full or associate membership, and affiliated telescopes. These three categories are separated in table 1 by dividing lines. The table also indicates the diameters of the individual telescopes and their system noise in Jy (at the main EVN observing wavelengths).

2.1. EVN SENSITIVITY

The obvious advantage that the EVN has over other networks is the enormous collecting area it can routinely draw upon across a broad range of frequencies. It is worth remembering that in terms of collecting area, the larger EVN telescopes such as the Effelsberg 100-m, Westerbork tied array and Lovell 76-m, are either individually larger or comparable to the combined collecting area of the ten 25-m antennas that comprise the VLBA. In practical terms, this large collecting area permits much fainter sources

Institute	Telescope	Diam (m)	SEFD (Jy) $\lambda\lambda$21|18|6|5|4|1|0.7 cm
MPIfR (DE)	Effelsberg	100	20|19|20|55|20|140|600
ASTRON (NL)	WSRT	14 × 25	30|30|60|–|120|–|–
JBO (UK)	Lovell	76	35|35|25|25|55| -
	Mk2	25	350|320|320|910|–|910|–
	Cambridge	32	220|212|136|136|–|720|–
IRA (IT)	Medicina	32	390|582|296|900|270|1090|2800
	Noto	32	820|784|260|–|770|2500|3000
	Sardinia	64	Not yet known
OSO (SE)	Onsala-85	25	450|390|600|1500|–|–|–
	Onsala-60	20	–|–|–|–|1630|1380|1310
SHAO (CN)	Shanghai	25	–|1090|520|–|590|1606|–
UAO (CN)	Urumqi	25	1068|1068|353|–|396|2950|–
TCfA (PL)	Torun	32	250|230|250|300|–|–|–
OAN (ES)	Yebes 14-m	14	–|–|–|–|3300|–|4160
	Yebes 40-m	40	Not yet known
MRO (FI)	Metsahovi	14	–|–|–|–|–|2608|4500
NAIC (USA)	Arecibo	305	3|4|6|9|–|–|–
HRAO (ZA)	Hartebeesthoek	26	–|450|700|800|940|–|–
IfAG (DE)	Wettzell	20	–|–|–|–|750|–|–
DSN (USA|ES)	Robledo 70-m	70	–|42|–|–|23|100|–
	Robledo 34-m	34	–|–|–|–|88|–|–
CGS|ASI (IT)	Matera	20	–|–|–|–|900|–|–
NMA (NO)	Ny-Alesund	20	–|–|–|–|1255|–|–
CrAO (UA)	Simeiz	22	–|1600|–|3000|–|1200|3000|–

TABLE 1. EVN member telescopes including those currently under construction at Sardinia and Yebes. System Equivalent Flux Density (SEFD) values for new or refurbished telescopes/receivers (e.g. the re-surfaced Lovell 76-m) are current best estimates or as yet unknown. The table is divided into 3 parts: full, associate members and affiliated observatories.

to be detected, imaged and self-calibrated with the EVN. This advantage applies particularly to spectral line observations (both emission and absorption studies). In addition, another key advantage of the EVN is its capability to perform *sustained* VLBI observations at very high data rates (currently 512 Mbit/sec). Uniquely, the EVN is able to observe at these sustained data rates for 12 hours or longer.

Figure 2 shows the 1σ r.m.s. (image) noise level of a standard EVN array (excluding the DSN telescopes and Arecibo) at λ18 cm for typical spectral line single channel widths of 6 and 30 kHz, dual polarisation continuum bands of 16 MHz (128 Mbits/sec), 32 MHz (256 Mbits/sec) and 64 MHz

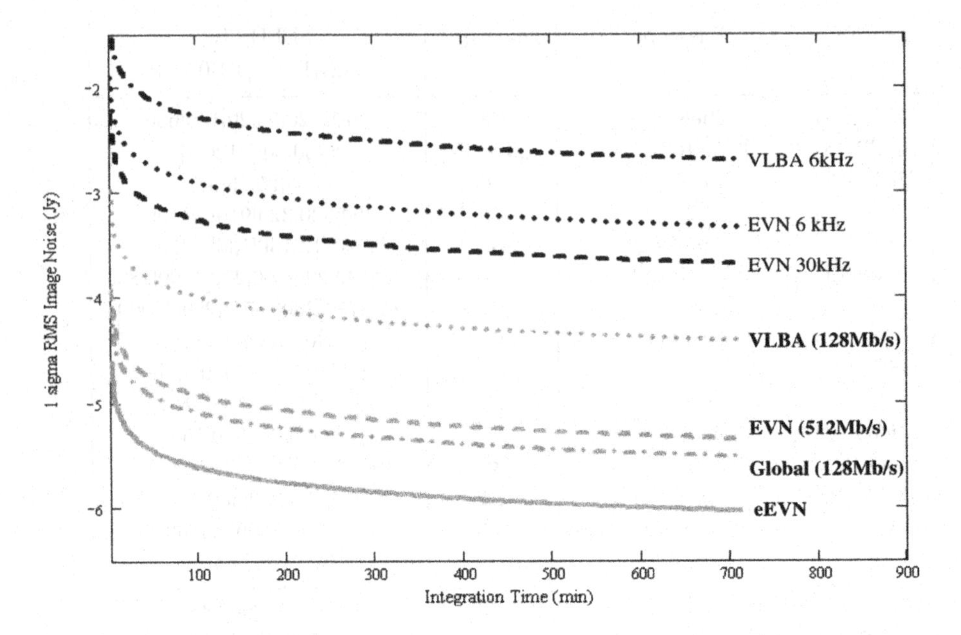

Figure 2. The 1-σ image noise level achieved by the EVN at λ18 cm as a function of time (minutes) for typical spectral line channel widths (6 and 30 kHz), dual-polarisation continuum data rates of 128, 256 and 512 Mbit/sec. For comparison the noise levels achieved by the VLBA for a channel width of 30 kHz and its maximum sustained data rate of 128 Mbits/sec are also included. The noise level achieved by a global VLBI array operating at a sustained data rate of 128 Mbits/sec is also presented. Finally we present the anticipated performance of the *e*EVN – the current EVN telescopes connected together by optical fibres.

(512 Mbits/sec). As a point of reference I plot the corresponding noise levels of the VLBA at 6 kHz, and 16 MHz (the latter being equivalent to the VLBA's maximum sustainable data rate of 128 Mbits/sec over 12 hours). In the same figure I also include an extremely sensitive Global VLBI array including the EVN, VLBA, VLA, GBT, DSN and Arecibo. Casting an eye towards the future, I have plotted the noise level that would be achieved by the EVN at λ6 cm, assuming fibre connections (*e*EVN) and a bandwidth of $\sim$ 2 GHz per polarisation (see section 4). In the latter case, the *e*EVN can be expected to reach impressive sub-microJy noise levels in a typical 12 hour (on-source) observing run.

2.2. UNIQUE EVN FREQUENCIES

Another important feature of the EVN is the availability of observing frequencies that are essentially unique, at least in the northern hemisphere. These include UHF band observations ($\sim 800 - 1300$ MHz) that have been used to search in relatively distant extra-galactic systems for (redshifted) neutral hydrogen in absorption, and the λ5 cm receivers that have been used to infer the presence of circumstellar discs around massive stars located within star forming regions in our own galaxy (see Booth these proceedings for a summary of the latest extra-galactic HI absorption and Methanol maser results). The fact that these receivers were constructed quickly, and then rapidly deployed across a substantial fraction of the network, emphasises the EVN's ability to respond flexibly to "bottom up", user driven demand.

2.3. COMBINED JOINT EVN-MERLIN OBSERVATIONS

The EVN often co-observes with the UK's MERLIN radio telescope network. Two of the MERLIN telescopes conduct both VLBI and MERLIN observations simultaneously – usually the Cambridge 32-m telescope and one of the Jodrell Bank "home" telescopes (either the Lovell 76-m or Mk2 telescope). The advantage of joint EVN-MERLIN observations is the excellent uv-coverage that can be obtained from the combined data set. MERLIN provides baselines on scales ranging from 6 to 217 km, thus providing overlap with the shortest (projected) EVN baselines (Jodrell-Cambridge and Effelsberg-Westerbork in particular). The combined data set therefore includes a range of baseline lengths, from a few to several thousand kilometers. This is of course ideal for detecting and imaging large extended sources that might otherwise be resolved-out or be extremely difficult to image accurately with the EVN alone.

Usually EVN-MERLIN observations take place during every session at one of the main EVN observing frequencies (usually $\lambda\lambda$18 or 6 cm, although joint observations at λ1 cm are also possible). The inclusion of the common Jodrell-Cambridge baseline in both the EVN and MERLIN arrays, ensures that the data sets can be combined together in a consistent fashion.

2.4. THE EVN AS PART OF THE GLOBAL VLBI NETWORK

VLBI is an international effort. The very longest baselines available via the ground require collaborations between various VLBI networks. In the northern hemisphere a particularly strong collaboration exists between the EVN and the VLBA. Both networks employ very similar recording systems (MkIV and VLBA respectively) which provide a wide range of compatible observing modes. Global VLBI observations usually involve the participation of the most sensitive VLBI telescopes in the world, including those that provide the longest baseline lengths – it's not uncommon for up to 20 VLBI telescopes to participate in a single 12 hour observing run. In addition, many Global VLBI projects are also made together with simultaneous MERLIN observations. This high resolution, Global VLBI Network currently provides the ultimate in terms of both sensitivity and uv-coverage (see Figure 3). Snapshot observations of a large number of sources, or of galactic sources that evolve quickly also becomes feasible with such a 20 station global array.

2.5. ON-GOING ENHANCEMENTS TO THE EVN

So far we have focussed on the areas where the EVN is strongest, *viz.* unmatched sensitivity and the ability to observe at several unique frequencies. There is a continuous and vigorous EVN programme of development to maintain and enhance these capabilities. General enhancements to the network (e.g. the recent upgrade to 2-head recording and 512 Mbit/sec data rates) are coordinated via the EVN Technical & Operations Group (see `www.evlbi.org/tog/tog.html` for more details).

Several significant events are expected to take place in the short-term, in particular the addition of two large telescopes to the EVN over the course of the next two to three years. These are the 64-m Sardinia Radio Telescope (the SRT – to be built at San Basilio, near Cagliari and operated by the IRA) and the OAN-Yebes 40-m telescope (to be built alongside the current 14-m OAN antenna at Yebes, near Madrid). In 2005 the Miyun 50-m mesh telescope (located near Beijing, China) should be complete, and this may also participate in EVN observations (up to and including λ3.6 cm). In addition, several other major upgrades of EVN telescopes have just been completed. These include the recent upgrade of the WSRT array, and the installation of an active surface for the Noto 32-m telescope. The replacement of the existing reflecting surface of the Lovell 76-m telescope is on-going and is expected to be complete by the end of 2002. These will permit the Lovell telescope to observe usefully at frequencies up to 10 GHz – boosting its sensitivity by a factor of 5 at λ6 cm. This major engineering development (which includes the upgrade of the drive and pointing control system) will transform the Lovell's capability as a VLBI and MERLIN antenna. Progress with the Lovell telescope upgrade and the construction of the new 40-m telescope at Yebes (as of summer 2001) is presented in Figure 4.

It must also be noted that there are certainly areas where the performance of a heterogeneous network such as the EVN might be considered less than optimal, at least in comparison to a homogeneous, full-time network such as the VLBA. Certainly the EVN is a more difficult instrument to calibrate, and only recently have a significant number of telescopes achieved frequency flexibility. In addition, the geographical location of the majority of the antennas is also not optimal for high frequency observations ($>$ 20 GHz). The EVN's ability to react to "target-of-opportunity" observations is also more limited than the VLBA – at least outside of network sessions. Similarly it is difficult for the EVN to adequately monitor sources with evolving radio structure, at least in comparison to the uniform temporal coverage that the VLBA can provide.

Nevertheless, progress is being made in all these areas. Automatic pipelining of EVN (and global) VLBI data (see section 3.1 and Reynolds, Paragi & Garrett 2002) now largely hides the intricacies of EVN calibration from the user. Experiments requiring fast frequency switching are beginning to become more common in network sessions, and the addition of the new 40-m Yebes and 64-m Sardinia telescopes (capable of operating at frequencies up to 115 GHz) will enhance the EVN's sensitivity at higher frequencies. Vigorous efforts to move the EVN towards real-time operations (see section 4) will also provide increased flexibility to conduct more uniform monitoring campaigns or to respond to "target-of-opportunity" events.

Figure 3. Left: The upgrade of the Lovell Telescope surface – a factor of 5 improvement in sensitivity is expected at λ6 cm. Right: Construction continues of the 40-m telescope at Yebes – the pedestal is complete, work continues on the backing structure and reflector panels. The new 40-m telescope (inset) will operate at frequencies up to 115 GHz

3. The EVN MkIV Data Processor at JIVE

The construction and development of a VLBI correlator entirely dedicated to EVN activities is one of the great achievements of the last decade. The EVN MkIV Data Processor at JIVE (Casse 1999, Schilizzi et al. 2002b) was developed as part of an international collaboration, the primary contributors being the European Consortium for VLBI and the MIT Haystack Observatory, in the USA. The EVN MkIV Data Processor is operated by JIVE and is now the main-stay of EVN data correlation (including global projects which it shares with the NRAO-VLBA).

The Data Processor is capable of handling data from 16 telescopes simultaneously (more via multiple pass correlation) and can handle MkIV, VLBA and MkIII data formats. Standard correlation of the vast majority of EVN and global VLBI continuum and spectral line experiments are now routinely processed at JIVE. The capacity of the correlator is continually being enhanced, and new capabilities introduced (see `www.jive.nl` for the most up-to-date information). On-going projects include: recirculation (in order to provide superb spectral resolution in excess of 8192 channels per baseband), Pulsar Gating (to optimise Pulsar detection limits) and the PCInt (Post-correlator Integrator) that will permit high-speed read-out of the correlator at data rates of up to 160 MBytes/second).

The PCInt development is expected to see "first-light" by the end of 2002 (Parsley 2001a) – it will transform the capability of the EVN (and Global VLBI arrays) providing the possibility to image dozens of faint sub-mJy radio sources within the primary beam of the individual VLBI antennas (see Garrett these proceedings). To take advantage of the fantastic output data rates the PCInt can generate, significant off-line computing resources will be required. Off-line computing resources are likely be the main bottleneck in the new system, at least for the first few years of operation.

3.1. EVN USER SUPPORT

As well as operating the EVN Data Processor, JIVE is largely responsible for EVN user support. JIVE support scientists (and in addition other JIVE staff) are involved in providing a level of user support that was previously unknown within the EVN, and rivals or surpasses that provided by other instruments. In particular, the following services are routinely provided:

- advice regarding the technical content of proposals (e.g. cover-sheet specifications, choice of mode, observing strategy etc)
- scheduling assistance and maintenance/development of NRAO's Sched (for specific EVN requirements)
- absentee correlation and data quality check-out
- automatic calibration of EVN and Global VLBI data correlated at JIVE (via a Pipeline process)
- direct assistance with VLBI data and image analysis.

In addition, the support scientists also contribute to monitoring the reliability and performance of the EVN (via special Network Monitoring Experiments) and also conduct network tests aimed at extending the capabilities of the network.

Financial support is available to those EVN users that wish to visit JIVE in order to avail themselves of these services (in particular scheduling and data analysis). Indeed the EVN is in receipt of a substantial award from the European Commission in Brussels (*Access to Research Infrastructures*), that comprehensively supports EVN users that are not affiliated to the EVN Consortium institutes but are located within the European Union or Associated States. In addition, there is also internal EVN support for users

Figure 4. The EVN MkIV Data Processor at JIVE – the 16 station Data Processor is now the main-stay of EVN data correlation.

that are directly affiliated to EVN Consortium institutes. Both programmes support not only visits to JIVE but also to other members of the EVN. For example, users frequently visit Jodrell Bank Observatory in order to take advantage of the local expertise in combining joint EVN-MERLIN data sets.

4. The Future of the EVN

Across the globe radio telescopes and interferometer arrays are involved in significant efforts to improve their overall performance and their continuum sensitivity in particular. These developments are desperately needed in order for radio astronomy to maintain its competitiveness with other next generation instruments – especially sub-mm and IR telescopes. Anticipated improvements largely rely on the possibility of observing and processing much larger continuum bandwidths than was previously possible. The use of optical fibre technology now permits the digital transport of many GHz of bandwidth over 1000's of km. Next generation correlators are now being designed and constructed to handle the associated Gbit/sec input data rates and subsequent processing requirements. The EVLA, e-MERLIN and LOFAR telescopes will be the first radio telescopes to take advantage of these developments – permitting huge areas of sky to be mapped-out with sub-arcsecond resolution and microJy sensitivity.

The consequences for VLBI, and the EVN in particular, are crystal clear. In order to remain competitive with, and complementary to these upgraded or new radio instruments, the EVN *must* be able to observe and process several GHz of bandwidth too. The connection of the EVN telescopes to commercial "λ-networks" (fibre data transport utilising wavelength-division multiplexing techniques) is now being vigorously explored in both Europe and the USA - the first connections and fringe tests are expected to occur within the year (Schilizzi 2002a, Whitney 2002, Parsley 2002). Perhaps all VLBI telescopes (in particular those located nearby densely populated areas) will be connected to such networks in 5-10 years time, the exact timescale depending on local circumstances. Trans-continental connections also appear feasible too. Meanwhile the new disk-based MkV (Whitney 2001) and PC-EVN (Parsley 2001b) recording systems are set to replace the current generation of tape recording systems. These disk-based PC systems can already record data at rates that are similar to current MkIV or VLBA systems. In addition, the same systems are poised to take advantage of the expected expansion in the capability of PC hardware over the next few years. While the investment in both fibre or disk-based technologies is substantial, it will provide VLBI networks such as the EVN with sensitivity levels that are similar or even better than that anticipated for either the EVLA or e-MERLIN. A natural consequence of employing observing bandwidths that span several GHz is the almost complete uv-coverage that accompanies it. Figure 5 shows the uv-coverage of a fibre connected EVN (*e*EVN). The transparent, real-time combination of the *e*EVN and e-MERLIN will also result in a significant enhancement in imaging capabilities of the combined array.

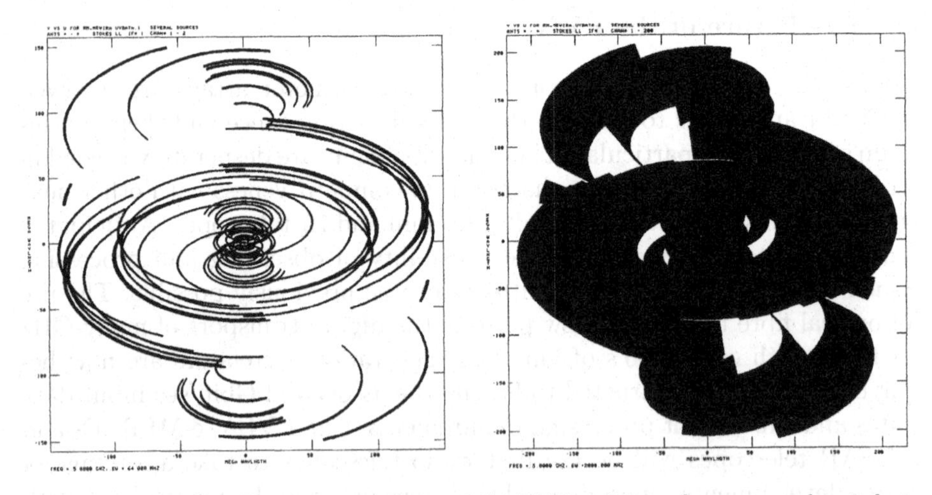

Figure 5. Left: The current uv-coverage of the EVN at λ6 cm for a source located at $\delta = 30°$. Right: the extended (almost full) uv-coverage of the *e*EVN for the same source, assuming a total bandwidth of 2 GHz.

As a consequence of all these developments, a replacement for the EVN Data Processor will also be necessary. The new correlator will need to be capable of handling a global array of $\sim$ 30 telescopes (each generating 10-30 GBits/sec of data) and the phenomenal output data rates that will enable the natural field of view (then set by the primary beam of individual VLBI elements) to be imaged out in its entirety. New broad-band receivers and a new generation of VLBI data acquisition electronics will also be required at the telescopes, in order to take full advantage of the available bandwidth. There are (not surprisingly) severe implications for ("off-line") data processing requirements too. This not only concerns raw processing power but also the development of new calibration and image algorithms. In a very real sense, the *e*EVN with baselines on the scales of several thousand km, "fantastic" data rates and microJy sensitivity, will be the natural test-bed to investigate some of the problems and possible limitations that might be relevant to next generation instruments such as the SKA (see Kus these proceedings).

References

Casse, J.L. 1999, NewAR **43**, 503.

Mujunen, A. et al. 2002, "The EVN Gbit/s e-VLBI Data Acquisition and Playback System", *Proc. e-VLBI Workshop*, `web.haystack.mit.edu/e-vlbi/`.

Parsley, S. 2001a, "PCInt Design Specification", EVN Document 112.

Parsley, S. 2001b, "PC-EVN Project Plan", EVN Document 113.

Parsley, S. 2002 "VLBI participation in next -generation network development. ", *Proc. e-VLBI Workshop*, `web.haystack.mit.edu/e-vlbi/`.

Reynolds, C., Paragi, Z. & Garrett, M.A. 2002 "Pipeline Processing of VLBI Data" presented at the URSI General Assembly, Maastricht (astro-ph/0205118).

Schilizzi, R.T. 2002a " Networking in Europe – status and plans", *Proc. e-VLBI Workshop*, `web.haystack.mit.edu/e-vlbi/`.

Schilizzi, R.T. et al. 2002b, Exp. Astron (in press).

Whitney, A.R. 2001 " Concept for an Affordable High-Data-Rate VLBI Recording and Playback System", Mark V Memo series No. 1.

Whitney, A.R. 2002 "Mark 5 and e-VLBI", *Proc. e-VLBI Workshop*, `web.haystack.mit.edu/e-vlbi/`.

Zensus, A., Diamond, P.J. Napier, P.J. 1995, *Very Long Baseline Interferometry and the VLBA*, ASP Conf. series, volume 82 (San Francisco, ASP).